Proteins
Concepts in Biochemistry

Proteins
Concepts in Biochemistry

Paulo Almeida

GS Garland Science
Taylor & Francis Group

LONDON AND NEW YORK

Garland Science
Vice President: Denise Schanck
Senior Editor: Summers Scholl
Editorial Assistant: Michael Roberts
Production Editor: Natasha Wolfe
Illustrator: Nigel Orme
Cover Design: AM Design
Copyeditor: John Murdzek
Typesetter: Nova Techset Private Limited, Bengaluru & Chennai, India
Proofreader: Susan Wood
Indexer: Indexing Specialists (UK) Ltd.

About the Author
Paulo Almeida is a Professor in the Department of Chemistry and Biochemistry at the University of North Carolina Wilmington. Prior to his academic appointment at UNCW, he was an Assistant Professor of Chemistry at the University of Coimbra and at the University of Algarve, Portugal. His research seeks to understand the mechanism of antimicrobial, cytolytic peptides, and cell-penetrating peptides, as well as study domain formation in lipid membranes in terms of the molecular interactions between lipids, and how they are influenced by proteins and peptides bound or incorporated into the membrane.

ISBN 978-0-8153-4502-2

Library of Congress Cataloging-in-Publication Data

Names: Almeida, Paulo, author.
Title: Proteins : concepts in biochemistry / Paulo Almeida.
Description: New York, NY : Garland Science, Taylor & Francis Group, [2016] |
Includes bibliographical references.
Identifiers: LCCN 2015046366 | ISBN 9780815345022
Subjects: | MESH: Proteins | Protein Conformation | Protein Stability |
Protein Folding | Biochemical Processes
Classification: LCC QP551 | NLM QU 55 | DDC 572/.6–dc23
LC record available at http://lccn.loc.gov/2015046366

Published by Garland Science, Taylor & Francis Group, LLC, an informa business, 711 Third Avenue, New York, NY 10017, USA, and 3 Park Square, Milton Park, Abingdon, OX14 4RN, UK.

Printed in the United States of America
15 14 13 12 11 10 9 8 7 6 5 4 3 2 1

Garland Science
Taylor & Francis Group

Visit our web site at http://www.garlandscience.com

To Antje

PREFACE

Overburdening necessarily leads to superficiality. Teaching should be such that what is offered is perceived as a valuable gift, and not as a hard duty.

Albert Einstein

This is a textbook on concepts in biochemistry: concepts from physics, from chemistry, from biology. I have chosen proteins as the subject of the book because they provide the richest illustrations of those concepts. However, the book is not a description of the physical properties of proteins and their biological functions. Rather, my goal is to communicate a way of thinking about problems in biochemistry—in science, really. In this sense, I use proteins as a vehicle. But proteins are much more than a vehicle for this process: They are the most fascinating molecules that exist.

This book was born from teaching. Having taught biochemistry for about 20 years, I came to appreciate that students struggle with the sheer volume of material in introductory courses. They are asked to know an ever-growing amount of information (yet what a human being can reasonably learn and comprehend has not increased) at the expense of conceptual understanding. Over the past decade, more than 30 undergraduate students have worked in my research laboratory. I have noticed that the amount of material learned in the classroom that is actually retained in usable form is quite small. This is true even of students who completed the courses with excellent results. Why?

Biochemistry textbooks have become increasingly concentrated on biological aspects and on large amounts of information. Conversely, understanding chemical, physical, and mathematical aspects of biochemical topics have received less attention. Yet, already about 10 years ago the National Research Council (NRC) of the National Academies reached the conclusion that *"life science majors must acquire a much stronger foundation in the physical sciences... than they now get,"*[1] and recommended increasing connections between biological and physical sciences as a means to bridge the communication gap that exists between researchers in different disciplines. The NRC also emphasized the importance of raising the interest for research in undergraduate students as early as possible. I wanted to attempt something to help change this state of affairs.

---

[1] National Research Council (2003). Bio 2010: Transforming Undergraduate Education for Future Research Biologists. National Academic Press.

I believe this book will be helpful for all those who seek a foundation in more physical aspects of biochemistry. The book contains a manageable amount of material for a one-semester course on proteins for advanced undergraduates and beginning graduate students in chemistry, biochemistry, biophysics, and biology. At the PhD level, courses on proteins often have a heterogeneous population of students, with varied backgrounds. Those courses are often taught on the basis of journal articles from the scientific literature. In these cases, I think this textbook will provide the necessary foundation for many students. Furthermore, chemistry departments whose concentration in biochemistry is limited may find this text appropriate for a course that fulfills their biochemistry requirement. The prerequisites for the course are one semester of calculus and a knowledge of introductory physical chemistry. A basic undergraduate general biochemistry course is helpful but not essential, because the book covers the basics of protein and DNA composition and structure. I have taught this course, with some variations, to beginning graduate and advanced undergraduate students in chemistry and biology. My hope is to endow students with a frame of thought that will help them tackle problems they will encounter in the course of their careers.

I designed this book to present essential concepts in biochemistry without overburdening students with too many facts or too much formalism. The book covers protein structure, evolution, stability, folding, ligand binding, and enzyme kinetics. I wanted to approach these topics from a perspective of probabilistic (statistical) and quantitative thinking. Therefore, the book begins with an introduction to statistical thermodynamics, applied to biological macromolecules. The concept of the partition function is introduced early and used throughout. With the exception of Chapter 3 (evolution), all chapters use the partition function method to solve various problems. Mathematics is essential in science, and many topics in biochemistry require mathematics. However, I have tried to keep the mathematical formalism to a minimum. When mathematics is used, I have tried to explain the meaning of formulas and procedures in plain English. Mathematics tends to scare biochemistry students. The barrier, however, is mainly one of language not of concepts; I have found that most students can follow a mathematical argument if it is explained in words. I have also sacrificed formalism to intuition. In so doing, I have taken approaches that purists may frown upon. I think this is a risk worth taking.

The book starts simple and becomes progressively more complex. Later chapters build on the previous ones: the book is meant to be read in sequence; it is not a reference text to look things up. Because of that, some sections may be assigned as homework reading, rather than covered in class. Each chapter ends with a set of problems of variable degrees of difficulty, not necessarily in order. The overall complexity of the problems tends to increase as the book progresses. Finally, each chapter includes a list of references, organized by topics. The references are listed under the topic for which they are most relevant, but some references are relevant for more than one topic. The references are not cited in line, to facilitate reading, but it should be fairly evident which references were used for a particular topic.

Some additional resources are highly recommended. One is the Protein Data Bank (PDB), which is the source of the protein structures that illustrate this book. The second is a means of visualization and rendering of those structures, such as the PyMol software. All protein structures in this book were produced with Open Source PyMol, which is free for everyone, and which I am pleased to acknowledge.

The third tool is the Kinemage web site, organized by Jane and David Richardson at Duke University. This site contains various tools that make the study of protein structure a wonderful experience. Students are strongly encouraged to visit those sites, download the programs, and begin exploring proteins.

I would like to acknowledge all the people who have helped me in this endeavor. I thank my editor at Garland, Summers Scholl, for her constant support and understanding during the writing of this book, and editorial assistant Michael Roberts for his help in the last stretch. I thank the reviewers of the manuscript, or of several of its chapters—namely, (in alphabetical order), Charles E. Bell (Ohio State University), David Case (Rutgers University), Athel Cornish-Bowden (CNRS, Marseille, France), Paul Raymond Gooley (University of Melbourne, Australia), Andrew Herr (University of Cincinnati College of Medicine), Anne Hinderliter (University of Minnesota, Duluth), Matt Junker (Kutztown University of Pennsylvania), Jim Morton (Lincoln University), Bill Pearson (University of Virginia), Gordon Rule (Carnegie Mellon University), Brian Shilton (University of Western Ontario), Nicolas Silvaggi (University of Wisconsin, Milwaukee), Jennifer Surtees (University at Buffalo School of Medicine and Biomedical Sciences), Christopher Taylorson (University College London, UK), Darren Thompson (University of Sussex, UK), Peter Tipton (University of Missouri, Columbia), and Honggao Yan (Michigan State University). All of their criticisms, major and minor, were extremely helpful.

In addition, I want to especially thank all those who accepted to read, and sometimes re-read, some of the chapters at my request. Some were also formal reviewers but others were not. Again in alphabetical order, I thank Robert Baldwin (Stanford University) for his comments and many discussions on the hydrophobic effect and protein stability; Steve Benkovic (Pennsylvania State University) for answering my questions on DHFR; Arieh Ben-Naim (Hebrew University of Jerusalem, Israel) for his comments on the hydrophobic effect; Alan Fersht (University of Cambridge, UK) for his comments and suggestions on protein folding, and for his encouragement; Karen Fleming (Johns Hopkins University) for an important suggestion; Martin Gruebele (University of Illinois at Urbana-Champaign) for his many comments and suggestions on protein folding; Phil Hanoian (Pennsylvania State University) for his comments on the DHFR section of enzyme kinetics; Michael Harms (University of Oregon) for his comments on protein evolution; Andrew Herr (University of Cincinnati College of Medicine) for many excellent suggestions; Anne Hinderliter (University of Minnesota, Duluth) for her criticism on so many chapters; Alesia McKeown (University of Utah) for her review of the protein evolution chapter, many discussions, suggestions, and papers on evolution; Bill Pearson (University of Virginia) for his criticism of the first version of the protein evolution chapter and for his comments on the last version; George Rose (Johns Hopkins University) for many discussions and suggestions on protein stability and folding; Gordon Rule (Carnegie Mellon University) for reviewing several chapters, some repeatedly, and for his excellent suggestions; and Shozo Yokoyama (Emory University) for kindly answering my questions on the evolution of opsins. One of the most delightful aspects of writing this book was the opportunity to meet, in person or by e-mail, many scientists whom I have admired throughout my career. What is more, my requests to them have been met with kindness and helpfulness. Their criticism and suggestions were essential to improve the book.

The views presented here, however, and any mistakes or errors remaining, are of my exclusive responsibility. I welcome any corrections from readers, which can be sent to science@garland.com.

I would also like to thank the colleagues in my department who have read and commented on various chapters: Antje Pokorny Almeida, on ligand binding; Chris Halkides, on enzyme kinetics; Mike Messina, on statistical thermodynamics; Pam Seaton, on NMR; and especially Hee-Seung Lee, who read and provided feedback on almost the entire manuscript. Their comments, suggestions, and corrections are deeply appreciated.

One of my motivations to write this book was to pass to students what I have learned. Therefore, I want to take this chance to thank the people from whom I myself have learned the most: my first mentor in science, Winchil Vaz (University of Coimbra, Portugal), for the many hours and years of great discussions; my PhD advisor, Tom Thompson (University of Virginia), from whom I learned so many things, among which is not to be afraid of starting to work on something I knew nothing about; and my post-doctoral advisor, Rod Biltonen (University of Virginia), from whom I learned to use the partition function in biochemical problems, and to whom, in this and many other senses, this book is a tribute.

Finally, I thank Antje for her support and patience during all the spoiled weekends and vacations for the past three years, while the most intense writing took place.

Paulo Almeida
Wilmington, NC, March 2016

RESOURCES FOR INSTRUCTORS

The instructor's teaching resources on the Garland Science website are password protected and available only to adopting instructors. We hope these resources will make it easier for instructors to prepare dynamic lectures and activities for the classroom. Instructor Resources are available on the Garland Science Instructor Resource Center, located at www.garlandscience.com and selecting the "instructor" tab. The Website provides access not only to the teaching resources for this book but also to all other Garland Science textbooks. Adopting instructors can obtain access to the site from their sales representative or by emailing science@garland.com.

Art of *Proteins: Concepts in Biochemistry* The images from the book are available in two convenient formats: PowerPoint® and JPEG. They have been optimized for display on a computer. Figures are searchable by figure number, by figure name, or by keywords used in the figure legend from the book.

Solutions Manual This PDF contains solutions to all of the problems in the book.

CONTENTS

DETAILED CONTENTS

Statistical Thermodynamics of Biological Molecules

1.1 BIOLOGICAL MOLECULES HAVE SPECIAL CHEMICAL PROPERTIES

As your study of science progresses, you will realize that questions are at least as important as answers. A good question can open a new field of research. A poor question is usually too vague to lead to anything. The question of what is biochemistry is a poor question. A common answer is that it is the chemistry of the reactions in living cells, but this gives no information beyond that already contained in the word biochemistry. Let us attempt to answer, instead, a different question: what sets biochemistry apart from the rest of chemistry? Like chemistry in general, biochemistry deals with molecules and their assemblies. However, biological molecules combine a set of properties that make them unique. Their functions are shaped by water, structure, evolution, weak interactions, and communication. In this book we will concentrate on proteins, because they provide the richest illustrations of the most important concepts in biochemistry.

The number of pages of this book is exactly infinite. None is the first; none, the last. I don't know why they are numbered in this arbitrary manner. Perhaps to indicate that the terms of an infinite series can take any number.

Jorge Luis Borges

Water Shapes the Properties of All Biological Structures

Let us examine each of the five aspects in our list. We begin with water. The Greek philosopher Thales of Miletus (ca. 550 BCE) is quoted by Aristotle (*Metaphysics, 983b*) as having observed that the nature of all living things is moist and that water, as the origin of the moist things, must be the principle from which everything originated. This statement is essentially correct: life as we know it would not exist without water. The structure of the major cellular components, especially proteins and lipid membranes, is a consequence of the structure of water to a great extent. Proteins fold into globular structures to hide their hydrophobic residues from the contact with water. Biological membranes are based on lipid bilayers. A lipid bilayer (**Figure 1.1**) adopts its structure to hide the nonpolar acyl chains of the lipids from water. Interactions in the cell, which is an aqueous medium, can only be understood if the role of water is taken into account.

Figure 1.1 A lipid bilayer membrane forms to shield the hydrocarbon chains of the lipids from contact with water.

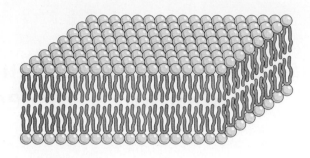

The Structures of Biological Macromolecules are Key to their Function

Second, structure plays an enormous role in biochemistry. Understanding protein structures is key to understanding their function. These structures are complex, like those of no other molecules. And yet protein structures are regular and highly symmetric (**Figure 1.2**).

In a protein, the polypeptide chain is organized in local regular structures, such as helices and strands, which in turn assemble in a three-dimensional arrangement with very high symmetry. Why has nature favored such regular structures? The answer, in essence, is that those structures are the best to ensure that all hydrogen bonds in the protein interior are satisfied, since they cannot be formed with water.

DNA, too, has a symmetric structure. It also forms a helix—a double helix (**Figure 1.3**). That structure told us how DNA works, how it is replicated.

The Structures of Proteins that Exist Today have been Selected by Evolution

Third, the function and structure of biomolecules have arisen through evolution. One of the most amazing processes in biochemistry is enzyme catalysis. Enzymes catalyze chemical reactions with a remarkable efficiency compared to regular chemical catalysts. The active site of an enzyme—the center of catalytic action—is usually located close to the protein surface. The strange thing is, the structure of the entire protein is needed to host a relatively small active site. However, it is difficult to appreciate the importance of protein structures without understanding *how* they became what they are today. Those structures were shaped by evolution. Evolution selects for protein structure, not for amino acid sequence—because it is the structure that determines the function.

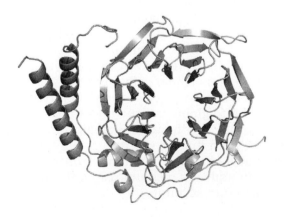

Figure 1.2 Structure of the β-γ complex of the GTP-binding protein transducin (PDB 1TBG).

Proteins evolve with the organisms that host them. At some point, early in time, an organism existed that contained a certain protein (**Figure 1.4**). Today, that organism—the common ancestor—no longer exists, but organisms that evolved from it do. These organisms contain proteins that evolved from that ancestor protein. By *divergent evolution*, the ancestor protein has changed in different ways, and the extant forms of this protein are found, for example, in the mouse and in humans. The way those changes occurred, in each of the branches of the evolutionary tree that led to today's species, tells a story about the protein and how it works.

Interactions in Biomolecules are Weak

The fourth aspect in our list relates to *interactions*—interactions of a protein with other molecules and interactions within the protein. What may be surprising in molecules with such a high level of organization and symmetry is that these interactions are *weak*. A process that occurs with a large negative Gibbs energy change (ΔG) is irreversible, and therefore not controllable. In most biochemical processes ΔG values are small, ensuring reversibility.

What do we mean by weak interactions? We mean that the ΔG involved are not larger than about 10 times the thermal energy, or the average kinetic energy of the "heat bath." The heat bath is the environment where the reaction takes place, the medium with which heat is exchanged, by collisions between molecules. The temperature is simply a measure of the average kinetic energy of the heat bath. The thermal energy is kT (per molecule) where k is the Boltzmann constant ($k = 1.38 \times 10^{-23}$ J K^{-1} per molecule) and T is the temperature in kelvin (K). More often we will write the thermal energy as RT (per mole) where R is the gas constant ($R = 1.987$ cal K^{-1} mol^{-1}, using the the conversion 1 cal = 4.184 J). The constants R and k are simply related by $R = N_A k$, where $N_A = 6.022 \times 10^{23}$ is Avogadro's number (molecules per mole). At room temperature, $RT = 0.6$ kcal/mol. This is our *reference energy*, relative to which other energies are large or small.

Consider, for example, protein unfolding. In its native state, a protein has a well-defined, regular, three-dimensional structure; but this structure is lost at high temperatures. The protein is denatured by heat: it unfolds. In the denatured state the protein is mainly disordered. However, denaturation is reversible. The Gibbs energy difference between the native and the denatured states of the protein is small, about $\Delta G = 5$ to 10 kcal/mol at room temperature. Moreover, this energy is not concentrated in one bond or interaction, but comprises many small interactions, each one not much larger than RT.

Communication Occurs through Interactions between Molecules or within Molecules

Finally, the fifth aspect in our list is communication. Communication happens through physical interactions. In higher organisms, communication with the exterior is essential to maintain the function of the cell in harmony with the rest of the organism. In microorganisms, communication with the exterior is essential to obtain information for orientation toward a food supply, for example. Communication between the outside and the inside of a cell occurs at its membrane. The information obtained is transmitted to the cell interior by complex mechanisms that we call signal transduction. However, the

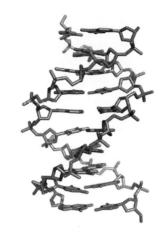

Figure 1.3 Double-helical structure of the DNA molecule (PDB 1D49).

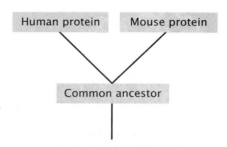

Figure 1.4 A simple evolutionary tree for a protein that exists in humans and in the mouse.

primary communication takes place through binding of a molecule to a membrane protein.

Another example of communication occurs in the control of protein function or enzyme activity. The reactions catalyzed by enzymes are regulated by interactions with protein inhibitors or activators. The active site of an enzyme "is informed" that an allosteric regulator (an inhibitor or an activator) is bound at a different (regulatory) site through changes in the protein structure, which are communicated to the active site by interactions within the protein. For example, binding of an oxygen molecule at one of the four heme sites in hemoglobin increases the affinity of another site in the same protein for a second oxygen molecule. In these cases we are dealing with molecular communication. In the case of hemoglobin, this communication leads to *cooperative* binding of oxygen, which is essential for the function of the protein. Cooperative interactions are a consequence of communication within a protein molecule.

1.2 STATISTICAL THERMODYNAMICS RELATES MICROSCOPIC INTERACTIONS TO MACROSCOPIC PROPERTIES

In the study of biochemical systems, proteins in particular, we will need to make frequent use of thermodynamics. This introduction to the basic concepts of thermodynamics and their relation to statistical mechanics is not meant to be complete. Rather, we will learn the concepts of statistical thermodynamics while often sacrificing formalism for intuition. These concepts are needed to understand the rest of the book. More details and refinements will be added as we go along.

States with the Same Energy are Equally Probable

Thermodynamics is concerned with the properties of macroscopic systems that can be measured experimentally. Those systems contain an enormous number of molecules (N). Even systems that you may usually think of as small contain many molecules. Consider the example shown in **Figure 1.5**. You have 1 milliliter of a polypeptide solution in water, with a concentration of 1 micromolar. This system contains 1 nanomole or $N \sim 10^{15}$ polypeptide molecules. More exactly, we have $N = 1 \text{ mL} \times 1 \text{ μM} \times 6.02 \times 10^{23} = 6.02 \times 10^{14}$ molecules, which is *of the order* of magnitude of 10^{15}. The symbol $\sim$ indicates an order-of-magnitude estimate; the actual value is within a factor of 10 of the number indicated. If we want to indicate a slightly more accurate, but still approximate estimate, we use the symbol $\approx$. We will also use the symbol $\sim$ to indicate the main mathematical form of a certain function, when we want to omit less important factors or numerical constants. For example, $f(t) \sim e^{-kt}$ indicates that the function f varies essentially as an exponential function of time (t); constant factors and less important ones, such as those linear in t, are omitted.

We call the collection of polypeptide molecules in our system an *ensemble*. The polypeptides in the ensemble can have different conformations, with different energies (**Figure 1.6**). (The term ensemble is used in a somewhat different sense in the Gibbs formulation of statistical mechanics; see the book by Hill [1980].) The ensemble has certain macroscopic properties, such as energy. The thermodynamic system

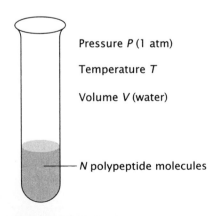

Pressure P (1 atm)

Temperature T

Volume V (water)

N polypeptide molecules

Figure 1.5 A system consisting of an aqueous polypeptide solution in a test tube, with N polypeptide molecules, volume V, at a temperature T and pressure P.

Figure 1.6 Polypeptides in the ensemble can have different conformations, with different energies.

includes the ensemble of N polypeptide molecules, but also the solvent (water), and is further characterized by the temperature (T), and the volume (V) or the pressure (P).

It may also be possible to measure individual properties of the molecules, such as the conformation of a particular polypeptide, but thermodynamics does not tell us about those. Thermodynamics tells us about the properties of the whole ensemble of molecules, not about individual ones, except in an average sense. It is statistical mechanics that tells us how the thermodynamic properties of the system are related to the molecular properties of its components. Whereas thermodynamics provides relations between macroscopic properties of the system, statistical mechanics provides a way to *interpret* those properties and the relations between them.

If we measure the energy of all the polypeptides in our system, and divide it by their number, we obtain the average energy of a polypeptide molecule in the ensemble. This is a thermodynamic property of the system at equilibrium, measured at a certain instant in time. How does it relate to the energy of an individual molecule in the ensemble? Suppose you could follow one particular polypeptide molecule in the solution and measure its energy as a function of time. You would see that the energy of this molecule varies—it *fluctuates*. However, if you measure the energy of this molecule for a very long time and calculate its average energy (by adding all the energy values you measured and dividing by the number of measurements), you will obtain the same value as for the average energy that you had determined for the entire ensemble of molecules (by dividing the total energy by the number of polypeptides). The *ergodic hypothesis* tells us that, in a system at equilibrium, the *time average* of a property of an individual molecule is the same as the *ensemble average* of that property, over all molecules at any given time. In other words, to measure an equilibrium property, you can watch one molecule all of the time or you can watch all of the molecules at one time. The ergodic hypothesis tells us that the two averages are equivalent.

The polypeptide molecules in solution can have different conformations; each different conformation is a different state. Now suppose that all those conformations have the *same energy* ε. You isolate the system, so no energy or molecules can enter or leave it. This ensemble has N polypeptide molecules in a certain volume V, and an energy E. (For simplicity we are not including the energy of the solvent (water) molecules in E. That additional energy E_w is part of the energy of the system, but not of the ensemble of polypeptides. This, however, makes no difference for the present argument.) We say that the system has characteristic variables N, E, V. Each polypeptide molecule has energy ε, so their total energy is $E = N\varepsilon$. Each polypeptide constantly changes conformation over time. The question is, how likely is a polypeptide molecule to adopt one particular conformation instead of another? The answer is that, as long as they have the same energy, all conformations are equally likely. This is the *principle of equal a priori probabilities*. The polypeptide molecule samples the conformational space uniformly.

States with Lower Energies have a Higher Probability

Now let us go back to our actual system, 1 mL of an aqueous polypeptide solution in a test tube in the laboratory. The polypeptide molecules do *not* all have the same energy. Rather, their temperature T is fixed by the heat bath. The water, where the polypeptides are dissolved, is primarily responsible for providing the heat bath. (Of course, the temperature of the water is itself set by the laboratory temperature, or by an external water bath where the test tube is placed.) The polypeptides still cannot escape from the test tube, so their number N is fixed. And assume for now that the volume V is also fixed (this is not strictly true, but we will correct it shortly). In this case the characteristic variables are N, T, V. The difference to the previous case is that now the individual polypeptide molecules in the ensemble have *different energies*. Each polypeptide conformation, which we designate by i, has a certain energy ε_i. The average energy $\bar{\varepsilon}$ is what we measure experimentally in the macroscopic system. How likely is a certain molecule to have a particular energy ε_i?

Suppose that the values of the energy are discrete (they change in steps) as shown in **Figure 1.7**. There are a certain number of conformations, each with a certain energy ε_i. We want to find the probability p_i that a molecule has an energy ε_i. This probability is simply the number of molecules N_i with energy ε_i divided by the total number of molecules,

$$p_i = \frac{N_i}{N}. \tag{1.1}$$

The average energy of each molecule is the total energy divided by the number of molecules,

$$\bar{\varepsilon} = \frac{E}{N}. \tag{1.2}$$

The total number of molecules is the sum of the numbers of molecules in each state. Using the symbol $\sum$ to indicate summation over all states i, we write,

$$\sum_i N_i = N \tag{1.3}$$

and the sum of all their energies is the total energy of the ensemble,

$$\sum_i N_i \varepsilon_i = E. \tag{1.4}$$

Therefore, the average energy is

$$\bar{\varepsilon} = \frac{\sum_i N_i \varepsilon_i}{N}. \tag{1.5}$$

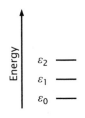

Figure 1.7 A system with three conformational states corresponding to the discrete energy states ε_0, ε_1, and ε_2.

Using Equation 1.1, we can also write the average energy of a molecule as

$$\bar{\varepsilon} = \sum_i p_i \varepsilon_i. \tag{1.6}$$

This sum is a *weighted average* of the energy, where each value ε_i is weighted by its probability p_i. The probabilities must add to 1,

$$\sum_i p_i = 1. \tag{1.7}$$

As you see from Equation 1.6, to calculate the average energy, all you need to know is the probability p_i of finding a molecule with energy ε_i.

The probability that a molecule is in conformation i depends only on its energy ε_i and on the temperature T, but not on any details of the conformation, and is proportional to the *Boltzmann factor*,

$$e^{-\varepsilon_i/kT}. \tag{1.8}$$

The Boltzmann factor is a relative probability. The sum of the absolute probabilities of all states must equal 1, but the sum of the Boltzmann factors is not necessarily 1. Therefore, we need to normalize the Boltzmann factor, dividing each one by their sum for all states (conformations). This sum is called the *partition function*,

$$Q = \sum_{\text{states } (i)} e^{-\varepsilon_i/kT}. \tag{1.9}$$

Now suppose that there are a certain number of conformations, and each conformation has a certain energy, but there may be more than one conformation with the same energy, as shown in **Figure 1.8**. We call *microstates* the various conformations belonging to the same energy level, ε_i.

If there are a number W_i of conformations with the same energy ε_i, the corresponding Boltzmann factor appears W_i times in the sum of Equation 1.9. We can also group the microstates W_i that have the same energy ε_i, and write the partition function as a sum over energy levels,

$$Q = \sum_{\text{energy levels } (i)} W_i e^{-\varepsilon_i/kT}. \tag{1.10}$$

The number of microstates W_i belonging to the energy level ε_i is called the *degeneracy* or the *multiplicity* of that state. Now the probability that a molecule has energy ε_i is

$$p_i = \frac{W_i e^{-\varepsilon_i/kT}}{\sum_i W_i e^{-\varepsilon_i/kT}}. \tag{1.11}$$

As it should be, the sum $\sum_i p_i = 1$. We will see shortly that, much more than a mere factor to normalize relative probabilities, the partition function is a fundamental quantity, with extremely important and useful properties.

If there are many, closely spaced energy values, then in practice the energy varies continuously. Then it is not practical to count numbers of conformations. Instead, we speak of continuous distributions of the energy and of their associated density of states $W(\varepsilon)$. The probability density represents the distribution of probabilities as a function of energy. It tells us how likely it is to find a molecule with a certain energy ε. In a discrete distribution, the probability that a molecule has energy ε_i is $p_i = \frac{N_i}{N}$. In a continuous distribution, we write the probability of finding a molecule with an energy very close to ε as

Figure 1.8 A system with conformational states distributed over discrete energy levels. The number of microstates increases with energy. Here, there is one microstate (conformation) in the zero energy level (ε_0), four microstates in level 1 (ε_1), 10 microstates in level 2 (ε_2).

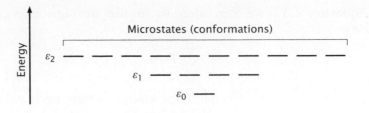

$$p(\varepsilon) = \frac{N(\varepsilon)}{N}, \tag{1.12}$$

where $N(\varepsilon)$ is the number of molecules with energy very close to ε.

The Boltzmann factor $e^{-\varepsilon_i/kT}$ decreases exponentially with energy, so the probability of high energy states decreases sharply. However, in a macromolecule with many degrees of freedom (in a polypeptide, the degrees of freedom are essentially the number bonds over which rotations are possible), the number $W(\varepsilon)$ of possible conformational states with similar energy increases sharply as the ε increases (see Figure 1.8). Therefore, because it is the product of an increasing factor ($W(\varepsilon)$) and a decreasing factor ($e^{-\varepsilon_i/kT}$), the probability distribution of the energy is a bell-shaped curve. For large polypeptides, we might approximate this probability by a Gaussian distribution,

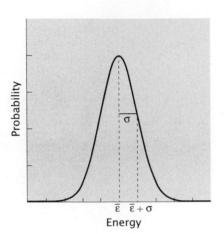

Figure 1.9 A Gaussian probability distribution, with mean $\bar{\varepsilon}$ and standard deviation σ.

$$p(\varepsilon) = \frac{1}{\sqrt{2\pi}\sigma} e^{-(\varepsilon - \bar{\varepsilon})/(2\sigma^2)}, \tag{1.13}$$

where $\bar{\varepsilon}$ is the average value of the energy, which corresponds to the maximum in a Gaussian probability distribution. The Gaussian probability distribution, $p(\varepsilon)$, is plotted in **Figure 1.9**. The standard deviation (σ) is a measure of the width of the energy distribution.

1.3 THE ENERGY OF AN ISOLATED SYSTEM IS CONSTANT; THE ENTROPY INCREASES TOWARD A MAXIMUM

The Energy is a Measure of Motion and Interactions in a System

The temperature is a measure of the kinetic energy of a system. The kinetic energy expresses the motion of molecules: how fast their translational motion is in a solution, how fast they rotate or tumble as a whole, how fast internal rotations are around single bonds, and how fast the vibrations of those bonds are. The potential energy measures interactions. Ultimately interactions are a consequence of electron distributions. When a favorable interaction is established, such as a covalent bond, an ionic pair, or a hydrogen bond, energy is released. To break interactions you need to provide energy, usually in the form of heat. In a system with high energy, fewer favorable interactions exist.

The *first law of thermodynamics* tells us that the energy of an isolated system is constant. The energy can change by the amount of work (w) done and heat (q) exchanged with the surroundings:

$$\Delta E = q + w. \tag{1.14}$$

The energy depends only on the *state* of the system, and not on how the system got there. Because of this, the energy is a *state function*. However, the heat absorbed or released, and the work done by the system or on the system are not state functions. They depend on how the process is performed. For example, if a gas expands slowly against a fixed pressure (provided by a piston) the system (gas) does more work than if it expanded quickly. In the end, however, the energy is the same in both cases, provided the state is the same (if the temperature, the volume, and the number of molecules are the same). The property of being a state function is extremely important. We can calculate the difference in energy between a certain equilibrium state and another, without having to worry about how the system got from one to the other: it doesn't matter, because the energy is a state function. However, the energy of a system does not tell us how the system is going to evolve, how it will change until it reaches equilibrium. The energy just is.

The Entropy is a Measure of the Number of States

What causes a system to evolve is the change of another state function called the entropy (S). Entropy means "transformability." The *second law of thermodynamics* says that the entropy of an isolated system increases toward a maximum. Then, equilibrium is reached. Clausius summarized the first and second laws in a famous statement, "Die Energie der Welt ist konstant; die Entropie der Welt strebt einem Maximum zu" (the translation is essentially the title of this section). But what is entropy? The best way to understand entropy is through a simple experiment. Take a glass beaker from the laboratory and fill it with glass marbles: first put in a layer of clear marbles, then a layer of black marbles, as shown in **Figure 1.10**A.

Then cover the beaker and shake it. After a while, if you keep shaking, your beaker eventually looks like that shown in **Figure 1.10**B. The glass marbles are randomly mixed. Now, if you shake it even more, will the system ever go back to the separated state? No, you *know* it won't. Why not? There is no difference in energy between the states in Figure 1.10A and B. The energy here is purely steric, hard-core repulsion of two marbles if they try to occupy the same space. So what makes the marbles mix in the first place and never separate again? It's entropy. There are just many more ways of having the marbles mixed than separated. The state that we call randomly mixed (B) contains many more *microstates*, or possible arrangements of the marbles

(A)   (B)

Figure 1.10 A beaker with glass marbles separated in two layers (A) and completely (randomly) mixed (B). (Courtesy of Antje Almeida.)

Figure 1.11 Tombstone of Ludwig Boltzmann in *Zentralfriedhof* (central cemetery) in Vienna. In the formula $S = k \log W$, the "log" is the natural logarithm, which we write as "ln." (Courtesy of Herbert Pokorny.)

than the state that we call separated (A). Because it has many more microstates, there are many more ways of achieving the mixed state, and it is *more probable*. By shaking the glass marbles you allow them to sample the available microstates, and you are more likely to obtain an arrangement (microstate) that belongs to the mixed state.

Ludwig Boltzmann (**Figure 1.11**) called the number of ways of obtaining a state *Wahrscheinlichkeit*, a German word that means probability. Because of this, we use the capital letter W to indicate the number of microstates belonging to a state (or macrostate). The entropy is simply related to the number of such microstates by the Boltzmann formula,

$$S = k \ln W, \qquad (1.15)$$

where k is the Boltzmann constant.

The second law of thermodynamics tells us that the entropy of an isolated system increases until it reaches a maximum, and then it stops changing. Statistical mechanics tell us that the system changes until it reaches its most probable macrostate, which is the state with the largest number of microstates. If we require that the number of microstates be the largest possible under the constraints of a certain overall energy E and number of molecules N, we obtain *the Boltzmann distribution*. In a Boltzmann distribution, the probability of each microstate i with energy ε_i is proportional to its *Boltzmann factor*, $e^{-\varepsilon_i / kT}$.

The Entropy can be Explicitly Related to Probabilities

Consider again the experiment with marbles. You have a total of N marbles, of which n_B are black and n_C are clear,

$$N = n_B + n_C.$$

There was only one arrangement in the separated state (see Figure 1.10A); we can say that the number of possible arrangements $W = 1$. Then you shook the beaker, and eventually the system reached the mixed state (see Figure 1.10B), because the marbles exchanged positions, or *permuted*. The total number of such *permutations* is $N!$ (read "N factorial"), which is the product

$$N! = N \times (N - 1) \times (N - 2) \times (N - 3) \times \cdots \times 2 \times 1. \qquad (1.16)$$

To see how this number arises, imagine you have a bag with all the N marbles and you have a box with N slots as shown in **Figure 1.12** (we could make the same argument with the positions in the beaker, but the slots in the box are easier to see).

You want to place a marble in each of the N slots in the box. You can place the first marble (black or clear) in *any* of the N slots. Thus, there are N possible arrangements just from the position of the first marble; this brings in the factor N in Equation 1.16. Then you can place the second marble at any of the *remaining* $N - 1$ positions; this brings in the factor $N - 1$. You can place the third at any of the remaining $N - 2$ slots, which brings in the factor of $N - 2$, etc. Until you get to the last marble, and there is only one place for it, which brings in the factor of 1.

Thus $N!$ would give the total number of different ways of arranging the marbles in the mixed state. However, we are overcounting because

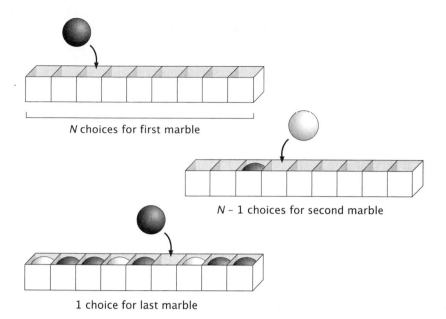

N choices for first marble

$N - 1$ choices for second marble

1 choice for last marble

permutations of black marbles (exchanges of black with black) or permutations of clear marbles do not produce new arrangements. So, to obtain the total number of arrangements or microstates W in the mixed case, we must divide $N!$ by the numbers of permutations among the black marbles and the permutations among the clear marbles. This division corrects for the initial overcounting, and we obtain

$$W - \frac{N!}{n_B! n_C!}. \tag{1.17}$$

This is a very large number if N is large. Even if you have only 100 black marbles and 100 clear marbles in the beaker, the factorials of these numbers are so large that if you enter 100! in your calculator you will get an error message. A useful formula to calculate factorials of large numbers is *Stirling's approximation*,

$$\ln N! = N \ln N - N. \tag{1.18}$$

If we take the natural logarithm of both sides of Equation 1.17, use Stirling's approximation for the factorials, and use $N = n_B + n_C$, we obtain (recall that $\ln(1/x) = -\ln x$)

$$\ln W = N \ln N - N - (n_B \ln n_B - n_B + n_C \ln n_C - n_C)$$
$$= (n_B + n_C) \ln N - n_B \ln n_B - n_C \ln n_C$$
$$= n_B \ln \frac{N}{n_B} + n_C \ln \frac{N}{n_C}. \tag{1.19}$$

The probability p_B of picking a black marble by chance out of the N marbles is just n_B/N, and similarly for clear marbles,

$$p_B = \frac{n_B}{N} \tag{1.20}$$

and

$$p_C = \frac{n_C}{N}. \tag{1.21}$$

So, if we divide Equation 1.19 by N and replace the ratios n_B/N and n_C/N by p_B and p_C, we obtain

$$\frac{1}{N} \ln W = -p_B \ln p_B - p_C \ln p_C. \tag{1.22}$$

Now, the Boltzmann equation tells us that $\ln W = S/k$. Therefore, the entropy S of the mixed state in our marble experiment is

$$S = -Nk(p_B \ln p_B + p_C \ln p_C). \tag{1.23}$$

Since $W = 1$ for the initial (separated) state, we have found that the entropy change upon mixing is

$$\Delta S = -Nk(p_B \ln p_B + p_C \ln p_C). \tag{1.24}$$

This is a very important result. We derived it for mixing marbles of two colors, but we could easily extend it to any number of different colors of marbles. In fact, this formula is valid in general, not only for marbles, but for any number N of molecules of L different kinds, or states. It is also valid, not only for mixing processes, but for any probabilities p_i. It is as general and as important as $S = k \ln W$. So we will write it once more in a more general form. The entropy per molecule (N molecules or systems of $i = 1, \ldots, L$ states or kinds) is

$$S/N = -k \sum_{\text{states } (i)} p_i \ln p_i. \tag{1.25}$$

The Second Law is a Statement about Probability

The second law of thermodynamics is a statement about probability in systems with a large number of atoms or molecules. It is not valid for one molecule or a few molecules. In a very small isolated system, a decrease in entropy *could* be observed. For example, suppose you have a solution of N protein molecules in a volume V of water (1 mL of 1 µM solution), and place it in a chamber, separated from another identical chamber filled with water by a removable partition, as shown in **Figure 1.13**A.

Now you remove the partition, allowing the proteins to occupy the entire volume ($2V$). Eventually you will have a homogeneous solution, with a uniform protein concentration half the original one, as shown in Figure 1.13B, just as in the experiment with marbles. What is the probability that all molecules, by chance, move back to the original chamber? The probability that one protein is found in the original chamber is $1/2$. The probability that all N molecules independently move to the original chamber (by chance) is the product of the probabilities that each molecule is found in that chamber—that is, $(1/2) \times (1/2) \times (1/2) \ldots \times (1/2) = (1/2)^N$. If you have $\sim 10^{15}$ molecules, the probability that they are all found in only one of the two chambers is $(1/2)^{10^{15}} \sim 10^{-10^{14}}$, which is zero for all practical purposes. However, if you only had four molecules, then $(1/2)^4 \approx 0.06$ is not negligible: there is a 6% chance that all molecules will be found in the original compartment. According to the ergodic hypothesis, all four molecules will be found in the original compartment 6% of the time.

How is this probability related to the entropy change? Consider again the mixing process of Figure 1.13. There is only one arrangement with all proteins in the original chamber, so $W = 1$, just like in the case of the separated black and clear marbles. Thus, $S = k \ln 1 = 0$ for the initial

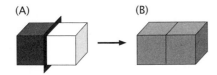

(A)   (B)

Figure 1.13 (A) A protein solution is initially in one chamber of volume V (left), separated by a partition from another chamber with an identical volume V of water. (B) When the partition is removed, the protein equilibrates over the two compartments, producing a homogeneous solution with half the concentration in the volume $2V$.

state. After the proteins spread over both chambers, each protein has a probability $p_L = 1/2$ of being in the left chamber and a probability $p_R = 1/2$ of being in the right chamber. To find the entropy of the final state, we use Equation 1.25, where the p_i are now p_L and p_R (N molecules and two states, in the left chamber or in the right chamber), to obtain,

$$S = -Nk \left(\frac{1}{2} \ln \frac{1}{2} + \frac{1}{2} \ln \frac{1}{2} \right)$$
$$= -Nk \ln \frac{1}{2}$$
$$= Nk \ln 2. \tag{1.26}$$

Thus, the entropy change is

$$\Delta S = Nk \ln 2, \tag{1.27}$$

which is $\sim Nk$, a huge number because N is very large ($N \sim 10^{15}$ and $\ln 2 = 0.69$, which is of the order of magnitude of 1). Now consider the opposite process, by which all proteins would spontaneously move back to the left chamber, by chance. The entropy change is the same, but with the opposite sign,

$$\Delta S = -Nk \ln 2. \tag{1.28}$$

Now, you *know* this process will *not* happen. However, suppose again you only have four proteins ($N = 4$). Then the entropy change is $\Delta S = k \ln(1/2)^4 = k \ln 0.06$, or $-2.8k$. This ΔS is *negative* and *could* be observed. However, in an isolated macroscopic system, even as small as 1 nanomole ($N \sim 10^{15}$), that probability is so tiny that a spontaneous negative entropy change *never* happens. This is what the second law of thermodynamics tells us.

1.4 THE GIBBS ENERGY IS A MINIMUM AT EQUILIBRIUM

The Enthalpy is a Thermodynamic Function More Useful than the Energy in the Laboratory

Usually in the biochemistry laboratory, we cannot control the volume of our system. In the example of Figure 1.5, to prepare an aqueous solution of a polypeptide in a test tube, you weighed a certain mass of polypeptide and measured a certain volume of water, to obtain the required polypeptide concentration (N/V). However, if the temperature varies or the pressure varies, the volume of your solution will change—and it is not easy to control. Therefore, in biochemistry it is much more convenient to use as our system variables the number of molecules, the temperature, and the pressure (N, T, P). The number of molecules and the temperature are easy to control, and the pressure is usually fixed at 1 atm by the atmospheric pressure. The energy, because it is a function of the volume, which we do not control, is not a very convenient thermodynamic property in the biochemistry laboratory. Instead, we use another thermodynamic function, the enthalpy (H), which is related to the energy by

$$H = E + PV. \tag{1.29}$$

The enthalpy is also a state function. The difference between the energy and the enthalpy is that, if the pressure is constant but the volume changes, the enthalpy change (ΔH) includes the work $w = P\Delta V$ done by the system against the fixed external pressure ($P = 1$ atm). Most important, the enthalpy has a practical meaning: ΔH is the heat exchanged at constant pressure (the symbol H for enthalpy comes from "*heat function*"). The heat absorbed or released in a process can easily be measured experimentally. The change in enthalpy is almost always larger that the change in energy (which is the heat exchanged at constant volume). Most substances expand on heating (the melting of ice to liquid water is a notable exception). When you heat a substance at constant volume, the heat is used to increase the temperature or to break molecular interactions—for example in the vaporization of water or in protein denaturation by heat, at constant temperature. If you heat the substance at constant pressure but let the volume vary, the heat supplied is used to increase the temperature or to break molecular interactions, as happened at constant volume, but in addition the system can expand, doing work equal to $P\Delta V$. Because of this additional capacity to take in heat by expansion, $\Delta H > \Delta E$.

In biochemical systems, which are in the liquid or in the solid states for the most part, the volume does not change much with pressure. We say that condensed phases (solids and liquids) are essentially incompressible. Therefore, the term PV is very small in practice, and the enthalpy is almost equal to the energy. It is usually easier to think in terms of energy, because the concept is more familiar and the energy is a more fundamental thermodynamic function, but the enthalpy is more useful in practice.

The Partition Function is Related to the Gibbs Energy

The partition function appropriate for a system at constant N, T, P, such as our polypeptide solution, is similar to that at constant N, T, V, but the enthalpy occupies the place of the energy. We will designate this partition function by the same letter, Q. No confusion should result because this is the partition function that we will use from now on. The partition function is still a sum of Boltzmann factors, now of the form $e^{-H_i/kT}$. If the sum is over all the states, or different conformations of our polypeptide, we have

$$Q = \sum_{\text{states } (i)} e^{-H_i/kT}, \tag{1.30}$$

where terms corresponding to different states with the same enthalpy H_i appear several times in Q. The probability of a state i is then given by

$$p_i = \frac{e^{-H_i/kT}}{Q}. \tag{1.31}$$

Now, in addition to the enthalpy, there is another thermodynamic *state function* that is especially important in systems with constant N, T, P. This function is the Gibbs energy, defined by the combination of the state functions H and S (and T),

$$G = H - TS. \tag{1.32}$$

The Gibbs energy is a *free energy* because it contains an energy component (H) and an entropy component (S).

It turns out that there is a fundamental relation between the partition function and the Gibbs energy of a system with constant N, T, P. Let us see what that relation is. We begin with Equation 1.25 for the entropy in terms of the probability

$$S = -k \sum_{\text{states }(i)} p_i \ln p_i$$

and substitute in it the expression for the probability p_i from Equation 1.31, to obtain

$$S = -k \sum_{\text{states }(i)} p_i \ln \frac{e^{-H_i/kT}}{Q}$$

$$= -k \sum_{\text{states }(i)} p_i(-H_i/kT - \ln Q)$$

$$= \frac{1}{T} \sum_{\text{states }(i)} p_i H_i + k \ln Q \sum_{\text{states }(i)} p_i. \qquad (1.33)$$

The first sum in Equation 1.33 is just the average enthalpy per molecule, which we call H,

$$\sum_{\text{states }(i)} p_i H_i = H, \qquad (1.34)$$

and the second sum is just equal to one,

$$\sum_{\text{states }(i)} p_i = 1.$$

Therefore, we can simplify Equation 1.33 and write the average entropy per polypeptide molecule as

$$S = \frac{H}{T} + k \ln Q. \qquad (1.35)$$

If we multiply both sides by the temperature and rearrange we get

$$-kT \ln Q = H - TS. \qquad (1.36)$$

Now compare Equations 1.36 and 1.32. The right-hand sides are identical, so we have found the general relation between the Gibbs energy and the partition function,

$$G = -kT \ln Q \qquad (1.37)$$

per molecule, or (with $N_A k = R$)

$$G = -RT \ln Q \qquad (1.38)$$

per mole. In either case, Q is the partition function of the system. It could be the partition function of a molecule or an entire system. In the example that we have considered, where the partition function is a *molecular* partition function (the states are the conformations of the molecules, each state i having enthalpy H_i) and the polypeptide molecules are independent of each other, the Gibbs energy of the entire system (G_{system}) of N independent molecules (at constant T and P) is

$$G_{\text{system}} = -NkT \ln Q. \qquad (1.39)$$

We could also have written the partition function Q as

$$Q = \sum_{\text{enthalpy levels }(i)} W_i e^{-H_i/kT}, \qquad (1.40)$$

where W_i is the multiplicity, or the number of microstates, or conformations, that have the same enthalpy H_i. Now, since the entropy is given by the Boltzmann formula $S = k \ln W$, we can invert this equation and express the multiplicity as an entropy by writing $W_i = e^{S_i/k}$. Then the partition function becomes

$$Q = \sum_{\text{levels } (i)} e^{-H_i/kT + S_i/k}$$

$$= \sum_{\text{levels } (i)} e^{-G_i/kT}, \tag{1.41}$$

where $G_i = H_i - TS_i$. The function G_i is the Gibbs energy of a molecule in the enthalpy level H_i. Finally, if we choose the lowest Gibbs energy state as the reference (G_0) and express all Gibbs energies in relation to this state ($\Delta G_i = G_i - G_0$), we can write the partition function as

$$Q = 1 + \sum_{\text{levels } (i)} e^{-\Delta G_i/kT}, \tag{1.42}$$

where the term 1 is the relative probability of the lowest Gibbs energy state, which is now excluded from the summation.

The Gibbs Energy Provides the Criterion for Equilibrium at Constant T and P

When you first studied thermodynamics, you learned that the Gibbs energy decreases in a favorable reaction at constant pressure and temperature,

$$\Delta G < 0. \tag{1.43}$$

This reaction can be a chemical transformation or it can be a physical change, such as a change in protein conformation. Now you can understand why this is so. The Gibbs energy change is the difference between the Gibbs energy of the products and that of the reactants. Since $G = -RT \ln Q$, the Gibbs energy decreases as the partition function increases. This means that $\Delta G < 0$ if the partition function of the products is larger than that of the reactants. What makes the partition function larger? Look at Equation 1.40. The partition function increases with the availability of lower enthalpy (energy) states (because the Boltzmann factor $e^{-H_i/kT}$ increases as the enthalpy H_i decreases) and a large number (or density) of states W_i, especially in the lower enthalpy levels H_i. When a reaction is spontaneous, the system will change in a way that the Gibbs energy decreases until it reaches a minimum, at equilibrium. At that point, any change in the system will leave the Gibbs energy unchanged. Thus, when equilibrium is reached,

$$\Delta G = 0 \tag{1.44}$$

for any possible change in the system.

The Gibbs energy change is the difference between the enthalpy and entropy changes, the latter weighted by the temperature,

$$\Delta G = \Delta H - T\Delta S. \tag{1.45}$$

Therefore, in a process at constant T and P, a negative ΔG can arise from a sufficiently negative ΔH or a sufficiently positive ΔS. A positive ΔS is favorable because it corresponds to an increase in the number of available states. A negative ΔH arises if the new states available (not present in the reactants) lie at lower enthalpy levels. Then, because the probability p_i is proportional to $e^{-H_i/kT}$, those enthalpy levels become preferentially occupied. On average, the enthalpy of the molecules decreases ($\Delta H < 0$). This happens, not because there is anything inherently advantageous in a lower enthalpy (as if the system "would like to give off heat"), but because a greater occupancy of the lower enthalpy levels corresponds to the distribution of molecules with the largest number of states (W).

Conversely, a positive ΔH arises if the new states available lie on average at higher enthalpy levels. Even though those enthalpy levels are higher, they will be occupied to some extent in the products because that results in a distribution with a larger number of states. After the reaction occurs, the molecules have on average a higher enthalpy ($\Delta H > 0$). Now, for this process to occur spontaneously ($\Delta G < 0$), the unfavorable ΔH must be more than compensated by the increase in the number of microstates, corresponding to $\Delta S > 0$.

Finally, keep in mind that the criterion for the increase in entropy as indicative of a spontaneous process applies to isolated systems. If a system can exchange energy with the surroundings, what is important is the change in entropy in the combined system + surroundings. For example, in an endothermic reaction, the system absorbs an amount of heat ($\Delta H > 0$) identical in magnitude to the amount released by the surroundings ($\Delta H < 0$). This process is spontaneous if the entropy increases in the combined system + surroundings.

1.5 THE GIBBS ENERGY DETERMINES THE CONCENTRATIONS AT EQUILIBRIUM

The Chemical Potential of a Component is its Contribution to the Gibbs Energy of the System

Consider the binding reaction of a small molecule X (a ligand) to a protein P in water to form a protein–ligand complex C,

$$P + X \rightleftharpoons C. \tag{1.46}$$

The chemical potential μ_P of the protein is measured by determining how the Gibbs energy of the system changes when the number of moles of protein (n_P) varies,

$$\mu_P = \left(\frac{\partial G}{\partial n_P} \right)_{n_X, n_C, T, P}. \tag{1.47}$$

All other variables are kept constant: the temperature (T), the pressure (P), and the numbers of moles of X (n_X) and C (n_C). The variables held constant are indicated in the subscript of the derivative in Equation 1.47. The symbol ∂ indicates this is a *partial* derivative, which means that only one variable is changed, n_P in this case, while the others are fixed.

In a pure system, which contains only one component, the chemical potential is simply the Gibbs energy per mole:

$$\mu = \frac{G}{n}. \tag{1.48}$$

In the system we are considering, which contains several components, μ_P is given by Equation 1.47, which expresses the contribution of the proteins to the Gibbs energy of the system. To put it simply, μ_P is the Gibbs energy contribution of a protein molecule per mole of protein in this system. The total Gibbs energy is the sum of the contribution of each component per mole multiplied by the number of moles of each component,

$$G = n_P\mu_P + n_X\mu_X + n_C\mu_C. \tag{1.49}$$

The Gibbs energy change (per mole) when one protein binds one ligand to produce one complex is the difference between the chemical potentials of the product and the reactants:

$$\Delta G = \mu_C - \mu_P - \mu_X. \tag{1.50}$$

In the reaction of Equation 1.46 the stoichiometric coefficients are all equal to 1. If they are not all 1, as for example if the protein binds two ligands,

$$P + 2X \rightleftharpoons C, \tag{1.51}$$

these coefficients need to be taken into account in ΔG. The chemical potentials are then multiplied by the corresponding stoichiometric coefficients, 2 in the case of μ_X,

$$\Delta G = \mu_C - \mu_P - 2\mu_X. \tag{1.52}$$

The Chemical Potential Increases with Concentration

In a general reaction

$$aA + bB \rightleftharpoons cC + dD, \tag{1.53}$$

the Gibbs energy change is given by (products minus reactants),

$$\Delta G = c\mu_C + d\mu_D - a\mu_A - b\mu_B. \tag{1.54}$$

The chemical potential of component A is

$$\mu_A = \mu_A^\circ + RT \ln a_A, \tag{1.55}$$

where μ_A° is the standard chemical potential of A and a_A is the *activity* of component A. Let us take a moment to understand the meaning of the two terms in this equation.

What is the activity? The activity is an apparent, or effective concentration. In an ideal solution, the activity a_A of solute A is identical to its concentration [A]. There are two kinds of ideal solutions: dilute ideal and symmetric ideal. In dilute ideal solutions, the concentration of the solute is so low that there are no interactions between solute molecules and no changes to the properties of the solvent due to the presence of solute. In symmetric ideal solutions, the solute is so similar to the solvent that their mutual interactions are identical to those in their pure states. An example is a mixture of two very similar molecules, such as benzene and toluene, or two isotopes of the same molecule. The ideal solution behavior is only approximated in those two kinds of real solutions.

In a real solution, the activity is the product of the concentration and the activity coefficient γ_A,

$$a_A = \gamma_A[A]. \tag{1.56}$$

The activity coefficient measures deviations from ideality; $\gamma_A = 1$ in an ideal solution. You can think of the activity in this way: if molecules of A attract each other (a favorable interaction exists between A molecules), they are *less available* to react with B molecules. If you did not know about this favorable interaction among A, you might think that the concentration of A was *smaller* than it actually is. The activity is this *apparent concentration*. In this case $a_A < [A]$, or $\gamma_A < 1$. Conversely, if the A molecules repel each other (the interaction between A molecules is unfavorable), they are more *prone to react* with B molecules. The concentration of A appears to be *higher* than it really is. The activity of A is higher than its concentration: $a_A > [A]$, or $\gamma_A > 1$. Most often in this book, we will deal with dilute solutions, and make the approximation that $a_A \approx [A]$. Therefore, for dilute solutions we write

$$\mu_A = \mu_A^\circ + RT\ln[A]. \tag{1.57}$$

Activities are unitless. If the activity of a solute can be approximated by its concentration, the activity has the same numerical value of the concentration, but no units. For example, if a 1 μM solution is close to ideal, then the activity of the solute is 10^{-6} (no units).

Now we turn to the meaning of chemical potential μ. The concept, as a *potential*, is easiest to understand through its relation to concentration. The more concentrated a reactant is, the more likely it is to react—simply because there is more of it. The *potential* for reaction increases with concentration. The chemical potential is the Gibbs energy of a component per mole. Therefore, if two solutions s_1 and s_2 containing a component A are in equilibrium, μ_A must be the same in both solutions. Transfer of a molecule of A between the two solutions must have $\Delta G = 0$ at equilibrium. That is, $\Delta\mu_A = 0$, which implies that $\mu_A^{s_1} = \mu_A^{s_2}$. Similarly, if a solution of A is in equilibrium with pure solid A, μ_A must be identical in the solid and in the solution. Whether the activities (defined by Equation 1.55) are identical depends on the definition of the standard states, to which we turn next.

The Standard State must be Clearly Defined

When a substance is in the standard state its activity equals 1 (if you set $a_A = 1$ in Equation 1.55 you get $\mu_A = \mu_A^\circ$). But what is the *standard state*? The standard state can be defined as a matter of convenience for the problem at hand, and it can change depending on the problem, but it must be consistent for each substance within each problem. However, certain conventions are commonly used, and certain definitions of standard states have definite advantages. The following are the most common definitions. The standard pressure is 1 atm. (In 1982 the IUPAC changed the standard pressure to 100 kPa, or 0.987 atm, but in practice, reported standard thermodynamic data in biochemistry refer to 1 atm.) There is no standard temperature; it has to be specified. For solvents and pure liquids the standard state is the pure liquid and it is assigned an activity $a_A = 1$. Thus, the activity of pure water is 1 (the concentration of water is 55.5 M). For pure solids the standard activity is also 1.

For solutes in water, the usual standard state is a *hypothetical state*, obtained by extrapolating to 1 M the beginning of the curve that describes the dependence of solute activity on its concentration, as indicated in **Figure 1.14**. This means that in the standard state the solute concentration is 1 M but the solution behaves as if it

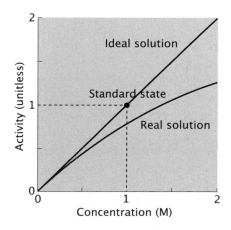

Figure 1.14 Definition of the usual standard state of a solute. The standard state is a hypothetical state, obtained by extrapolating the concentration dependence of the activity to 1 M.

were infinitely dilute. Namely, each solute molecule is completely surrounded by solvent and does not interact with other solute molecules. This standard state does not correspond to any real state, but it is a convenient way of relating Gibbs energies to a common reference.

If the reaction of Equation 1.53 takes place under standard conditions, the Gibbs energy change is $\Delta G°$, given by Equation 1.53,

$$\Delta G° = c\mu_C° + d\mu_D° - a\mu_A° - b\mu_B°. \tag{1.58}$$

For example, in the case of binding of a ligand to a protein (shown in Equation 1.46),

$$P + X \rightleftharpoons C,$$

the standard Gibbs energy change is

$$\Delta G° = \mu_C° - \mu_P° - \mu_X°. \tag{1.59}$$

This means that $\Delta G°$ is the Gibbs energy change per mole when a molecule of ligand X is combined with a molecule of protein P to produce a molecule of complex C, in a solution where the activities (concentrations) of P, X, and C are $a_P = 1$, $a_X = 1$, and $a_C = 1$, and the pressure is $P = 1$ atm. This reaction provides the *meaning* of $\Delta G°$. However, this is not how $\Delta G°$ is *measured*. To measure $\Delta G°$ you need to determine the equilibrium constant for the reaction (there are other ways, but this is the simplest to explain at this point). We turn next to that problem.

The Equilibrium Constant Provides a Measurement of the Standard Gibbs Energy Change

Let us begin again by writing the Gibbs energy change for the the general reaction

$$aA + bB \rightleftharpoons cC + dD,$$

not necessarily under standard concentrations, and replace each chemical potential by its expression in terms of the activity ($\mu = \mu° + RT \ln a$). We obtain

$$\begin{aligned}
\Delta G &= c\mu_C + d\mu_D - (a\mu_A + b\mu_B) \\
&= c\mu_C° + cRT \ln a_C + d\mu_D° + dRT \ln a_D \\
&\quad - (a\mu_A° + aRT \ln a_A + b\mu_B° + bRT \ln a_B) \\
&= c\mu_C° + d\mu_D° - (a\mu_A° + b\mu_B°) + RT \ln \left(\frac{a_C^c a_D^d}{a_A^a a_B^b} \right)
\end{aligned} \tag{1.60}$$

and finally

$$\Delta G = \Delta G° + RT \ln \left(\frac{a_C^c a_D^d}{a_A^a a_B^b} \right). \tag{1.61}$$

At equilibrium $\Delta G = 0$. Therefore,

$$\Delta G° = -RT \ln \left(\frac{a_C^c a_D^d}{a_A^a a_B^b} \right)_{eq}, \tag{1.62}$$

where the subscript *eq* indicates that all the activities (concentrations) are those that exist at equilibrium. The ratio of equilibrium activities of each reactant and product raised to the power of their stoichiometric coefficients (a, b, c, d) is defined as the *equilibrium constant K* for the reaction,

$$K = \frac{a_C^c \, a_D^d}{a_A^a \, a_B^b},$$

(1.63)

where we did not explicitly indicate, but it is understood, that these are activities (concentrations) at equilibrium. Thus, we write Equation 1.62 as

$$\Delta G^\circ = -RT \ln K.$$

(1.64)

This equation provides a way of determining ΔG°, by measuring the equilibrium constant. Note, however, that ΔG° refers to the Gibbs energy change of the reaction when reactants and products are in their standard concentrations—namely all their activities are equal to 1. We can see this directly by substituting $a = 1$ for all the activities on the right-hand side of Equation 1.61. Then you get $\Delta G = \Delta G^\circ$. However, those activities are *not* equilibrium activities. We just used equilibrium to measure ΔG°. Nothing in Equation 1.64 depends on concentration. Equilibrium *constants* depend on the temperature and pressure, but not on concentrations (assuming the approximation of activities by concentrations is valid). The concentrations at equilibrium will vary, depending on the initial concentrations used, but their *ratio K* will not: this is why K is a constant. The equilibrium constant tells us what are the differences in interactions between molecules in their standard states.

Equilibrium Constants are Often Reported with Units (as a Reminder of the Standard State)

Equilibrium constants are unitless because they are ratios of activities, which are themselves unitless. Consider again the binding reaction of a ligand X to a protein P in water to form a protein–ligand complex C (shown in Equation 1.46),

$$P + X \rightleftharpoons C.$$

The equilibrium constant K for this reaction is an *association constant* or *binding constant*, written as a function of the equilibrium activities of C, P, and X (but we omit the subscript eq because it is obvious this is an equilibrium constant),

$$K = \frac{a_C}{a_P \, a_X}.$$

(1.65)

Let us suppose that you determined this equilibrium constant by measuring the equilibrium concentrations of C, P, and X in a very dilute solution, so that the approximation of activities by concentrations is valid, and you obtained $K = 10^6$. Strictly speaking this number is unitless because the activities in Equation 1.65 have no units.

In practice, however, you measure concentrations. If you make the approximations $a_C \approx [C]$, $a_P \approx [P]$, and $a_X \approx [X]$, and use these approximations in Equation 1.65, you get

$$K = \frac{[C]}{[P][X]}.$$

(1.66)

Now suppose you obtained the concentrations [X] = 1 mM, [P] = 1 μM, and [C] = 1 mM. As written in Equation 1.66, if all concentrations are expressed in molar units (M = mol/L), the equilibrium constant K has units of M^{-1}. Hence you can write $K = 10^6$ M^{-1} because the units in Equation 1.66 do not cancel out. You could also write the reaction as a dissociation,

$$C \rightleftharpoons P + X \tag{1.67}$$

and the corresponding equilibrium constant would be a *dissociation constant*, which we abbreviate as K_d,

$$K_d = \frac{[P][X]}{[C]}. \tag{1.68}$$

$K_d = 1/K$. In this case you would write $K_d = 10^{-6}$ M, or more commonly, $K_d = 1$ μM. This way of writing is convenient because the dissociation constant has units of concentration, to which biochemists can relate more intuitively than to a number like 10^6. Therefore, dissociation constants are usually written with units. However, we must keep in mind that, strictly, they have no units because they really are ratios of activities. This is why we can write

$$\Delta G° = -RT \ln K,$$

where RT has units of energy, the same as $\Delta G°$. Thus, the argument (K) of the logarithm cannot have units. Yet writing equilibrium constants with units serves as a reminder that we are expressing concentrations in molar units (or their submultiples) and are using a 1 M standard state. This concludes our summary of thermodynamics. We now move to applications.

1.6 THE PARTITION FUNCTION IS A POWERFUL TOOL TO STUDY DISTRIBUTIONS OF MOLECULES

Throughout this book we will learn a way of thinking, a manner of approaching problems in biochemistry (in chemistry, really). That approach is introduced in this section. The most we can know about a system in equilibrium is the distribution of its molecules over their different conformations or states—that is, how the molecules are *partitioned* among those various states. This knowledge goes beyond knowing the average value of some physical property. For example, we may know that the average degree of native structure, or degree of folding, of the protein ribonuclease A in aqueous solution is one-half. This, however, does not tell us if those protein molecules are all about half folded or if half of them are completely folded (with a structure like that shown in **Figure 1.15**), while the other half are completely unfolded. This information—the *distribution* of ribonuclease A between folded and unfolded conformations—would tell us much more about protein behavior.

We now reintroduce the concept of the *partition function* and its usage in a few simple cases. The partition function tells us how the population of molecules is distributed among the possible states in which they can exist. We call those the accessible states of the molecule. For example, we could consider the partitioning of protein molecules among the possible conformations they can adopt; or the partitioning of 2-butene molecules between *cis* and *trans* configurations; or even the

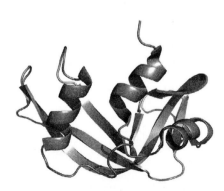

Figure 1.15 The structure of ribonuclease A in its folded state (PDB 7RSA).

partitioning of hydrogen nuclei between their two different spin states. In fact, any system—not only molecules, but also atoms, or nuclear spins—has a partition function.

In the simple cases that we will study first, using the partition function does not have a great advantage over the more common approach that uses equilibrium constants and a little algebra to calculate the populations of the different states present in a sample, but as the situations become more complicated, the power and simplicity of the partition function approach will become evident. In addition to being conceptually simpler, it will allow us to avoid almost all the algebra and concentrate on the physical problem.

Small Proteins are either Folded or Unfolded

Let us first consider protein denaturation. The native structure of a protein, which we know from X-ray crystallography or nuclear magnetic resonance (NMR), is unique. A protein in its native conformation has a single, well defined, three-dimensional structure. This is called the *native* or *folded state* of the protein, shown for ribonuclease A in Figure 1.15. This structure can be lost by a number of processes, such as an increase in temperature, the addition of urea or detergents, or changes in pH. All these processes preserve the covalent bonds of the protein, but lead to a complete transformation of the three-dimensional arrangement of its polypeptide chain to such a degree that there is no apparent order, or regularity. This new conformational state is called the *denatured* or *unfolded state* of the protein.

Protein denaturation is *reversible*. This apparently simple statement means, first, that all the information required for folding is contained in the protein amino acid sequence. No other cellular machinery is required. Second, because the protein always finds the same structure, the native state probably corresponds to the Gibbs energy minimum of the protein conformation. Third, we can apply to protein folding the principles of equilibrium thermodynamics.

Later we will treat protein thermodynamics in greater depth. For now, however, we will introduce the subject as simply as possible. We begin by writing the equilibrium,

$$N \rightleftharpoons D \qquad (1.69)$$

where N represents the native and D the denatured state. (The abbreviations F for folded and U for unfolded will also be used.) At this point, this equilibrium is essentially an assumption. It turns out that the assumption is valid to a very good approximation for small proteins. The equilibrium constant for this reaction can be written as

$$K = \frac{[D]}{[N]}. \qquad (1.70)$$

The implications of this equation are probably not obvious to you at this point. It would not be valid if there were well-populated intermediates between the folded and unfolded states. Under physiological conditions, the equilibrium $N \rightleftharpoons D$ is shifted toward the native state (left). However, $K = 10^{-7}$ to 10^{-3} for most small proteins, which corresponds to a Gibbs energy difference between the folded and unfolded states of only $\Delta G° \approx 5–10$ kcal/mol at room temperature. Thus, the folded state is only favored by a small amount: proteins are only *marginally stable*.

The Probabilities of Each State can be Obtained from the Partition Function

Most likely, you are used to thinking about Equation 1.70, $K = [D]/[N]$, in terms of concentrations. Concentrations, however, are proportional to *probabilities*. Thus, we would like to start thinking about Equation 1.70 in terms of probabilities. Let p_N and p_D be the probabilities of the native and denatured states. This means that if you were able to pick a protein molecule at random from an ensemble in solution, the probability that this protein molecule is folded is p_N and the probability that it is unfolded is p_D. Thus,

$$K = \frac{p_D}{p_N}. \tag{1.71}$$

We can also say that K is the probability of observing the denatured state *relative* to the native state. Now, thermodynamics only tells us about *changes*, not about absolute values. For example, changes in ΔG are meaningful, but the absolute value of the Gibbs energy of a molecular state (such as a folded protein) is not—because it *cannot be measured*. This restriction, rather than a disadvantage, is actually very convenient. It means that we can choose our *reference* state, and we can arbitrarily assign its Gibbs energy. So, let us choose the folded state as the reference and set its Gibbs energy to zero. You could also have chosen the unfolded state—it doesn't matter—but it is wise to choose as reference the state that is most convenient in each problem, the one that makes calculations easiest.

Since we have the freedom to choose the folded state as our reference, it would be convenient to just set $p_N = 1$. Then we would get simply $p_D = K$. However, we cannot set $p_N = 1$ because probabilities must vary between 0 and 1 *and* they must add to 1. If $p_N = 1$, there is nothing left for p_D. Out of a set of complete and mutually exclusive possibilities, one of them *must* occur, and only one *can* occur. That is, the sum of all the probabilities of the various states of a certain observable, in this case the folded and unfolded states of a protein, must equal 1,

$$\sum_i p_i = p_N + p_D = 1. \tag{1.72}$$

However, we *can* define the probabilities of each state relative to our reference state (N) by

$$\frac{p_N}{p_N} = 1 \tag{1.73}$$

and

$$\frac{p_D}{p_N} = K. \tag{1.74}$$

Thus, the numbers 1 and K represent the *relative probabilities* of the native and denatured states. You can think of them as "nonnormalized probabilities." These nonnormalized probabilities are called *statistical weights*. They represent the weight of each state in the distribution. They tell you how likely a state is relative to the reference state.

The partition function (Q) of a system is simply the sum of the statistical weights for all the accessible states of that system. In the case of a small protein, there are only two states: folded and unfolded. The statistical weights are defined relative to the probability of the folded state, which we have chosen as reference. The partition function is the sum of the two relative probabilities. That's all:

$$Q = 1 + K. \tag{1.75}$$

Now we can establish the following correspondences, which we will indicate by an arrow ($\rightarrow$), between the protein states and their statistical weights:

$$N \rightarrow 1 \tag{1.76}$$

and

$$D \rightarrow K. \tag{1.77}$$

In the end, we want to know the absolute probabilities of each state, not just the statistical weights. The partition function allows us to calculate those absolute probabilities very easily. Since it is the sum of all nonnormalized probabilities, the partition function is also the normalizing factor to convert the relative probabilities to absolute ones. To obtain the probability of each state, simply take the term that corresponds to that state in the partition function and divide it by the entire partition function,

$$p_N = \frac{1}{1 + K} \tag{1.78}$$

and

$$p_D = \frac{K}{1 + K}. \tag{1.79}$$

Thus, $p_N + p_D = 1$, as required.

1.7 TWO-STATE SYSTEMS ARE COMMON IN CHEMISTRY

Now let us use this approach to study three simple examples. In each case, the system only has two states. Two-state systems are simple, but very common in chemistry, as illustrated by the following cases.

Ribonuclease A is Almost Entirely Folded at Room Temperature

Our first example is ribonuclease A, a small protein of 124 amino acid residues. To a very good approximation it exists only in either the folded or the unfolded state, but not in any intermediate state. We want to find out how the ribonuclease molecules are distributed between the native and denatured states at room temperature (25°C). That is, what fraction of the protein is folded and what fraction is unfolded?

When trying to understand a biochemical system (or any system), it is very helpful to sketch its accessible states and indicate how those states are "connected," or what the path is to go from one state to the next. This type of diagram is shown in **Figure 1.16** for ribonuclease A. On the left side, we have the folded state (reference), with its associated statistical weight of 1. On the right, we have the unfolded state, with statistical weight K. We go from the folded to the unfolded state by multiplying 1 by K. We write K over the branch connecting the two states to indicate the path.

Figure 1.16 Diagram for unfolding of a protein. The probability of the unfolded state is obtained from that of the folded by multiplying 1 by the factor K written over the branch that connects the two states. (Left structure: PDB 7RSA.)

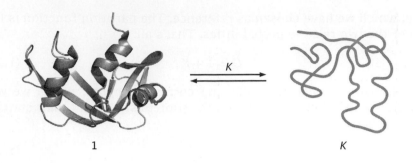

1   K

To proceed, let us write our approach as a recipe to obtain the probabilities of the native (p_N) and denatured (p_D) states:

1. From the equations of the equilibrium constant(s), write the probabilities of all states relative to the reference state by dividing each concentration by that of the reference. This yields $[N]/[N] = 1$ for the native state and $[D]/[N] = K$ for the denatured state. Then the nonnormalized probabilities are

$$N \to 1$$

and

$$D \to K.$$

2. Add the nonnormalized probabilities to obtain the partition function,

$$Q = 1 + K.$$

3. Divide each term (1 and K) by their sum, Q, to obtain the absolute probabilities, or the fractions, of each state

$$p_N = \frac{1}{1 + K}$$

and

$$p_D = \frac{K}{1 + K}.$$

(Note: If you are mathematically inclined, you can verify that you can obtain the probabilities directly by taking the derivative of the natural logarithm of the partition function with respect to the natural logarithm of the equilibrium constant, and making use of $d\ln x = dx/x$ in the second equality below.

$$p_D = \frac{d\ln Q}{d\ln K}$$

$$= \frac{K}{Q}\frac{dQ}{dK}$$

$$= \frac{K}{1 + K}.$$

This procedure is generally valid and saves much work in complicated cases. You should think about why it works.)

At pH 5, ribonuclease A has a melting temperature $T_m \approx 61°C$ and an enthalpy of unfolding $\Delta H° \approx 110$ kcal/mol. With this information you can calculate the equilibrium constant $K = [D]/[N]$. The T_m is the temperature at which there are equal numbers of the two states: $[N] = [D]$ or $p_N = p_D$. Thus, from $K = p_D/p_N$ (shown in Equation 1.70), $K = 1$ at $T = T_m$. (This is a good demonstration that equilibrium constants depend on temperature.) Substituting $K = 1$ in the relation between the Gibbs energy change and the equilibrium constant (shown in Equation 1.64),

$$\Delta G° = -RT \ln K,$$

we obtain $\Delta G° = 0$ at T_m. Also, under standard conditions Equation 1.45 reads

$$\Delta G° = \Delta H° - T \Delta S°. \tag{1.80}$$

In particular at T_m,

$$\Delta G° = 0 = \Delta H° - T_m \Delta S°. \tag{1.81}$$

Now substitute the numerical values of $\Delta H°$ and T_m in this equation:

$$0 = 110 - (61.3 + 273.15)\Delta S°.$$

We obtain $\Delta S° = 0.33$ kcal/mol/K. Next, calculate $\Delta G°$ at room temperature ($T = 298.15$ K) from Equation 1.80 (assuming that $\Delta H°$ and $\Delta S°$ do not depend on temperature),

$$\Delta G° = 110 - 298.15 \times 0.33 \approx 12 \text{ kcal/mol.}$$

The equilibrium constant at room temperature is obtained by inverting Equation 1.64,

$$K = e^{-\Delta G°/RT}$$
$$\approx 10^{-9}.$$

This is a very small number, which means that the protein is almost entirely folded at room temperature. Finally, we can calculate the probabilities of the folded and unfolded states:

$$Q = 1 + K \approx 1,$$

$$p_N = \frac{1}{Q} \approx 1,$$

and

$$p_D = \frac{K}{Q} \approx 10^{-9} \approx 0.$$

We found that there is essentially only one state (folded) populated at room temperature.

The *cis* and *trans* Isomers of 2-butene are Populated to Different Extents at Equilibrium

As our second example, we study the distribution of *cis* and *trans* isomers of 2-butene using the partition function method. The *cis–trans*

Figure 1.17 2-Butene isomerizes in the presence of an acid catalyst.

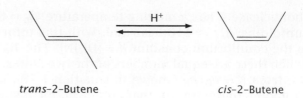

trans–2–Butene   *cis*–2–Butene

isomerization of 2-butene is familiar to most students who took introductory organic chemistry (**Figure 1.17**). You probably learned that there is no free rotation around double bonds. Therefore, *cis* and *trans* are not *conformations* of the same molecule, but different *configurations*. However, in the presence of an acid catalyst, the isomerization reaction takes place.

The equilibrium constant is

$$K = \frac{[cis]}{[trans]}.$$

The equilibrium constant for this reaction is $K = 0.33$. What are the probabilities of each state? Let us choose the *trans* isomer as the reference state. Then, we follow the procedure we just outlined.

1. From the equilibrium constant, write the probabilities of all states relative to the reference state. For *trans*, we have $[trans]/[trans] = 1$; for *cis*, we have $[cis]/[trans] = K$. So we have the following correspondences,

$$trans \rightarrow 1$$

and

$$cis \rightarrow K.$$

2. Add the nonnormalized probabilities to obtain the partition function,

$$Q = 1 + K = 1.33.$$

3. Divide each term (1 and K) by their sum, Q, to obtain the absolute probabilities (fractions) of each state

$$p_t = \frac{1}{1 + K} = 0.75$$

and

$$p_c = \frac{K}{1 + K} = 0.25.$$

Again, there was no need for much algebra. This time, both *cis* and *trans* states are appreciably populated, but one is more probable than the other.

Nuclear Spins in a Magnetic Field have Almost Identical Populations

As our third example, consider a population of hydrogen nuclei (protons). Protons have a nuclear spin of 1/2. You can think of these nuclei as little magnetic dipoles, which we represent by an arrow with a direction given by the spin state. When placed in an external magnetic field B_0 the nuclear spins align with the magnetic field, as shown in **Figure 1.18**. Now there are two possible orientations, which are the two spin states: state α is oriented in the direction of the field, or parallel; state β is oriented opposite to the field, or antiparallel. State α corresponds to spin $+1/2$ and has the lowest energy; state β corresponds to spin $-1/2$, and has higher energy.

The energy difference (ΔE) between the two spin states increases in proportion to the magnetic field B_0,

$$\Delta E = \gamma \hbar B_0 \qquad (1.82)$$

where γ is the gyromagnetic ratio and $\hbar = h/2\pi$, where h is Planck's constant. The dependence of ΔE on B_0 is depicted in **Figure 1.19**.

Transitions between the two spin states, from α to β, are induced by an applied radiofrequency field. The transition occurs when the *resonance condition* is met—that is, when the applied frequency ν is such that $\Delta E = h\nu$. The angular frequency ω is related to ν by $\omega = 2\pi\nu$. Thus, $\hbar\omega = h\nu$, and from Equation 1.82,

$$\hbar\omega = \gamma \hbar B_0$$

$$\omega = \gamma B_0$$

$$\nu = \frac{\gamma B_0}{2\pi}. \qquad (1.83)$$

Let us calculate the energy difference ΔE in an 600-MHz NMR spectrometer. For protons, the gyromagnetic ratio is $\gamma_H = 2.675 \times 10^8$ rad s^{-1} T^{-1} (1 Tesla (T) = 10,000 gauss). In this spectrometer, $\nu = 600$ MHz (the proton frequency); so, from Equation 1.83, the magnetic field is $B_0 = 14.1$ T. The Planck constant is $h = 6.62517 \times 10^{-34}$ J s per particle, or multiplying by Avogadro's number, $h = 3.99 \times 10^{-10}$ J s mol^{-1}. (Note: for energy in calories, $h = 9.54 \times 10^{-11}$ cal s mol^{-1} or $\hbar = 1.518 \times 10^{-11}$ cal s mol^{-1}.)

The result of the calculation is $\Delta E = 0.0572$ cal/mol. Note that this value is in calories, not kilocalories. This is a *very small* energy difference between the two spin states. There are no internal degrees of freedom in a spin. In a molecule, there are internal rotations and vibrations, so different isomers may have different entropies; one may have more microstates than the other, and therefore a higher entropy. There is no such "inner machinery" in a spin state. Therefore, the equilibrium constant between the spins states is determined by the energy difference alone. At room temperature, with $\Delta E = 0.0572$ cal/mol, $K = e^{-\Delta E/RT} = 0.999903$.

We can now calculate the partition function of the spins and determine the probabilities of the states α and β. Again, let us choose the lowest energy state (spin α) as the reference. We simply apply the recipe.

1. The probabilities of the two states relative to the reference are $p_\alpha/p_\alpha = 1$ for α, and $p_\beta/p_\alpha = K$ for β. Thus, we have the correspondences,

$$\alpha \to 1$$

and

$$\beta \to K.$$

Figure 1.18 Orientations of spins in states α $(+1/2)$ and β $(-1/2)$ in an external magnetic field B_0.

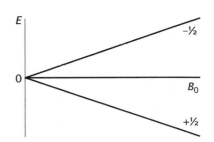

Figure 1.19 The energy difference ΔE between spins $-1/2$ and $+1/2$ increases linearly with the external magnetic field (B_0).

2. The partition function is

$$Q = 1 + K = 1 + 0.999903 \approx 2.$$

3. Divide each term (1 and K) by their sum, Q, to obtain the absolute probabilities, or the fractions, of each state:

$$p_\alpha = \frac{1}{1 + K} = 0.500025$$

and

$$p_\beta = \frac{K}{1 + K} = 0.499975.$$

In this case, both spin states are almost equally populated, with both probabilities $\approx 1/2$. This is because the energy difference ΔE between them is so small. There is almost no reason for a spin to be in one state over the other, so both have about identical populations.

The Partition Function Gives the Number of Occupied States

Let us take a moment to consider the results we obtained in the three cases we studied. There were only two accessible states in each case. In protein unfolding almost all proteins were in the folded state. Only one state was appreciably occupied; the partition function was $Q \approx 1$. In butene isomerization, both states were appreciably populated, but one more than the other; the partition function was $Q \approx 1.3$. In the proton spin system, the two states were equally occupied; the partition function was $Q \approx 2$. In each case, we chose the lowest Gibbs energy state as our reference and assigned its Gibbs energy to zero ($G_0 = 0$); its relative probability is 1. Then, if we do that, the value of the partition function gives the number of states that are *statistically* occupied, or effectively accessible, at equilibrium. (By "statistically occupied" we mean that the probabilities of those states are significant, typically not much less than $1/10$ of the most probable state.) This is another important property of the partition function. As the temperature increases and becomes much larger than the energy differences between the accessible states (all the Boltzmann factors $\to 1$ as T becomes very large compared to the energy levels), all states become occupied and $Q =$ number of states. This is the case already at room temperature for the proton spins because $\Delta E \ll RT$.

1.8 SUMMARY

Interactions in biological macromolecules, and in proteins in particular, are usually weak. A variety of conformations are therefore accessible to biological macromolecules. We can group those conformations into thermodynamic states, such as the folded and the unfolded states of a water-soluble protein, or the states of a membrane protein receptor with and without a bound hormone molecule. Defined in this manner, those states may comprise many microstates, such as the enormous number of conformations belonging to the unfolded state of a protein. However, those microstates can be grouped according to their energy or free energy. Once the states are clearly defined, we can apply the methods of thermodynamics and statistical mechanics to solve problems involving biological macromolecules, just as in much simpler physical chemical systems.

The first law of thermodynamics tells us that the energy of an isolated system (or the universe) is constant. In practice this means that, in any transformation, only the changes (but not the absolute value) of the energy matter. The energy of a system can change by the work done and by the heat exchanged with it surroundings, but it cannot be created or destroyed. The first law in itself, however, does not tell us how a system will evolve. It is the second law of thermodynamics that tells us that a system will change until it reaches the most probable state compatible with its energy. That most probable state is the one that can be obtained in most ways. The Gibbs energy (G) is a free energy because it is a combination of an energetic term, the enthalpy (H), and an entropic term ($-TS$). It is the combination $G = H - TS$ that determines the equilibrium state reached at constant pressure and temperature. However, these laws do not tell us how the macroscopic properties of the system relate to molecular interactions.

It is statistical thermodynamics (statistical mechanics) that relates the microscopic interactions in a system to its macroscopic, observable properties. In this chapter, we began to develop a systematic approach to establish this connection, using the partition function (Q). The partition function is related to the Gibbs energy of the system by $G = -RT \ln Q$. Each term in Q represents the statistical weight of an accessible state of the system. The statistical weights are nonnormalized (relative) probabilities—relative to a state chosen as reference. The ratio of each term to the sum Q is the absolute probability of each state (normalized). Those probabilities are just the fractions of each state in the total population. The partition function also gives the average number of statistically occupied states at equilibrium.

1.9 PROBLEMS

1.1 A system contains eight molecules distributed over three energy states (nondegenerate) with energies $E_0 = 0$, $E_1 = 1kT$, $E_2 = 2kT$. The total energy of the molecules is $E_t = 4kT$. The system is *isolated*; therefore, no heat exchange can occur across its boundary and its total energy is fixed.
(a) Find all possible distributions (how many molecules are in each state) of the eight molecules over the three energy states consistent with the fixed total energy.
(b) Calculate the number of microstates or the number of arrangements W of the eight molecules in each distribution.
(c) Note that the distribution of the molecules changes in time, between the limited set of distributions that you determined. The system is always found in one of those distributions, but not always the same. However, some distributions are more likely than others. If you were to take a snapshot of the system, how likely would it be to be found in each distribution? What is the probability of each distribution?
(d) The observed distribution is the average number of molecules in each energy state. You can obtain it by calculating the *weighted average* of the number of molecules in each state. In the weighted average, the number of molecules in a given energy state in a particular distribution is multiplied (weighted) by the probability of that distribution. Calculate this *average distribution*.
(e) Calculate the energy of the average distribution using the average numbers of molecules in each state and the

energy that each molecule has. (Hint: you must obtain $E_t = 4kT$.)
(f) Now suppose the system still has the same energy states, with energies $E_0 = 0$, $E_1 = 1kT$, $E_2 = 2kT$, and the same number of molecules (eight), but now its *energy is not fixed*. Instead, the *temperature T is fixed*. This is now a *closed* system, which follows a Boltzmann distribution. Calculate this distribution (average number of molecules in each state). (Hint: begin by writing the partition function.) You should find a distribution similar to that obtained in the isolated system, but not identical.
(g) Calculate the energy of the system in the closed system. This time, you will not obtain $E_t = 4kT$. Note that the energy is now an average value, not fixed, because heat can be exchanged with surroundings. The energy is determined by the temperature. For the small number of molecules in our example, the isolated and closed systems have slightly different distributions. However, if the system were very large, the two distributions would become identical.

1.2 A molecule has three states with energies $\varepsilon_0 = 0.3$ kcal/mol, $\varepsilon_1 = 1.0$ kcal/mol, and $\varepsilon_2 = 2$ kcal/mol.
(a) Assume first that none of the energy states is degenerate as shown in Figure 1.7. Calculate the partition function at room temperature using the lowest energy as the reference state. What does the result tell you?

(b) Now suppose you have 100 molecules in the system. Calculate the distribution of molecules at room temperature using the Boltzmann distribution.
(c) Do the same calculation at 70°C.
(d) Finally, suppose you have a molecule with the same energy levels, but this time there are a number of conformations (states) with the same energy for levels 1 and 2, as specified in the energy diagram of Figure 1.8. What is the distribution now at room temperature and at 70°C?

1.3 Suppose you did the marbles experiment we discussed using 100 black marbles and 100 clear marbles in the beaker.
(a) What is W in the separated state?
(b) Calculate W in the mixed state (use Stirling's approximation, $\ln N! = N \ln N - N$).
(c) Now calculate the entropy change from the separated to the mixed state.

1.4 Butane provides a simple example in which the equilibrium constant can be easily calculated from first principles. This is an instructive example of the concept of degeneracy.
(a) Write the partition function for the conformational equilibrium of butane. (Hint: butane molecules are partitioned between two thermodynamic states: *anti* and *gauche* conformations, but there are two distinct *gauche* conformations with the same energy.) See Figure 2.36 for the energy difference $\Delta E°$ between the two states.
(b) Use the partition function to calculate the equilibrium constant K_{eq} between *anti* and *gauche* states at room temperature. Assume $\Delta S° = 0$ between *each gauche* and

the *anti* conformation. Note that we want the equilibrium constant between the *anti* state and *any gauche* state. It can be *gauche*(+) or *gauche*(−), we don't care. K_{eq} is the equilibrium ratio of the probabilities of the *gauche* to *anti* states.
(c) Using the partition function and the numerical values you obtained, calculate the fractions of *anti* and *gauche* conformation of butane at room temperature.
(d) If there *were* no energy difference between the *anti* and *gauche* states ($\Delta E° = 0$), what would be the equilibrium constant K_{eq}?

1.5 At a pressure $P = 1$ atm, the molar volume of ice is 0.0196 L/mol (at 0°C); the molar volume of water is 0.0180 L/mol (between $T = 0°C$ and 100°C); and the molar volume of water vapor is 30.6 L/mol at 100°C and $P = 1$ atm. The heat of melting of ice at 0°C is $\Delta H = 1.435$ kcal/mol, and the heat of vaporization of water at 100°C is $\Delta H = 9.73$ kcal/mol. (Note: 1 atm = 1.013×10^5 Pa (pascal); Pa = N/m^2; J = N·m; 1 m^3 = 10^3 L; 1 cal = 4.184 J.)
(a) What is the energy change ΔE when one mole of ice melts to water? ($T = 0°C$, $P = 1$ atm.)
(b) What is ΔE for vaporization of 1 mole of water? ($T = 100°C$, $P = 1$ atm.)
(c) Compare the difference between ΔE and ΔH in the melting of ice (solid → liquid) and in the vaporization of water (liquid → vapor). Explain the similarity or the difference between ΔE and ΔH in these two cases.

1.6 In a protein solution at 37°C, 95% of the proteins are folded. What is $\Delta G°$ of unfolding at this temperature?

1.10 FURTHER READING

Dill K, & Bromberg S (2011) Molecular Driving Forces, 2nd ed. Garland Science.

Hill TL (1980) An Introduction to Statistical Thermodynamics. Dover.

McQuarrie DA, & Simon JD (1999) Molecular Thermodynamics. University Science Books.

Protein Structure

2.1 PROTEINS ARE POLYMERS OF A SPECIAL KIND

The history of the discovery of protein structures is a history of a search for beauty. Beauty, as we perceive it, derives in great measure from regularity and symmetry. Max Perutz and John Kendrew sought it in the structures of globins (**Figure 2.1**). Linus Pauling sought it in the secondary structure of α-keratins, and found it in an unlikely moment. In this chapter, we will explore protein structures, one of the most important aspects of biochemistry. To do that, we must toil for a moment with the details of what proteins are made of. Pauling knew that detail was important: it was crucial for him in discovering the α-helix.

You'll say that reality is under no obligation to be interesting. To which I'd reply that reality may disregard the obligation but that we may not.

Jorge Luis Borges

Polypeptides are Formed by Amino Acids Connected by Peptide Bonds

Proteins are a special kind of polymer in that they have a unique three-dimensional structure. Synthetic organic polymers are not like that. However, like any other polymer, proteins are composed of units, or monomers, connected by covalent bonds. In proteins, the units are 20 different amino acids. The chemical bond that links two amino acids is an *amide bond*, which biochemists usually call a *peptide bond*. A peptide bond is formed by linking two amino acids, with release of a water molecule (**Figure 2.2**).

Here a dipeptide is formed. If a third amino acid is added, we form a tripeptide, then a tetrapeptide (**Figure 2.3**), and so on. In general,

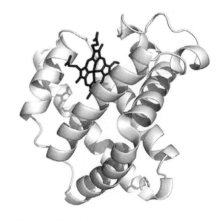

Figure 2.1 Three-dimensional structure of myoglobin in a ribbon representation. The heme group is shown in black (PDB 1A6N).

Figure 2.2 A peptide bond, or amide bond, is a covalent bond formed between two amino acids, with release of a water molecule. Amino acids differ in the groups designated by R_1 and R_2. The α carbon is attached to the NH_2 and COOH groups.

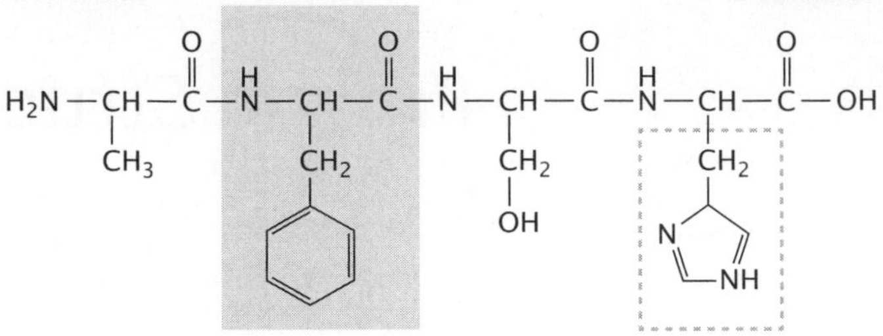

Figure 2.3 A tetrapeptide. The phenylalanine residue is indicated by the solid box. The side chain of the histidine residue is indicated by the dashed box.

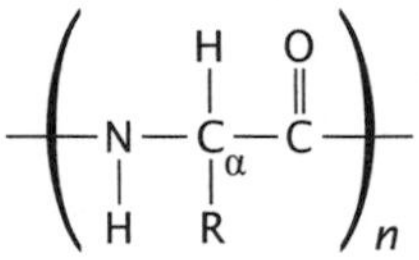

Figure 2.4 The repeat unit of a polypeptide. The sequence N–C_α–C forms the backbone. The side chains are indicated by R.

a polymer of amino acids is called a *polypeptide*. A polypeptide is *not* a protein: it is only called a protein when it acquires a unique *three-dimensional structure*. When incorporated in a polypeptide, the amino acid units are called *residues*. For example, when the amino acid phenylalanine is incorporated into a polypeptide chain, we speak of a phenylalanine residue, shown in the gray box in the tetrapeptide of Figure 2.3. Note this is no longer the *amino acid* phenylalanine. It has lost a proton from the amino group and a hydroxide from the carboxylic acid group—that is, a water molecule.

Any number of amino acids can be joined in this manner, forming a polypeptide chain. The *main chain* is the polypeptide *backbone*, constituted by a repeat of the amide nitrogen (N), the α carbon (C_α), and the carbonyl carbon (C=O or C'). The *side chain*, shown inside the dashed box in Figure 2.3, is the amino acid moiety that is not part of the backbone. The repetitive nature of the polypeptide chain is emphasized in **Figure 2.4**, where the side chains are abbreviated as R.

The Primary Structure of Proteins is the Sequence of Amino Acids

The sequence of a polypeptide is written from the amino terminus, or N-terminus, to the carboxylic acid terminus, or C-terminus. For example, the sequence of the tetrapeptide in Figure 2.3 could be written as H_2N-Ala-Phe-Ser-His-COOH, where Ala, Ser, Phe, and His are three-letter abbreviations of the amino acid names. Here we wanted to emphasize the nature of the two terminal groups (NH_2, COOH). This is sometimes important if the peptide has modified termini. For example, if the N-terminus was modified by acetylation, and the C-terminus was modified by amidation, the sequence would be written as CH_3COHN-Ala-Phe-Ser-His-$CONH_2$, or as Ac-Ala-Phe-Ser-His-amide. For unmodified terminal groups, the chemical nature of the termini need not be specified. By convention, a polypeptide sequence is always written *from the N- to the C-terminus*; in this case, we would write Ala-Phe-Ser-His. For longer sequences, a single-letter notation is used for each amino acid; in the case of this tetrapeptide, we would write the sequence as AFSH.

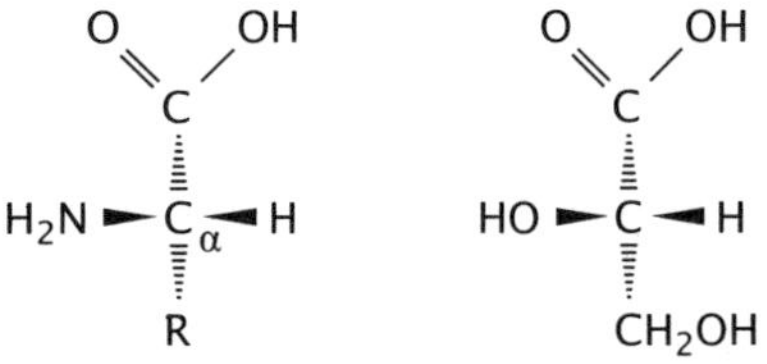

Amino Acids are the Building Blocks of Proteins

L amino acid   L–glyceraldehyde

Figure 2.5 General structure of an L amino acid and L-glyceraldehyde for comparison. The configuration is L if the NH_2 (or OH) group is on the left when the carbon chain is drawn vertically, pointing into the plane of the page, and it is D if the NH_2 (or OH) is on the right.

The naturally occurring α-aminocarboxylic acids, or amino acids for short, have the general structure shown in **Figure 2.5**. The R group is different for each of the 20 amino acids. The central carbon (C_α) is *chiral*, because it has four different substituents attached. This means there are two enantiomers for each amino acid, which are mirror images of each other. The two enantiomers differ in the configuration of the central carbon. These enantiomers are designated by D and L, from the

Latin *dexter*, which means *right*, and *laevus*, which means *left*. The D,L configuration is assigned by comparison with L-glyceraldehyde: if you draw the carbon chain vertically, with the substituents on C_α pointing into the plane of the page, the configuration is L if the NH_2 (or OH) group is on the left, and it is D if the NH_2 (or OH) is on the right. In organic chemistry you learned how to assign the absolute configurations *R* and *S*, and you could do the same here, but there is no universal correspondence between two systems. Only the L enantiomers of the amino acids are found in the proteins synthesized in the ribosomes. However, some of the L enantiomers are *R*, whereas others are *S*. (Nature recognizes the D,L rather than the *R,S* system.)

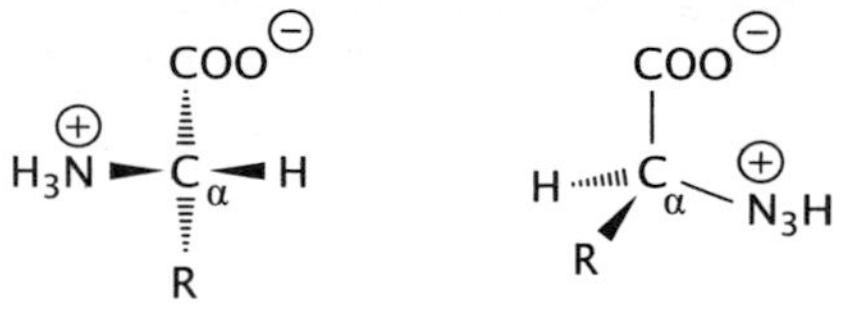

The amino acid form shown in Figure 2.5 is not the major ionization state at pH 7. Rather, in a neutral solution, the amino acids occur in their *zwitterionic* form (**Figure 2.6**). A zwitterion is a chemical species that contains charged groups but has no *net* charge. The ionization state of any chemical group depends on its *acidity constant* K_a, usually expressed as the pK_a, and on the pH of the solution. Recall the definitions of pH and pK_a,

$$pH = -\log[H^+] \qquad (2.1)$$

and

$$pK_a = -\log K_a, \qquad (2.2)$$

where [H$^+$] is the proton concentration (strictly, its chemical activity). We will consider this topic in much greater detail shortly, but for now it is useful to keep in mind that an ionizable group is *mostly protonated*—has a proton (H$^+$) attached—if its pK_a is larger than the pH (pK_a > pH). In the zwitterion, both the amino and the carboxylic acid groups are ionized. At pH 7, the amino group is *protonated* because its conjugate acid, the ammonium group (NH_3^+) has a $pK_a \approx 9.5$ > pH; the carboxylic acid group is *deprotonated* because its $pK_a \approx 2$ < pH. Note, however, that when a polypeptide is formed the α-amino and the α-carboxylic groups of a residue are no longer ionizable (see Figure 2.3).

There are Twenty Amino Acids in Proteins

We are now going to take a tour of the 20 amino acids found in proteins. We will make a few first remarks on some of their properties, such as polarity and ionization (pK_a), but not in too much detail for now. First, however, perhaps some of these questions occurred to you: Why are there *20* amino acids? Why *these* 20? Why the *same* 20 in all living organisms? For example, in the course of evolution, some organisms might have changed the structure of the amino acids in their proteins. But this did not happen. These 20 amino acids have been selected very early during evolution and have remained the same until now. We should look at them with some respect.

The names of the amino acids can be abbreviated by a three- or one-letter code. The simplest amino acid is glycine (Gly, G), which is the only one that is not chiral. Its side chain is just a hydrogen (H) atom (**Figure 2.7**).

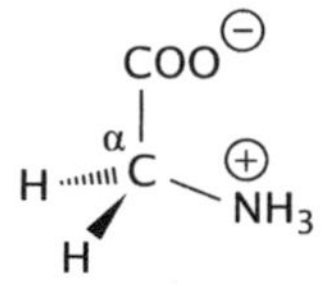

Figure 2.6 The zwitterionic form of the L amino acids shown in two different stereochemical perspectives.

Glycine
Gly

Figure 2.7 Glycine.

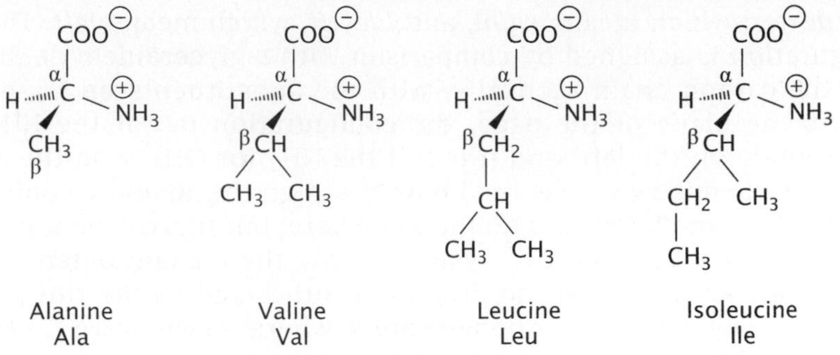

Figure 2.8 The aliphatic amino acids: alanine, valine, leucine, and isoleucine.

Alanine
Ala

Valine
Val

Leucine
Leu

Isoleucine
Ile

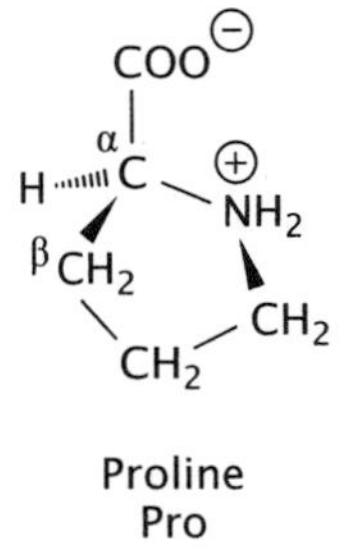

Proline
Pro

Figure 2.9 Proline.

Next, consider the amino acids with *aliphatic*, nonpolar side chains: alanine (Ala, A), valine (Val, V), leucine (Leu, L), and isoleucine (Ile, I) shown in **Figure 2.8**. The carbon atoms are designated sequentially by the Greek letters α, β, γ, δ, ϵ. The α carbon is the first of the chain; it is attached to the NH_3^+ and the COO^- groups.

These amino acids can be seen as constructed by sequentially replacing hydrogens by methyl groups. Replacing H by CH_3 in glycine yields alanine. Then replacing two H in alanine by CH_3 yields valine. Insertion of $-CH_2-$ in the middle of the valine side chain yields leucine, whereas addition of CH_3 at the end yields isoleucine.

Proline (Pro, P) is also aliphatic, but is unique because its side chain connects back to the nitrogen of the amino group, to produce a five-membered ring (**Figure 2.9**).

Now we move on to amino acids with functional groups on their side chains. They come in pairs. There are two amino acids that contain a hydroxyl (–OH) group: serine (Ser, S) and threonine (Thr, T) (**Figure 2.10**). Their side chains are polar but usually not ionizable, with a pK_a similar to that of ethanol ($pK_a = 15.5$). However, serine is often found at the active sites of enzymes, where it reacts as a nucleophile, as in the case of serine proteases.

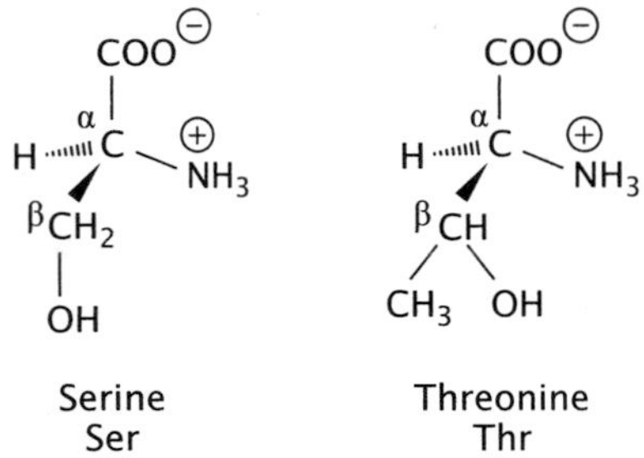

There are two amino acids that contain sulfur: cysteine (Cys, C) and methionine (Met, M) (**Figure 2.11**). They have very different properties. Methionine is a large, nonpolar amino acid, which is normally not reactive. Cysteine is nonpolar in its protonated form, but has a $pK_a \approx 8$ and

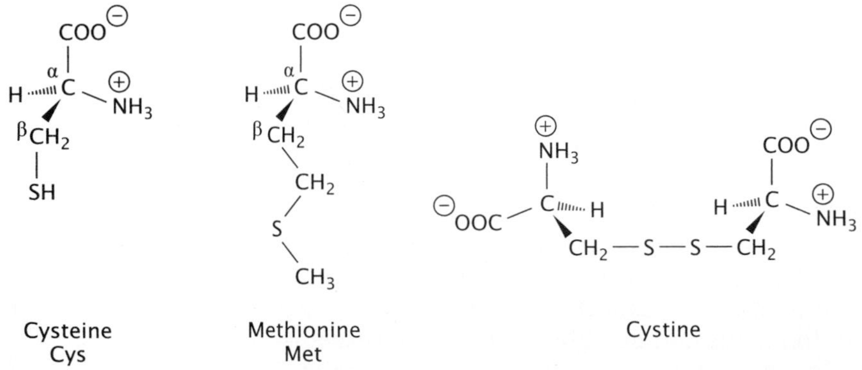

Serine
Ser

Threonine
Thr

Figure 2.10 Serine and threonine.

Cysteine
Cys

Methionine
Met

Cystine

Figure 2.11 Cysteine, methionine, and cystine. Cystine is not an additional amino acid, but is formed when two cysteines link via a disulfide bond.

is therefore appreciably ionized at pH 7. Like serine, it is often a nucleophile in the active sites of enzymes. In addition, cysteine is important for protein structure because it makes *disulfide bonds*, in which two cysteine residues are linked together covalently, forming a cystine (see Figure 2.11).

Next, there are two basic amino acids, lysine (Lys, K) and arginine (Arg, R) (**Figure 2.12**). Both carry a positive charge at pH 7, at the end of a long aliphatic side chain, and are thus strongly polar. The ϵ-amino group of lysine has $pK_a \approx 10$, so that at moderately basic pH it can lose a proton and become a good nucleophile (NH_2). The δ-guanidinium group ($C(NH_2)_3^+$) of arginine, however, has $pK_a \approx 12$ and is therefore always positively charged in the entire pH range of interest for proteins.

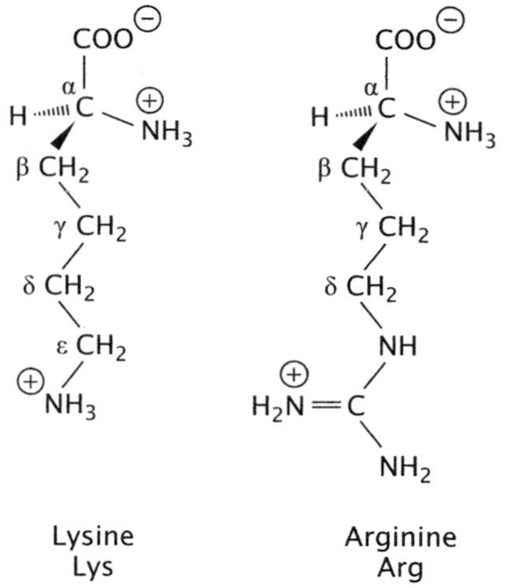

As if to keep the balance, there are also two acidic amino acids, aspartic acid (Asp, D) and glutamic acid (Glu, E), both of which carry a carboxylic acid group at the end of their side chains. Because the carboxylic acid group is fully ionized (deprotonated) at pH 7, these two amino acids are often named as the corresponding carboxylate ions, aspartate and glutamate, which are the forms shown in (**Figure 2.13**). Aspartic and glutamic acids differ only by a methylene group, and have similar acidity constants, with $pK_a \approx 4$. Therefore, they carry a negative charge at pH 7 under most circumstances and are very polar residues found on the surface of proteins. Then there are two amides corresponding to the two carboxylic acids: asparagine (Asn, N) and glutamine (Gln, Q) (see Figure 2.13). They are not charged but are very polar, because the amide group at the end of their side chains makes very favorable interactions with water—namely dipolar interactions and especially hydrogen bonds.

Now we come to histidine (His, H), which occupies a unique position among the amino acids (**Figure 2.14**). Histidine has a weakly basic side chain, the aromatic heterocycle imidazole. It has such special properties that it is better treated in a class by itself, not with the other basic or aromatic side chains. The imidazole group of histidine has a $pK_a \approx 6.5$ in proteins (**Figure 2.15**), which means that at pH 7 significant fractions exist of both the neutral and the positively charged species. The reason for the much lower pK_a compared to aliphatic amines is that, when protonated, the positive charge is delocalized by resonance between the two nitrogen atoms of the ring.

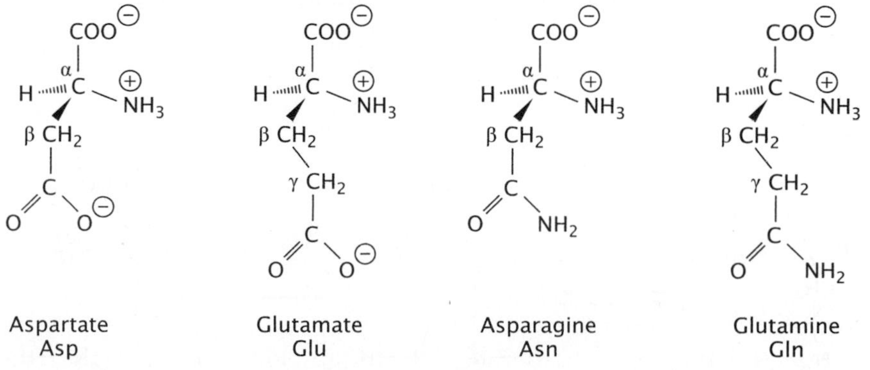

Figure 2.12 Lysine and arginine.

Figure 2.13 Aspartate and glutamate, and their corresponding amides, asparagine and glutamine.

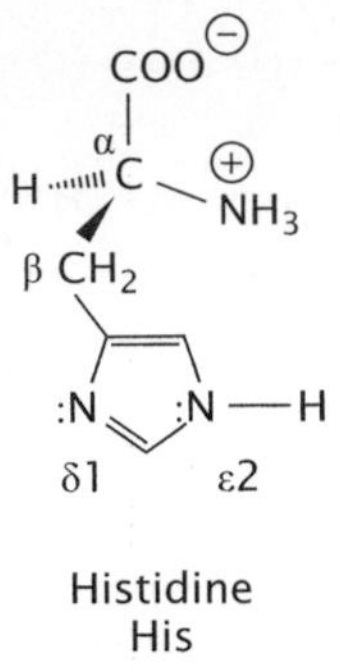

Histidine
His

Figure 2.14 Histidine.

The two nitrogen atoms on the imidazole ring of histidine are not equivalent. They are designated alternatively by $\delta 1$ and $\epsilon 2$, N-1 and N-3, or π and τ. We will use $\delta 1$ and $\epsilon 2$. In model compounds, the pK_a of N-$\delta 1$ is about 0.6 pH units lower than that of N-$\epsilon 2$, which means that $\epsilon 2$ binds the proton slightly better. Thus, the proton spends more time on the nitrogen $\epsilon 2$, but there is a rapid equilibrium (called a tautomerism) between the two forms of the imidazole ring (**Figure 2.16**). In proteins, however, this small difference is obliterated by the effect of the local environment of the histidine residue, especially the location of its hydrogen-bonding partners.

In the neutral form, each nitrogen atom in the imidazole ring has a lone pair of electrons (see Figure 2.14). Why is it that, when *another* proton binds to the ring, it always attaches to the deprotonated nitrogen, never to the nitrogen that already has a proton? That is, the two protons never end up on the same nitrogen. The reason is that in the structure shown the lone pair of electrons on N-$\epsilon 2$ is part of the *aromatic system*. Protonation of N-$\epsilon 2$ would destroy the aromaticity of the ring, and can only occur if the proton concentration is very large, at extremely low pH.

In principle, the neutral form of histidine could also lose its proton to produce a negatively charged species (**Figure 2.17**). The corresponding $pK_a \approx 14$, which means that proton dissociation will not happen in normal conditions. But it can occur if the histidine is located in the very special environment of the active site of an enzyme, as in triose phosphate isomerase.

Finally we come to the aromatic amino acids phenylalanine (Phe, F), tyrosine (Tyr, Y), and tryptophan (Trp, W) (**Figure 2.18**). The aromatic amino acids are difficult to classify in terms of polarity. The hydrocarbon rings are essentially nonpolar, but the π electrons are very polarizable. Phenylalanine is probably the most nonpolar, the one most excluded from contact with water. Tryptophan is also nonpolar, but is very often found on protein and membrane interfaces with water. In addition, the N–H group of its indole ring is a hydrogen-bond donor.

Tyrosine is the most polar of the three. Its side chain is a phenol and, as such, $pK_a \approx 10$ for the –OH proton, instead of $pK_a \sim 15$ in alcohols. The reason for the much lower pK_a is resonance delocalization of the negative charge throughout the ring (**Figure 2.19**).

The 20 amino acids do not occur with the same frequency in proteins (**Table 2.1**). Some are very common, others are rare. For example, alanine and leucine are very common, whereas histidine, cysteine, and especially tryptophan are much rarer. Very often, there is not more than one tryptophan residue in a protein.

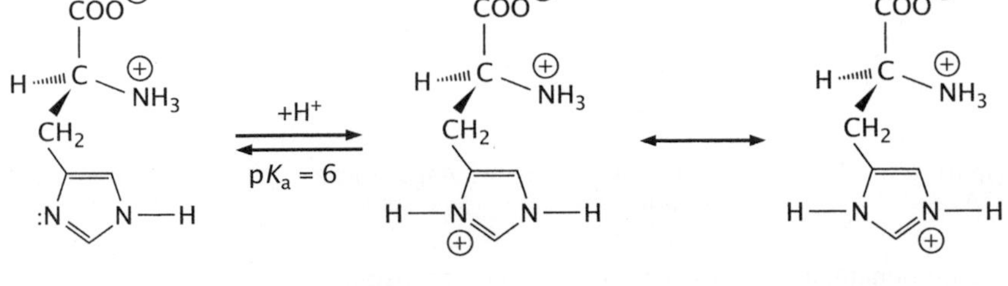

Figure 2.15 Ionization of histidine and stabilization of the cationic form by resonance.

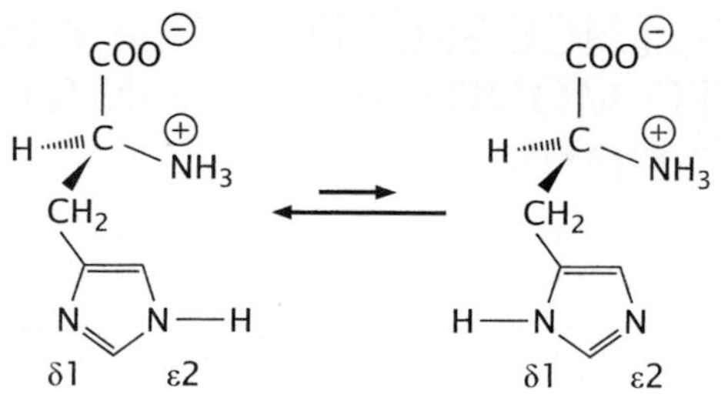

Figure 2.16 Equilibrium between the two protonation sites of the imidazole ring of histidine.

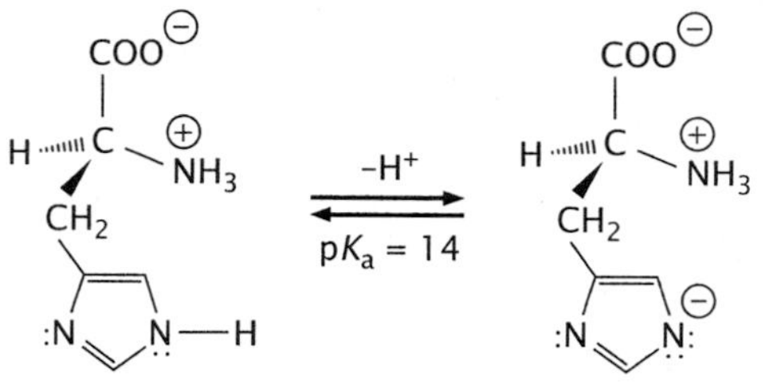

Figure 2.17 Loss of the second proton of the histidine ring ($pK_a \approx 14$) would normally require an extreme pH, but it can happen in the active site of enzymes, as in the case of triose phosphate isomerase (pK_a value from Richard JP [2012] *Biochemistry* 51:2652–2661.)

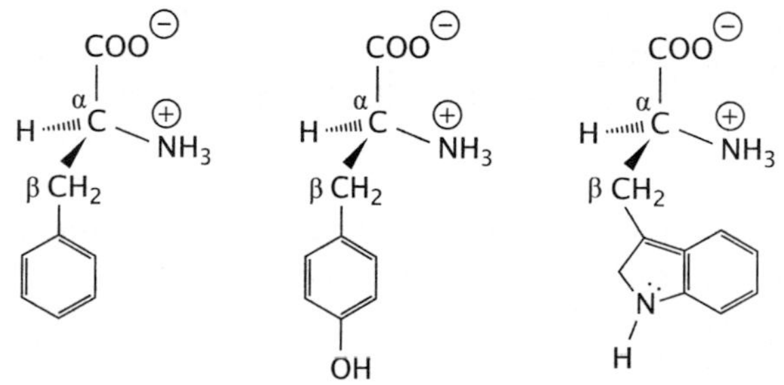

Phenylalanine
Phe

Tyrosine
Tyr

Tryptophan
Trp

Figure 2.18 Phenylalanine, tyrosine, and tryptophan

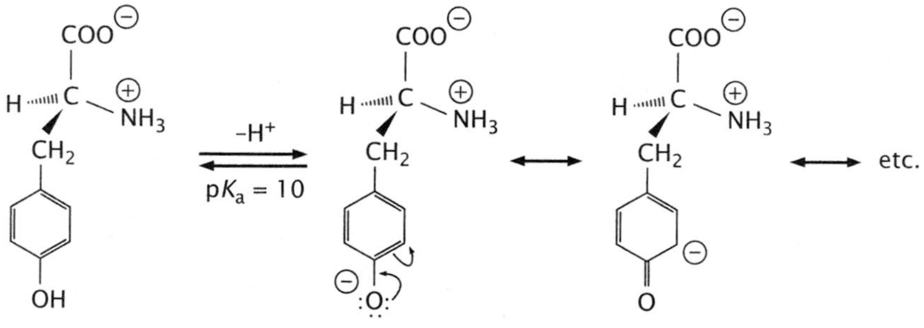

Figure 2.19 Ionization of tyrosine and stabilization of the negative charge in the phenol ring by resonance delocalization.

Table 2.1 Average frequency (%) of occurrence of amino acid residues in the genomes of archaea, bacteria, and eukaryotes[a]

Amino acid	%	Amino acid	%	Amino acid	%
Gly	6.8	Asp	5.2	Ser	6.5
Ala	7.8	Glu	6.7	Thr	5.1
Val	7.0	Asn	4.4	Cys	1.1
Leu	10.2	Gln	3.5	Met	2.3
Ile	7.0	Arg	5.2	Phe	4.4
Pro	4.3	Lys	6.3	Tyr	3.3
		His	2.0	Trp	1.1

[a]Data from Gilis D, Massar S, Cerf NJ & Rooman M [2001] *Genome Biol* 2:research0049.

2.2 FLUORESCENCE SPECTROSCOPY IS A USEFUL TECHNIQUE TO MONITOR PROTEIN STRUCTURE AND INTERACTIONS

Aromatic residues, especially tryptophan (but also tyrosine, albeit to a lesser extent) are very useful probes of protein structure. They absorb ultraviolet (UV) light and, most important, they emit light. Absorption and emission of light by a molecule result in transitions between its electronic states. The energy difference ΔE between the electronic ground state, the state with lowest energy (called S_0, for ground-state *singlet*), and the first excited state (S_1) is large ($\Delta E \approx 100$ kcal/mol). $\Delta E = h\nu$, where h is the Planck constant ($h = 9.54 \times 10^{-11}$ cal s mol^{-1}), and ν is the frequency, which can be written as $\nu = c/\lambda$, where c is the speed of light ($c = 3.0 \times 10^8$ m s^{-1}) and λ is the wavelength ($\lambda \approx 300$ nm for UV light absorbed by aromatic amino acids). If you calculate the distribution of molecules between the states S_0 and S_1, at room temperature, using the partition function, you find that nearly all molecules are in the electronic ground state. Most are also in the vibrational ground state ($\lambda \approx 10$ μm).

When a molecule absorbs a photon of UV or visible light, an electron is promoted from the S_0 to the S_1 state, as indicated by the vertical upward arrows in **Figure 2.20**. In the process, the molecule is also excited to higher vibrational states within S_1. The molecule then quickly relaxes (in $\sim 10^{-12}$ s) to the lowest vibrational state of S_1 by *internal conversion*, a *radiationless* energy transfer to lower vibrational states with concomitant release of heat to the solvent. Finally, the molecule relaxes to the S_0 state (vertical downward arrows) with emission of a light photon: this emission is called *fluorescence*.

Now if you look carefully at Figure 2.20, you see that the transitions in absorption correspond in general to larger ΔE than in emission because of the vibrational levels involved. Therefore, on average, the energy of the emitted photon is lower than that of the absorbed photon: Fluorescence occurs at a longer wavelength than absorption. This shift to longer wavelength in emission is called the *Stokes shift*. A Stokes shift always happens to longer wavelengths (lower energy).

Figure 2.21 shows the absorption (A) and emission (B) spectra of the aromatic amino acids. Recall that, according to the Beer–Lambert law, the absorbance is given by $A = \epsilon Cl$, where ϵ is the molar absorptivity or extinction coefficient, C is the molar concentration, and l is the path length (in cm). Tryptophan has an absorption maximum $\lambda_{max} = 280$ nm and a fluorescence emission $\lambda_{max} \approx 350$ nm in water, which is a large Stokes shift.

Figure 2.20 Absorption and emission (fluorescence) of light. When a photon of the correct frequency is absorbed by a molecule (left), an electron is promoted from the ground state (singlet S_0) to the excited state (singlet S_1). The decay from higher vibrational states of the S_1 electronic state occurs by internal conversion. Fluorescence emission (right) then occurs from the lowest vibrational state of S_1, and is on average of lower energy than the absorption.

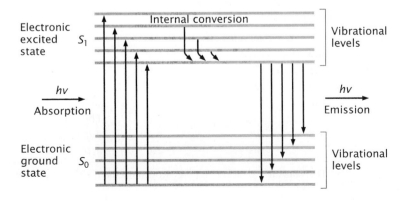

(A)

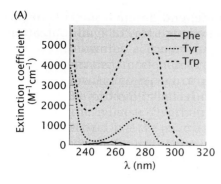

(B)

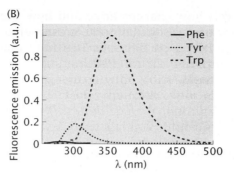

Figure 2.21 (A) Absorption and (B) emission spectra of the aromatic amino acids in aqueous solution, 0.1 M phosphate buffer, pH 7: tryptophan (dashed), tyrosine (dotted), and phenylalanine (solid line). (Data from Du H, Fuh R-CA, Li J et al. [1998] *Photochem Photobiol* 68:141–142.)

Table 2.2 Spectral properties (absorption and emission) of the aromatic amino acids at pH 7[a]

Amino acid	Absorption		Emission	
	λ_{max} (nm)	ϵ (M^{-1}cm^{-1})	λ_{max} (nm)	Quantum yield
Tryptophan	280	5.6×10^3	350[b]	0.2
Tyrosine	275	1.4×10^3	303	0.2
Phenylalanine	258	2.0×10^2	282	0.04

[a]Data from Cantor CR & Schimmel PR [1980] Biophysical Chemistry. Freeman; Lakowitz JR [2006] Principles of Fluorescence Spectroscopy. Springer; and Creighton TE [1993] Proteins. Structure and Molecular Properties. Freeman.
[b]$\lambda_{max} \approx 350$ in water, $\lambda_{max} \approx 320$ in a nonpolar medium.

The main spectral properties of the aromatic amino acids are listed in **Table 2.2**. The quantum yield is the ratio of the number of emitted to absorbed photons. Tryptophan and tyrosine have similar and moderate quantum yields (~0.2), but the extinction coefficient (ϵ) of tyrosine is four times smaller than that of tryptophan. The emission of tyrosine depends on pH, because of ionization of its phenol side chain. Phenylalanine, however, absorbs very weakly and has a very low quantum yield; for all practical purposes, it does not fluoresce.

The emission wavelength is sensitive to the environment, which is why fluorescence is a very useful probe of protein structure. Tryptophan fluorescence occurs at $\lambda_{max} \approx 350$ nm in water but at ≈ 320 nm in a nonpolar solvent. We say that an emission shift to a longer wavelength is a *red shift*, because red light has a longer wavelength ($\lambda \approx 680$ nm) than blue light ($\lambda \approx 480$ nm). Conversely, if the emission wavelength is shorter in a different environment, we say there is a *blue shift*. The side chain of tryptophan is nonpolar. When a protein is folded, this residue is usually in a nonpolar environment and emits at $\lambda \approx 320$ nm; but when the protein unfolds, tryptophan becomes exposed to water and its emission is red-shifted to $\lambda \approx 350$ nm.

2.3 CHARGES IN A PROTEIN ARE DETERMINED BY THE IONIZATION OF ITS AMINO ACIDS

Charges in proteins exist as a consequence of ionization of amino acid residues. Charges play an essential role in molecular recognition—for example, in protein–protein interactions, such as antigen–antibody interactions, or in substrate binding to the active site of an enzyme. Charges are also essential for protein solubility in water, and in their interactions with membranes. Therefore, we need to understand very well how charges arise and how they depend on environment. In this section we concentrate on the effect of pH on amino acid ionization. We will also begin to see how using the partition function method simplifies our calculations.

The pK_a of a Chemical Group is a Measure of Its Proton (H$^+$) Binding Affinity

Amino acids contain at least two ionizable groups, attached to their α carbon. These are the α-carboxylic acid group (COOH) and the α-amino group (NH$_2$). In a carboxylic acid, the acidic or protonated form (COOH) is neutral and the basic form (conjugate base) is negatively charged (COO$^-$). In an amine, the acidic or protonated form is an ammonium ion (NH$_3^+$), which is positively charged, and the basic form (conjugate base) is neutral (NH$_2$). The concept involved, however, is the same: the acidic form loses a proton (H$^+$) as it is converted to the basic form. If the side chain is also ionizable, an amino acid has three ionizable groups. We will define pK_a rigorously in the next section, but it is useful to keep in mind that the pK_a of an ionizable group is a measure of its affinity for H$^+$: the larger the pK_a, the greater the proton affinity.

Table 2.3 lists the pK_a's of the α-carboxylic (pK_C), the α-amino (pK_N), and the side chain (pK_R) of the amino acids. Note that pK_N relates to the loss of a proton from the ammonium ion (NH$_3^+$) to produce the amino group (NH$_2$), *not* to the loss of yet another proton from the amino group (NH$_2$) to produce an amide ion (NH$^-$). However, we will refer to pK_N as the pK_a of the amino group (NH$_2$). The values listed in Table 2.3 for the side chains (pK_R) in peptides are the best estimates for the unperturbed pK_a's of the amino acid residues in proteins. The values of pK_R in proteins (last column) correspond to averages in folded globular proteins, which include values of anomalous pK_a's, for residues in special environments (active sites, for example).

One Ionizable Group has Two Forms, Depending on pH

Let us begin our study of protonation of amino acids with one ionizable group. We want to find out which of its chemical species exist as the pH is varied. To simplify notation, we will write HA for the protonated form (COOH, NH$_3^+$, or any other acidic group) and A for the deprotonated form (COO$^-$ or NH$_2$). Thus, HA and A may refer to neutral or charged species; we will usually omit charges in HA and A. The equation for dissociation of the proton (H$^+$) is

$$HA \rightleftharpoons A + H^+ \tag{2.3}$$

and the corresponding *acidity constant* (K_a) is the equilibrium constant for the *proton dissociation* reaction,

$$K_a = \frac{[H^+][A]}{[HA]}. \tag{2.4}$$

Note that because K_a is a dissociation constant, the smaller the K_a, the better the binding of the proton to A, and the more stable is the bound state HA.

Now, when solving any problem, the first question you need to ask is, "What exactly do I want to know?" Here, the answer is that we would like to describe the fractions of the different forms of the ionizable

Table 2.3 The pK_a values of the ionizable groups of amino acids and amino acid residues: pK_N is the pK_a of the α-NH_3^+ group, pK_C is the pK_a of the α-COOH group, and pK_R is the pK_a of the side chain[a]

	Free amino acid			Peptides[b]	Proteins[c]
	pK_C	pK_N	pK_R	pK_R	pK_R
Gly	2.3	9.6			
Ala	2.3	9.7			
Val	2.3	9.5			
Leu	2.3	9.6			
Ile	2.3	9.6			
Pro	2.0	10.5			
Ser	2.1	9.1			
Thr	2.2	9.0			
Cys	1.9	10.3	8.1	8.6	6.8 ± 2.7
Met	2.2	9.1			
Asp	2.0	9.7	3.7	3.9	3.5 ± 1.2
Glu	2.2	9.6	4.2	4.3	4.2 ± 0.9
Asn	2.0	8.7			
Gln	2.2	9.0			
Arg	2.0	9.0	12.1	12.3	
Lys	2.2	9.2	10.7	10.4	10.5 ± 1.1
His	1.7	9.1	6.0	6.5	6.6 ± 1.0
Phe	2.2	9.1			
Tyr	2.2	9.0	10.1	9.8	10.3 ± 1.2
Trp	2.4	9.3			
N-terminus				8.0	7.7 ± 0.5
C-terminus				3.7	3.3 ± 0.8

[a]Data for free amino acids from the CRC Handbook of Chemistry and Physics, 88th ed. CRC Press. Data for peptides and proteins from Grimsley GR, Scholtz JM & Pace CN [2009] *Protein Sci* 18:247–251, and Pace CN, Grimsley GR & Scholtz JM [2009] *J Biol Chem* 284:13285–13289.
[b]Values of pK_a's in aqueous solutions of pentapeptides of Ac-Ala-Ala-X-Ala-Ala-NH_2, where X is the residue of interest, and the C- and N-termini are blocked (uncharged) by acetylation (Ac, CH_3CO) and amidation (NH_2).
[c]Average and standard deviations of pK_a's in 78 folded, globular proteins.

groups of the amino acids, the deprotonated (A) and the protonated (HA) states, as a function of pH. That is, we want

$$f_0 = \frac{[A]}{[A_T]} \tag{2.5}$$

$$f_1 = \frac{[HA]}{[A_T]} \tag{2.6}$$

where $[A_T]$ is the total concentration of the amino acid,

$$[A_T] = [HA] + [A]. \tag{2.7}$$

Second, before attempting to solve the problem, we need to know if that is even possible. Here we have two unknowns, so we need two relations between them. Which brings us to the second question you need to ask: "What do I know?" Well, we know the total amino acid concentration in solution, $[A_T]$, because presumably we made the solution correctly. We just don't know how much of A_T is in the form HA and how much is in the form A. Those two concentrations are further constrained by the equilibrium constant (shown in Equation 2.4).

So we have two unknowns, the concentration [HA] and [A], and two equations (shown in Equations 2.4 and 2.7). Apparently, there is one more unknown: the proton concentration $[H^+]$. However, $[H^+]$ is related to the pH by Equation 2.1,

$$pH = -\log[H^+]$$

and we know the pH because we can *measure* it. Strictly speaking, we measure the proton chemical *activity* (a_H), but we will use the approximation $a_H \approx [H^+]$ in dilute solutions. Recall, however, from Chapter 1, that the activity of pure water is 1. Therefore, if we write the acidity constant for water K_w, corresponding to Equation 2.4, as

$$K_w = \frac{[H^+][OH^-]}{[H_2O]},\tag{2.8}$$

we must use the activity of water ($a_w = 1$) instead of the concentration $[H_2O] = 55.5$ M in Equation 2.8, because 55.5 M is *not* dilute. Since the *ionic product* of water is $K_w = [H^+][OH^-] = 10^{-14}$, the pK_a of water is 14 (not 15.7, as we would obtain by using $[H_2O] = 55.5$ M in Equation 2.8).

Now we are ready to calculate f_0 and f_1. This is just algebra. Using the definition of $[A_T]$ (shown in Equation 2.7) we can write the fractions f_0 and f_1 (shown in Equations 2.5 and 2.6) as

$$f_0 = \frac{[A]}{[A] + [HA]} = \frac{1}{1 + [HA]/[A]}.\tag{2.9}$$

The ratio $[HA]/[A]$ can be obtained from Equation 2.4 for K_a as

$$\frac{[HA]}{[A]} = \frac{[H]}{K_a}.\tag{2.10}$$

Substituting this expression in Equation 2.9 we obtain for f_0

$$f_0 = \frac{1}{1 + [H]/K_a}.\tag{2.11}$$

Similarly, the protonated fraction f_1 is given by

$$f_1 = \frac{[HA]}{[A] + [HA]} = \frac{[HA]/[A]}{1 + [HA]/[A]}\tag{2.12}$$

and finally,

$$f_1 = \frac{[H^+]/K_a}{1 + [H^+]/K_a}.\tag{2.13}$$

Take a moment to appreciate the symmetry in Equations 2.11 and 2.13. You will encounter this type of symmetry any time you deal with various populations (fractions) of chemical species as a function of an external variable, such as pH or temperature. There is a certain correspondence of terms. If $[A] \leftrightarrow 1$, then $[HA] \leftrightarrow [H]/K_a$. If we choose $[A]$ as our *reference state*, then, *relative* to the concentration of A, the concentration of HA is given by $[H]/K_a$. You probably recognize the same type of reasoning we used when we wrote partition functions in Chapter 1. It is the same. The denominator is just the partition function. We will use that reasoning from the beginning in the next section.

The Partition Function Allows a Rapid Calculation of the Fractions of Acidic and Basic Forms

Now let us use the partition function to solve the same problem, the calculation of the fractions of acidic and basic forms of an ionizable group as a function of pH. The first step is to sketch the accessible states and their connection, the path from one state to the other. The two states are the protonated and the deprotonated forms, HA and A. The diagram for this case is shown **Figure 2.22**.

Recall how we derived the probabilities. First we chose the *reference state*; let us choose A. Then we used the equation of the equilibrium constant to write the probabilities of all states *relative* to the reference state, by dividing each concentration by that of the reference state. Thus, for the deprotonated state we get [A]/[A] = 1 (reference); for the protonated state, we rearrange Equation 2.4 to get

$$\frac{[HA]}{[A]} = \frac{[H^+]}{K_a}. \tag{2.14}$$

Note well that this equation tells us that to obtain the probability of HA we multiply the probability of A by $[H^+]/K_a$. Thus, the factor $[H^+]/K_a$ connects the probability of A to that of HA; we write it over the branch in the diagram of **Figure 2.23**.

Therefore, we have the following correspondences between each state and its relative probability.

$$A \to 1$$

$$HA \to [H^+]/K_a$$

The only significant difference from the case of protein unfolding is that now the proton concentration appears as a factor in the probability of HA. This makes sense because the greater $[H^+]$, the more *likely* it is that the group will be protonated. It is just mass action. Also, because K_a is the equilibrium constant for proton *dissociation*, it appears in the denominator: the smaller the K_a, the more *likely* it is to find the proton bound to A.

Now we write the *relative probabilities* below each state to complete the diagram (**Figure 2.24**). To obtain the partition function, we just add the relative probabilities (statistical weights).

$$Q = 1 + [H^+]/K_a. \tag{2.15}$$

To obtain the absolute probabilities (fractions), divide each term by their sum.

$$f_0 = \frac{1}{1 + [H^+]/K_a}$$

$$f_1 = \frac{[H^+]/K_a}{1 + [H^+]/K_a}.$$

In the diagram (see Figure 2.24), the statistical weights (1 and $[H^+]/K_a$) are indicated below each state, and the factor connecting the two probabilities ($[H^+]/K_a$) is written above the branch that connects those two states. We arrived at the same result as before, but we used reasoning instead of algebra.

The Fractions of Protonated and Deprotonated States Change Sharply When the pH Approaches the pK_a

We are now ready to write the fractions of deprotonated and protonated states, f_0 and f_1, as explicit functions of pH and pK_a. Equations 2.1

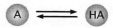

Figure 2.22 Diagram for binding of a proton to the basic form (A) of an ionizable group, producing the acidic form (HA).

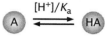

Figure 2.23 Diagram for one ionizable group, with two states. The factor $[H^+]/K_a$, written over the branch, connects the probability of the deprotonated state to that of the protonated state.

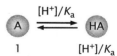

Figure 2.24 Diagram for binding of a proton to the basic form (A) of an ionizable group, producing the acidic form (HA). The relative probability of the protonated state is obtained by multiplying the relative probability of the deprotonated state, 1, by the factor $[H^+]/K_a$ written over the branch that connects them.

Figure 2.25 Fractions of ionization states of the side chains of (A) Glu, (B) His, and (C) Tyr as a function of pH.

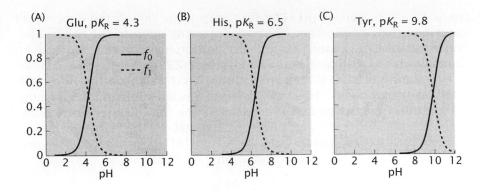

and 2.2,

$$pH = -\log[H^+]$$

$$pK_a = -\log K_a$$

can be inverted to yield,

$$[H^+] = 10^{-pH} \tag{2.16}$$

$$K_a = 10^{-pK_a}. \tag{2.17}$$

Now we substitute 10^{-pH} for $[H^+]$ and 10^{-pK_a} for K_a in the ratio $[H^+]/K_a$ in Equations 2.11 and 2.13 (the sign of pK_a in the exponent switches because 10^{-pK_a} appears in the denominator of the ratio $[H^+]/K_a$). We obtain

$$f_0 = \frac{1}{1 + 10^{(pK_a - pH)}} \tag{2.18}$$

and

$$f_1 = \frac{10^{(pK_a - pH)}}{1 + 10^{(pK_a - pH)}}. \tag{2.19}$$

Let us plot those fractions against pH for a few amino acid side chains, using the values of pK_R listed in Table 2.3. **Figure 2.25** shows the effect of pH on the ionization of a glutamic acid residue (A), a histidine residue (B), and a tyrosine residue (C). In all cases, the solid line represents f_0 (A) and the dashed line f_1 (HA). The lines cross at $pH = pK_a$ when both fractions equal 1/2.

The two populations change very sharply about pK_a; f_0 increases and f_1 decreases. Mathematically, this is because of the exponential dependence of f_0 and f_1 on $pK_a - pH$ (shown in Equations 2.18 and 2.19). Physically, this is because of the *buffering capacity* of the ionizable group in the region about $pH = pK_a \pm 1$. An excess of H^+ is absorbed by A to produce HA, whereas an excess of OH^- is absorbed by HA, which donates a proton to the hydroxide to produce water. In neither case does the acid (H^+) or base (OH^-) appear in solution; therefore, close to pK_a, the pH does not change much when we add an acid or a base.

Observable Properties of Molecular Ensembles are Weighted Averages of Individual Contributions

How are the properties of individual molecules related to observable physical properties? Suppose you have an aqueous solution of the polypeptide ASALHTS. You want to know its *average* charge at pH 6.0.

The first thing to do is to find out what ionizable groups are present in this peptide. In this case, there are the N- and C-termini, and a histidine residue. The pK_a's of the termini, however, are very far from the pH. If you look at Figure 2.25, you see that the fractions of the different states of ionization only begin to change when the pH is less than 2 units away from the pK_a. As the pH increases, approaching pK_a from below, the fraction of deprotonated form, f_0, begins to rise from zero, and the fraction of protonated form, f_1, begins to fall from 1. But the fraction of the deprotonated form only becomes appreciable ($>10\%$) when $pK_a - pH \leq 1$.

In a polypeptide, $pK_a = 8$ for the N-terminus and $pK_a = 3.7$ for the C-terminus. At pH 6, we are 2 pH units away from the pK_a of the N-terminus and 2.3 pH units away from the pK_a of the C-terminus. Thus, the N-terminus is completely protonated, with a $+1$ charge, and the C-terminus is completely deprotonated, with a -1 charge. Those two charges cancel out, and the charge of our peptide is determined only by the charge of the histidine residue; that is the only one we need to worry about.

For the histidine side chain, $pK_a = 6.5$. Thus, pH = 6 $< pK_a$, so we know that in a solution of this peptide most histidine residues will be protonated—but not all. The *average* charge will not be $+1$, but some *fractional* value between 0 and 1. That is, the observable charge of the ensemble of histidine residues in solution is an average of the charges of the individual molecules (either 0 or $+1$) *weighted* by the fractions of the two forms that exist, f_0 with charge 0, and f_1 with charge $+1$. Now this problem is easy, because we just learned how to calculate those fractions from the partition function. All we need is to use Equations 2.18 and 2.19 with the values of pH = 6.0 and $pK_a = 6.5$; we obtain $f_0 = 0.24$ and $f_1 = 0.76$. Finally, our *observable*, the average charge, is $0.24 \times 0 + 0.76 \times (+1) = +0.76$. This observable charge is the result of *fast exchange* of protons between species: protonation and deprotonation are extremely rapid events.

Now, you may say, "Wait a minute, why, or how, is this average charge *observable*?" It actually *is* observable. If we run an electrophoresis experiment, where the migration of the polypeptide in a solution under the influence of an electric field is observed, we find that it is the *average* charge that determines how much the peptide moves. When pH = pK_a the imidazole has a mean charge of $+0.5$. If you do the experiment at pH $\approx 5 \ll pK_a$, the imidazole group has almost a full positive charge ($+1$) and the polypeptide will move a lot in the electrical field. But at pH 7 its average charge is only about $+0.24$, and the peptide will move little from the origin.

The Problem with Two Ionizable Groups is Easy to Solve Using the Partition Function

Let us now consider two ionizable groups in an amino acid. For example, the amino acid alanine has a carboxylic acid group (COOH) and an amino group (NH_3^+) attached to its α carbon, but does not have an ionizable side chain. We use pK_C for COOH and pK_N for NH_3^+. Note that $pK_C \approx 2$ and $pK_N \approx 9.5$ for *all* amino acids (see Table 2.3). This means that the proton binds much better to the amino group (NH_2) than to the carboxylate group (COO^-). Conversely, as dissociation takes place, the first proton that comes off is *almost exclusively* that of the COOH group. Thus, to a very good approximation, the first dissociation constant measured experimentally is identical to the

Figure 2.26 Diagram for an amino acid with two ionizable groups.

dissociation constant of the COOH group (pK_C), and the second experimental dissociation constant is identical to the dissociation constant of the NH_3^+ group (pK_N). To simplify notation, we write H_2A for the diprotonated form of the amino acid (both COOH and NH_3^+ protonated), HA for the monoprotonated form (COO^-, NH_3^+), and A for the fully deprotonated (COO^-, NH_2). Again, we will omit the charges H_2A^+ and A^-. The equilibria for proton dissociation from the di- and monoprotonated forms are

$$H_2A \rightleftharpoons H^+ + HA \tag{2.20}$$

$$HA \rightleftharpoons H^+ + A \tag{2.21}$$

and the corresponding acidity constants are

$$K_C = \frac{[HA][H^+]}{[H_2A]} \tag{2.22}$$

$$K_N = \frac{[A][H^+]}{[HA]}. \tag{2.23}$$

This time, we are going to use the partition function method from the beginning. We want to know how the amino acid molecules are *partitioned* among the possible ionization states they can adopt as a function of pH. But there are now *three states*: A, HA, and H_2A. We begin by sketching the diagram of the problem to indicate those three states, in **Figure 2.26**.

As previously, we choose the completely deprotonated state (A) as our reference. Then we rearrange Equations 2.22 and 2.23 to yield the relative probabilities of the three states. But instead of trying to write them all immediately relative to the reference we write each one relative to the *preceding state*. Rearranging Equations 2.22 and 2.23, we obtain the probability of H_2A relative to HA, and the probability of HA relative to A,

$$\frac{[H_2A]}{[HA]} = \frac{[H^+]}{K_C} \tag{2.24}$$

$$\frac{[HA]}{[A]} = \frac{[H^+]}{K_N}. \tag{2.25}$$

The factors $[H^+]/K_N$ and $[H^+]/K_C$ *connect* the probability of each state to that of the preceding one. Now we write those factors over the branches of the diagram in **Figure 2.27**. But note that the proton binds much better to the amino than to the carboxylate group. Therefore, the first H^+ binds to NH_2 and $[H^+]/K_N$ comes first; the second H^+ binds to COO^- and $[H^+]/K_C$ comes second.

Finally, we take advantage of the diagram to complete the calculation. To get to HA from A, multiply the relative probability of A (which is 1) by $[H^+]/K_N$; this yields $[H^+]/K_N$ for HA. To get to H_2A from HA, multiply the relative probability of HA (which is $[H^+]/K_N$) by $[H^+]/K_C$; this yields $[H^+]^2/(K_N K_C)$ for H_2A. We can now write the correspondences,

Figure 2.27 Diagram for an amino acid with two ionizable groups. The factors over the connecting branches relate the probability of each state to that of the preceding one.

$$[A] \to 1$$

$$[HA] \to \frac{[H^+]}{K_N}$$

$$[H_2A] \to \frac{[H]^2}{K_N K_C}.$$

These three terms are the relative probabilities of the three states—relative to the reference state A. We write them under each state in the diagram, shown in **Figure 2.28**.

The partition function is obtained by adding the relative probabilities (nonnormalized),

$$Q = 1 + [H^+]/K_N + [H^+]^2/(K_N K_C). \qquad (2.26)$$

Finally, the absolute probabilities (fractions) of each protonation state are calculated by dividing each term by Q,

$$f_0 = \frac{1}{1 + [H^+]/K_N + [H^+]^2/(K_N K_C)}$$

$$f_1 = \frac{[H^+]/K_N}{1 + [H^+]/K_N + [H^+]^2/(K_N K_C)}$$

$$f_2 = \frac{[H^+]^2/(K_N K_C)}{1 + [H^+]/K_N + [H^+]^2/(K_N K_C)}.$$

Now we can replace $[H^+]$, K_C, and K_N, using Equations 2.1 and 2.2, to obtain f_0, f_1, and f_2 as explicit functions of pH and the pK_a's,

$$f_0 = \frac{1}{1 + 10^{pK_N - pH} + 10^{pK_C + pK_N - 2pH}}$$

$$f_1 = \frac{10^{pK_N - pH}}{1 + 10^{pK_N - pH} + 10^{pK_C + pK_N - 2pH}}$$

$$f_2 = \frac{10^{pK_C + pK_N - 2pH}}{1 + 10^{pK_N - pH} + 10^{pK_C + pK_N - 2pH}}.$$

As an example, we plot the fractions f_0, f_1, and f_2 against pH for valine in **Figure 2.29**, which has p$K_C = 2.3$ and p$K_N = 9.5$ (see Table 2.3).

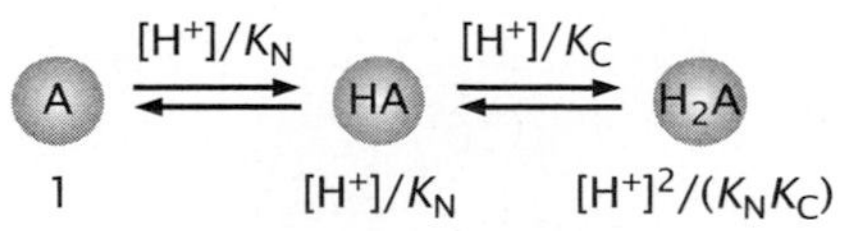

2.4 THE SECONDARY STRUCTURE IS THE LOCAL SPATIAL ARRANGEMENT OF THE POLYPEPTIDE CHAIN

In this section we begin to explore protein structure. Intuitively, most of us think of structure as arising from an organization of elements in space, the connections between those elements appearing from favorable, or attractive, interactions. We will see, however, that the concept of structure is much broader. Although favorable interactions do indeed occur in proteins, structure also arises from *constraints*. That is, by forbidding some conformations, others become the only possible ones for the polypeptide chain. The secondary structure of proteins is the result of *restrictions* on possible conformations and the requirement for *hydrogen-bond* formation.

The Peptide Group is Planar

The first restriction that we will encounter is on the peptide group itself. The peptide bond, between the nitrogen atom and the carbonyl carbon, is usually represented as a single bond (N—C=O). This is misleading. Electrons are delocalized from the nitrogen atom to the carbonyl group by resonance, and the real structure is a hybrid of two resonance forms (**Figure 2.30**). The two resonance forms do not contribute equally to the real structure. The form with an N—C single bond

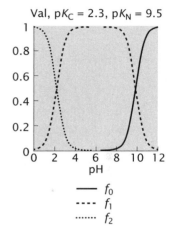

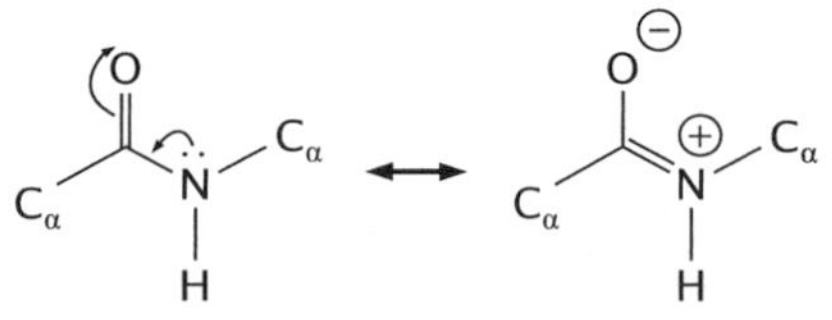

Figure 2.28 Complete diagram for an amino acid with two ionizable groups and three states. The probability of each state is obtained by multiplying that of the preceding one by a factor of the form $[H^+]/K_a$, written over each connecting branch.

Val, pK_C = 2.3, pK_N = 9.5

— f_0
---- f_1
······ f_2

Figure 2.29 Fraction of the different ionization states of valine as a function of pH: f_0 (A, solid line), f_1 (HA, dashed), and f_2 (H$_2$A, dotted).

Figure 2.30 The peptide group is a resonance hybrid of two chemical structures, which contribute unequally to the real structure.

Figure 2.31 Geometry of the peptide group. The atoms in the rectangle are in the same plane. The bond lengths are given in Å. (Data from Ramachandran GN, Kolaskar AS, Ramakrishnan C & Sasisekharan V [1974] *Biochim Biophys Acta* 359:298–302).

contributes about 60%, whereas the form with an N═C double bond contributes about 40%. Therefore, the peptide bond has approximately 40% double-bond character.

As a consequence of the partial double-bond character of the peptide bond, the two C_α's, the N and its H, and the carbonyl C and its O are all in the same plane. The geometry of the peptide bond is shown in **Figure 2.31**. The bond angles are close to 120°. The atoms inside the rectangle, which constitute the *peptide group*, are all in the same plane. The maximum distance between two consecutive C_α's is 3.80 Å. Note that the peptide bond length is 1.33 Å, which is shorter than a C—N single bond.

The Configuration About the Peptide Bond can be *cis* or *trans*

The partial double-bond character of the peptide bond leads to the existence of *cis* and *trans* isomers (as in alkenes) defined by the positions of the C_α's (**Figure 2.32**).

For all amino acid residues, with one exception, the equilibrium strongly favors the *trans* isomer ($K = 10^3$), mainly because of steric hindrance. The exception occurs if proline is the residue following the peptide bond (**Figure 2.33**). In proline, the side chain closes onto the α-NH_2 group, forming a five-membered ring. Therefore, the nitrogen is attached to two carbon atoms and no hydrogen. The two carbon atoms are different; one, the α carbon, is connected to the rest of the polypeptide chain, whereas the other is part of the side chain. Nevertheless, the repulsive energy difference between the two isomers is not very large and, although still favoring the *trans* isomer, the equilibrium constant is much smaller ($K = 4$). Proline isomerization determines the rate of folding of many proteins.

Conformations are Changed by Rotations Around Single Bonds

The polypeptide chain can adopt a very large number of conformations. Each rotation about a single bond corresponds to a new conformation. We have seen that, because of its partial double-bond character, there is no rotation about the CO—NH bond. But there are still rotations about the N—C_α bond and about the C_α—CO bond. We will see, however, that rotations about those two bonds are much more restricted than it may appear at first glance. Further, we can describe the conformations of each amino acid residue independently of the other residues (with a few exceptions). This renders the study of polypeptide conformations much simpler.

Let us first define what a conformation is and distinguish the concepts of *conformation* and *configuration*. A molecule can change its conformation by rotations about single bonds. The configuration of a molecule, however, is fixed by the spatial orientation of the covalent bonds and cannot be changed by rotations, but only by breaking and remaking covalent bonds. The simplest example is the configuration at

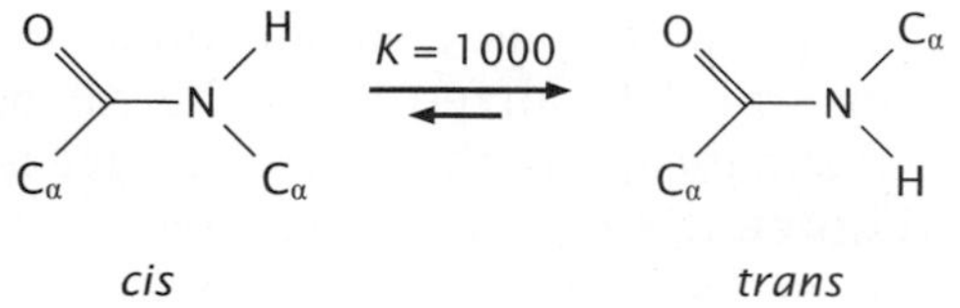

Figure 2.32 The *cis/trans* equilibrium of the peptide bond lies strongly on the *trans* side, with an equilibrium constant $K = 1000$.

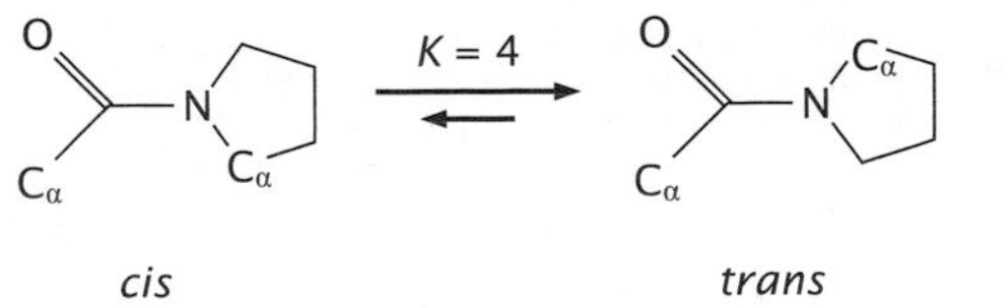

cis   *trans*

Figure 2.33 The *cis/trans* equilibrium of the peptide bond preceding a proline residue favors *trans* marginally, with an equilibrium constant $K = 4$.

a *chiral center*, such as the α carbon of alanine. The naturally occurring enantiomer has the L configuration (*S* absolute configuration in the case of alanine) and that configuration cannot be changed by any rotation. Another example is the *cis* or *trans* configuration across alkene double bonds, as in 2-butene.

Conformations are defined by *dihedral angles*. We will introduce the concept of dihedral angles with another example from organic chemistry. Butane has one *anti* conformation, two *gauche* conformations, and several eclipsed conformations. (In the biochemical literature the *anti* conformation is often called *trans*, although strictly the latter term applies only to configurations across double bonds.) The structure of butane ($CH_3CH_2CH_2CH_3$) is shown in **Figure 2.34** as a Newman projection, looking along the bond between the two central carbons: C_2, shown as a point in front, and C_3, as a circle in the back. The *anti* and *gauche* conformations interconvert by rotation about the central carbon–carbon bond.

In the projections of Figure 2.34, the angles subtended by bonds from C_2 to its attached atoms and C_3 to its attached atoms, are called *dihedral angles* or *torsional angles*. These are not bond angles, but angles between *planes* in space. For example, when projected onto the paper, the angle between the C_2-CH_3 and the C_3-CH_3 bond is $\pm 60°$ in the *gauche* conformations, and $180°$ in the *anti* conformation. This is the dihedral angle between the two methyl groups. The dihedral angle is defined as zero for the conformation with the two methyl groups eclipsed.

Now how do we know the *sign* of the dihedral angles in Figure 2.34? Consider the four-atom group shown in **Figure 2.35**. A dihedral angle is *positive* if a *clockwise* rotation (to the right) of the bond to the front atom (A) is required to bring it into eclipse with the rear atom (D). The angle is *negative* if a *counterclockwise* rotation (to the left) of the bond to the front atom is required. In this definition it does not matter if we look at the bond from one end or the other. If you look at the bond in the C–B direction, D becomes the front atom and A the rear atom, but the sign of the angle does not change.

The energy of butane as a function of dihedral angle is shown in **Figure 2.36**. The *anti* and *gauche* conformations correspond to the

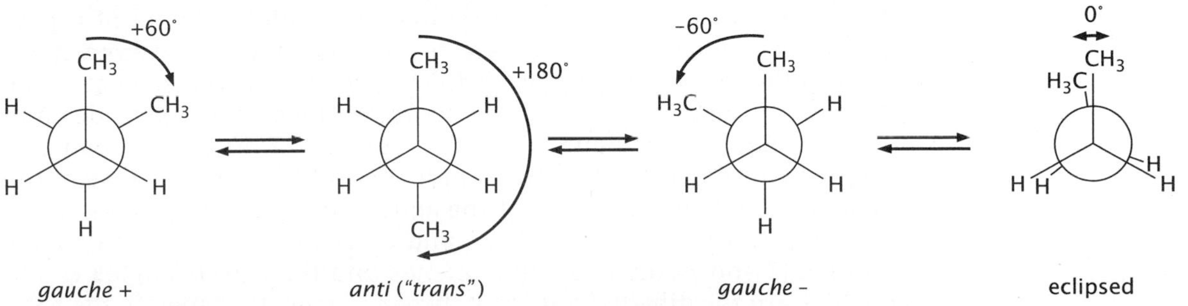

gauche +   *anti ("trans")*   *gauche –*   *eclipsed*

Figure 2.34 Conformations of butane obtained by rotation about the two central carbons. The structure of butane is shown as a Newman projection, viewed along the bond between the two central carbons. The dihedral angle that defines the conformation is indicated.

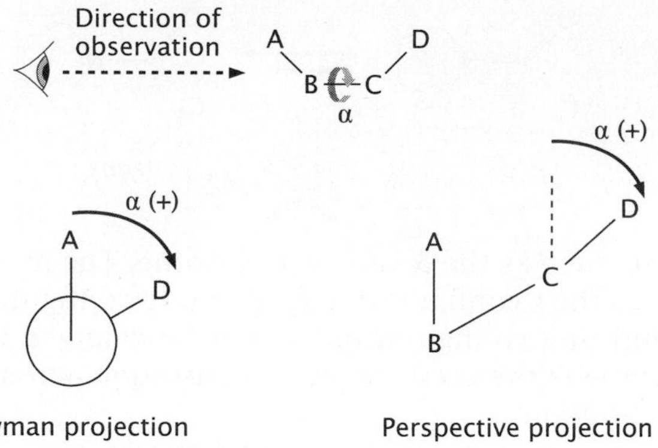

Figure 2.35 Assignment of positive and negative values of angles according to the IUPAC-IUB Commission on Biochemical Nomenclature [1970] *Biochemistry* 9:3471–3479.

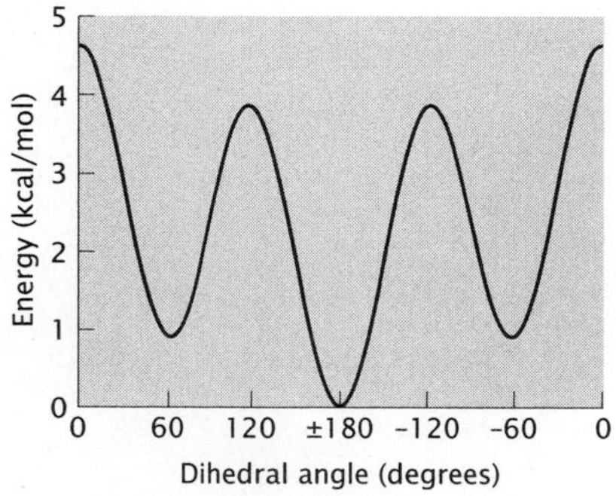

Newman projection   Perspective projection

Figure 2.36 Energy of the butane molecule as a function of rotation around the C_2–C_3 bond.

energy minima in the diagram. The eclipsed conformations are of much higher energy, and therefore essentially not populated. Thus, there are only two stable (thermodynamic) states, *anti* and *gauche*, but there are two different *gauche* microstates (+ and −). There are three eclipsed conformations, which correspond to the peaks in the energy diagram. There are also all the conformations in between. However, *statistically*, butane has only three conformations: one *anti* and two *gauche*.

We have reduced the conformational space of the rotations around the C_2–C_3 bond in butane from a continuum of states to three states. We can do this because the potential energy function is steep, so that we have three well-defined minima separated by three maxima (the eclipsed conformations). The molecule spends its time moving between the three minima, and only very transiently goes through the maxima (on its way to the next minimum). We say that there are only three *thermodynamic states*, which lie at the bottom of energy wells, shown in Figure 2.36. We could also say that there are two states, *anti* and *gauche*, and that the *gauche* state has two microstates (conformations). Only those three conformations are appreciably populated at equilibrium. In other words, >99.9 % of the butane molecules in a sample are found in one of those three conformations, within small angular variations about each one. The *gauche* conformations have about 0.9 kcal/mol higher energy than the *anti* conformation. At room temperature, there is an equilibrium between the *anti* and *gauche* conformations with a distribution of ~70 : 30 in favor of *anti* (depending on the temperature).

The Conformations of Residues in Peptides are Defined by the Torsion Angles φ and ψ

We will now take a closer look at the conformations allowed for a polypeptide chain. At this point, you should build a model of a peptide segment including two amide planes, as shown in **Figure 2.37**. Use it in the remainder of this and the following sections. That is the only way to really understand the conformations of polypeptides and the plots that we are about to discuss. Consider a residue at position *i* in a polypeptide, flanked by two amide (peptide) bonds. Because of the partial double bond character of the amide bond, the six atoms around it (C_i^α, N_i, H_i, C_{i-1}' (carbonyl), O_{i-1}, and C_{i-1}^α) are all in the same plane. Figure 2.37 shows the two amide planes and the *torsion* angles φ and ψ, which are the dihedral angles (between planes) that specify the rotations around the N_i–C_i^α and the C_i^α–C_i' bonds respectively. The picture shows the conformations with torsion angles φ and ψ equal to 180° (A) and 0° (B).

Why are the actually observed regions somewhat different? First, minor changes in the standard backbone geometry—that is, slight deviations from the standard *bond lengths* and *bond angles*—result in significant expansion in the allowed regions of the diagram. This is particularly evident in the regions of regular secondary structure. It happens because hydrogen bonds within the polypeptide are favorable and compensate for some moderately unfavorable φ and ψ combinations and slight deviations from the standard backbone geometry. The case of pre-proline is especially clear: with strictly standard backbone geometry, the α-helical region is forbidden for a residue that precedes a proline, but deviations from standard backbone geometry allow this region to be well populated in Figure 2.41D. In addition, some conformations that are allowed on the basis of steric hindrance prevent the establishment of hydrogen bonds with water. The conformations on the right side of the "bridge region" between the α (bottom) and β (top) regions of the Ramachandran plot belong to this class (see Figure 2.40A). They are observed only when the hydrogen bonds can be satisfied internally, within the polypeptide.

2.5 MOST OF THE SECONDARY STRUCTURE IN PROTEINS CONSISTS OF α-HELICES AND β-SHEETS

The α-Helix is Stabilized by Hydrogen Bonds Between Residues *i* and *i* + 4.

The discovery of the α-helix (**Figure 2.42**) by Linus Pauling is one of those truly inspired moments in the history of science. Pauling tells the story in the book by Horace Judson. In 1948 Pauling went to Oxford, then fell sick. "In Oxford, it was April, I believe, I caught a cold. I went to bed, read detective stories for a day, and got bored, and thought why don't I have a crack at that problem of the alpha keratin. I took a sheet of paper—I still have this sheet of paper—and drew, rather roughly, the way I thought a polypeptide chain would look if it were spread out into a plane." He took scissors and cut out the chain of connected amide bonds from the paper. "And then I looked to see if I could form hydrogen bonds from one part of the chain to the next." Folding the strip of paper to a helix, Pauling realized that hydrogen bonds could be established between the amide C=O group of a residue (*i*) and the amide N—H of another (*i* + 4), four residues away along the polypeptide chain (see Figure 2.42).

Pauling had discovered the α-helix. The simplicity of his paper model, however, hides Pauling's deep knowledge of chemistry, and the importance of structure, of which he was well aware. Later, he told Judson, "We got the alpha helix, you know, and Bragg and Perutz and Kendrew were trying to do the same thing, and they—ah, failed—because of, really, a lack of knowledge of the principles of chemistry, of structural chemistry."

The α-helix, shown in several representations in **Figure 2.43**, is one of the most common types of secondary structure found in proteins. On average there are 3.6 residues per turn of the helix and the rise per residue is 1.5 Å, which means that the rise per turn, or the *pitch* of the helix, is 5.4 Å. The α-helix is stabilized by hydrogen bonds between the amide C=O group of residue *i* and the amide N—H group of residue *i* + 4. The hydrogen bond length in the helix is about 2.8 Å (distance between the heavy atoms, N and O). The common L amino acids preferentially form a right-handed α-helix. Its mirror

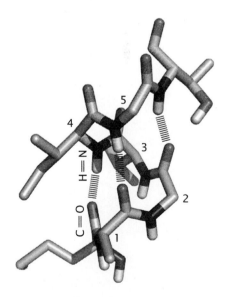

Figure 2.42 Hydrogen bonds (dashed lines) in a segment of the right-handed α-helix, between the C=O of a residue (*i* = 1) and the N—H of the fourth residue from that one (*i* + 4 = 5) on the polypeptide chain. The carbonyl groups (C=O) are pointing up, whereas the amide protons (N—H bonds) are pointing down. (PDB 1MSR.)

Figure 2.43 The right-handed α-helix, in three different representations. The N-terminus is at the bottom and the C-terminus on top. (A) Stick model; (B) sticks and cartoon (ribbon); (C) ribbon structure only. The carbonyl groups of the peptide groups are pointing up, whereas the amide protons (NH) are pointing down. Model: glycophorin A (PDB 1MSR), a membrane protein of the red blood cell.

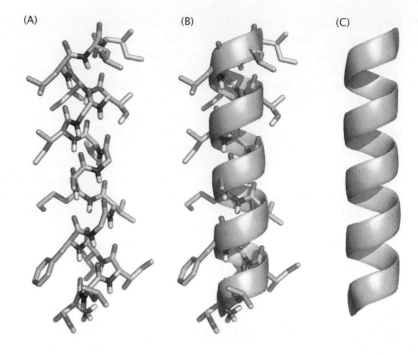

(A)   (B)   (C)

image, the left-handed α-helix, would be preferred by D amino acids. A polypeptide made of L amino acids *can* form a left-handed α-helix (see Figure 2.40A) but it only occurs in very short stretches in natural proteins (see Figure 2.41A). A proline residue does not have an amide N–H; therefore, it cannot be a hydrogen-bond donor and tends to break the helix at the point where it occurs in the chain. It is remarkable that the structure of the right-handed α-helix was proposed by Pauling based only on the structure of the amino acid residue and on the requirement that all hydrogen bonds be satisfied by a repetitive structure, much before the crystal structure was available.

Figure 2.44A shows a top view of the α-helix. The side chains protrude *outward* from the helix (see also Figure 2.43). Contrary to what the figure suggests, however, there is no hole in the center of the α-helix, as a space filling model would show. Figure 2.44B shows a schematic top view called a *helical wheel*. This representation is particularly useful to examine the distribution of residues along the helix. Note that residue 5 is close to residue 1 (approximately one turn of the helix) and residue 8 is almost on top of residue 1 (two turns). Since the are 3.6 residues per turn, after seven residues from the beginning (residue 8) we have almost two complete turns.

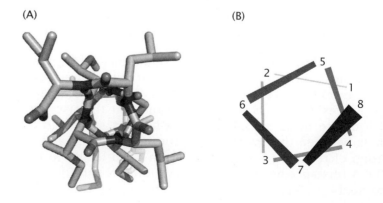

(A)   (B)

Figure 2.44 Top view of the α-helix from the C-terminus. (A) Stick model; (B) helical wheel projection of the backbone. (PDB 1MSR.)

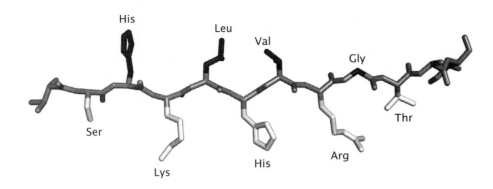

Figure 2.45 Side view of a β-strand from the satellite tobacco necrosis virus (PDB 2BUK) showing the alternate, up-and-down orientation of the side chains.

The β-Sheet is Stabilized by Hydrogen Bonds Between Adjacent Strands

The β-pleated sheet, proposed by Pauling and Corey, is the other major type of secondary structure and is about as common as the α-helix in globular proteins. The element that forms β-sheets is the β-strand, where the backbone atoms form an almost fully extended chain. The dihedral angles are approximately $\phi \simeq -130°$ and $\psi \simeq 120°$, whereas $\phi = \psi = \pm 180°$ in the fully extended polypeptide chain. Viewed from the side, the β-strand has a rippled structure, with the α carbons at kinks, alternately up and down. As a consequence, the side chains protrude alternately above and below the main chain (**Figure 2.45**).

An isolated β-strand would not be stable because the amide hydrogen bonds are not satisfied, but there are two structures, based on the β-strand, that allow the hydrogen bonds to be made. They correspond to the two "flavors" of β-sheets: the *parallel* and the *antiparallel* β-pleated sheets. In parallel β-sheets, two strands run side by side in a parallel orientation—that is, the polypeptide chain runs from N → C in both strands (**Figure 2.46**). The hydrogen bonds between N—H···O=C are somewhat distorted from the ideal orientation, which makes them slightly weaker. The pitch is 6.5 Å. In antiparallel β-sheets the two strands run in opposite directions (**Figure 2.47**). The hydrogen bonds N—H···O=C between the amide groups are satisfied in an ideal orientation. The pitch is 7.0 Å.

In both cases, because of the kinked, up-and-down structure of the β-strands, the sheet is not flat but pleated. The side chains protrude outward, above and below the sheet (not shown in Figures 2.46 and 2.47). Each strand has a right-handed twist, which causes further deviation from a flat structure. In the representation of **Figure 2.48** an arrow indicates the C-terminus and the blunt end indicates the N-terminus.

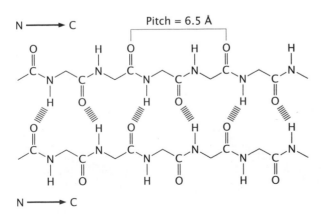

Figure 2.46 The parallel β-pleated sheet viewed from top. Note that the H_α and the R groups were omitted for clarity.

Figure 2.47 The antiparallel β-pleated sheet viewed from top. The H$_\alpha$ and the R groups were omitted for clarity.

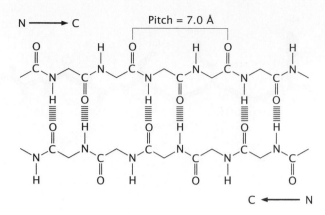

Circular Dichroism is a Simple Method to Identify α-Helices and β-Sheets

In organic chemistry you learned that chiral molecules are optically active: they change the plane of polarization of light. Light is an electromagnetic perturbation, periodic in time and space. In plane-polarized light, the electrical component of light is a vector that oscillates in a plane. The interaction of this electrical field with the electrons in a molecule is what gives rise to optical activity. Circular dichroism (CD) is one of the expressions of optical activity. Because the amino acids that constitute proteins are chiral, the secondary structures adopted by polypeptides are also chiral. Therefore, CD is especially useful to determine the secondary structure of proteins or polypeptides that contain only (or mainly) one type of (chiral) secondary structure. α-Helices, β-sheets, and "random coils" (disordered polypeptides) yield characteristically different CD spectra (**Figure 2.49**).

To understand the physical basis of CD, we need to view plane-polarized light as the result of the superposition of left- and right-circularly polarized light (**Figure 2.50**). Suppose we have light polarized in the vertical plane. Consider two electrical vectors of light, one rotating clockwise, or to the right (R), and the other counterclockwise, or to the left (L). Their sum (gray vector in Figure 2.50A) varies from a positive value to zero to negative, and then back up again as the two individual vectors rotate; but it is always in the vertical plane. Because chiral molecules have an asymmetric electronic structure, they *absorb differently* the left- and right-circularly polarized components of light. Thus, after interacting with chiral molecules, the sum of the two

(A)   (B)

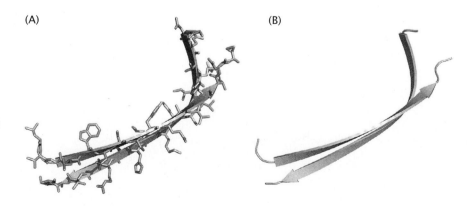

Figure 2.48 The antiparallel β-sheet showing the right-handed twist of the strands. (A) Sticks and ribbon representation. The side chains protrude above and below the sheet. (B) Simplified ribbon representation. An arrow indicates the C-terminus; the blunt end indicates the N-terminus of the strand. Model: satellite tobacco necrosis virus (PDB 2BUK).

electrical vectors is no longer in the vertical plane (see Figure 2.50B). Indeed, the light is changed in two ways: first, there is a shift of the polarization by angle φ relative to the original plane; and second, the resultant electrical vector now describes an ellipse. Circular *dichroism* is the *difference* in absorbance of the two circular components. The molar ellipticity [θ] (units of deg M^{-1} cm^{-1}) is related to the difference in molar absorptivities for left- and right-polarized light by [θ] = $3300(\epsilon_L - \epsilon_R)$. The CD spectrum is a plot of [θ] as a function of the absorption wavelength λ.

Loops and Turns are Needed for Proteins to Fold to Compact Structures

If a polypeptide adopts the structure of an α-helix or a β-sheet for an extended length, it can form a fibrous protein. In order to fold into a compact globular structure, the polypeptide chain must turn around and fold several times upon itself. Thus, a long chain can be packed into a small region of space. The polypeptide chain turns back by means of *loops* and *reverse turns*. Loops are flexible stretches of the polypeptide chain without a regular or repetitive structure. They often provide the reversal of the direction of the chain, connecting α-helices or β-strands. In other cases, the reversal of the direction of a β-strand occurs at a sharp turn. Those are called β-turns (or β-bends). The most common β-turns in proteins are type I and type II (**Figure 2.51**). Types I' and II', which are the corresponding mirror images (for backbone atoms), are much rarer.

Four amino acids are needed to make a β-turn. The torsional angles are, for type I, $\phi_2 = -60°$, $\psi_2 = -30°$, $\phi_3 = -90°$, $\psi_3 = 0°$, and for type II, $\phi_2 = -60°$, $\psi_2 = 120°$, $\phi_3 = 90°$, $\psi_3 = 0°$. The main difference between types I and II is the central amide bond, between C_2^α and C_3^α: it is turned by a 180° difference in the two types of turns. In both cases the C=O of residue 1 makes a hydrogen bond to the N—H of residue 4. Residue 2 of both types is often proline, because it can assume the required conformation for the turn without losing much in terms of rotational freedom, since it is *already constrained*. Furthermore,

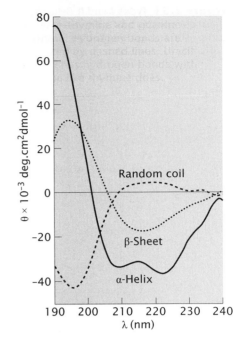

Figure 2.49 CD spectra of the α-helix (solid line), the β-sheet (dotted), and the random coil (dashed) structures. (Data from Greenfield N & Fasman GD [1969] *Biochemistry* 8:4108–4116.)

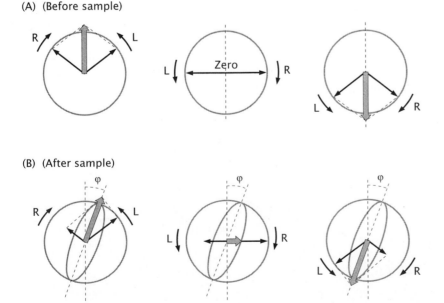

(A) (Before sample)

(B) (After sample)

Figure 2.50 (A) Plane-polarized light (before entering a sample) is the result of the sum (gray arrow) of left-(L) and right-(R) circularly polarized components. (B) After interacting with a sample that absorbed the right-circularly polarized component more than the left-, the light is elliptically polarized and the major axis of the ellipse is shifted by an angle φ relative to the original plane of polarization.

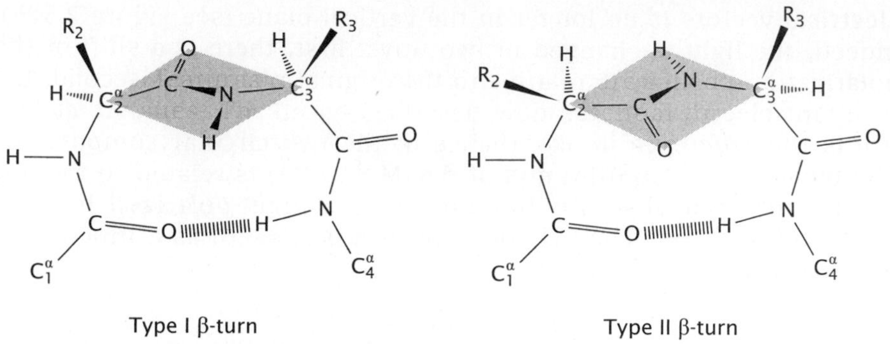

Figure 2.51 Types I and II β-turns, the most common kinds of reverse turns. The H_α are omitted for clarity.

Type I β-turn

Type II β-turn

proline automatically solves the problem of satisfying the hydrogen bond to the N—H of residue 2, because there is *no amide proton* (N—H) in a proline residue. Turns that do not contain glycine are mostly of type I. Conversely, in type II β-turns, residue 3 is usually glycine because the oxygen of the C=O group of residue 2 makes it difficult to accommodate any other side chain in position 3. Residues in the turn that are able to make hydrogen bonds with the central amide bond are favored.

Even tighter, with only three amino acids in the turn, are γ-turns (**Figure 2.52**). The central amino acid is in a sharp kink.

Globular Proteins Also Contain Another Type of Helix: the 3_{10}-Helix

Not all helices in proteins are α-helices. About 10% of the residues in helices adopt another regular structure, called the 3_{10}-helix (**Figure 2.53**). The 3_{10}-helices are usually short, about five residues. They are most often found as connections between β-strands in all-β proteins and less frequently as N- or C-terminal extensions of α-helices. Some small peptides also form a 3_{10}-helix in solution, such as the gp41 fusion protein of HIV-1. The 3_{10}-helix has three residues per turn and 10 atoms along the polypeptide chain between the hydrogen bond donor and the acceptor. (Hence, its name. In this notation, an α-helix, with 3.6 residues per turn and 13 atoms along the chain between the hydrogen bond donor and the acceptor, is a 3.6_{13}-helix.) The 3_{10}-helix is more stretched than the α-helix (see Figure 2.53A). Hydrogen bonds are made between the amide C=O of residue i and the amide N—H of residue $i + 3$. These hydrogen bonds, however, are a little too long, and therefore weaker than those in α-helices. A helical wheel representation of a 3_{10}-helix shows that residues 1, 4, and 7 are on top of each other (see Figure 2.53B,C).

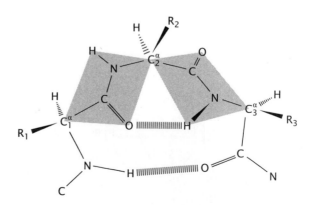

Figure 2.52 The γ-turns, the sharpest of all reverse turns.

(A)

(B)

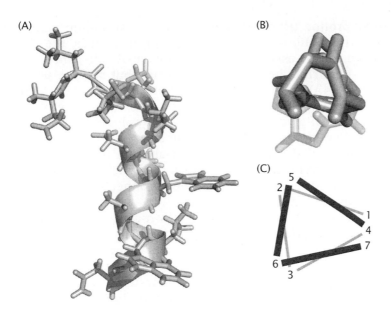

(C)

Figure 2.53 The 3_{10}-helix. (A) side view, modeled with a 13-residue peptide from the gp41 fusion protein of HIV-1 (PDB 1LB0); (B) end view of same helix; (C) helical wheel representation.

Collagen Adopts a Special Type of Helical Structure

To complete this survey of protein secondary structure, we describe a very different type of structure, the collagen helix (**Figure 2.54**). This helix is so named because it is the structure adopted by collagen, the main protein in connective tissue. Collagen is a fibrous, not globular, protein. The sequence of collagen is special: it consists of a repetition of the Gly-X-Y motif, where X is usually proline and Y is usually proline

(A)   (B)

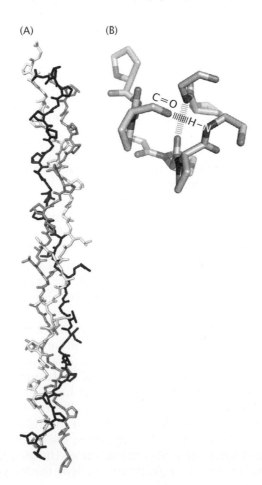

Figure 2.54 The collagen triple helix. (A) Side view. The Pro and Hyp residues are recognized by their five-membered rings (PDB 1BKV). (B) Hydrogen bonding in the collagen triple helix. A detail of the inside of the helix shows two hydrogen bonds between the C=O of Pro or Hyp and the N—H of Gly (PDB 1A3J).

or hydroxyproline (Hyp). Hydroxyproline is an amino acid resulting from post-translational modification of proline, after the collagen is synthesized, and is essential for the stability of collagen.

Collagen is a triple helix of three distinct, but similar, polypeptides (see Figure 2.54A). Each of these three polypeptide chains forms a left-handed helix, whose structure is similar to a type of helix formed by homopolymers of glycine or proline, called polyglycine II and polyproline II. In collagen, the three chains coil around each other, forming an overall right-handed triple helix. In the absence of the modification of proline to hydroxyproline, the collagen triple helix opens, and the collagen fibers unravel. The result is a weak connective tissue, ultimately giving rise to scurvy.

Unlike the α-helix and the 3_{10}-helix, which are stabilized by *intrachain* hydrogen bonds, the collagen triple helix is stabilized by *interchain* hydrogen bonds, made between the amide N—H of glycine and the C=O of proline or hydroxyproline (see Figure 2.54B). In the interior of the triple helix, only the side chain (H) of glycine fits. That is why glycine is always the first residue of the Gly-Pro-Hyp repeat. As we have seen, proline is unique in that both *cis* and *trans* peptide bonds to the preceding residue are common, unlike the other amino acids, which almost always make *trans* peptide bonds. In polyproline II, all peptide bonds are *trans*. Polyproline can also exist in another secondary structure, polyproline I, in which all peptide bonds are *cis*. (Polyglycine I also exists, but it is a β-sheet structure.)

2.6 THE PROTEIN TERTIARY STRUCTURE IS THE GLOBAL ARRANGEMENT OF ITS POLYPEPTIDE CHAIN

We have learned what *local* conformations the polypeptide chains can adopt in proteins. How are those elements of secondary structure organized in the protein? The overall arrangement of the polypeptide chain in space is the protein *tertiary structure*. We will learn that most protein three-dimensional structures are highly symmetrical, organized in a regular way.

The Search for the Three-dimensional Structure of Proteins is the Story of a Search for Beauty

The story of the search for the three-dimensional structure of proteins is a story of the search for beauty, for something simple, an "ultimate truth," as Max Perutz called it. It is also a story of frustration and disappointment. Perutz (1964) said, "In 1937, a year after I entered the University of Cambridge as a graduate student, I chose the X-ray analysis of hemoglobin, the oxygen-bearing molecule of the blood, as the subject of my research."

It is hard to imagine the leap of faith that Perutz was accomplishing. Hemoglobin is a tetrameric protein, each monomer consisting of about 150 amino acid residues. At that time, X-ray analysis had been used to determine molecular structures, but only of molecules not much larger than amino acids. The most complex that had been solved was the structure of phthalocyanine (**Figure 2.55**). Perutz himself wondered, "How could I hope to locate the thousands of atoms in the molecule of hemoglobin?" As he later wrote, "Fortunately, the examiners of my doctoral thesis did not insist on a determination of the structure, otherwise I should have had to remain a graduate student for 23 years."

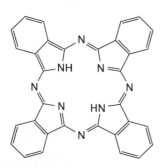

Figure 2.55 The structure of phthalocyanine, the largest molecule whose structure had been determined by X-ray analysis when Perutz decided to tackle hemoglobin.

(A)

(B)

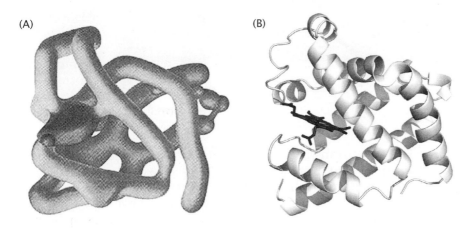

Figure 2.56 (A) The low-resolution structure of myoglobin obtained by Kendrew. (B) A modern high-resolution structure (PDB 1A6N). The heme group (in black) is shown on the left side of the structure. (A, from, Perutz MF [1964] *Sci Am* 211:64–76. With permission from Scientific American, Inc.)

In the meantime, John Kendrew had decided to choose his own molecule. He settled on the slightly more modest goal of tackling myoglobin, with *only* a single chain of about 150 amino acids. The initial structure was determined at very low resolution (**Figure 2.56**A). Kendrew described it like this, as related by Perutz (1964): "Perhaps the more remarkable features of myoglobin are its complexity and its lack of symmetry. The arrangement seems to be almost totally lacking in the kind of regularities which one instinctively anticipates, and it is more complicated than has been predicted by any theory of protein structure."

Looking at the myoglobin structure, Max Perutz was outright shocked: "Could the search for the ultimate truth really have revealed so hideous and visceral-looking an object? Was the nugget of gold a lump of lead?" What Kendrew had seen at low resolution would one day look like the symmetric arrangement of helices we now know, shown in Figure 2.56B. As Perutz concluded, "Fortunately, like many other things in nature, myoglobin gains in beauty the closer you look at it."

Simple Folding Motifs are Recurrent in Protein Structures

The global arrangement of the polypeptide is described by the tertiary structure, but within that structure there is an intermediate level of organization. Certain types of local arrangements of the secondary structure elements occur frequently and are present even in tertiary structures that are not similar overall. We call these local combinations *motifs* of supersecondary structure. In globular proteins, the polypeptide chain must make many turns in space to fold into a compact structure. The motifs are the ways in which the chain turns and the combinations of secondary structure it contains. We will examine a few of these motifs before studying in more detail the different types of tertiary structures and how they are determined.

β-Hairpin motifs connect antiparallel β-strands. In order to assemble an antiparallel β-sheet, a β-strand needs to turn around. The simplest manner in which this can happen is through a β-turn. This motif is called a β-hairpin (**Figure 2.57**A). Figure 2.57B shows a set of three

(A)

(B)

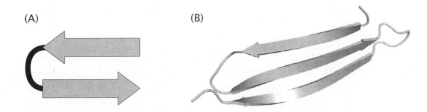

Figure 2.57 (A) The β-hairpin motif. (B) β-Hairpin connections in an antiparallel β-sheet of human brain hexokinase (PDB 1HKC).

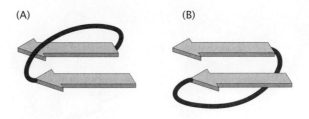

Figure 2.58 Right-handed (A) and left-handed (B) crossover connections between β-strands. The left-handed connection is almost never found.

Figure 2.59 A β-α-β motif, from alcohol dehydrogenase (PDB 1E3E).

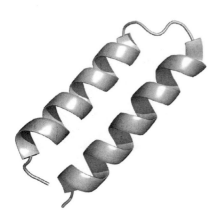

Figure 2.60 A helix-turn-helix motif from cytochrome b$_{562}$ (PDB 1QPU).

β-strands (two hairpins), forming an antiparallel β-sheet. Note that the β-strands are not flat, but have a right-handed twist.

Crossover connections link parallel β-strands (**Figure 2.58**). α-Helices in proteins are right-handed; β-strands have a right-handed twist. Right-handedness of the connecting elements is also the rule. Parallel β-strands cannot be connected as β-hairpins. Instead, the β-strands can be linked by a crossover connection, which is almost always right-handed (see Figure 2.58A); left-handed connections (see Figure 2.58B) are extremely rare.

Often, the crossover connection between two β-strands is an α-helix. This is called a β-α-β motif (**Figure 2.59**); it is one of the most common motifs of supersecondary structure. Some proteins, such as those containing a triose phosphate isomerase (TIM) type of barrel, consist essentially of a series of this motif.

Turns often connect antiparallel helices, forming a *helix-turn-helix* motif (**Figure 2.60**). This structure has the same topology of a β-hairpin, but with helices instead of strands. The turn shown in Figure 2.60 makes an angle of 180°, but other angles occur frequently. The helix-turn-helix motif is especially common in all-α proteins, such as four-helix bundles, globins (for example, myoglobin), DNA binding proteins (for example, the tryptophan and λ repressors), and calcium-binding proteins (calmodulin, for example).

2.7 PROTEIN STRUCTURES CAN BE DETERMINED BY X-RAY CRYSTALLOGRAPHY OR NUCLEAR MAGNETIC RESONANCE

X-ray crystallography and nuclear magnetic resonance (NMR) are the only methods that can produce a full three-dimensional structure of a protein at atomic resolution. Here we will learn the essential concepts on which the two methods are based and how the protein structures are obtained. We will also learn that neither method is an imaging technique. Microscopy is an imaging technique: it just amplifies a microscopic object to a size that we can see. NMR and X-ray crystallography generate sets of information about the positions of the atoms in the protein from which the structure can be reconstructed. The structure that we get in the end is the best model of the protein compatible with the sequence and the X-ray or NMR data.

Periodic Functions of Space and Time are Fundamental in X-ray and NMR

Both X-ray crystallography and NMR use electromagnetic radiation, although from very different regions of the spectrum. To understand how these methods work, we need to briefly review the basic physics and mathematics of waves. We will keep things simple because the goal is to understand the physical concepts, not the mathematical complexity.

An electromagnetic wave is an oscillatory perturbation in time and space. It changes *periodically*, oscillating between a maximum positive amplitude A and its negative $-A$ (**Figure 2.61**). The rate of change in time t (see Figure 2.61A) is given by the *frequency* ν (in s^{-1}) or by the *angular frequency* ω (in radians per second, rad s^{-1}); the two are related by $\omega = 2\pi\nu$. The wave contains a *phase* ϕ, which represents the shift of the periodic function relative to the origin ($t = 0$). After a certain time τ called the *period*, the wave is identical to what it was at $t = 0$, corresponding to a complete cycle (2π); thus, $\omega\tau = 2\pi$. We can represent this wave mathematically by

$$f(t) = A\cos(\omega t - \phi). \tag{2.27}$$

The product ωt corresponds to an angle θ. The cosine repeats itself in periods of 2π because $\cos(\theta + 2\pi) = \cos\theta$. The phase shift corresponds to making zero the argument of the cosine in Equation 2.27 ($\omega t - \phi = 0$). Thus, the phase shift in time is (see Figure 2.61A)

$$t = \frac{\phi}{2\pi}\tau. \tag{2.28}$$

We can also represent a periodic function of the angle θ by a complex number

$$f(\theta) = e^{i\theta}, \tag{2.29}$$

which is given by Euler's equation

$$e^{i\theta} = \cos\theta + i\sin\theta, \tag{2.30}$$

where i is the imaginary number ($i = \sqrt{-1}$). This form is convenient because it is much easier to perform calculations with exponentials than with sines and cosines. The real part of the complex number, in this case $\cos\theta$, is what corresponds to a physical quantity.

A wave is also a function of space (distance) at a fixed moment in time (see Figure 2.61B). This wave repeats itself after a distance λ, which is the *wavelength*. Mathematically we can represent this periodic function of distance (x) as

$$f(x) = A\cos(kx - \phi), \tag{2.31}$$

where k is the *wave number* (in rad m^{-1} or rad cm^{-1}). After a full wavelength λ, an oscillatory cycle (2π) is completed; thus, $k\lambda = 2\pi$. Again, the phase shift corresponds to making zero the argument of the cosine,

(A)

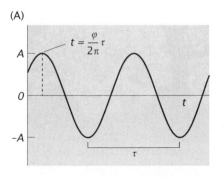

(B)

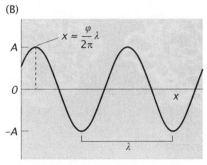

Figure 2.61 An electromagnetic wave in time (A) and in space (B). The period τ and the phase shift are indicated in (A). The wavelength λ and phase shift in space are indicated in (B).

now in Equation 2.31 ($kx - \phi = 0$). Thus, the phase shift in space is (see Figure 2.61B)

$$x = \frac{\phi}{2\pi}\lambda. \tag{2.32}$$

For photons, which move at speed c ($c = 3.0 \times 10^8$ m s^{-1}), the time and space waves are related by $\omega = 2\pi c/\lambda$.

The Structure in X-ray Crystallography is the Fit of the Molecular Model to the Experimental Electron Density of the Protein in a Crystal

The first step in obtaining an X-ray structure is to crystallize the protein. This is no easy task. The conditions to obtain crystals that diffract well vary from protein to protein. Only well-ordered crystals produce high-quality diffraction patterns, which give rise to high-resolution structures. Finally, once you obtain a good crystal, it needs to be mounted on a diffractometer. Protein crystals are fragile because they contain a lot of solvent: they are squishy, not solid like a NaCl crystal. To prevent damage from the radiation, the crystals are often cooled to $< -150°$C. If all this is successfully accomplished, we are ready to collect X-ray diffraction data.

The positions of objects can only be determined by photons to a resolution of $\lambda/2$, where λ is the wavelength of the radiation. X-rays used in protein structure determination have wavelengths of 0.5 Å to 2.5 Å, so they are ideal to determine structures at atomic resolution (distances $\sim$1 Å). X-ray determination of protein structure is based on two physical principles: first, electrons in the atoms scatter X-rays; second, the waves scattered by the atoms in the molecule *interfere*, giving rise to a *diffraction pattern*. The positions of the atoms become encoded in the diffraction pattern, and our task is to decode it, to reveal the original electron density that produced it. In the end, a molecular model of the protein is fitted to this experimental electron density.

What is diffraction? Any atom will scatter X-rays. However, the diffraction pattern of a single protein molecule would be continuous. You could measure some degree of scattering at any angle from the beam. A crystal is a three-dimensional *periodic* lattice composed of repeating unit cells. The periodic array of atoms in the crystal only allows us to view scattering at *discrete* angles. To understand how this works, consider the sketch in the left side of **Figure 2.62**.

An incident beam with a certain angle θ is deflected by the electrons in the atom and comes off at the same angle θ (see Figure 2.62, bottom ray). If the atoms are placed at regular intervals in a crystal lattice with

Figure 2.62 Diffraction of an X-ray beam by an ordered crystal (spheres). A delay (thick lines, bottom ray) is introduced in the scattered beams, with the result that their waves are not in phase. The interference of the electromagnetic waves results in a diffraction pattern, shown for lysozyme on the right. (Courtesy of Wikimedia.)

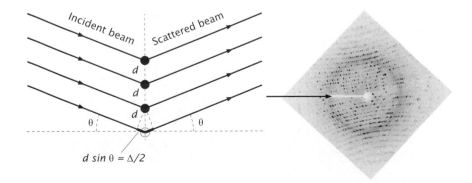

spacing d, a relative delay $\Delta = 2d\sin\theta$ is introduced in the path length (the factor of 2 appears because there is a delay of $d\sin\theta$ before and after hitting the atom). As a consequence of this delay there is a *phase shift*, and the waves of the X-rays scattered by successive atoms in the crystal *interfere*. If the waves are *in phase*, they interfere positively: they add. If *out of phase*, they interfere negatively: they subtract from each other. In order for the scattered beams to be in phase, Δ must equal an *integer number n* of wavelengths λ. The angles θ at which positive interference occurs are given by Bragg's law:

$$n\lambda = 2d\sin\theta. \qquad (2.33)$$

Thus, we will only measure scattering from our protein for angles that obey Bragg's law. Any other angles will produce no scattering. The result of the sum of the scattered X-ray waves is therefore a pattern of dots—the *diffraction pattern* (see Figure 2.62, right). It contains the information on the positions of the atoms in the crystal, but is *not* an image of the protein. Rather, the diffraction process creates a representation of the protein structure in *reciprocal space* (m^{-1}): the structure is there, but it is encrypted in a code that we still need to decipher.

The X-ray diffraction experiment generates a physical "print" of what is called mathematically a Fourier transform $F(k)$ of the electron density $f(x)$ into the diffraction pattern,

$$F(k) = \frac{1}{\sqrt{2\pi}}\int f(x)e^{ikx}dx. \qquad (2.34)$$

Here the integral is over the entire space and $1/\sqrt{2\pi}$ is just a normalization factor. In this case, the transform is from a periodic function of space $f(x)$, the crystal, to a periodic function of reciprocal space $F(k)$, the diffraction pattern. The inverse Fourier transform gives back the protein electron density,

$$f(x) = \frac{1}{\sqrt{2\pi}}\int F(k)e^{-ikx}dk. \qquad (2.35)$$

In X-ray diffraction, x and k are actually vectors with three dimensions. Thus, instead of the simple product kx, we should write the dot product $\vec{k}\cdot\vec{x}$, but we will use the simpler version because the concept is the same. Also, in protein structure determination by X-ray diffraction, the radian part of the function is explicitly written outside, so that k itself has units of reciprocal space (m^{-1} instead of rad m^{-1}). $F(k)e^{i\phi}$ is called the *structure factor*. The inverse Fourier transform of the structure factor is usually written as,

$$f(x) = \sum_k F(k)e^{i\phi}\, e^{-i2\pi kx} \qquad (2.36)$$

where the sum (over all the diffraction spots) replaces the integral of Equation 2.35 because we are dealing with discrete spots.

The scattering intensity of a spot in the diffraction pattern gives $F(k)$, but *not* the phase ϕ in Equation 2.36. This is called the *phase problem* in X-ray crystallography. Thus, before the inverse transform can be applied, the phases have to be determined separately. The phases are actually more important than the intensities. If a similar structure is available, the corresponding molecular model can be used as a starting point to reconstruct the electron density of our molecule. Essentially we use the available model, *replacing* the real molecule, to calculate the diffraction pattern and find initial values for the phases. This procedure

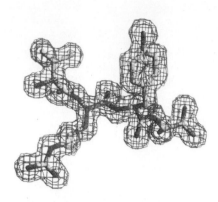

Figure 2.63 The polypeptide chain is fitted into the electron density (shown as a mesh) obtained from the X-ray diffraction pattern. (Adapted from Rupp B [2010] Biomolecular Crystallography. With permission from Garland Science.)

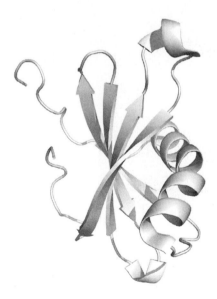

Figure 2.64 Crystal structure of human cystatin A obtained by X-ray diffraction at 1.7 Å resolution (PDB 3KSE).

is called *molecular replacement.* If a similar structure is not available, additional phasing experiments are necessary. Anomalous scattering can be used if the wavelength can be varied. Alternatively, heavy-atom substitution can be used. For example, a marker atom may be introduced in the crystal, by soaking it in a solution of a heavy atom. The positions of those heavy atoms then allow us to determine the phases.

Once the electron density is calculated, we are ready to fit to it a molecular model of the protein. We take a model of the polypeptide at atomic detail, with correct bond lengths and bond angles, and fit it to the electron density (**Figure 2.63**), like you would fit a hand into a glove.

Finally, the best-fit model is subjected to energy minimization, to correct steric residual clashes and incorrect angles. The molecular model of the protein thus obtained is what we call the crystal structure. But it is still a model, not a photograph of the protein at very high magnification. **Figure 2.64** shows a representation of the X-ray crystal structure of the human protein cystatin A, displayed to emphasize its secondary structure (α-helices, β-strands, and loops).

The resolution of the X-ray diffraction pattern determines the level of detail of the model. Thus, structures obtained at 3.5 Å resolution show the secondary structure; at 2.5 Å resolution, we can see most amino acid side chains; at 2.0 Å resolution, good ϕ and ψ angles are obtained; and at 1.5 Å, we can begin to resolve hydrogen atoms.

NMR Yields the Structure of a Protein in Solution from a Simulation of the Molecular Model Under Constraints Derived from Experimental Spectra

Let us now see how a protein structure is obtained with nuclear magnetic resonance (NMR), and how this structure differs from that obtained by X-ray crystallography. NMR requires chemical elements with a nonzero nuclear spin. Nuclei with either an unpaired proton or an unpaired neutron have spin 1/2. Those are the nuclei used for protein structure determination. For example $^{1}_{1}H$ (called *protons* in NMR work; not to be confused with the protons in the nuclei of heavier elements), $^{15}_{7}N$, and $^{13}_{6}C$ have spin 1/2, whereas $^{2}_{1}H$ (or D, for deuterium) and $^{14}_{7}N$ have spin 1. The most common isotope of carbon, $^{12}_{6}C$ has no unpaired protons or neutrons and therefore has zero spin. In practice for most structure determinations by NMR, proteins need to be synthesized in the laboratory (by recombinant DNA techniques) uniformly labeled with ^{15}N and ^{13}C isotopes.

Nuclei with spin 1/2 behave like little magnets, which align themselves in an external magnetic field B_0 (see Figure 1.18). Under the influence of B_0, a small energy difference ΔE exists between the spin states $+1/2$ and $-1/2$. Radiation with the resonance frequency ν, given by $\Delta E = h\nu$, induces a transition of the nuclear magnetic dipole from parallel to antiparallel to B_0. The spins of the excited nuclei change from $+1/2$ to $-1/2$, with an increase in energy ΔE, resulting in a change in the Boltzmann distribution equivalent to an increase in temperature. In NMR, the energy difference ΔE is very small, corresponding to the radiofrequency region of the electromagnetic spectrum: $\nu \sim 300$ MHz (1 Hz = 1 s^{-1}), or $\nu \sim 10^9$ s^{-1} ($\lambda \sim 1$ m). For comparison, fluorescence (UV/visible light) has $\nu \sim 10^{15}$ s^{-1} ($\lambda \sim 500$ nm) and X-rays have $\nu \sim 10^{18}$ s^{-1} ($\lambda \sim 1$ Å), several orders of magnitude larger in energy.

This small ΔE means that the populations of $+1/2$ and $-1/2$ spins are almost identical (see Chapter 1). Therefore, the net number of spins that can be promoted to the $-1/2$ state is very small, resulting in an intrinsic low sensitivity of NMR.

The energy difference ΔE, and thus the frequency ν, depends on B_0, which varies with the NMR spectrometer. Therefore, the resonance frequencies are reported in a dimensionless scale called the *chemical shift* δ, which is calculated relative to, and normalized by, a reference frequency ν_0,

$$\delta = \frac{\nu - \nu_0}{\nu_0} \times 10^6. \qquad (2.37)$$

The δ scale is expressed in parts per million (ppm), because the differences $\nu - \nu_0$ are very small. The normalization removes the field dependence of NMR resonances and allows direct comparison of spectra obtained in different instruments.

NMR has two major advantages, which more than compensate for its low sensitivity. First, the chemical shift of a specific nucleus is sensitive to its local chemical environment, because the electron distribution alters the magnetic field effectively felt by the nucleus. Thus, identical nuclei have slightly different δ values, which report on the local structure. Further, the chemical shifts of the protein backbone atoms, especially those of C_α and $C{=}O$, depend on the ϕ and ψ dihedral angles, and thus provide information on the local secondary structure. The second major advantage of NMR is that the excited nuclear spin state has a *very long lifetime*, of milliseconds to seconds. For comparison, fluorescence lifetimes are of the order of nanoseconds. The long-lived spin state allows the excitation, or magnetization, in NMR, to be transferred from one nucleus to another, if the nuclei are close in the sequence or in space, before it decays.

The magnetization of an excited nuclear spin state can be transferred to other nuclei in two ways: through chemical bonds, called *scalar* or *J-coupling*, or through space, called *dipolar coupling*. **Figure 2.65**A shows some of the most common types of magnetization transfer used in NMR experiments.

In scalar coupling, the transfer of magnetization takes place *through* bonds (typically 1–3) before it decays considerably, and thus identifies, or *correlates*, atoms close to each other in the polypeptide sequence. *J*-coupling between the amide nitrogen and its proton (J_{NH}) and between the amide proton and the α proton ($J_{\mathrm{H_N H_\alpha}}$) of the same amino acid residue are particularly strong. Scalar coupling is used to assign the

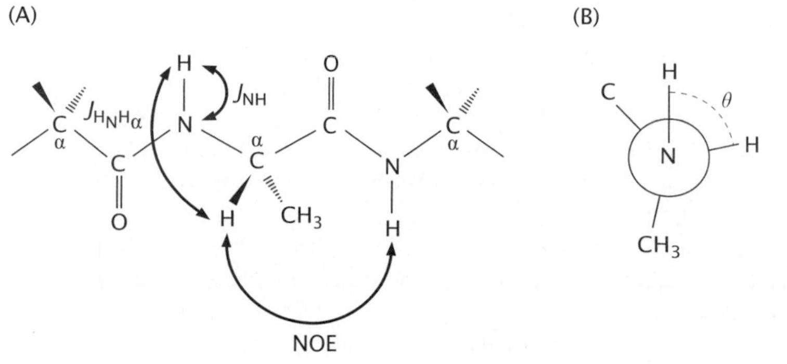

Figure 2.65 (A) Transfer of magnetization in a polypeptide. The transfer can occur through bonds (scalar or *J*-coupling) or through space (dipolar coupling, NOE). Shown are *J*-couplings between the amide nitrogen and its proton (J_{NH}) and between the amide proton and the α proton ($J_{\mathrm{H_N H_\alpha}}$), and the NOE between the α proton and the amide proton of the following residue. (B) Newman projection looking down the central N—C bond shown in (A), with N in front and the C_α in the back (represented by a circle). The coupling constant $J_{\mathrm{H_N H_\alpha}}$ depends on the dihedral angle θ between the two protons, which is related to the dihedral angle ϕ.

NMR signals to specific atoms, by determining which atoms are chemically bonded to each other. In particular, the coupling between the $^{13}C{=}O$ of one residue and the ^{15}N of the adjacent residue allows us to determine which set of signals belong to adjacent residues in the polypeptide chain. Because J-couplings depend on the ϕ (especially) and ψ dihedral angles (see Figure 2.65B), they can be used to determine secondary structures and torsional angles of side chains.

In dipolar coupling, on the other hand, the magnetization is transferred from one spin to another *through space*. This is called the nuclear Overhauser effect (NOE). The NOE is analogous to the transfer of heat from a "hot" (excited) spin to a neighbor close in space. Figure 2.65A indicates an NOE between the α proton on a residue and the amide proton of the next residue in the polypeptide sequence. Because NOEs can only be observed over distances ≤ 6 Å, and their intensity decreases with distance, they are powerful indicators of *spacial proximity* and thus fundamental in protein structure determination by NMR.

When magnetization is transferred between two or more nuclei, either by J- or dipolar coupling, their resonance frequencies become *correlated*. For example, if the magnetization is passed between an amide nitrogen and its attached proton (J_{NH}, see Figure 2.65), the signal (excited state) evolves in time (t) as a wave of the form

$$s(t) \sim A\, e^{i\omega_N t} e^{i\omega_H t} \tag{2.38}$$

where ω_N is the nitrogen frequency, ω_H is the proton frequency, and A is the amplitude (intensity) of the signal. The frequencies of both nuclei are *encoded* in the signal $s(t)$. This signal is then subjected to a Fourier transform from the time to the frequency domain (ω), analogous to the transform between x and k (shown in Equation 2.34), to generate the NMR spectrum $S(\omega_H, \omega_N)$. The spectrum is the intensity as a function of the frequencies of the coupled nuclei—in this case, ^1H and ^{15}N. We plot this spectrum as a two-dimensional surface, where the x- and y-axes are the chemical shifts of the nitrogen and the proton, as shown in **Figure 2.66**. The intensity (third dimension) is represented by level curves, like a topographical map. Now, since the signal encodes both frequencies (ω_H, ω_N), this NMR spectrum includes a *cross peak* at the intersection of the frequencies of the two correlated nuclei (see Figure 2.66).

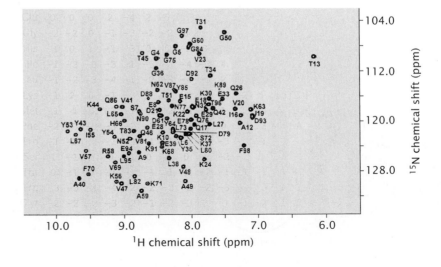

Figure 2.66 A two-dimensional NMR spectrum (^{15}N-HSQC) of the human protein cystatin. The labels indicate the corresponding amino acids. (Adapted from Tate S, Ushioda T, Utsunomiya-Tate N et al [1995] *Biochemistry* 34:14637–14648. With permission from American Chemical Society.)

There are many types of multidimensional NMR experiments that use scalar couplings to correlate the frequencies of ^{1}H, ^{15}N, and ^{13}C atoms. They are known by acronyms, such as COSY, HMQC, and HSQC, which we will not need to explain here. The important point is that these spectra allow us to *assign the chemical shifts* to the corresponding atoms in the protein (see Figure 2.66). Similarly, the NMR experiment that uses the nuclear Overhauser effect (through-space couplings) is called a NOESY. Once we have assigned the chemical shifts of the amino acid residues, and know their through-space distances and the dihedral angles formed by their bonds, we know the *constraints* that the protein structure must satisfy. Most often these constraints consist of inter-atomic distances, determined from NOESY type spectra; dihedral (or torsional) angles, determined by *J*-couplings over three bonds; chem-ical shifts of the backbone atoms; and hydrogen bonds, determined from H/D exchange experiments. In H/D exchange experiments, a pro-tein is placed in a solution in which D_2O replaces H_2O as the solvent. What is measured is the degree of protection of amide protons (N—H) from exchange with the solvent. Amide protons involved in hydrogen bonds (secondary structure) are *protected from exchange*. As a result, these N—H protons exchange *slower* with D from the solvent D_2O.

The final step in protein structure determination by NMR is a com-puter simulation of the protein conformation under the effect of the intermolecular forces acting on each atom. The molecular model used in the simulation is constructed with correct bond lengths, bond angles, and torsional angles. Those structural properties are supple-mented by a "force field" provided by the constraints derived from NMR. The best NMR models are those that agree most closely with the constraints. Unlike X-ray crystallography, which produces a single best crystal structure, the solution structures determined by NMR are ensembles of the conformations that best conform to the constraints. **Figure 2.67** shows the solution structure of the human protein cys-tatin A obtained by NMR. The ensemble of best models is shown in Figure 2.67A, and one model only is shown in Figure 2.67B. The N- and C-termini of the protein, visible on the left side of the structure, are typically the regions least well determined. This may reflect a lack of structural constraints in the NMR data or an actual larger conforma-tional disorder in those regions of the protein. If you compare the NMR

(A)

(B)

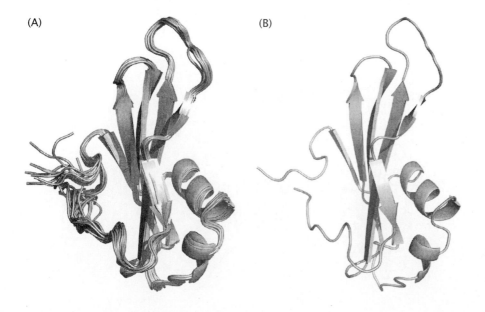

Figure 2.67 Solution structure of human cystatin A obtained by NMR (PDB 1CYU). (A) Overlay of the ensemble of 15 best models. (B) One of the models (model 1).

with the X-ray structure (see Figures 2.64 and 2.67), you can see that the two are very similar but not exactly identical.

2.8 PROTEIN STRUCTURES BELONG TO A FEW TYPES OF FOLDS

In this section, we will describe the major types of protein folds. The first detailed classification of protein structures is due to Jane Richardson in her classic 1981 paper. Two other classifications are now commonly used: CATH, whose acronym is derived from levels of protein organization (class, architecture, topology, and homologous superfamily), and SCOP (structural classification of proteins). We will use a simplified classification, based mainly on Richardson and CATH. However, at the highest level of classification, SCOP is not very different. There are several major types of folds, which are grouped in classes. These are really the folds of single domains in proteins, not necessarily of the entire structure. Some proteins contain only one domain, but many contain several, and each of those domains may have a different fold. We divide the main folds in four classes: α, β, α/β, and other.

Proteins in the α Class Contain Little or no β Structure

The most typical folds in the α class consist of repeats of helix-turn-helix motifs. A large fraction of these proteins have the structure of an antiparallel, up-and-down α-helical bundle. An example is the four-helix bundle of cytochrome b_{562} from *E. coli* (**Figure 2.68**A). This type of up-and-down helical bundle is very common in membrane proteins, such as the aquaporin water channel (see Figure 2.68B) and rhodopsin, the central protein in vision in vertebrates, which is found in the rod cells of the retina and which transduces the absorption of a light photon into molecular information.

The helix-turn-helix motif is also the building block of Ca^{2+}-binding proteins such as calmodulin (**Figure 2.69**A), the most important transducer of Ca^{2+} signals in the cell, parvalbumin, a protein that functions as a Ca^{2+} buffer system in muscle and neuronal cells (see Figure 2.69B), and the annexins (**Figure 2.70**), a family of Ca^{2+}-binding peripheral membrane proteins that constitute about 1% of the cytoplasmic protein (yet their function remains unknown).

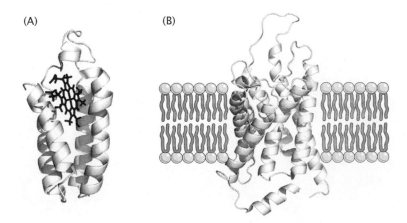

(A)　　　　(B)

Figure 2.68 (A) The four-helix bundle of cytochrome b_{562} with its heme cofactor attached (PDB 1QPU). (B) The up-and-down α-helical bundle of the bovine AQP1 aquaporin water channel, an integral membrane protein (PDB 1J4N).

(A)

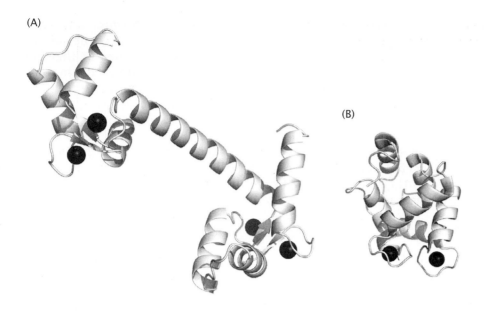

(B)

Figure 2.69 Calcium-binding proteins with the EF-hand motif. (A) Calmodulin (PDB 1CLM) and (B) parvalbumin (PDB 4CPV). Ca^{2+} ions are shown in black.

Similarly, the structures of many DNA-binding proteins, also called transcription factors, are based on the helix-turn-helix motif. The tryptophan repressor is shown in **Figure 2.71**.

The globin fold, which is the structure of myoglobin (**Figure 2.72**) and the various hemoglobin subunits, contains several pairs of antiparallel α-helices that pack onto each other at approximately 90-degree angles.

Coiled coils are a different type of all-α structure that occurs in diverse protein families. It is obtained when two α-helices coil around each other, forming a left-handed superhelix. The superhelix is stabilized by interactions between hydrophobic residues in the contact regions between the helices. Coiled coils are easily recognized by their sequence because of the characteristic *heptad repeat* of residues **a**-*b*-*c*-**d**-*e*-*f*-*g*, where *a* and *d* are hydrophobic (**Figure 2.73**).

A type of coiled coil, common in DNA-binding proteins, is the *leucine zipper*. An example is the transcription factor GCN4 from yeast shown in **Figure 2.74**. It is called a leucine zipper because the hydrophobic residues in the middle of the coil, mainly leucines, lock into each other like the teeth of a zipper. The two α-helices run *parallel* to each other and wrap around each other making a left-handed superhelix. Another example is the MAX transcription factor, which is shown binding to a DNA molecule in **Figure 2.75**. The coiled coil portion of the protein forms a long stalk, while the DNA binding is achieved by means of the shorter α-helices, which grab the DNA like a pair of pliers. The best examples of coiled coils, however, are the α-keratins. These *fibrous* proteins constitute hair, nails, and horns in mammals. The coiled coil is also the structure of the light chains of myosin, a protein involved in muscle contraction and molecular motors.

Proteins in the β Class Contain Little or no α Structure

There are several types of all-β folds. Let us begin with antiparallel β folds. The simplest antiparallel folds are up-and-down β meander barrels. Imagine a planar antiparallel β-sheet; a strand runs up, then turns around, comes back down, turns, goes back up, and so on, repeating the β-strand–turn–β-strand motif many times. Now bend this β-sheet, until the first and last strands meet sideways and establish hydrogen

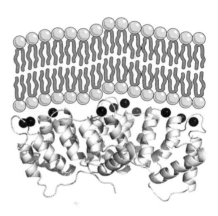

Figure 2.70 Annexin V (PDB 1A8A), a calcium-binding peripheral membrane protein. Ca^{2+} ions are shown in black.

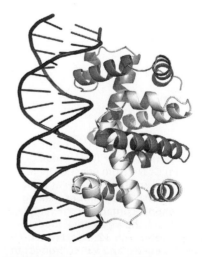

Figure 2.71 The tryptophan repressor (gray) bound to its DNA (black) binding site (PDB 1TRO).

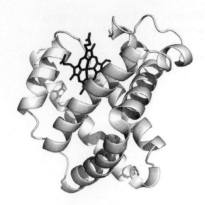

Figure 2.72 Antiparallel α-helices: the globin fold from sperm whale myoglobin (PDB 1A6N).

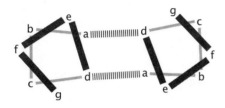

Figure 2.73 Interactions stabilizing a coiled coil: the residues in positions *a* and *d* of the heptad pseudorepeat **a**-*b*-*c*-**d**-*e*-*f*-*g* are hydrophobic, often leucine or valine, and their contacts stabilize the coil.

bonds with each other; the sheet closes on itself, forming a β-barrel. This fold is present in small proteins, such as domain 2 of the sulfidryl protease papain (**Figure 2.76**A). The clearest examples of this fold, however, are the very large integral membrane proteins of the porin family (see Figure 2.76B). Porins form water-filled pores in the outer membrane of Gram-negative bacteria (see Figure 2.76C).

A different type of antiparallel β-sheets are β-propeller folds (**Figure 2.77**). They are so called because when viewed from the top the arrangement of the sheets suggests a propeller (Figure 2.77A). Note how the chain connects one "blade" of the propeller to the next, moving from the top β-strand of one blade to the bottom β-strand of the following one, and making its way up again to the top of the next blade. In the end, the β-sheet also closes on itself, like in the β-barrel folds, satisfying the hydrogen bonds of the last and the first strands. Note also that the barrel is *not flat*; all its β-strands have about the same length. (The strands closer to the center of the propeller just appear shorter because they are further away from the observer.) The number of blades of the propeller varies, but 6–8 are common. A seven-blade propeller is found in galactose oxidase (see Figure 2.77B,C) and a six-blade propeller is found in sialidase (or neuraminidase) (see Figure 2.77D,E).

Our third and final type of antiparallel β-sheets are *Greek key* β-barrels. These barrels owe their name to the resemblance between the topology of the β-strands in the central barrel and a decorative motif frequently found in ancient Greek pottery (**Figure 2.78**).

In Greek key barrels (**Figure 2.79**A), the polypeptide chain also runs up and down, as do antiparallel β-strands connected by β-turns or longer loops (which can include helices), but it does not simply meander. The topology is more complicated. Chains adjacent to each other in the structure are not necessarily contiguous in the sequence. And the shape of the barrel is distorted, as if the barrel were somewhat collapsed. Because of that, they are sometimes called "jellyrolls" or "β-sandwiches."

There are two types of Greek key β-barrels, simple and complex. Simple Greek key barrels (see Figure 2.79A) have *one long connection* at each end of the barrel (one on top and one at the bottom), which link the front side to the back side of this collapsed barrel. Thus, there are two long connections in total. All the other links between the β-strands are direct connections between strands adjacent in the sequence. A typical example of simple Greek key barrels is the immunoglobulin fold, thus named because of its abundance in immune system globulins.

Figure 2.74 Coiled coil motif of the transcription factor (DNA binding protein) GCN4 from yeast (PDB 2ZTA). (A) Top view. (B) Side view, at right angles to (A). The residues making the hydrophobic contacts in the middle of the coil are shown as stick models (Leu-5, Val-9, Leu-12, Asn-16, Leu-19, Val-23, Leu-26, and Val-30 from both α-helices). The N-termini of both helices are on the left side and the C-termini are on the right.

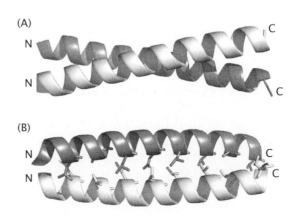

(A)

(B)

(A)

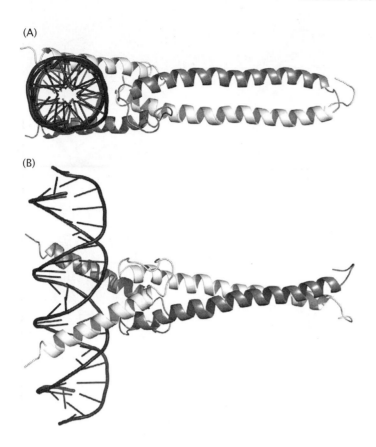

Figure 2.75 The MAX transcription factor, a coiled coil, bound to DNA. (A) Top view (DNA as hollow cylinder) and (B) side view. The coiled coil portion is on the right side of the protein structure. The two shorter helices grab the DNA (PDB 1AN2).

(B)

What do these barrels have in common with decorative keys in ancient Greek pottery? To answer, we need to draw a diagram of the *topology* of the protein chain. Imagine you look at the structure from the left (see Figure 2.79A). Now "pry" open the barrel between the first strand (N-terminus) and the last (C-terminus), as indicated by the curved arrows in Figure 2.79A. Then, flatten out the barrel. Draw the strands, starting from the N-terminus, and indicate the direction of the chain (N-to-C) by an arrow (see Figure 2.79B). Next, follow the turns of the polypeptide chain *in the actual structure* (A), and connect the strands in the diagram (B) in the same way they are connected in the structure. The result (see Figure 2.79C) is a Greek key with the same topology as the polypeptide chain in this β-barrel.

(A)   (B)   (C)

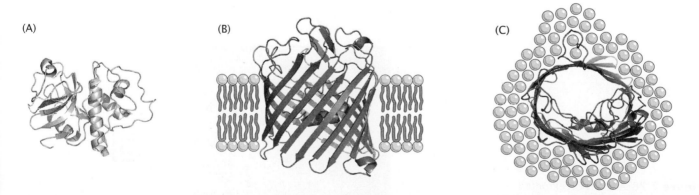

Figure 2.76 Antiparallel up-and-down β-barrel folds. (A) Domain 2 of papain, on the left side of the protein (PDB 9PAP). (B) Side view of a *porin*, the outer membrane protein F of *E. coli* (PDB 2OMF). (C) Same as (B), top view; the center of the barrel is filled with water.

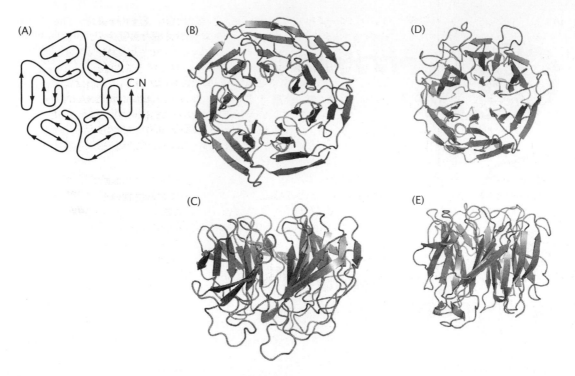

Figure 2.77 The β-propeller fold. (A) An idealized six-bladed β-propeller. (B, C) Top and side views of the seven-bladed β-propeller domain of galactose oxidase from *Dactylium dendroides* (PDB 1GOG). (D, E) Top and side views of the six-bladed β-propeller domain of sialidase (or neuraminidase) from *Micromonospora viridifaciens* (PDB 1EUU).

Complex Greek key β-barrels (**Figure 2.80**A) differ from the simple ones because they have *two long connections* at each end of the barrel, which link the front face to the back face of the barrel. Thus, there are four connections in total, two on top and two at the bottom of the barrel. To draw the topological diagram of this barrel (how each strand is connected to the next in the structure) is just as easy as in the case of simple Greek key barrels. The procedure follows the same steps, and is illustrated in Figure 2.80.

In all-β proteins, most β-sheets are antiparallel. However, the next type of fold in our survey, β-solenoids (or β-helices), consist essentially of parallel β-sheets (**Figure 2.81**). In these structures, a solenoid of up to 16 helical turns is formed; its faces consist of parallel β-sheets. A right-handed β-solenoid with one parallel β-sheet face occurs in the structure of the antifreeze protein from the beetle *Tenebrio molitor* (see

Figure 2.78 Ancient Greek pottery. The interior of a kylix, which is wide cup (bowl) used for drinking, shows the Greek key motif on the rim band. (Courtesy of Marie-Lan Nguyen.)

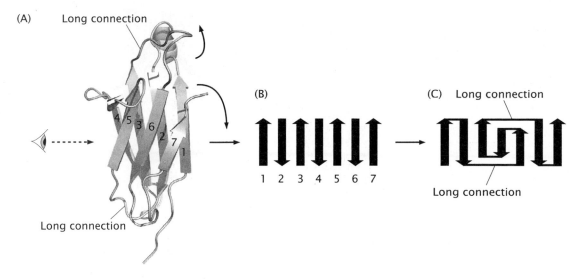

Figure 2.79 Drawing the Greek key topology of a simple β-barrel. (A) Simple barrel, from the human β-2-microglobulin (PDB 1LDS), with the β-strands numbered from the N- to the C-terminus. The curved arrows indicate how we would "pry" open the barrel. The dashed arrow indicated the direction of observation. (B) The β-strands after opening the barrel. (C) The connections between strands.

Figure 2.81A). The regularity of the solenoid is essential for the function of this protein: its period exactly matches that of an ice crystal, to which the protein binds, preventing crystal growth and consequent freezing of the beetle.

Left-handed β-solenoids are also common. The enzyme *N*-acetylglucosamine 1-phosphate uridyltransferase contains one such domain (see Figure 2.81B). Another example is the antifreeze protein from spruce budworm.

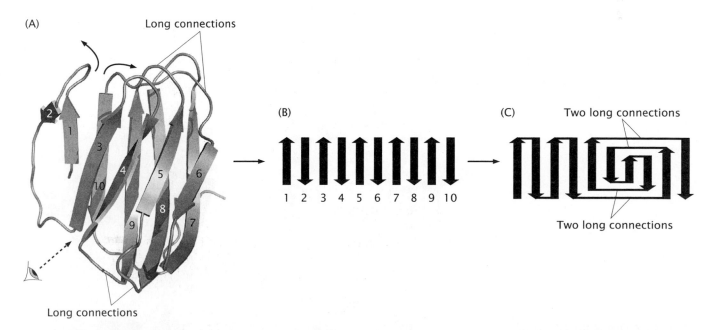

Figure 2.80 Drawing the Greek key topology in a complex β-barrel. (A) Complex barrel, from domain 3 of the tomato bushy stunt virus (PDB 2TBV), with the β-strands numbered from the N- to the C-terminus. The curved arrows indicate how we would "pry" open the barrel. The dashed arrow indicates the direction of observation. (B) The β-strands after opening the barrel. (C) The connections between strands.

Figure 2.81 β-Solenoids (or β-helices). (A) Right-handed β-helix of the antifreeze protein from the beetle *Tenebrio molitor* (PDB 1EZG).(B) Left-handed β-helical domain of uridyltransferase from *E. coli* (PDB 1FXJ).

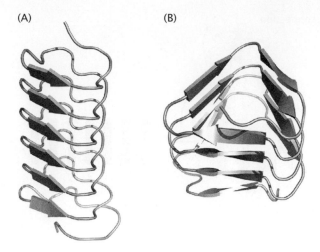

Proteins in the α/β Class Contain Significant Amounts of Both Types of Secondary Structures

Proteins in the α/β class contain significant amounts of α-helix and β-sheet structure. However, in the most important folds in this class, the *parallel* α/β folds, the overall fold is dominated by the spatial arrangement of the β-sheet; the α-helices form the connections between the β-strands. The two most common α/β folds are singly wound β-barrels and doubly wound β-sheets.

The singly wound β barrel (**Figure 2.82**) is characterized by a central, parallel β-sheet barrel, whose strands are linked by α-helical, right-handed crossover connections. This is a repetitive assembly of the β-α-β motif (see Figure 2.59). It is called *singly wound* because the connections (mostly α-helices) stay on *one side* of the β-strands they link—namely, on the outside of the barrel. The classical example is the triose phosphate isomerase (TIM) barrel (also called a TIM barrel fold). A very large number of proteins with unrelated functions share this fold.

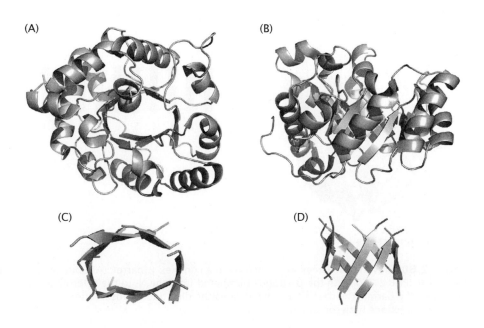

Figure 2.82 One of the TIM barrel domains of triose phosphate isomerase (PDB 1MO0). (A) Top view and (B) side view. (C) Top view of the central β-barrel; (D) side view.

(A)   (B)

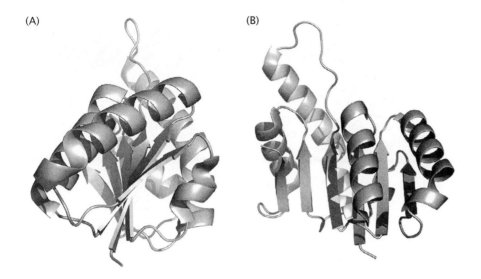

Figure 2.83 Parallel β-sheets, doubly wound: domain 1 of L-lactate dehydrogenase viewed from two different perspectives at right angles to each other (PDB 1LLD).

The doubly wound β-sheet fold (**Figure 2.83**) consists of an open, parallel (or sometimes mixed) β-sheet, coated on each side by right-handed crossover connections, which are mostly α-helices. Because of the right-handedness of these crossovers, the connections occur alternately on each side of the β-sheet. This fold is also known by several other names: β-saddle, because the shape of its β-sheet resembles a riding saddle when viewed from the side (see Figure 2.83A); dinucleotide binding domain, because proteins that bind dinucleotides (for example, nicotinamide adenine dinucleotide, NADH) have this fold; or Rossmann fold, after who described it. An example is the first domain of L-lactate dehydrogenase (see Figure 2.83).

Proteins Often Have More than one Domain

The types of folds that we have discussed are really folds of *domains*, not folds of entire proteins. A single polypeptide chain may fold to a single-domain protein, but often proteins have more than one domain. According to the CATH classification, about 70% of proteins have only one domain, 24% have two domains, and 4% have three. Proteins with more than three domains are rare. Each domain may have a different fold, but a protein may have two domains with identical folds, as in the case of triose phosphate isomerase. **Figure 2.84** shows examples of multidomain proteins.

How do we define a domain? Domains are usually identified in the protein structure using a mixture of visual intuition and computational algorithms. If we inspect the multidomain proteins shown in Figure 2.84, identifying different domains in each protein may seem fairly evident. However, if we are asked to define what a domain is, the question is not so simple. Like Saint Augustine wrote about time in his *Confessions*, "What then is time? If no one asks me, I know what it is; if I wish to explain it, I don't know."

In the early days of protein chemistry, domains were defined by proteolytic cleavage. Proteases, such as trypsin, for example, are not able to cleave the polypeptide chain inside the compact structure of a folded globular domain. However, they often cleave the chain at the loops that connect the domains. (The reason is interesting from a biochemical point of view. The binding site for the substrate polypeptide chain that is to be hydrolyzed is an extended ridge on the protease surface.

Figure 2.84 Multidomain proteins. (A) Alcohol dehydrogenase, with a Rossmann fold domain (an NAD$^+$ molecule is shown), and other α and β domains (PDB 1E3E); (B) sialidase, with β-propeller domains (a galactose molecule bound at the active site), and two β-barrel domains (PDB 1EUU); (C) uridyl transferase, with three distinct domains: β-solenoid, a single α-helix, and Rossmann fold (PDB 1FXJ); (D) tomato bushy stunt virus, with two β-barrel domains (PDB 2TBV); (E) triose phosphate isomerase, with two TIM barrels (PDB 1MO0); (F) pyruvate kinase, with three different domains: β-barrel, TIM barrel, and Rossmann fold (PDB 1PKN).

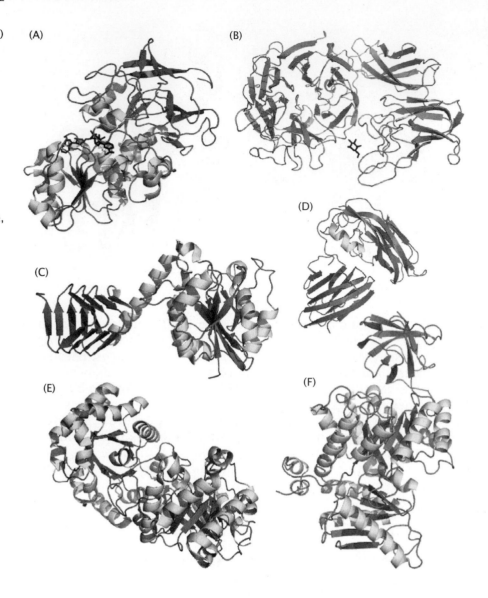

The chain segment that will undergo cleavage must be flexible to adapt to the binding site. This is usually true of loops connecting domains, but not of the polypeptide chain in a compact domain.) Often, but not always, the domains obtained by proteolytic cleavage coincide with those defined on the basis of structure.

Domains can also be defined on the basis of thermodynamics, by examining protein unfolding. One approach to identify folding domains is by calorimetry. In a differential scanning calorimetry (DSC) experiment, the temperature is slowly increased and the heat absorbed upon protein unfolding is measured. If a protein has two domains that unfold separately, two melting transitions are observed in the DSC experiment (**Figure 2.85**). However, there is a significant caveat. If the domains unfold *cooperatively* (that is, if the two domains interact, so that the unfolding of one renders *more likely* the unfolding of the other domain), then only one transition may be observed even if the protein has two structural domains. Another approach to unfold the protein is to add a denaturant, such as urea. Domains with different stabilities will unfolded at different urea concentrations and can therefore be identified. This method, again, identifies domains as cooperative

(A) 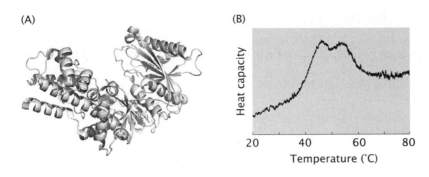 (B)

Figure 2.85 Domains in hexokinase. (A) The structure of hexokinase has two distinct domains: an all-α domain (left) and an α/β domain, a doubly wound β-sheet (right). It also contains another, small, doubly wound β-sheet in the hinge regions between the two larger domains (PDB 3O6W). (B) Differential scanning calorimetry (DSC) of hexokinase reveals the separate melting (unfolding) of the two major domains as two peaks in the heat capacity curve. (Adapted from Takahashi K, Casey JL & Sturtevant JM [1981] *Biochemistry* 20:4693–4697. With permission from American Chemical Society.)

segments of the polypeptide chain. This definition of domains coincides with the structural domains defined by CATH in most cases, but differs in about one-third of the cases.

Most Proteins form Oligomers in Solution

In the cell, the majority of proteins are not monomeric. Rather, they associate with other protein molecules, identical or different, to form *oligomers*. Oligomers are noncovalent complexes, but are stable in solution, under physiological conditions. They usually represent the functional form of the protein, although there are also inactive complexes; in those cases, the functional form is the dissociated protein.

When part of the oligomer, each protein molecule is called a *protomer*. The spatial organization of the protomers in the oligomer is called the *quaternary structure* of the protein. It is estimated that about half of the protein oligomers in cells are homomers (complexes of identical protomers). The classical example of a protein with quaternary structure is hemoglobin (**Figure. 2.86**), which is a tetramer of two types of protomers ($\alpha_2\beta_2$). The quaternary structure of hemoglobin is of fundamental importance for its function.

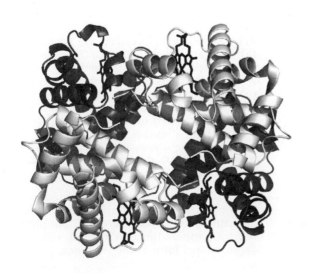

Figure 2.86 The quaternary structure of hemoglobin, a tetramer ($\alpha_2\beta_2$) of two types of protomers, α (white) and β (dark gray). The four heme groups are shown as stick models, in black (PDB 1A3N).

2.9 SUMMARY

Proteins are composed of 20 amino acids, linked by peptide (amide) bonds. The ionization of amino acid side chains (and the end groups) determines the electrostatic charges in a protein. The ionization states depend on the pK_a's of the amino acid residues and change with pH. The partition function approach provides a simple method to describe and understand those changes.

The possible conformations of the polypeptide chain are constrained by van der Waals (steric) repulsions. Conformations that allow formation of hydrogen bonds between peptide groups (CONH) of the chain are highly favored. Regular conformations, in which the φ and ψ angles of consecutive residues are similar, constitute the secondary structure of proteins. Most of the secondary structure in globular proteins consists of α-helices and β-sheets, but the polypeptide chain must make frequent turns in order to fold into a compact structure. In native proteins, the chain adopts a well-defined and unique tertiary structure. The tertiary structures of proteins can be classified under a limited number of different folds, many of which have a distinctive symmetry.

Many techniques can be used to study various aspects of protein structure and conformation, such as absorption and fluorescence spectroscopy, circular dichroism, or calorimetry, but only NMR and X-ray crystallography allow the determination of the complete three-dimensional structure of proteins. Most proteins consist of only one domain, but many have two domains, or even three. In multidomain proteins, each domain can have a competely different fold. Finally, most proteins associate reversibly under physiological conditions to form oligomers, which often represent the functional form of the protein. The spatial arrangement of the protomers in the oligomer is the protein quaternary structure.

2.10 PROBLEMS

2.1 Draw the structure of the most important ionization state of glutamine at pH 7. Is the pK_a of the acid that is obtained by protonating the side chain of glutamine less than zero or greater than 14?

2.2 Consider the amino acids cysteine and lysine. Indicate one pH at which the major forms of these two amino acids are in accordance with the following conditions: the α-amino groups are protonated (+ charge), the α-carboxylic groups are deprotonated (−), the side chain of cysteine is deprotonated (−), and the side chain of lysine is protonated. Hint: start by drawing the structures with the stipulated properties.

2.3 Using the pK_a values in Table 2.3, answer the following questions.
(a) Draw the structures of the major forms of cysteine and aspartic acid at pH 3, 7, and 9.
(b) What are approximately the net charges of these two amino acids at each of those pH's?

2.4 The side chain of the amino acid tryptophan is a heterocycle called indole. The nitrogen of the indole ring has a lone electron pair, but it does not get protonated unless the pH is extremely low. Explain why.

2.5 What is the net charge of a histidine *residue* at pH 2? Use Table 2.3. Think first whether you need a calculation or not.

2.6 What is, approximately, the net charge of the polypeptide EKLADRANTWKKGRR at pH 3 and at pH 7? No elaborate calculation is needed. Estimate the charges from Table 2.3.

2.7 The following questions are about cysteine.
(a) What are the structures of the major and the second most important form of the amino acid cysteine at pH 7? What are fractions of those forms at pH 7?
(b) Write the proton binding isotherm—that is, the fraction of the protonated state, as a function of pH for a cysteine *residue* that is part of a polypeptide chain. Think carefully! This is about a residue, not an amino acid.
(c) What is the *average* charge of this *residue* at pH 7.0? (Hint: You have to do a calculation, similar to the one in Section 2.3.)

2.8 Derive the fraction of each protonation state of cysteine as a function of pH. Note that there are now *three* ionizable groups, and, therefore, four protonation states. Use the partition function method, with the completely deprotonated form as the reference.

2.9 Using the result from the previous question, plot the fractions of all the ionization states of glutamic acid, histidine, and lysine as a function of pH, in the pH range of 0–14. (Use the pK_a values from Table 2.3.)

2.10 Consider the following plot of the ionization of an amino acid. Which amino acid is this?

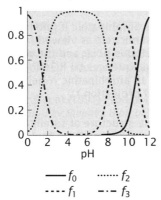

— f_0 ······ f_2
---- f_1 —·— f_3

2.11 Consider the *cis/trans* isomerization of a peptide bond. For a bond between two amino acids (different from proline), the equilibrium constant is $K = 1000$ in favor of *trans*. If proline is the residue following the peptide bond, $K = 4$.
(a) What is ΔG° between the two isomers in the two cases above, at room temperature?
(b) Use the partition function to calculate the fraction of *cis* bonds in both cases.

2.12 Build short peptides, say a pentapeptide, and set the dihedral angles to each of the values given in Table 2.5, so as to generate all the types of secondary structure and familiarize yourself with these structures.

2.13 Go to the Kinemage web site (http://kinemage.biochem.duke.edu/). Download and install the program King on your computer. Then consider a central residue (n) and the two flanking residues preceding ($n - 1$) and following it ($n + 1$) in a polypeptide chain. For example, designate the peptide carbonyl of the central residue by CO(n), the amide nitrogen group of the following residue by NH($n + 1$), and the second carbon of the side chain of the central residue by Cβ(n) (this is also sometimes called Cb or C2). Use the kinemage c1Basics.kin provided at the Kinemage web site, under kinemages from Branden and Tooze (http://kinemage.biochem.duke. edu/kinemage/kinlist.php), or molecular models to explore the Ramachandran diagram. In King, use the *measure* tool to determine distances between atoms, so that you find out which distances are too short compared to the limits given in Table 2.4.
(a) The upper right quadrant of the Ramachandran diagram is mainly forbidden because of steric hindrance between which two groups?
(b) What forbidden overlaps block the lower right quadrant of the diagram? (Hint: compare any regular amino acid with glycine.)
(c) Why is there so much more accessible space on the top left than on the bottom left quadrant? (Hints: think of the differences between glycine and the other amino acids. Use the computer molecular models in this kinemage file or real models to understand the problem.)
(d) The central line at $\phi = 0$ in the Ramachandran diagram is forbidden mainly because of steric hindrance between which groups?

2.14 Use the kinemage c2Motifs.kin provided at the Kinemage web site, under kinemages from Branden and Tooze (http://kinemage.biochem.duke.edu/kinemage/ kinlist.php), to do the following:
(a) Do the tutorial for kin 1 (α-helix). Measure the ϕ and ψ angles as described in the tutorial for the residues of the α-helix shown there. Record your values and calculate the mean and standard deviation.
(b) Do the same with kin 3 (β-sheet). Again, measure the ϕ and ψ angles for strands b2, b3, and b4 for the β-sheet shown. Record your values and calculate the mean and standard deviation.
(c) Produce a Ramachandran diagram: plot the individual values of ϕ and ψ that you measured for both the helix and the sheet as points in the Ramachandran plot. Include the plot in your answer.

2.15 Why does proline break an α-helix at the point where it occurs in the polypeptide chain?

2.16 Calculate the populations of the electronic states S_0 and S_1 involved in the absorption of light by tryptophan ($\lambda = 280$ nm), at room temperature, using the partition function method. See Section 2.2 for the values of constants.

2.17 Suppose you are studying protein denaturation by the fluorescence emission of a Trp residue of the protein. You obtained a fluorescence spectrum of the protein at 20°C and another at 65°C, and you know that the protein is folded at 20°C and unfolded (denatured) at 65°C. Note that the spectrum of the unfolded protein is red-shifted. Then you recorded a fluorescence spectrum of this protein at 45°C. The three spectra, all normalized to the fluorescence of the highest value recorded, are shown below. The spectrum in gray (45°C) appears intermediate between those at 20°C and 65°C.

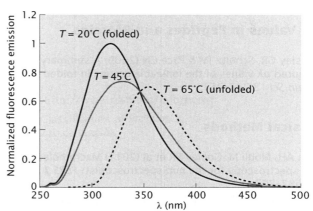

What is the fraction of folded protein at 45°C? (Hint: choose one wavelength to do your calculation. This problem is similar to the calculation in Section 2.3.)

2.18 Suppose you are following the denaturation of protein A, but now you measure its circular dichroism (CD) spectrum as you increase the temperature. This protein, when folded, is entirely α-helical; when unfolded, it loses all structure (random). The graph below shows the CD spectrum of protein A at a certain temperature T (dashed line) together with the characteristic spectra for an α-helix (solid) and a random coil (dotted). What percent of the proteins are denatured at this temperature? (Hint: the

Pauling L, Corey RB & Branson HR (1951) The structure of proteins: two hydrogen-bonded spiral configurations of the polypeptide chain. Proc Natl Acad Sci USA 37:205-211.

Pauling L & Corey RB (1951) The pleated sheet, a new layer configuration of polypeptide chains. Proc Natl Acad Sci USA 37:251-256

Ramachandran GN & Sasisekharan V (1968) Conformation of polypeptides and proteins. Adv Protein Chem 23:283-437.

Protein Tertiary Structure

Orengo CA & Thornton JM (2005) Protein families and their evolution—a structural perspective. Annu Rev Biochem 74:867-900.

Evolution of Protein Sequence & Structure

3.1 NOTHING IN BIOLOGY MAKES SENSE EXCEPT IN THE LIGHT OF EVOLUTION

The title of this section is that of a famous article by Theodosius Dobzhansky. Evolution is a unifying concept in biochemistry. The structures of the proteins that exist today (*extant* proteins) have been shaped by evolution. Amino acid interactions depend on the context of the overall three-dimensional structure of the protein. To study those interactions, it is not possible to prepare protein mutants with all possible combinations of residue substitutions. However, if we follow the history of the amino acid changes in a protein, we can learn about the effects of amino acid residues on each other, as those changes actually happened. We can also understand how the amino acid substitutions that occurred historically in a certain protein shaped the future changes that the protein could and could not undergo. Interacting systems co-evolve. Studying the changes that actually occurred in the course of evolution provides a better understanding of the contributions of individual amino acids to protein structure and function.

3.2 THE TREE OF LIFE SHOWS EVOLUTIONARY RELATIONSHIPS BETWEEN SPECIES

Life appeared probably about 4 billion years ago. The major divergence between bacteria, archaea, and eukaryotes occurred about 2 billion years ago. Mammals began to diverge from each other about 100 million years ago. **Table 3.1** summarizes the landmarks in evolution.

The fossil record provides strong evidence for evolution. Yet the evidence at the molecular level, in the composition, sequence, and structure of biomolecules, is even more compelling. The genetic code is universal. The same 20 amino acids are used as building blocks for the proteins of all life forms. All amino acids in proteins (synthesized in the ribosomes) are of the L stereoisomer. Extant living organisms do not have *different* amino acids in their proteins because this set of amino acids was already used by their common ancestors and has remained the same.

Charles Darwin used the image of the *tree of life* to convey the idea of the relation between species that share a common ancestor and the

These examples made it possible for a librarian of genius to discover the fundamental law of the Library. This thinker observed that all the books, no matter how diverse they might be, are made up of the same elements: the space, the period, the comma, the twenty-two letters of the alphabet.

Jorge Luis Borges

Table 3.1 Important dates in evolution[a]

Event	Time (billion years ago)
Origin of the universe	14
Formation of the solar system	4.6
First self-replicating organism	3.5
Bacterial/archaeal/eukaryotic divergence	~2
Animal/plant/fungal divergence	~1
Vertebrate/invertebrate divergence	0.5
Beginning of mammalian radiation	0.1

[a]Based on Doolittle RF, Feng DF, Johnson MS & McClure MA [1986] *Cold Spring Harb Symp Quant Biol* 51:447–455.

limbs of a tree that share a common branch point from the main trunk. Two drawings of the tree of life, or evolutionary tree, are shown in **Figure 3.1**. An evolutionary tree of living organisms represents the relationships between species in time. It shows their phylogeny: the order in which different living organisms branched out from their common ancestors. Therefore, these trees are also called phylogenetic trees or organismal trees. The numbers indicated on the trees represent the times when branching occurred (in billion years) counted backward from the present. In Figure 3.1A, the lengths of the tree branches are drawn (roughly) in proportion to these times.

The type of tree in Figure 3.1B is called a *cladogram*. It contains the same information, but is better at conveying the organization of the tree. This tree organizes the species in groups, called clades, that are nested hierarchically, from the outer limbs down to the main trunk. Starting from the top left, the first clade consists of human and horse; then, there is the human-horse-fish clade; then, the human-horse-fish-insect clade. Each clade ends at a horizontal branch, which is "rooted" on the next vertical branch. What the human-horse clade says is that there was a common ancestor from which both humans and horses descend. What the human-horse-fish clade says is that there was an *older* common ancestor from which humans, horses, and fish descend.

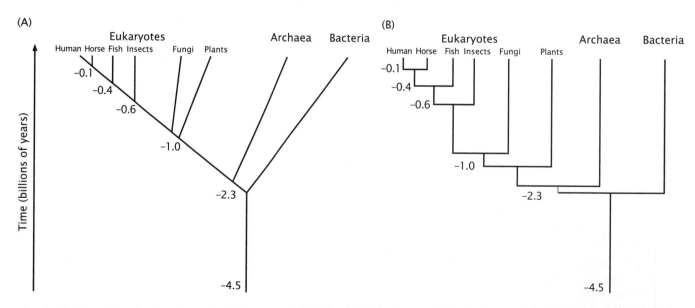

Figure 3.1 Evolution landmarks in the divergence of living organisms, in two equivalent representations: (A) evolutionary tree, (B) cladogram. The times at the points of divergence are indicated in billion years.

Humans are more closely related to horses than to fish because they share a common ancestor (which existed 0.1 billion years ago) that is more recent than the common ancestor between humans and horses and fish (which existed 0.4 billion years ago).

The trees show how species are related in the sense of how long ago they diverged from their most recent common ancestor. Also, the tree does not say that humans descended from horses, but that they share a common ancestor, from which they both evolved in divergent directions. In a cladogram the time refers only to the length of the vertical branches, which are drawn (roughly) in proportion. Typically, only two vertical branches are attached to each horizontal one.

3.3 MUTATIONS IN PROTEINS ARISE FROM MUTATIONS IN DNA

Protein sequences change over time because of mutations that are a result of mutations in the genetic material. In most extant organisms, the genetic material is deoxyribonucleic acid (DNA). Some viruses use ribonucleic acid (RNA) instead. The *central dogma* of molecular biology establishes the flow of information from the primary genetic record, DNA, to an intermediate record, RNA, to its final expression, the proteins (**Figure 3.2**).

To understand protein evolution, we need to briefly review essential aspects of replication (copying of the DNA code), which is written in the language of nucleotides, transcription of this code from DNA to RNA, and translation of the nucleotide code to the language of proteins (amino acids).

DNA is Composed of the Same Four Nucleotides in all Living Organisms

The composition of DNA tells a similar story as that of proteins about evolution. The structure of DNA was proposed by Watson and Crick (1953) in one of the most important papers in the history of science. DNA forms a right-handed double helix (**Figure 3.3**). The helix is periodic along its axis: it repeats itself after each stretch of a certain length, called the pitch, which in DNA is 34 Å (see Figure 3.3A). The pitch corresponds to a complete turn of the helix, and it takes 10 residues (or base pairs) for a turn. Thus the rise per residue along the helical axis is 3.4 Å. The diameter of helix is 20 Å. DNA has two grooves: the major groove, between two consecutive turns, and the minor groove, between the two strands (see Figure 3.3B).

The two strands of DNA are polymers of nucleotides. Each nucleotide consists of a phosphate, a carbohydrate, and a base (**Figure 3.4**). What is remarkable with regard to evolution is that all carbohydrate units in DNA are of the D isomer, and they are the same carbohydrate (deoxyribose) in all living organisms. What is more, the same four bases, adenine (A), thymine (T), guanine (G), and cytosine (C) (**Figure 3.5**) are used in the DNA of all life forms.

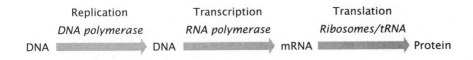

Figure 3.2 The central dogma of molecular biology: the genetic information in DNA is transcribed to RNA, which is then translated to the language of proteins.

Figure 3.3 The double-helix structure of the DNA molecule (PDB 1BNA). (A) Stick model. (B) Space-filling model. The pitch of the helix is 34 Å, with 10 residues (or base pairs) per turn, so the rise per residue is 3.4 Å. The diameter is 20 Å.

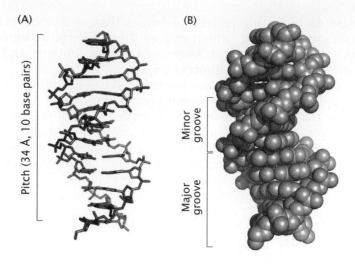

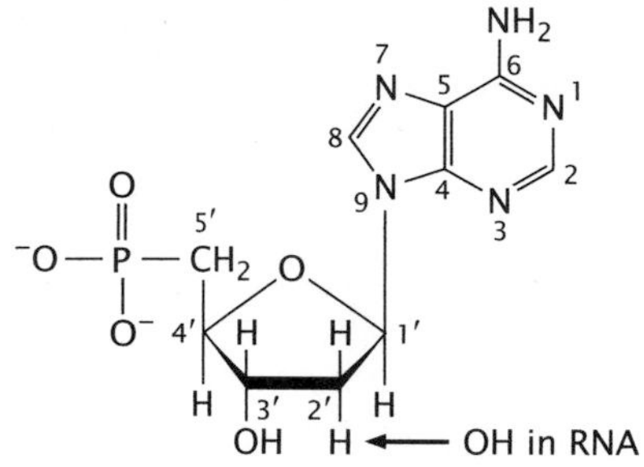

Figure 3.4 Structure of a DNA nucleotide, deoxyadenosine-3'-monophosphate (dAMP). In RNA there is a hydroxyl group (OH) at position 2'.

The double-helix of DNA is stabilized by hydrogen bonds between the two strands. Two bases, one from each strand, form a base pair. Adenine and thymine establish two hydrogen bonds, guanine and cytosine, three (**Figure 3.6**). Each hydrogen bond contributes only ~1 to 2 kcal/mol of Gibbs energy to the stability of the helix. This is a weak interaction (~$2RT$). However, there are many such hydrogen bonds in each DNA molecule, making a large overall contribution to DNA stability. When DNA is replicated or transcribed it must open locally and *reversibly*, so that one strand can serve as template for a new DNA or RNA strand. In order for this to happen, *each* bond must be weak.

In addition to hydrogen bonds, the DNA helix is stabilized by *base stacking*. There are two components to base stacking: the *hydrophobic effect*, which tends to reduce contact between aromatic rings and water; and *polarization* of the π electrons of the aromatic rings of the bases, which leads to strong induced dipole–induced dipole interactions between adjacent bases along the length of the DNA molecule.

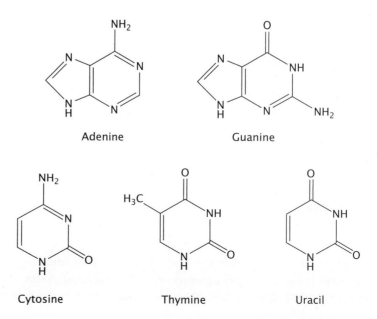

Figure 3.5 Structure of four DNA bases and the RNA base, uracil. Two of those bases, adenine and guanine, are purines; the other two, cytosine and thymine, are pyrimidines. In RNA, uracil (U) replaces thymine.

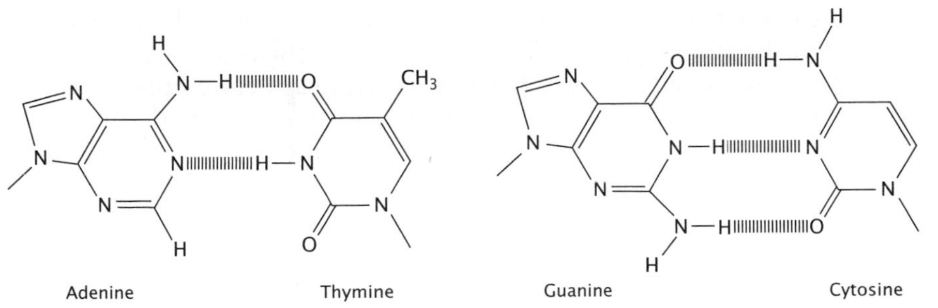

Figure 3.6 Watson–Crick base pairs: adenine–thymine and guanine–cytosine. Hydrogen bonds are indicated by dashed lines. Uracil establishes hydrogen bonds with adenine like thymine does.

DNA Replication and Transcription Take Place on a Template Strand

When DNA replication occurs, a buldge opens in the double strand, unpairing a few bases, and a replication fork is assembled. The central protein is a DNA polymerase, which catalyzes the synthesis of a new DNA strand on each of the parent strands. The DNA polymerase reaction takes place in a mandatory order, from the 5' to the 3' end of the nascent DNA strand (**Figure 3.7**A). To begin replication, DNA polymerase requires a primer, a short piece of nucleic acid assembled on the template strand (the primer is made of RNA). In addition, it requires the triphosphate deoxynucleotides of the four bases (dATP, dGTP, dCTP, and dTTP) and the metal ion Mg^{2+}.

The first step to produce a protein is the transcription of DNA into messenger RNA (mRNA), catalyzed by RNA polymerase. This enzyme uses triphosphate ribonucleotides (ATP, GTP, CTP, and UTP) and does not require a primer. It works faster, but also makes more mistakes that DNA polymerase. However, this is not as important because a transcription error only affects the batch of protein synthesized from this mRNA, and will not be passed to the offspring.

The sequences of the two strands of DNA are different—they are complementary. For a given sequence, only one of the strands is translated into RNA. The RNA polymerase uses as template the DNA strand complementary to the coding strand; thus, the sequences of the mRNA and the coding strand are the same, except for U instead of T (see Figure 3.7B).

DNA Contains Regulatory Elements in Addition to Genes

Not all DNA is transcribed; most of it is not. A significant amount of DNA has no active function, but some of the DNA that does not constitute genes has well-defined functions. Regulatory signals and protein

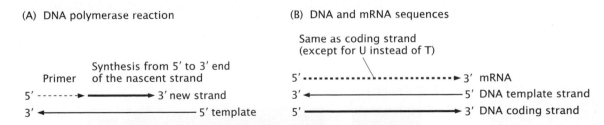

Figure 3.7 Relation between DNA and mRNA strands. (A) Replication, catalyzed by a DNA polymerase. (B) Transcription, catalyzed by RNA polymerase. The mRNA molecule has the same sequence as the DNA coding strand, which is complementary to the template strand.

binding sites are located in the DNA sequences close to the transcribed genes. The beginning and ending of the sequences to be transcribed have special features that identify them. These are called promoter and terminator sites. The promoter regions are essential for the assembly of the machinery required for mRNA synthesis, of which the RNA polymerase is only one element. The DNA sequence before the gene itself (the protein-coding stretch of the DNA) is called the upstream region, and the sequence after the gene is called the downstream region. DNA regions that do not code for protein sequences or serve as binding sites for gene transcription and regulation elements are highly variable.

A Promoter Controls an Operon of Functionally Related Genes

Often, particularly in prokaryotes, genes involved in the same metabolic pathway are grouped together in the DNA in a continuous stretch called an *operon*. The expression of the operon is usually controlled by a single promoter where the RNA polymerase binds. The region of the promoter that controls binding of the RNA polymerase is called the operator. Transcription factors are proteins involved in regulation of gene expression. Some, like the ρ protein, are involved in transcription termination. Others activate or inhibit the activity of RNA polymerase at the promoter site.

For example, the tryptophan operon in *E. coli* comprises the genes that code for the enzymes necessary for the synthesis of tryptophan from chorismate (**Figure 3.8**). The two subunits of anthranilate synthase are encoded by the genes trpE and trpD. The synthase complex catalyzes the formation of anthranilate from chorismate. It also contains the transferase enzymatic activity (also encoded by trpD) that produces the next intermediate, phosphoribosyl anthranilate. TrpC codes for a protein with two domains, which contain the next two enzymatic activities: phosphoribosyl anthranilate isomerase (PRAI), which produces 1-(*o*-carboxyphenylamino)-1-deoxyribulose-5-phosphate, and indole glycerol phosphate synthase (IGPS), which converts it to indoleglycerol-3-phosphate. Finally, *trpB* and *trpA* code for the β and α subunits of the complex tryptophan synthase, which produces indole (from indoleglycerol-3-phosphate) and transfers it to serine to make tryptophan.

The tryptophan repressor (see Figure 2.71) is a DNA-binding protein of the helix-turn-helix motif; its active form is a dimer. It binds two tryptophan molecules, one on each subunit. In the tryptophan-associated state, the repressor binds to the operator of the Trp operon, inhibiting its transcription. The repressor thus functions as a *sensor* for the abundance of tryptophan in the cell. When the tryptophan level is high, there is no need for the cell to synthesize tryptophan; thus, tryptophan associates with the repressor, which in turn binds to the DNA and switches off expression of the genes needed for tryptophan synthesis.

Figure 3.8 The Trp operon in *E. coli*. The set of genes specific for the Trp biosynthesis, *trpE, D, C, B*, and *A*, are contiguous in the DNA and transcribed as a single mRNA. Their expression is controlled at the promoter site, which includes the operator.

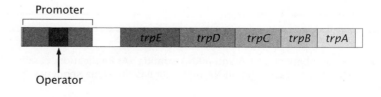

The Genetic Code is (Almost) Universal

The genetic code consists of triplets of nucleotides on the mRNA, called *codons*. Each codon codes for an amino acid. The genetic code is almost universal (**Table 3.2**). This in itself is strong evidence for evolution. There are some variations in a few elements of the code, between very distant organisms, but these are minor exceptions. The most interesting variations are those found in mitochondrial DNA. Mitochondria evolved from primitive organisms that associated symbiotically with other species; therefore, mitochondrial DNA originated much before the divergence of the species that exist today, and some differences have remained. Nevertheless, the universality of the code across species as different as humans and bacteria is much more impressive.

The code is *degenerate*: there are several codons for each amino acid. Many amino acids have four different codons; some have six (Leu, Arg). Tryptophan, the rarest amino acid, has only one codon (UGG), and cysteine has only two (UGU, UGC). The degeneracy of the code is a must because there are 64 codons and only 20 amino acids. You might think that one could just use 20 of the 64 codons, leaving the others as nonsense codons. The problem is, what would happen if a mutation converted one of the 20 used codons to a nonsense codon?

The genetic code in the mRNA is "read" on the ribosomes by adapter RNA molecules called transfer RNA or tRNA (**Figure 3.9**). Each tRNA molecule has a specific amino acid covalently attached to its 3' end, and binds to the codon on the mRNA through an *anticodon*. The *anticodon* is a triplet of nucleotides in the tRNA that is *complementary* to the codon on the mRNA and recognizes it by base pairing (see Figure 3.9A). The three-dimensional structure of tRNA resembles the letter L, with the amino acid at one end and the anticodon at the other end of the L (see Figure 3.9B).

Table 3.2 The standard genetic code

UUU	Phe	UCU	Ser	UAU	Tyr	UGU	Cys
UUC	Phe	UCC	Ser	UAC	Tyr	UGC	Cys
UUA	Leu	UCA	Ser	UAA	Stop[a]	UGA	Stop
UUG	Leu	UCG	Ser	UAG	Stop	UGG	Trp
CUU	Leu	CCU	Pro	CAU	His	CGU	Arg
CUC	Leu	CCC	Pro	CAC	His	CGC	Arg
CUA	Leu	CCA	Pro	CAA	Gln	CGA	Arg
CUG	Leu	CCG	Pro	CAG	Gln	CGG	Arg
AUU	Ile	ACU	Thr	AAU	Asn	AGU	Ser
AUC	Ile	ACC	Thr	AAC	Asn	AGC	Ser
AUA	Ile	ACA	Thr	AAA	Lys	AGA	Arg
AUG	Met[b]	ACG	Thr	AAG	Lys	AGG	Arg
GUU	Val	GCU	Ala	GAU	Asp	GGU	Gly
GUC	Val	GCC	Ala	GAC	Asp	GGC	Gly
GUA	Val	GCA	Ala	GAA	Glu	GGA	Gly
GUG	Val	GCG	Ala	GAG	Glu	GGG	Gly

[a]There are four differences in the genetic code of mitochondria: AUA codes for Met; UGA codes for Trp; AGA and AGG are stop codons.
[b]Met or start. AUG codes both for the start of the protein, with an initial Met, and for internal Met.

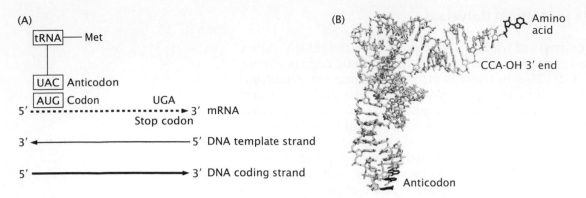

Figure 3.9 (A) Translation of mRNA into protein through the matching of codons in mRNA to anticodons in tRNA. (B) Structure of the adapter tRNA molecule (PDB 1EHZ).

During protein synthesis, two tRNA molecules bind to adjacent codons in the mRNA, which are located in the P-site (for peptidyl) and A-site (for aminoacyl) in the ribosome. A new peptide bond is formed when the growing polypeptide chain attached to the tRNA in the P-site is transferred to the amino acid on the tRNA in the A-site. Then, the tRNA/mRNA complex moves in the ribosome. The tRNA molecule that was originally in the P-site is released from the ribosome; the original A-site becomes the new P-site, containing the growing peptidyl-tRNA; and a new A-site opens, ready to accept a new aminoacyl-tRNA. This process is repeated until the entire mRNA sequence is translated. The termination codons are not recognized by tRNA molecules, but by proteins called release factors, which bind to the ribosome/RNA complex at those termination sites and release the protein from the ribosome.

Eukaryotic Genes are not Continuous

In eukaryotes, the primary transcript (messenger RNA) is further processed before being translated into the polypeptide. The primary mRNA contains sequences that are not translated, called intervening sequences, or *introns*, as well as regions that are translated, called expressed sequences, or *exons*. Maturation of the primary mRNA involves the removal of introns and the *splicing* of the exons together to produce a shorter mRNA, which is then translated (**Figure 3.10**).

Splicing occurs at precise locations in the mRNA: an intron begins usually with GU and ends with a pyrimidine-rich sequence ending in AG. Mutations that alter these sites result in proteins that are too long and usually do not fold properly. Split genes, consisting of alternating exons and introns, offer evolutionary advantages. Recombination of genes can occur in the course of evolution by rearranging exons, a process called *exon shuffling*. Further, split genes offer the possibility

Figure 3.10 Splicing of mRNA is a complex process that consists of cutting out the introns and connecting the exons.

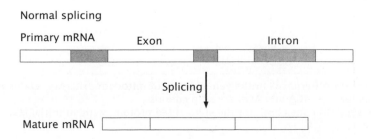

of alternative splicing: more than one protein can be produced from the same gene by splicing the mRNA in different manners.

The evolutionary origin of introns is a controversial question. For some time, the hypothesis that introns existed in primordial bacteria but were lost during evolution was favored. However, the dominant belief now is that introns never existed in bacteria. They seem to have invaded the eukaryotic genome, at some early stage of the appearance of eukaryotic cells.

3.4 EVOLUTION OCCURS BECAUSE OF MUTATIONS IN DNA

A Point Mutation Occurs when a Nucleotide is Replaced by Another in the DNA

Mutations in the DNA are the primary material on which evolution works. Why do mutations occur? Mutations occur naturally, by chance. (The rates of mutation are enhanced by exposure to external factors, such as chemicals and radiation.) The error rate of DNA polymerases is low, but sometimes an error occurs and the wrong base is incorporated in the nascent DNA strand. The substitution of a single base (nucleotide) by another on the DNA is called a *point mutation*. Purine ↔ purine or pyrimidine ↔ pyrimidine exchanges are called transitions; purine ↔ pyrimidine exchanges are called transversions. Point mutations are the simplest and most common kind of mutations.

A point mutation may result in the replacement of one amino acid by another in a protein sequence, but not always, because of the degeneracy of the genetic code. In most cases, a mutation in the third nucleotide does not result in a change of amino acid (see Table 3.2). For example, changing the third nucleotide in the CTT codon for leucine in the DNA does not result in a change of amino acid, because CTT, CTC, CTA, and CTG all code for leucine. DNA mutations that result in the same amino acid are called *synonymous substitutions* or *silent* mutations. Mutations that change the amino acid type are called *non-synonymous substitutions*.

Furthermore, if a mutation changes only one of the letters (nucleotides) of a codon, the resulting amino acid is usually not too different from the original one. For example, mutation of the first nucleotide (C) of the CUU codon (Leu) to A or G results in AUU (Ile) or GUU (Val), producing amino acids that are not very different from leucine. Mutation of the third nucleotide of the codons for glutamic acid, from GAA or GAG to GAT or GAC, respectively, is usually of no consequence, since the result is a similar residue, aspartic acid. These are examples of *conservative* mutations. The mutant protein is likely to function normally. This feature of the genetic code shields the organism from the effects of mutations, to a point.

DNA is constantly subject to repair by enzymes that detect and correct misincorporated bases. Still, some errors go undetected. In humans, there are $\sim 1 \times 10^{-8}$ mutations per nucleotide per generation. Since the human genome contains $\sim 3 \times 10^9$ base pairs (bp), there are on average ~ 70 mutations per individual per generation. As a result, two randomly selected individuals from the human population differ in $\sim 0.1\%$ of their genome (1 in every 1000 bp). These natural variations in the DNA, which result from point mutations, are called single nucleotide polymorphisms (SNPs).

Mutations in noncoding regions of the DNA have no deleterious consequences, for the most part. Hence, those DNA regions are highly variable. However, mutations in regions flanking genes may or may not be tolerated, depending on whether they affect nucleotides that are important for the regulation of gene expression.

Occasionally a single nucleotide substitution has serious consequences. Sickle-cell anemia is caused by a single nucleotide substitution in the gene encoding for the β chain of hemoglobin. The sixth residue of the β-globin sequence is glutamic acid, whose codon is GAA or GAG. If the middle A is mutated to T, the codon changes to GTA or GTG, which codes for valine. In most circumstances this mutation, although a change from a charged to a hydrophobic amino acid, would probably be unimportant. In fact, the mutant β-globin folds correctly. The problem is that Val-6 fits in a hydrophobic pocket on the surface of *another* β-globin molecule, and this interaction leads to hemoglobin *polymerization*. The filaments formed push on the ends of the red blood cell, which acquires a sickle shape. The pressure exerted on the membrane renders it susceptible to lysis. The loss of red blood cells results in anemia. The molecular origin of sickle-cell anemia seems to indicate that single amino-acid mutations have devastating effects. In fact, this is rather the exception than the rule. Protein structures are so adaptable that most amino acid substitutions have negligible effects on structure and function.

Insertions and Deletions Usually Change the Reading Frame of mRNA

More serious are the consequences of mutations that affect post-translational modification sites, such as proteolytic cleavage sites and glycosylation sites, or initiation or termination codons. Point mutations preserve the *reading frame* of the RNA. Small-scale insertions or deletions, however, most often change the reading frame. For example, the human hemoglobin α chain has 141 residues. In the Cranston mutant, the first 141 residues are the same as in the normal protein, but two nucleotides are inserted (AG in **Figure 3.11**). The reading frame changes after that codon and the original termination codon (UAA) is bypassed. The synthesis of the protein continues until another UAA codon appears, after residue 172. (This was not a termination codon in the original nucleic acid sequence because it only appears after the reading frame shift.) Such mutations may not be lethal if the new termination codon is not too far from the original one, but the RNA is translated into a longer protein.

Insertions or deletions of large DNA portions, and internal duplications or repeats occur as well. Large-scale rearrangements occur because of unequal crossing over of the DNA and because of exon shuffling.

Figure 3.11 Terminal sequence of the human β-globin. In the Cranston mutant, insertion of the two nucleotides AG occurred. (Based on Creighton TE [1993] Proteins. Structure and Molecular Properties, 2nd ed. Freeman.)

Normal β-globin

```
Lys Tyr His Stop
AAG UAU CAC UAA GCU CGC UUU CUU CGU GUC CAA UUU CUA UUA A
```

Cranston mutant

```
Lys Ser Ile Thu Lys Leu Ala Phe Leu Leu Ser Asn Phe Tyr Stop
AAG AGU AUC ACU AAG CUC GCU UUC UUG CUG UCC AAU UUC UAU UAA
```

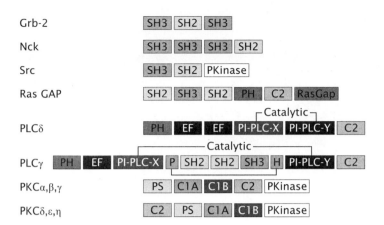

Grb-2   SH3 | SH2 | SH3

Nck   SH3 | SH3 | SH3 | SH2

Src   SH3 | SH2 | PKinase

Ras GAP   SH2 | SH3 | SH2 | PH | C2 | RasGap

PLCδ   PH | EF | EF | PI-PLC-X | PI-PLC-Y | C2

PLCγ   PH | EF | PI-PLC-X | P | SH2 | SH2 | SH3 | H | PI-PLC-Y | C2

PKCα,β,γ   PS | C1A | C1B | C2 | PKinase

PKCδ,ε,η   C2 | PS | C1A | C1B | PKinase

Figure 3.12 Domain structure of several signal transduction proteins. (Based on Gomperts BD, Kramer IM & Tatham PER [2002] Signal Transduction, Elsevier.)

Mosaic Proteins Result from Large-scale DNA Recombination

New proteins can be assembled by the recombination of genes from unrelated proteins. These types of proteins are called *mosaic proteins* (**Figure 3.12**). They seem to have been assembled from modules. Many proteins involved in signal transduction are built from a set of modular domains, combined differently, sometimes repetitively. Examples are the *src* homology domains (SH2 and SH3), the pleckstrin homology domain (PH), the pseudosubsrate domain (PS) in some protein kinases, the phosphoinositite-specific domain (PI), and the Ras GAP domain.

3.5 HOMOLOGOUS PROTEINS EVOLVED FROM A COMMON ANCESTOR

Proteins that are related by evolution have similar structures. This is because mutations act on the sequence, but evolution acts on structure and function. Evolution mainly selects *against* nonfunctional structures.

There are many different types of protein folds. Most proteins have very different structures (**Figure 3.13**); therefore, we have no reason to think they are related by evolution. The sequences of most proteins, even if of similar sizes, are not more similar to each other than what is expected by chance. In some cases, however, different proteins have similar sequences and similar structures. Consider for example trypsin and chymotrypsin, two serine proteases (**Figure 3.14**).

(A)   (B)   (C)   (D)

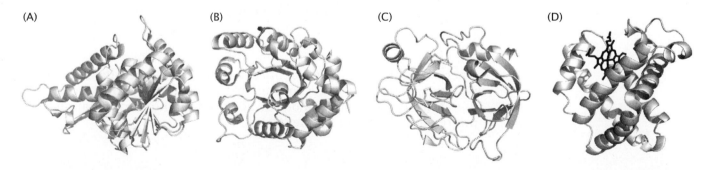

Figure 3.13 Examples of different protein structures belonging to various folds. (A) Lactate dehydrogenase (PDB 1LLD); (B) triose phosphate isomerase (one of its two similar domains, PDB 1MO0); (C) bovine trypsin (PDB 3OTJ); (D) myoglobin (PDB 1A6N).

Figure 3.14 Structures of (A) trypsin (PDB 3OTJ) and (B) chymotrypsin (PDB 1YPH). The residues of the catalytic triad are shown as stick models.

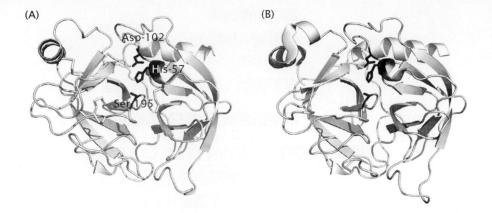

These enzymes owe their name to a serine residue in the active site, Ser-195, which is the primary nucleophile in the catalysis of the cleavage of polypeptide chains. Trypsin and chymotrypsin have different specificities. Trypsin cleaves the chain after a basic residue (Lys or Arg) and chymotrypsin after an aromatic residue (Phe, Tyr, or Trp), but all serine proteases share a common mechanism and their active site comprises the same *catalytic triad*: His-57, Asp-102, and Ser-195 (numbers of chymotrypsin).

If we compare the sequences of bovine trypsin and chymotrypsin, we find that 43% of their amino acids are identical. It is usually true that if two protein sequences are similar, their structures are also similar. The most plausible explanation for the sequence and structural similarity is that these proteins are related by evolution. They descend, by mutation, from a *common ancestor* protein. Over time, mutations in their genes have led to different sequences and to some variations in structure and specificity. But the structure of the common ancestor is still discernible in these proteins. Proteins that arise by *divergent evolution* from a common ancestor are called *homologous* proteins or homologs.

3.6 PROTEIN SEQUENCES DIVERGE EXPONENTIALLY IN TIME

Homology is the central concept in protein evolution. We can establish protein homology through the comparison of protein sequences and structures. Because we know many more sequences than structures, homology is most often inferred from sequence comparison. However, before we learn how to compare sequences and assess the significance of those comparisons, we must learn how protein sequences diverge in the course of evolution. Subject to random mutations, two sequences diverge from each other as an exponential decay in time,

$$\frac{n_0(t)}{N} = e^{-kt} \tag{3.1}$$

where n_0 is the number of identical residues, N is the total number of residues, and k is a rate constant that expresses the intrinsic rate of mutation. It gives the number of accepted amino acid mutations (m) per residue (N) per unit time (t),

$$k = \frac{m}{Nt}. \tag{3.2}$$

The exponential decay (shown in Equation 3.1) essentially describes the curve shown in **Figure 3.15**.

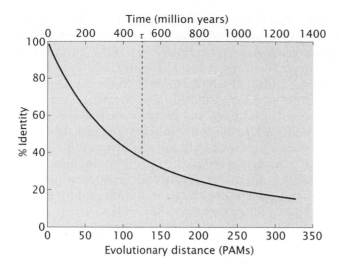

Figure 3.15 Decay of the percent identity of two protein sequences in time. The lower horizontal axis shows the evolutionary time in PAMs, or point accepted mutations per 100 residues. The curve shown is a plot of a calculation for a sequence of average composition. (Data from Dayhoff MO [1978] Atlas of Protein Sequence and Structure, vol. 5, suppl. 3. National Biomedical Research Foundation.) This curve is very similar to the expressions $e^{-t/\tau}$ or $e^{-m/N}$ derived in the text, but is not exactly identical because it is based on empirical mutability data. The upper horizontal axis shows an actual time scale for a protein with a characteristic time $\tau \approx 500$ million years. The relation between the two time scales is different for different proteins.

The dependent variable (y-axis) is the percent identity, $(n_0/N) \times 100$. The characteristic time (τ) of the decay,

$$\tau = \frac{1}{k},$$ (3.3)

is the mean time elapsed per mutation per residue and is indicated on the top horizontal axis. After a time τ each residue has been mutated once *on average*. Instead of using the real time as the independent variable (top horizontal axis in Figure 3.15), the evolutionary distance is often measured in units of the number of *point accepted mutations* or PAMs (bottom horizontal axis). Here, *accepted* means accepted by natural selection: the mutations occurred in the DNA and were incorporated in viable proteins. PAMs are expressed per 100 residues. For example, 200 PAMs corresponds to two accepted mutations per residue ($m/N = 2$). In Figure 3.15, at 200 PAMs each residue has been mutated twice on average, but the two sequences are still ~25% identical. Using Equation 3.2, we can write the exponential decay of the fractional identity (shown in Equation 3.1) explicitly in terms of the number of accepted mutations per residue ($m/N \approx$ PAMs/100):

$$\frac{n_0(m)}{N} = e^{-m/N}.$$ (3.4)

The Exponential Decay of Sequence Identity is a Consequence of Simple Statistics

The exponential decay expressed by Equations 3.1 and 3.4 can be easily derived. We could compare the mutated sequence with the original, or compare two homologous sequences that diverge from each other. In the latter case, m is the number of accepted mutations in one sequence *or* the other, because either produces a *change*. We will derive Equations 3.1 and 3.4 for divergence from the original ancestor because it is conceptually simpler.

Consider a protein with N amino acid residues. Over time, a number of residues $n(t)$ change as a consequence of mutations (at time zero $n(0) = 0$). At any time t, the number of original residues remaining is

$n_0(t) = N - n(t)$. Thus, the fractional identity is n_0/N. The probability that a residue *has* been mutated is n/N, and the probability that it has *not* been mutated is $1 - n/N = n_0/N$. Let m be the number of mutations that *occurred* at time t (PAMs). Not all these m mutations result in a *change* in the fractional identity: only those mutations that *change an original* amino acid do. A position in the sequence can be mutated several times (multiple hits), in which case mutations occur but the fractional sequence identity does not change, and reversal to the original amino acid can also occur (back mutations). Initially, the chance that mutations lead to *differences* in the two sequences is very high. Over time, however, there are fewer and fewer residues that have *not* been changed, and the chance that a mutation will hit one of those and produce a *difference* decreases.

Assuming that each of the N residues is equally likely to be mutated, the probability that a mutation hits any one residue is $1/N$. If the number of accepted mutations at a certain time is m, the fraction of mutated residues—that is, the mean number of mutations per residue (μ)—is

$$\mu = \frac{m}{N}. \tag{3.5}$$

Now if mutations occur stochastically (randomly) and the probability of mutation is small, the process follows *Poisson statistics*. In a Poisson distribution with mean number of mutations per residue μ, the probability that a given residue suffers s mutations is

$$P_s = \frac{\mu^s}{s!} e^{-\mu}. \tag{3.6}$$

P_0 corresponds to $s = 0$, for zero mutations. Thus, the probability that a residue (any residue) has *not* changed is

$$P_0 = e^{-\mu}$$
$$= e^{-m/N}. \tag{3.7}$$

Note that $P_0 = n_0/N$. The exponential decay expressed by Equation 3.7 is very similar to the line plotted in Figure 3.15 (in %). They are not exactly identical because the curve shown represents Dayhoff's calculation based on empirical mutability data. It includes the effects of back mutations (not included in Equation 3.7) and multiple hits (included).

The Change in Amino Acid Identity is Analogous to a Chemical Reaction

Very often, it helps our understanding of a problem to look at it from a different perspective. If the language of probability does not come intuitively to you, you may prefer to look at this problem as analogous to a chemical reaction. The amino acid sequence of a protein, AGWLVDK...K, is shown schematically in **Figure 3.16**. The first residue is alanine. A mutation, occurring with rate constant k_1, can change it to any of the other amino acids. A back mutation can change the new residue type to alanine with a rate constant k_{-1} (which is much smaller than k_1, because there are many more ways of changing A to *any* of the other amino acids than to change *one* of the other amino acids to A).

The rate equations that describe the change of sequence in time, for the number of changed residues (n) and those unchanged

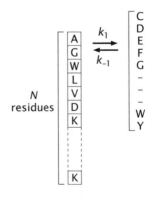

Figure 3.16 A protein of N residues with sequence AGWLVDK...K can undergo mutations in each of its residues to any other of the 20 amino acids with rate constant k_1, and back-mutations, to the original type, with rate constant k_{-1}.

$(n_0 = N - n)$, are

$$\frac{dn}{dt} = k_1 n_0 - k_{-1} n \qquad (3.8)$$

$$\frac{dn_0}{dt} = -k_1 n_0 + k_{-1} n. \qquad (3.9)$$

The solution of these equations, under the condition there are no mutations at time zero ($n(0) = 0$), is

$$\frac{n(t)}{N} = 1 - e^{-(k_1 + k_{-1})t} \qquad (3.10)$$

$$\frac{n_0(t)}{N} = e^{-(k_1 + k_{-1})t}. \qquad (3.11)$$

Now, since $k_{-1} \ll k_1$, we have $k_1 + k_{-1} \approx k_1 = k$. Thus, we obtain again Equation 3.1,

$$\frac{n_0(t)}{N} = e^{-kt},$$

but with a different reasoning.

3.7 PROTEIN SEQUENCE COMPARISON IS A POWERFUL METHOD TO ESTABLISH EVOLUTIONARY RELATIONSHIP

We are now ready to learn how to establish protein homology. If we can demonstrate that certain proteins are homologous, we know they share a common evolutionary ancestor. Then they have similar three-dimensional structures. Typically, if their sequences are at least ~30% identical over >100 residues, these proteins are homologous. However, many homologous proteins have substantially <30% sequence identity, and many proteins with apparently similar structures are not homologous. How can we determine if two proteins are homologous? Homology can be established from the comparison of two protein sequences if they have a statistically significant sequence similarity. To achieve this level of significance, however, we cannot simply determine the percent identity. Rather, we need to show that the two protein sequences have an *excess similarity* relative to what is expected from two random sequences, with a statistically significant degree of confidence.

Why do we compare protein sequences instead of DNA sequences? The reason is that a comparison of protein sequences is much more sensitive. In DNA there are only four bases (A,T,G,C). Therefore, *any* two unrelated DNA sequences of the same length are 25% identical on average. The chances of fortuitous matches between unrelated sequences are much smaller in proteins. In protein sequences we have 21 "words" (20 amino acid codons + terminator codon). If all amino acids occurred with the same frequency (they don't), two unrelated sequences would be about 5% identical on average. In addition, the reading frame is determined, further decreasing the chances of fortuitous matches. Thus, if possible, we should always compare protein sequences. If only the DNA sequence is available, we should translate it first into the protein sequence.

Protein Sequences are Compared by Finding the Best Alignment

To establish homology between protein sequences, we must perform a *pairwise alignment*. It is also possible to compare several sequences at once, with a *multiple alignment*. A multiple alignment may reveal important structural and functional residues, which are conserved across a number of homologous proteins, as well as information about their phylogenetic relationship. However, a pairwise alignment is necessary to determine the statistical significance of the sequence comparison. To align two protein sequences, we need a few tools: a *scoring matrix*, to decide on the importance of amino acid matches; an algorithm, to find the best match; and a way to decide on the statistical significance of the alignment. Some of the best reference materials and tools on protein sequence alignment are found on the Internet (see References). We will make use of these tools to understand the basic concepts and develop the ability to compare sequences.

Scoring Matrices Incorporate the Likelihood of Amino Acid Substitutions

A scoring matrix, or substitution matrix, consists of an ordered set of numbers that indicate how likely an amino acid is to be substituted by another in the course of evolution. The BLOSUM62 matrix, perhaps the most commonly used, is shown in **Figure 3.17**.

Other common matrices are the classic PAM (point accepted mutations) matrices introduced by Dayhoff. The PAM250 matrix was designed to reflect probabilities of substitution in proteins that differ by 250 PAMs (250 mutations per 100 residues). However, the PAM matrices were constructed by successive multiplications of the PAM1 matrix, which is derived from substitution frequencies in protein sequences that differ only in 1% of their residues (PAM250 = $(PAM1)^{250}$). Thus, because they are ultimately built from data for very close homologs, the original PAM matrices do not perform very well for

	Ala	Arg	Asn	Asp	Cys	Gln	Glu	Gly	His	Ile	Leu	Lys	Met	Phe	Pro	Ser	Thr	Trp	Tyr	Val
Ala	4
Arg	-1	5
Asn	-2	0	6
Asp	-2	-2	1	6
Cys	0	-3	-3	-3	9
Gln	-1	1	0	0	-3	5
Glu	0	0	0	2	-4	2	5
Gly	-2	-2	0	-1	-3	-2	-2	6
His	-1	0	1	-1	-3	0	0	-2	8
Ile	-1	-3	-3	-3	-1	-3	-3	-4	-3	4
Leu	-1	-2	-3	-4	-1	-2	-3	-4	-3	2	4
Lys	-1	2	0	-1	-3	1	1	-2	-1	-3	-2	5
Met	-1	-1	-2	-3	-1	0	-2	-3	-2	1	2	-1	5
Phe	-2	-3	-3	-3	-2	-3	-3	-3	-1	0	0	-3	0	6
Pro	-1	-2	-2	-1	-3	-1	-1	-2	-2	-3	-3	-1	-2	-4	7
Ser	1	-1	1	0	-1	0	0	0	-1	-2	-2	0	-1	-2	-1	4
Thr	0	-1	0	-1	-1	-1	-1	-2	-2	-1	-1	-1	-1	-2	-1	1	5
Trp	-3	-3	-4	-4	-2	-2	-3	-2	-2	-3	-2	-3	-1	1	-4	-3	-2	11
Tyr	-2	-2	-2	-3	-2	-1	-2	-3	2	-1	-1	-2	-1	3	-3	-2	-2	2	7
Val	0	-3	-3	-3	-1	-2	-2	-3	-3	3	1	-2	1	-1	-2	-2	0	-3	-1	4

Figure 3.17 The BLOSUM62 matrix.

distant homologs (however, modern matrices based on the PAM idea perform well). BLOSUM (blocks substitution matrix) matrices are now generally preferred. The entries in BLOSUM62 (see Figure 3.17) are derived from the frequencies of amino acid substitutions observed in blocks of *local alignments* of protein sequences with ≤62% identity. Similarly, BLOSUM50 is based on alignments of sequences with ≤50% identity.

The more positive the number in the scoring matrix of Figure 3.17, the more likely the substitution; the more negative, the less likely. Some amino acids are more likely to be replaced than others. For example, alanine is very common and well tolerated in environments with different polarities. Consequently, alanine is often replaced by other amino acids in different homologous proteins. Accordingly, a match between two alanines is not especially noteworthy, and it is assigned a low score in the BLOSUM62 matrix (4). On the other hand, cysteine and tryptophan are rare amino acids and tend to be conserved in homologous proteins. Thus, matches between two cysteine or two tryptophan residues are significant, and they are assigned the highest favorable (positive) scores (9 and 11). Conversely, it is unlikely that a charged residue, like aspartate, will replace a large hydrophobic one, like leucine, because placement of the charge in the protein interior (where most leucines are found) would be very unfavorable. Thus, this entry in the scoring matrix is very negative (−4).

Different Alignment Algorithms are Designed for Different Goals

Several kinds of algorithms are used for sequence alignment. Rigorous algorithms find the best alignment of two sequences. A global algorithm, such as the Needleman–Wunsch algorithm, produces the best overall alignment of two sequences; it is thorough but slow. Local algorithms search for *regions* of the sequences with high identity, and align the protein sequences from those regions. The Smith–Waterman algorithm produces the best local sequence alignments. But the fastest are heuristic algorithms (work by trial and error), such as FASTA or BLAST (basic local alignment search tool), which is the most commonly used to search databases for sequence matches.

Allowance for Gaps Increases the Match Between Two Sequences

If in the course of evolution an insertion or a deletion occurs in one of two homologous sequences, a block of residues is present in one but not in the other: a *gap* of missing residues exists in one of the sequences. Allowing for gaps in sequence alignments almost always improves the match between protein sequences. How much "gapping" should be allowed? With many gaps it may be possible to find good alignments between unrelated sequences. On the other hand, some gaps clearly make sense, because they originate from insertions and deletions that actually occurred in the course of evolution. There is no general answer to this question. In practice, what is done is to assign a penalty to each gap. Therefore, gaps that result in substantial improvement in amino acid matches will be favorable. But introduction of too many gaps will be penalized. With BLOSUM62, typical gap penalties are −10 points to open a gap and −1 to extend it.

The Alignment Score is Calculated from the Substitution Matrix and the Gap Penalties

We are now ready to align two protein sequences and calculate the alignment score. As an example, consider the two alignments of the sequences of bovine trypsin and chymotrypsin shown in **Figure 3.18**. Both alignments were obtained with the Needleman–Wunsch algorithm and the BLOSUM62 matrix, but alignment A was obtained with very high gap penalties (−100 to open a gap and −10 to extend it by one amino acid), whereas alignment B was obtained with smaller gap penalties (open gap, −10; extend, −1). Consequently, there are only two small gaps in A, but seven in B (initial gap not included). There are 74 identities in A (indicated by the vertical lines) and 99 identities in B. The corresponding percent identities are 33% and 43% (calculated relative to the regions aligned, 226 and 228 residues, respectively).

Calculating the alignment score (S) is simple. First, we add the scores given in the substitution matrix (see Figure 3.17) for all pairs of amino acids matched in the alignment. For example, in Figure 3.18B, we begin with the matches I/I (+4), V/V (+4), G/N (−2), and so on, adding all those scores. Then, we subtract a penalty for each gap open (−10), and a penalty for each residue by which a gap is extended (−1). The final scores for the alignments shown are $S = 93$ for A and $S = 449$ for B. Now, which alignment should we choose? We cannot simply compare the scores obtained with different gap penalties. Each one is the best alignment for the conditions chosen. These decisions are not always easy. A multiple alignment of related sequences may help in reaching the best decision. In this case, alignment B identifies similarities that are missed in A and does not introduce unreasonable gaps.

How Significant is the Alignment Score?

Now we have aligned two sequences and obtained a score and a percent identity. But we still don't know what the alignment means to protein homology unless we can estimate its significance. The sequences of bovine trypsin and chymotrypsin are 43% identical. Is this significant? Does it mean that those two proteins arose by divergent evolution from a common ancestor—that is, they are homologous? Or could this level of identity be obtained by chance? A usual rule of thumb is that sequence identity greater than 30% over the entire length indicates homology (in sequences > 100 residues). In this case, 43% is a high level of similarity and these proteins are homologous. The problem

(A)

```
Trypsin        1 --------------IVGGYTCGANTVPYQVSLNSGYHFCGGSLINSQWV    35
                               .......:|.......:...:|:|||||||||..||
Chymotrypsin   1 CGVPAIQPVLSGLIVNGEEAVPGSWPWQVSLQDKTGFHFCGGSLINENWV    50

Trypsin       36 VSAAHC-YKSGIQVRLGEDNINVVEGNEQFISASKSIVHPSYNSNTLNND    84
                 |:|||| ...:.|..||.:........:.:|....:.....|||.|:|||
Chymotrypsin  51 VTAAHCGVTTSDVVVAGEFDQGSSSEKIQKLKIAKVFKNSKYNSLTINND   100

Trypsin       85 IMLIKLKSAASLNSRVASISLPTSCASAGTQCLISGWGNTKSSGTSYPDV   134
                 |.|:||..:|||:...:::|.||::..........||.||||:||::.||.
Chymotrypsin 101 ITLLKLSTAASFSQTVSAVCLPSASDDFAAGTTCVTTGWGLTRYANTPDR   150

Trypsin      135 LKCLKAPILSDSSCKSAYPGQITSNMFCAGY--LEGGKDSCQGDSGGPVV   182
                 ||:....:|:||:||.:.|.||||.:.:|||| ...:.:||||||||||
Chymotrypsin 151 LQQASLPLLSNTNCKKYWGTKIKDAMICAGASGVSSCMGDSGGPLVCKKN   200

Trypsin      183 CSGKLQGIVSWGSGCAQKNKPGVYTKVCNYVSWIKQTIASN    223
                 |.:.:.|.||||||......:|||.:.|...:|||:|:|:|
Chymotrypsin 201 GAWTLVGIVSWGSSTCSTSTPGVYARVTALVNWVQQTLAAN    241
```

(B)

```
Trypsin        1 --------------IVGGYTCGANTVPYQVSL--NSGYHFCGGSLINSQWV    35
                               ||.|........:|:||| .:|:||||||||..||
Chymotrypsin   1 CGVPAIQPVLSGLIVNGEEAVPGSWPWQVSLQDKTGFHFCGGSLINENWV    50

Trypsin       36 VSAAHC-YKSGIQVRLGEDNINVVEGNEQFISASKSIVHPSYNSNTLNND    84
                 |:|||| ...:.|..||.:........:.:|....:.....|||.|:|||
Chymotrypsin  51 VTAAHCGVTTSDVVVAGEFDQGSSSEKIQKLKIAKVFKNSKYNSLTINND   100

Trypsin       85 IMLIKLKSAASLNSRVASISLPTSC--ASAGTQCLISGWGNTKSSGTSYP   132
                 |.|:||..:|||:...:::|.||::. :.........||.||||:||::|
Chymotrypsin 101 ITLLKLSTAASFSQTVSAVCLPSASDDFAAGTTCVTTGWGLTRYANT--P   148

Trypsin      133 DVLKCLKAPILSDSSCKSAYPGQITSNMFCAGYLEGGKDSCQGDSGGPVV   182
                 | .:||:....:|:||:....|.||||.:.:|||.:.:...||||||||||
Chymotrypsin 149 DRLQQASLPLLSNTNCKKYWGTKIKDAMICAG--ASGVSSCMGDSGGPLV   196

Trypsin      183 C----SGKLQGIVSWGSGCAQKNKPGVYTKVCNYVSWIKQTIASN    223
                 |    .:.:.|.||||||......:|||.:.|...:|||:|:|:|
Chymotrypsin 197 CKKNGAWTLVGIVSWGSSTCSTSTPGVYARVTALVNWVQQTLAAN    241
```

Figure 3.18 Alignment of cow trypsin and chymotrypsin, using the Needleman–Wunsch algorithm with the BLOSUM62 matrix, with different gap penalties. (A) Gap open, −100; gap extend, −10. (B) Gap open, −10; gap extend, −1. The alignments were obtained with the program NEEDLE in the EMBL-EBI Internet site.

is that percent identity is a poor criterion to judge homology. Some short sequences have more than 30% identity and are not homologous; conversely, many homologous sequences have much less than 30% identity.

We need to do better. First, we will define two quantities that will help us reach an informed decision on homology, the bit-score (S') and the expectation value (*E-value*). Second, we will use BLAST to examine a few examples and begin to build an intuition for these problems. We switch to BLAST because it calculates the bit-score and the expectation value for us. BLAST does not necessarily find the best global alignment, but finds optimal *local* alignments of two sequences called high-scoring segment pairs (HSPs). The bit-scores, *E*-values, and percent identities are calculated over each of these HSPs. Sometimes the best HSP is the entire sequence, but most often it is not. It may even be only a small stretch of the sequence. Thus, percent identities calculated with BLAST are not very meaningful unless they relate to a large fraction of the sequence.

Raw alignment scores obtained with different scoring systems or parameters cannot be compared. However, the bit-score S' reported by BLAST (as score, in bits) is normalized, taking the statistical properties of the scoring system into account. The bit-score is defined by

$$S' = \frac{\lambda S - \ln K}{\ln 2} \tag{3.12}$$

where S is the raw score and λ and K are statistical parameters. The important point is that $2^{S'}$ gives an estimate of the number of sequence segments that we would need to search in order to find a score at least as good. For example, if $S' = 50$, you would need to search $2^{50} = 10^{15}$ database entries. For a typical protein, a bit-score of 50 is significant. But the bit-score still depends on the database size. The same bit-score is more significant in a smaller database (the Swiss–Prot database has about 500,000 sequences; the nr (nonredundant sequences) database, default in BLAST, has about 60,000,000).

The E-value (called *Expect* in BLAST) corrects the bit-score for the database size,

$$\text{E-value} \sim \frac{N}{2^{S'}}, \tag{3.13}$$

where the search space N is proportional to the database size. The E-value tells you what is the probability of a given alignment score to be obtained by chance. Alignments of two random sequences yield E-values ~1, whereas E-values <0.001 are reliable indicators of protein homology. E-values in the "gray zone" of 0.001–0.01 tend to be inconclusive.

Excess Sequence Similarity Indicates Homology

Let us now examine some examples from the trypsin family and apply what we have just learned. (Note: we will use the term *family* in its broadest sense; it includes all proteins that are evolutionarily related to a particular one.) The trypsin family of serine proteases has many members, in organisms as different and as distant as bacteria and mammals, whose evolutionary branches diverged from each other more than 2 billion years ago (see Figure 3.1). Yet the three-dimensional structures of these enzymes are similar, as shown in **Figure 3.19**

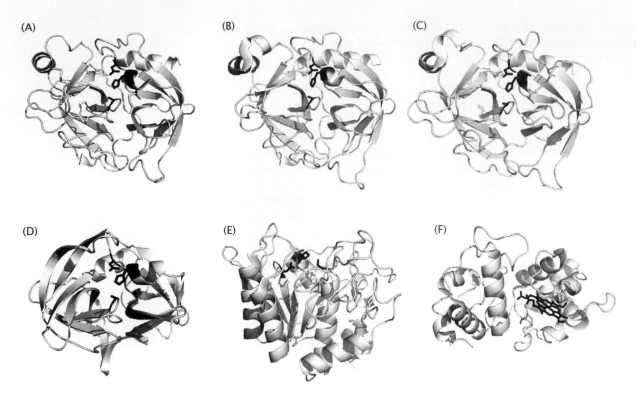

Figure 3.19 Structures of four homologous serine proteases and two unrelated proteins. (A) Bovine trypsin (PDB 3OTJ); (B) bovine chymotrypsin (PDB 1YPH); (C) *S. griseus* trypsin (PDB 1SGT); (D) *S. griseus* protease A (PDB 2SGA); (E) subtilisin (PDB 1SBT); (F) cytochrome c4 (PDB 1ETP). The residues of the catalytic triad (His, Asp, Ser) of the serine proteases, and the heme prosthetic group of cytochrome are shown as sticks.

for bovine trypsin (A) and chymotrypsin (B), and *Streptomyces griseus* trypsin (C) and protease A (D).

First, we compare the sequences of bovine trypsin and chymotrypsin (using BLAST with BLOSUM62 and default gap penalties -11 to open, -1 to extend). We obtain the same alignment as in Figure 3.18B, with the same 43% identity (the BLAST alignment is over the entire sequence in this case). But now we also obtain a bit-score $S' = 150$ bit and an E-value $\sim 10^{-48}$, which indicates the alignment score is statistically significant. The two proteins have *excess sequence identity* relative to what is expected by chance. It is this excess similarity that allows us to infer homology. Indeed the structures of bovine trypsin and chymotrypsin are extremely similar (see Figure 3.19A,B).

Second, we compare trypsins from cow and the bacterium *S. griseus*. The sequences have only 35% identity (over 96% of the sequence), the same as in a global alignment. But the E-value $\sim 10^{-29}$ ($S' = 99$), indicating that the proteins are homologous, as evident by their similar structures (see Figure 3.19A,C).

Now, the sequences of bovine trypsin and *S. griseus* protease A are only 24% identical in a global alignment. Using BLAST with the BLOSUM45 matrix, which is tailored for sequences with at most 45% identity (with default gap penalties -15 to open, -2 to extend), we obtain 21% identity (local, over $\approx$80% of the sequence). Two main segments are aligned by BLAST with E-values of 0.02 and 6 ($S' \approx 12$ and 20). Based on these results, we could *not* conclude that the proteins are homologous. Yet the similarity of their structures shows that they are (see Figure 3.19A,D). The important point is that we cannot conclude that the proteins are *not* homologous even if their sequence similarity is low and the E-values are greater than the significance threshold: after we do a statistical test, we still need to interpret it. In this case, we have *additional evidence* (the structures) that shows the proteins are homologous, despite the lack of statistically significant sequence similarity. The low level of identity arises because we are comparing bacterial and mammalian proteins, which have diverged more than 2 billion years ago, and have therefore undergone a very large number of mutations since their last common ancestor. Homologous proteins are not required to have *any* prescribed level sequence identity. Homology only means that they have the same ancestor. If they are homologous, their structures are similar, even without sequence similarity.

Sequences Unrelated by Evolution Score in a Pairwise Alignment Like Two Random Sequences

Now let us compare subtilisin from *Bacillus subtilis* with bovine trypsin (see Figure 3.19A,E). Subtilisin is not a member of the trypsin family. But its active site is also a catalytic triad, consisting of His-64, Asp-32, and Ser-221. Even though the residue positions in the sequence are different, their spatial organization in the active site is similar to that of trypsin, and the catalytic roles of the three amino acids are the same as in the other serine proteases. It could be thought that subtilisin, too, evolved by divergence from an ancestor common to the other serine proteases.

These two proteins have about 20% sequence identity in a global alignment, about the same as that between *S. griseus* protease A and bovine trypsin. Based only on sequence similarity, subtilisin and trypsin *could* be homologous. The local alignment with BLAST yields 23% identity (over ~40% of the sequence) and E-values of 2 and 7.6 for the major segments aligned (bit-scores ≈ 14). Thus, the statistical significance test is what is expected from two *random sequences*: these values could have been obtained by chance. They strongly suggest (but do not prove) that the two proteins are *not* homologous. The structures of subtilisin and trypsin, however, are very different (see Figure 3.19A,E) and show that these two proteins are not homologous in spite of the common active site and a marginal sequence similarity. Subtilisin and trypsin are *analogous* proteins: they have the same function but not a common ancestor. Their similarities are the result of convergent evolution.

Finally consider cytochrome c4, which also has about 20% sequence identity with bovine trypsin in a global alignment. The BLAST comparison also yields 20% similarity over the aligned region (~60% of the sequence) and E-values of 0.47 and 6.5 for the major segments aligned (bit-scores ~12–16). These E-values are of the order of magnitude of 1, which is what we expect to obtain by chance. The structure of cytochrome c4 (see Figure 3.19F) is very different from that of bovine trypsin. The two proteins are not homologous.

In conclusion, the sequences of homologous proteins often share significant similarity. However, the similarity may be almost unrecognizable by a simple comparison of amino acid identities. The percent identity depends on how long ago those proteins diverged from their common ancestor. It depends also on whether the function of those proteins changed. Evolution cares only about whether the protein is functional. However, the structure determines the function, and the sequence enters the problem because the structure is specified by the sequence alone. Usually, similar sequences lead to similar structures.

Thus, sequence comparison can be used to probe for evolutionary relationships and trace back 2 billion years of evolutionary history. But we must be careful in assigning homology: a similar mechanism, even a similar active site, is not enough to establish homology, as the example of subtilisin shows. Nor is a similar fold just by itself. Homology can only be established based on similarity of sequence or structure that is statistically significant, in excess of what is expected by chance.

3.8 EVOLUTIONARY TREES TRACE THE HISTORY OF HOMOLOGOUS PROTEINS

Evolutionary trees can be constructed to show how proteins diverged from a common ancestor. If a sequence match is found between human and yeast (fungi) proteins, for example, we may conclude that a common ancestral protein existed at least 1 billion years ago, because that is the time at which animals and fungi diverged from each other (see Figure 3.1). If proteins from fungi and bacteria are determined to be homologous, we can conclude that a common ancestor to these proteins must have existed about 2 billion years ago in an organism that was a common ancestor to bacteria and fungi, because those two branches diverged about 2 billion years ago. The protein must have existed before the divergence occurred.

Protein Evolutionary Trees Display the Distances to Common Ancestors

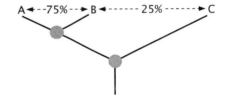

Figure 3.20 Evolutionary tree of proteins A, B, and C, with 75% sequence identity between A and B, and 25% between A and C, or B and C.

To understand the concept of constructing a protein evolutionary tree, let us consider a very simple example. Suppose we compare the sequences of three homologous proteins A, B, and C. Protein C has diverged from A and B earlier than A and B diverged from each other and functions as a reference. We call C an *outgroup sequence*, because it is evolutionarily more distant from those in the closer group (A and B). If A and B changed at the same rate, they should have about the same sequence similarity with C. Say A and B have 75% sequence identity, but each one of them has only 25% identity with C. Our protein evolutionary tree would look like that shown in **Figure 3.20**. The branch lengths reflect the evolutionary distance between proteins.

Let us construct an evolutionary tree of the trypsin family. In this case we use percent identities instead of bit-scores because we already know that these proteins are homologous. **Table 3.3** shows a comparison of trypsin sequences of humans, cow, pig, rat, salmon, and the bacterium *S. griseus*. We see that *S. griseus* trypsin has only ~30% sequence identity to *all* the vertebrate trypsins, which share much greater similarity among themselves. This indicates that *S. griseus*

Table 3.3 Percent identities among the sequences of several trypsins

Organism	Human (1TRN)[a]	Pig	Cow	Rat	Salmon
Pig (1S81)	79				
Cow (5PTP)	75	82			
Rat (1YKT)	79	81	74		
Salmon (1HJ8)	64	67	66	67	
S. griseus (1SGT)	30	33	35	31	33

[a]PDB identifiers in parentheses.

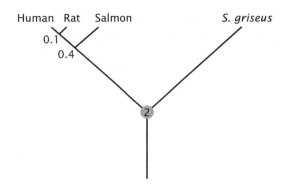

Figure 3.21 Evolutionary tree for trypsins. The numbers indicate the times of divergence in billion years.

diverged early from all the other species. Next, the salmon protein has the same percent identity (∼65%) with those of all mammals, indicating that the fish separated next from the mammalian branch. Finally, the mammalian trypsins diverged. The tree of **Figure 3.21** contains this information. In this example, *S. griseus* is an outgroup species.

The Time of Branch Points is Assigned from the Fossil Record

Now we want to put some real time information on the trypsin tree. To do so, we need to calibrate the bifurcations in the tree to geological time (fossil record), from the knowledge of the real time at which evolutionary bifurcations took place. The oldest fossils are those of prokaryotes, in rocks, which date from about 2.5 billion years ago. From the fossil record, we know that prokaryotes diverged from eukaryotes about 2 billion years ago (see Figure 3.1). The divergence of *S. griseus* trypsin from the other trypsins must have occurred at about that time. Therefore, we wrote 2 (billion years) on the branch point between *S. griseus* and the animal trypsins in Figure 3.21. Since then, mutations have accumulated in both types of trypsins, resulting in a sequence identity of only ≈30%. Further, we know from the fossil record that fish and mammals diverged about 0.4 billion years ago, and that the mammalian radiation occurred about 0.1 billion years ago. That information is also added to the tree in Figure 3.21.

Sophisticated Methods Allow the Reconstruction of Protein Sequences

To construct simple evolutionary trees, the information on sequence similarity (percent identity) may be sufficient. However, modern reconstruction of sequence evolution uses sophisticated statistical methods in combination with explicit models of evolution and known phylogenetic relationships. These methods allow us to determine the series of mutations that best represent the changes in sequence in the course of evolution. In this manner, we reconstruct the *evolution of the sequence* itself, and can more reliably infer the sequences of distant ancestors. There are three methods of sequence reconstruction. They begin with a multiple alignment of a set of sequences of closely related proteins, supplemented by outgroup sequences. The task is to find the best explanation for the differences in sequence on the basis of evolution.

The simplest approach is the *consensus* method. Here, the *most common* amino acid found at each position in extant proteins is assumed to be that present in the ancestor. The next two improvements use the phylogenetic tree to guide the reconstruction of sequence evolution.

The *maximum parsimony* method assumes that the most likely evolutionary path (tree) is the one that requires the *smallest number of changes* in the sequence, from the common ancestor to the extant proteins. Parsimony allows the reconstruction of sequences as ancient as ~50 million years.

The most reliable are *maximum likelihood* methods. Here, the probability of each possible ancestral sequence at each branch point in the tree is calculated. The maximum likelihood pathway is determined by the probability that the extant set of species could have evolved from that ancestor, assuming a certain tree and an evolutionary model. The model stipulates the transition probabilities for amino acid substitutions, which can be inferred from substitutions in known structures. The most likely ancestor at each node (branch point) is then chosen. Maximum likelihood methods have allowed the reconstruction of sequences that existed more than 1 billion years ago.

3.9 ORTHOLOGOUS PROTEINS DIVERGE WITH THE SPECIES THAT CARRY THEM

In some homologous proteins, the corresponding genes occupy equivalent positions (*loci*) in the genomes of the different organisms that carry those genes. These proteins are different because the organisms that host them evolved independently from each other, and their proteins accumulated random mutations in their amino acid sequences. Homologous proteins that are related in this manner are called *orthologous* proteins (or orthologs). They differ because of *speciation*. Usually they have similar functions.

Cytochrome c Evolved Because of Speciation

Cytochromes c from different species are orthologous proteins. These proteins diverged from a common ancestor and have similar structures and functions in different species. Under the effect of point mutations, insertions, and deletions, the sequences of cytochromes c have changed enormously. They have only ~15% sequence identity. **Figure 3.22** shows a *multiple alignment* of a few sequences from the cytochrome c family from several bacteria and eukaryotes. Only 18 residues are identical (conserved) across this set of sequences, which are indicated with an asterisk (*).

Yet the structures of all cytochromes c are very similar, as shown in **Figure 3.23** for human, tuna, baker's yeast, rice, *Rhodobacter capsulatus*, and *Paracoccus denitrificans*. Note that the three invariant (strictly conserved) residues Cys-17, His-18, and Met-80 (numbering of the mammalian proteins) interact directly with the heme group, which is viewed along its plane in the figure (stick models).

Orthologous proteins change because of speciation, as the species that carry them diverge from each other. The branch points in their evolutionary trees reflect the times of divergence of the species that host the protein. Thus, the branching pattern of the cytochrome c tree reflects the branching pattern of the phylogenetic trees of the corresponding species (see Figure 3.1). This happens necessarily if the proteins are orthologous.

```
Cytochrome c    Human          |PDB 1J3S  --------------------MGDVEKGKKIFIMKCSQCHTVEK-------GGKHKTGPN  32
Cytochrome c    Horse          |PDB 1GIW  --------------------MGDVEKGKKIFVQKCAQCHTVEK-------GGKHKTGPN  32
Cytochrome c    Tuna           |PDB 1I54  --------------------GDVAKGKKTFVQKCAQCHTVEN-------GGKHKVGPN  31
Cytochrome c1   Yeast          |PDB 2HV4  ----------------MTEFKAGSAKKGATLFKTRCLQCHTVEK-------GGPHKVGPN  37
Cytochrome c    Rice           |PDB 1CCR  -------------MASFSEAPPGNPKAGEKIFKTKCAQCHTVDK-------GAGHKQGPN  40
Cytochrome c2   R. capsulatus  |PDB 1C2N  MKISLTAAT-VAALVLAAPAFAGDAAKGEKEF-NKCKTCHSIIAPDGTEIV-KGAKTGPN  57
Cytochrome c2   R. viridis     |PDB 1CRY  MRKLVFGL--FVLAASVAPAAAQDAASGEQVF-KQCLVCHSIGP-------GAKNKVGPV  50
Cytochrome c2   R. sphaeroides |PDB 1CXC  --------------------QEGDPEAGAKAF-NQCQTCHVIVDDSGTTIAGRNAKTGPN  39
Cytochrome c550 P. denitrificans|PDB 1COT MKISIYATLAAITLALPAAAQDGDAAKGEKEF-NKCKACHMIQAPDGTDII-KGGKTGPN  58
                                                .   *   *  :* **:            * **

Cytochrome c    Human          |PDB 1J3S  LHGLFGRKTGQAPGYS-YTA-----ANKNKGIIWGEDTLMEYLENPKKYIP--------G   78
Cytochrome c    Horse          |PDB 1GIW  LHGLFGRKTGQAPGFT-YTD-----ANKNKGITWKEETLMEYLENPKKYIP--------G   78
Cytochrome c    Tuna           |PDB 1I54  LWGLFGRKTGQAEGYS-YTD-----ANKSKGIVWNNDTLMEYLENPKKYIP--------G   77
Cytochrome c1   Yeast          |PDB 2HV4  LHGIFGRHSGQAEGYS-YTD-----ANIKKNVLWDENNMSEYLTNPKKYIP--------G   83
Cytochrome c    Rice           |PDB 1CCR  LNGLFGRQSGTTPGYS-YST-----ANKNMAVIWEENTLYDYLLNPKKYIP--------G   86
Cytochrome c2   R. capsulatus  |PDB 1C2N  LYGVVGRTAGTYPEFK-YKDSIVALG--ASGFAWTEEDIATYVKDPGAFLKEKLDDKKAK  114
Cytochrome c2   R. viridis     |PDB 1CRY  LNGLFGRHSGTIEGFA-YSD-----ANKNSGITWTEEVFREYIRDPKAKIP--------G   96
Cytochrome c2   R. sphaeroides |PDB 1CXC  LYGVVGRTAGTQADFKGYGEGMKEAG--AKGLAWDEEHFVQYVQDPTKFLKEYTGDAKAK   97
Cytochrome c550 P. denitrificans|PDB 1COT LYGVVGRKIASEEGFK-YGEGILEVAEKNPDLTWTEADLIEYVTDPKPWLVKMTDDKGAK  117
                                          * *:.**   .    :  *    .    .*: :  *:  :*  :

Cytochrome c    Human          |PDB 1J3S  TKMIFVGIKKKEERADLIAYLKKATNE-------------  105
Cytochrome c    Horse          |PDB 1GIW  TKMIFAGIKKKTEREDLIAYLKKATNE-------------  105
Cytochrome c    Tuna           |PDB 1I54  TKMIFAGIKKKGERQDLVAYLKSATS--------------  103
Cytochrome c1   Yeast          |PDB 2HV4  TKMAFGGLKKEKDRNDLITYLKKACE--------------  109
Cytochrome c    Rice           |PDB 1CCR  TKMVFPGLKKPQERADLISYLKEATS--------------  112
Cytochrome c2   R. capsulatus  |PDB 1C2N  TGMAFK-LAKGGE--DVAAYLASVVK--------------  137
Cytochrome c2   R. viridis     |PDB 1CRY  TKMIFAGVKDEQKVSDLIAYIKQFNADGSKK---------  127
Cytochrome c2   R. sphaeroides |PDB 1CXC  GKMTFK-LKKEADAHNIWAYLQQVAVRP------------  124
Cytochrome c550 P. denitrificans|PDB 1COT TKMTFK-MGKNQA--DVVAFLAQNSPDAGGDGEAAAEGESN  155
                                          * *  :  .     :: :::  .
```

Figure 3.22 Multiple sequence alignment of species of the cytochrome c family from human, horse, tuna, yeast, rice, and several bacteria. An asterisk (*) indicates identical residues; a colon (:) indicates similar type; a period (.) indicates a lower degree of similarity.

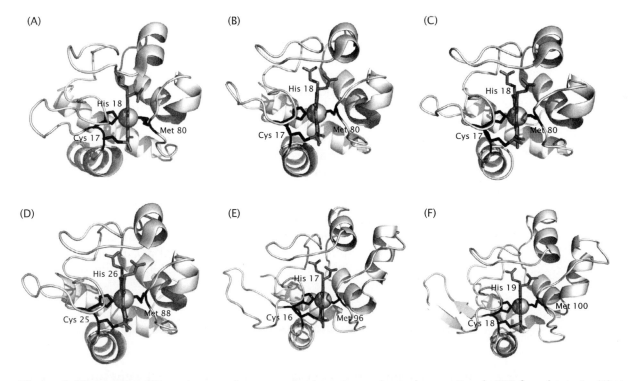

Figure 3.23 Structure of cytochrome c from several eukaryotes, and cytochrome c2 and c550 from bacteria. (A) Human (PDB 1J3S); (B) tuna (PDB 1I54); (C) baker's yeast (PDB 2HV4); (D) rice (PDB 1CCR); (E) *Rhodobacter capsulatus* (PDB 1C2N); (F) *Paracoccus denitrificans* (PDB 1COT). The residues that interact directly with the heme (Cys, His, and Met) are shown as stick models. The heme is viewed along its plane, with its iron metal ion (sphere) in the center.

3.10 PARALOGOUS PROTEINS DIVERGE FOLLOWING GENE DUPLICATIONS

Now suppose that that a *gene duplication* takes place. This could arise from an error in DNA replication or repair, or from DNA recombination during meiosis. The point is, there are now two copies of the same gene. If the function of the original protein is needed, one gene copy must be conserved. But the second copy can be transformed into a new protein, perhaps with a new function, by accumulation of mutations. These two proteins are homologous, but they do not occupy equivalent positions (loci) in the genome. Homologous proteins that result from a gene duplication are called *paralogous* (or paralogs). They often have different functions in the same or in different organisms. The three-dimensional structures of paralogous proteins are similar, but often the active sites are totally different because a new enzyme function has evolved, which catalyzes a different reaction. Paralogous proteins diverge because the mutations that occur in each sequence after the gene duplication are uncorrelated.

Myoglobin and Hemoglobin Diverged from a Primordial Globin

Genes that are the result of duplications form a *gene family*. A good example is the globin family, which includes myoglobin and the various hemoglobin chains. Myoglobin is monomeric, whereas hemoglobin is a tetramer (in vertebrates) of similar but not identical chains called globins. In adult humans, each tetramer has two α-globins and two β-globins. Each one of those chains is paralogous to myoglobin and to the other hemoglobin chains. The functions of these proteins are related, but not identical. They all bind oxygen. But whereas the function of myoglobin is to provide an oxygen buffer or reservoir in the muscles, the function of hemoglobin is to exchange oxygen between the lungs and the tissues, transporting it through the blood.

Figure 3.24 shows a multiple alignment of globin sequences from different eukaryotes and prokaryotes. Four residues, marked with an asterisk (*), are conserved among the sequences shown, but there are only two strictly invariant residues among all hemoglobins: His-87 (87 in human α-globin, 92 in β-globin) and Phe-43 (43 in α, 42 in β), which interact directly with the heme group.

The structures of myoglobin and hemoglobin chains are all very similar. They are shown in **Figure 3.25** for sperm whale myoglobin, human α-globin, lamprey hemoglobin, marine bloodworm *Glycera dibranchiata* hemoglobin, lupine leghemoglobin, and Hell's Gate globin I from *Methylacidiphilum infernorum*. The heme group is visible on the left side of the structures.

Branch Points in Paralogous Evolutionary Trees Mark Gene Duplication Events

In paralogous proteins, the branch points reflect gene duplications. The lengths of the branches give the evolutionary distance from the last common gene. In the globin family, a first gene duplication led to the separation of myoglobin and hemoglobin (**Figure 3.26**). Myoglobin has ~25% sequence identity with the α-and β-globins. A second gene duplication occurred later, which led to the divergence of the α- and β-globins from each other. Since this duplication is more recent, α- and

```
Sperm Whale Myoglobin PBD 4MBN ---------VLSEGEWQLVLHVWAKVEADV--AGHGQDILIRLFKSHPETLEKFDRFKHL  49
Human Hemoglobin A     PDB 1A3N ---------VLSPADKTNVKAAWGKVGAHA--GEYGAEALERMFLSFPTTKTYFPHFDLS  49
Lamprey hemoglobin     PDB 2LHB PIVDTGSVAPLSAAEKTKIRSAWAPVYSTY--ETSGVDILVKFFTSTPAAQEFFPKFKGL  58
Bloodworm hemoglobin   PDB 1HBG ---------GLSAAQRQVIAATWKDIAGADNGAGVGKKCLIKFLSAHPQMAAVFGFSGAS  51
Leghemoglobin          PDB 2GDM --------GALTESQAALVKSSWEEFNANI--PKHTHRFFILVLEIAPAAKDLFSFLKGT  50
Hell's Gate globin     PDB 3S1I ---------MIDQKEKELIKESWKRIEPNK--NEIGLLFYANLFKEEPTVSVLFQNPI--  47
                                      :    :    :   * .       .:     *     *

Sperm Whale Myoglobin PBD 4MBN KTEAEMKASEDLKKHGVTVLTAL--GAILKKKGH---HEAELKPLAQSHA--TKHKIPIK 102
Human Hemoglobin A     PDB 1A3N HGS------AQVKGHGKKVADAL--TNAVAHVDD---MPNALSALSDLHA--HKLRVDPV   96
Lamprey hemoglobin     PDB 2LHB TTADELKKSADVRWHAERIINAV--DDAVASMDDTEKMSMKLRNLSGKHA--KSFQVDPE  114
Bloodworm hemoglobin   PDB 1HBG --------DPGVAALGAKVLAQI--GVAVSHLGDEGKMVAQMKAVGVRHKGYGNKHIKAQ 101
Leghemoglobin          PDB 2GDM SE--VPQNNPELQAHAGKVFKLVYEAAIQLEVTGVVVTDATLKNLGSVHVSK---GVADA 105
Hell's Gate globin     PDB 3S1I ------------SSQSRKLMQVL--GILVQGIDNLEGLIPTLQDLGRRHKQY---GVVDS   90
                                      .  :    :       :    *         :  :.  *         :

Sperm Whale Myoglobin PBD 4MBN YLEFISEAIIHVLHSRHPGDFGADAQGAMNKALELFRKDIAAKYKELGYQG 153
Human Hemoglobin A     PDB 1A3N NFKLLSHCLLVTLAAHLPAEFTPAVHASLDKFLASVSTVLTSKYR------ 141
Lamprey hemoglobin     PDB 2LHB YFKVLAAVIADTVAA---------GDAGFEKLMSMICILLRSAY------- 149
Bloodworm hemoglobin   PDB 1HBG YFEPLGASLLSAMEHRIGGKMNAAAKDAWAAAYADISGALISGLQS----- 147
Leghemoglobin          PDB 2GDM HFPVVKEAILKTIKEVVGAKWSEELNSAWTIAYDELAIVIKKEMDDAA--- 153
Hell's Gate globin     PDB 3S1I HYPLVGDCLLKSIQEYLGQGFTEEAKAAWTKVYGIAAQVMTAEHHHHHH-- 139
                                    :    :    :             . .             :
```

Figure 3.24 Alignment of the sequences of globins from several organisms: Sperm whale myoglobin, human α chain hemoglobin, lamprey hemoglobin, marine bloodworm *Glycera dibranchiata* hemoglobin, lupine leghemoglobin, and globin I from *Methylacidiphilum infernorum*. The asterisk (*), colon (:), and period (.) indicate identity and decreasing levels of similarity between amino acids.

β-globins are more similar to each other (43% sequence identity) than they are to myoglobin.

Figure 3.27 shows the occurrence of the globin genes superimposed on the eukaryotic phylogenetic tree. Bony fish, which diverged from

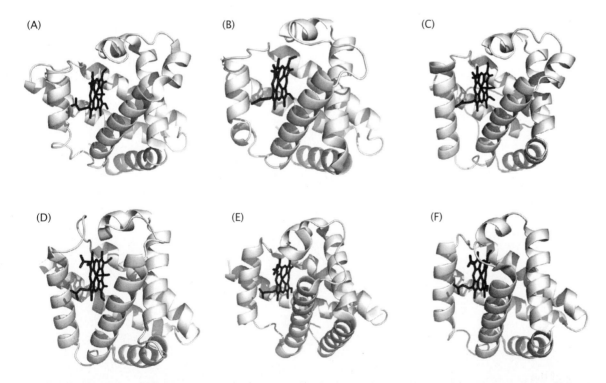

(A)   (B)   (C)

(D)   (E)   (F)

Figure 3.25 Structure of globins from several organisms. (A) Sperm whale myoglobin (PDB 4MBN); (B) human α-globin (PDB 1A3N); (C) lamprey hemoglobin (PDB 2LHB); (D) marine bloodworm *Glycera dibranchiata* hemoglobin (PDB 1HBG); (E) leghemoglobin, from root nodules of yellow lupine (PDB 2GDJ); (F) Hell's Gate globin I from *Methylacidiphilum infernorum* (PDB 3S1I). The heme group is shown as a stick model.

Figure 3.26 Evolutionary tree for the globin family. The numbers indicate the time since divergence in million years.

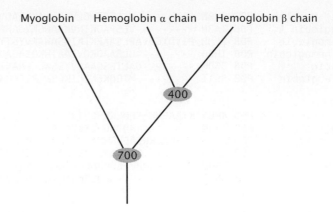

mammals ~400 million years ago, as well as all vertebrates that diverged later, have myoglobin and α- and β-globins. But jawless fish, which bifurcated before, have only one type of hemoglobin chain. Thus, we can place the globin gene duplication event that resulted in α and β chains at ~400 million years ago.

Insects and crustaceans have hemoglobin, but not the α and β chains found in mammals. Plants also have globins; for example, leghemoglobin exists in legume root nodules. Plants diverged from animals about 1 billion years ago, but the sequence similarities between plant and animal globins are still significant, and their structures are very similar (see Figure 3.25). The gene duplication that led to the separation of myoglobin and hemoglobin occurred after the bifurcation between plants and animals (~1000 million years ago), but before the separation between vertebrates and insects (~600 million years ago). It probably occurred ~700 million years ago. This is how the times shown in Figure 3.27 are derived. Some bacteria, such as *Methylacidiphilum infernorum*, also have a type of hemoglobin (see Figure 3.25F), which is similar to plant globins (24% sequence identity) and to animal globins (20% sequence identity).

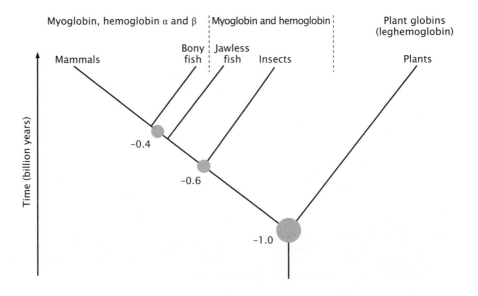

Figure 3.27 Appearance of the various globin genes superimposed on the eukaryotic phylogenetic tree. The numbers indicate the time since divergence in billion years.

3.11 PROTEIN SEQUENCES DO NOT CHANGE UNIFORMLY IN THE COURSE OF EVOLUTION

Now here is a puzzling observation. To a good approximation, the rate of DNA mutation is the same for all genes in a given organism, but in the proteins encoded by those genes the mutation rate varies enormously within an organism. For example, fibrinopeptides have changed much faster than hemoglobin, which has changed much faster than cytochrome c (**Figure 3.28**). The mean time (τ) elapsed per mutation per residue, which corresponds to $m/N = 1$ on the y-axis, is indicated in the figure.

Why are the rates so different in different protein families? The rate of mutation of DNA is much higher than that of the corresponding proteins. Most of the possible mutations at the DNA level do not result in protein mutations. Thus, a completely different factor determines the rate of change of protein sequences. The protein *structure* itself encodes the rate of change of its sequence.

The Strongest Constraints on Mutation are Due to Function

Cytochrome c has changed very slowly in time (see Figure. 3.28). Cytochrome c molecules from various species share only ~15% sequence identity. However, amino acids at some positions are highly conserved. The strongest constraints on the change of a residue are due to function. Cytochrome c is an essential protein for any organism. Thus, mutations that render it nonfunctional cannot be accepted. The invariant residues Cys-17, His-18, and Met-80 (**Figure 3.29**) interact directly with the heme group, which is central to the electron transport function of cytochrome c.

Hemoglobins have changed slightly faster than cytochrome c, but still contain several strongly conserved residues (see Figure 3.24), again due to function. The most important ones interact directly with the heme group (**Figure 3.30**). The strictly conserved residues are His F8, called the *proximal* histidine because it is bound directly to the iron

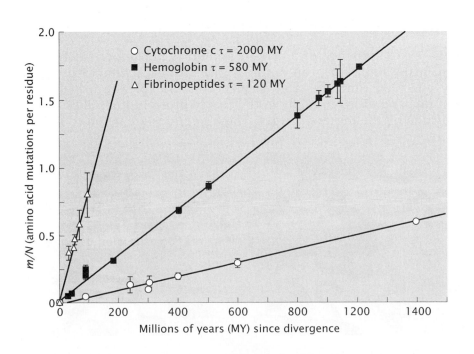

Figure 3.28 Rates of divergence of different proteins in different organisms. Time is counted backward from the present, which is the origin in the plot, to the time of the branch point for each pair of sequences compared. (Data from Dickerson RE [1971] *J Mol Evol* 1:26–45.)

Figure 3.29 Center of the cytochrome c molecule (human, PDB 1J3S) showing the heme group with its iron metal ion (sphere). The three invariant residues, Cys-17, His-18, and Met-80, directly interact with the heme group.

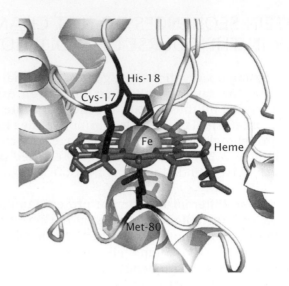

ion in the center of the heme, and Phe CD1. His E7 is also fairly conserved. It binds to the oxygen molecule on the other side of the heme. His E7 is called the *distal* histidine because it is further away from the heme than His F8.

Similarly, the residues of the catalytic triad in the serine proteases (His-57, Asp-102, and Ser-195) are strictly conserved. The regions of the heme groups in cytochrome c and the globins are analogous to the active sites of enzymes. Another set of conserved residues are those involved in interactions with other proteins. Cytochrome c is part of the electron transport chain. As such, it needs to interact with several proteins and those interactions are essential for its function.

Residues that form the inner hydrophobic core of the protein also tend to be conserved. They are essential to maintain the folded state. In other cases only the type of residue is conserved: an acidic replaces another (Asp/Glu), a basic replaces another (Lys/Arg), a nonpolar residue replaces another, or polar replaces polar. On the protein surface, switches are usually well tolerated.

Unconstrained Sequences Change Fast

If the exact sequence of a certain protein is not important for function, many mutations can occur in that sequence without producing any effect on function. This sequence would tolerate a high mutation rate and change rapidly in the course of evolution. An example is provided by the fibrinopeptides. These peptides are involved in blood

Figure 3.30 Center of the hemoglobin molecule (human, α chain): (A) deoxyhemoglobin, (B) oxyhemoglobin (with bound O_2). The heme group, with its iron metal ion, is visible in the center. His F8 directly coordinates the iron, whereas His E7 binds oxygen from the opposite side of heme in oxyhemoglobin. Phe CD1 interacts with the heme group (PDB 1A3N and 1HHO).

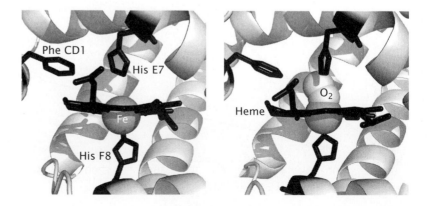

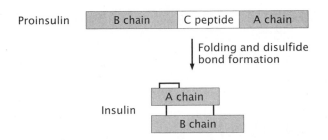

Figure 3.31 Maturation of insulin from proinsulin. The thick links indicate disulfide bonds.

clotting and are part of a larger precursor, fibrinogen. Without the fibrinopeptides, fibrinogen would aggregate. This is desirable if a blood clot is to form but not otherwise. The function of fibrinopeptides is to impart a negative charge to fibrinogen, to prevent its aggregation. Accordingly, they are rich in aspartate and glutamate. In addition, an arginine residue is required for proteolytic cleavage. Other than that, the sequence is not important. Therefore, fibrinopeptides have changed rapidly in evolution (see Figure 3.28).

The insulin C peptide is another example. The sequence of proinsulin (the precursor of insulin) is composed of three segments: chain B, the C peptide, and chain A. Proinsulin is converted to insulin upon folding, formation of disulfide bonds, and excision of the C peptide (**Figure 3.31**). The C peptide is not necessary for function. Once insulin is folded and stabilized by disulfide bonds, peptide C is proteolytically removed (**Figure 3.32**). Therefore, the detailed sequence of peptide C is not very important, and it has evolved at a rate seven times faster than the A and B chains of insulin.

The sequences of *pseudogenes* change even faster. Pseudogenes do not code for any protein. They are derived from functional genes (probably by reverse transcription from mRNA), but contain nonsense or frameshift mutations that render them nonfunctional. Since there are no constraints on mutations in pseudogenes, their evolutionary rate is about the same as the rate of DNA mutation.

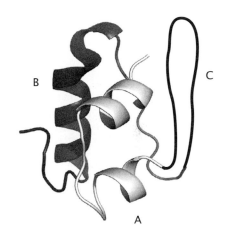

Figure 3.32 Structure of insulin: chains A (light) and B (dark), and the location of the C peptide (shown as a black loop) (PDB 1B2F).

Amino Acid Replacements in Homologous Proteins are not Random

Whereas genes undergo stochastic sequence mutations during evolution, physical and evolutionary contraints determine which amino acid changes are accepted. This is manifested in *how often* an amino acid is replaced, and *which* amino acid replaces it. Most often, similar amino acids replace each other. Further, proteins are not built from a uniform mixture of the 20 amino acids, but contain many more of some. For example, alanine is very common. Therefore, it is likely to replace other amino acids, and it is also more likely to be mutated. After correction for occurrence frequency, the probabilities of mutation of different amino acids, or the normalized mutabilities, are obtained (**Table 3.4**).

Table 3.4 Relative mutabilities of the amino acids during evolution, normalized to the value for alanine[a]

Asn	1.34	Thr	0.97	His	0.66	Phe	0.41
Ser	1.20	Ile	0.96	Arg	0.65	Tyr	0.41
Asp	1.06	Met	0.94	Lys	0.56	Leu	0.40
Glu	1.02	Gln	0.93	Pro	0.56	Cys	0.20
Ala	1.00	Val	0.74	Gly	0.49	Trp	0.18

[a]Data from Dayhoff MO [1978] Atlas of Protein Sequence and Structure, vol. 5, suppl. 3. National Biomedical Research Foundation.

Figure 3.33 Discrepancy between observed (*O*) and expected (*E*) amino acid replacements in closely related proteins. The discrepancy is defined here by $O/(E + \delta) - E/(O + \delta)$, ($\delta = 2$ in this plot). The two horizontal lines mark one standard deviation from the mean. (Data from Dayhoff MO [1978] Atlas of Protein Sequence and Structure, vol. 5, suppl. 3. National Biomedical Research Foundation, and Creighton TE [1993] Proteins. Structure and Molecular Properties. Freeman.)

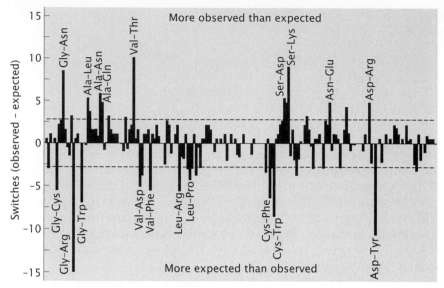

The Frequency of Amino Acid Replacements in Proteins Reflects Their Chemical Nature, not the Genetic Code

It might be thought that the frequency of amino acid replacements reflects the degeneracy in the genetic code (see Table 3.2). Some amino acids are encoded by many codons (for example, leucine), whereas others are coded for by few, or even only one codon (tryptophan). A conversion to leucine might appear more likely than a conversion to tryptophan. This is true, but the frequencies of those amino acid changes do not match those expected from the genetic code.

If we compare the distribution of probable mutations calculated on the basis of the genetic code with the distribution of mutations actually observed, we find that they are very different (**Figure 3.33**). If the observed amino acid replacements matched those expected, the plot in Figure 3.33 would consist of a horizontal line at zero. Instead, some exchanges occur much more often, and others, much less often than expected. Thus, replacement frequencies of amino acids in homologous proteins are *not* a consequence of the genetic code, but reflect the similarities in the chemical nature of the amino acids and their roles in protein structure and function.

3.12 PROTEIN SEQUENCES CHANGE MAINLY BECAUSE OF NEUTRAL MUTATIONS

Positive Selection is not The Major Cause of Sequence Evolution

Charles Darwin proposed that species evolve by *natural selection*. The idea of *positive selection* is that a distribution of phenotypical traits exists in a population, and those individuals with phenotypes that confer an advantage are favored. Over time, they win the competition, reproduce, and become dominant in the population. Their phenotypes are retained. The ability to survive and pass phenotypes (and genotypes) to the next generation is called *fitness*. Conversely, *negative selection* eliminates those individuals with deleterious traits. According to Darwinism, change in the dominant population phenotype is driven by positive selection.

The neo-Darwinistic view of evolution, or *selectionism*, maintains that natural selection is the main driving force for changes at the molecular level—namely, in protein sequences. The idea is that there is a sufficiently high genetic variability in any population on which positive selection acts. Mutations are the source of genetic variability, but which mutations are retained and passed to the offspring is determined by positive selection.

But is this true? We have seen that the number of amino acid replacements in protein sequences increases linearly with time since the species diverged. Substitutions occur most frequently in the parts of the sequence that are *not* important for structure or function, but rarely in the important regions. For example, fibrinopeptides change much faster than cytochrome c or hemoglobin because most of the sequence of fibrinopeptides is not important for function. And the residues in cytochrome c and hemoglobin that are most important for function, those interacting directly with the heme groups (see Figures 3.29 and 3.30), change much slower than residues on the protein surface. None of those observations could be explained if positive selection were the main means of sequence evolution.

Neutral Mutations are those that do not Affect Function

A much simpler explanation was proposed by Kimura (1968) and King and Jukes (1969) as the *neutral theory of evolution*. According to this hypothesis, most amino acid substitutions in proteins are the result of *neutral mutations*, which are those that do not significantly affect the structure and function of the protein. Neutral mutations are therefore "invisible" to natural selection. Natural selection acts mainly against deleterious mutations. Neutral mutations are tolerated, and over time become fixed in the population.

The mutation rate of an amino acid residue provides a measure of the importance of that residue for the protein structure. Some regions of the sequence undergo mutations much less frequently than others, because those regions are essential for structure or function, such as the active site, binding sites, or protein association surfaces. Most of the changes that do occur during evolution are those that *do not* affect function.

As we have seen, some proteins change slowly and others fast in the course of evolution. These differences do not arise because of different rates of mutation in some genes, or because some proteins are richer in amino acids that can be mutated more easily. Rather, differences in mutation rate arise because different proteins have different *fractions* of amino acid positions that can be mutated *without* affecting function.

Accordingly, proteins with a very low tolerance for mutations evolve very slowly. The typical example is cytochrome c. The reason for the low mutation tolerance is that this small protein is part of the electron transfer, respiratory chain. As such, cytochrome c must interact with several other proteins, accepting electrons from some and donating electrons to others. This significantly reduces the number of amino acids that can be mutated without affecting the interaction with other proteins. In addition, residues close to the heme binding site are also conserved because this is where the function of cytochrome c resides, its equivalent of an active site. Combined with the small size of the protein, this leaves little room for change. Therefore, the number of possible neutral mutations is small, and cytochrome c has changed very slowly in the course of evolution.

Not all mutations are neutral in this hypothesis. In fact, *purifying selection* is a key component of the neutral theory of evolution. By this we mean that deleterious mutations are purified out by natural selection, whereas neutral and beneficial mutations are retained. It is estimated that ~80% of non-synonymous substitutions are eliminated by purifying selection. Most mutations in the third base of a codon are synonymous; only 28% of those changes result in a different amino acid. In contrast, 95% of the changes in the first codon change the amino acid type (see Table 3.2). Synonymous substitutions do not change the protein sequence; therefore, they occur at a rate similar to the total mutation rate in the DNA, as predicted by the neutral mutation hypothesis.

Most Accepted Mutations are Neutral

The neutral mutation hypothesis explains why some proteins mutate much faster than others. If the requirement for the acceptance of a mutation is that it should not affect structure significantly, the bulk of accepted mutations are determined by the structure. Then, it does not matter how fast the DNA mutates. Proteins differ from each other in how much the sequence can be changed without producing noticeable effects—that is, without affecting function. The sequences of proteins whose functions are more affected by mutations change slower.

The constraints imposed by function and structure work as a slow step in an assembly line. One worker makes components of a robot and passes them to the next, who assembles the robot (**Figure 3.34**A). The first worker can produce parts faster, but there is a limit on the overall rate of assembly. If the second worker is much slower, he cannot cope with the flux of parts anymore (see Figure 3.34B).

(A)

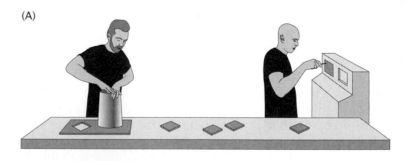

(B)

Figure 3.34 The rate-limiting step in an assembly line is the slower worker.

This situation is analogous to a two-step chemical reaction in which the first step is fast and the second is slow.

$$A \xrightarrow{\text{fast}} B \xrightarrow{\text{slow}} C.$$

If the second step is much slower than the first, so that the second is *rate-limiting*, the overall rate, which is the rate of formation of product C, is determined exclusively by the slow step. We can represent protein evolution by the "reaction"

$$\text{Original gene} \xrightarrow{\text{fast}} \text{Mutated gene} \xrightarrow{\text{slow}} \text{Protein.}$$

The second, slow step here is a "quality control" step: only some mutated proteins are viable. Thus, only some mutations in the genes find a "path" to functional proteins. This is why the overall rate is determined by the rate of neutral mutations.

In the neutral theory of evolution, the most important process through which mutations become accepted is *genetic drift*. What this means is that, in a finite population, some mutations happen to be fixed, by *random* sampling, even if they do not confer any advantage to the individuals that carry them. This is in stark contrast with selectionism. The neutral theory does not deny the existence of positive natural selection. There are many such examples, as we will see shortly. However, the *number* of mutations that result from positive selection is much smaller than the number of neutral mutations. This is why the overall rate of sequence divergence is determined by the rate of neutral mutations.

Important Functional Changes Result from Positive Selection

Whereas most mutations are neutral, some of the most important mutations in adaptation are the result of positive selection. The environment presents major constraints or driving forces for change. For example, if there are strong environmental constraints requiring a certain trait, no change can occur in that trait. Conversely, a change in environment, such as a modification of the habitat, or the appearance of new chemicals in it, leads to rapid evolution. Here are a few examples.

Crocodiles can stay under water for about one hour because their hemoglobin has undergone a functional change. Instead of binding organic phosphates (for example, bisphosphoglycerate), Cl^-, and CO_2 (as a carbamide by reaction with the N-terminus), it binds bicarbonate in the acidic environment of the tissues. This forces hemoglobin to release oxygen more completely, making more effective use of the oxygen loaded in the lungs. (In normal conditions, hemoglobin only releases ~40% of its bound oxygen in the tissues.) This adaptation is the result of only five amino acid changes out of 123 residues. Thus, most changes in crocodilian hemoglobin are neutral. Just those five are the most important for its specific function.

The hemoglobin of Andean camelids (llama, vicuña, alpaca, and guanaco) and bar-headed and Andean geese has adapted to the low pressure of oxygen at high altitude by increasing its oxygen binding affinity. The molecular mechanisms of affinity enhancement are different in geese and camelids, but the changes involve only a few amino acid substitutions in both cases. It is also remarkable that, in these and other species of birds, those changes occurred *independently*—that is, *in parallel.* Furthermore, sometimes those mutations occurred in the *same order* in time. This is strong evidence for positive selection.

The red and green opsins in human cone cells, which are responsible for color vision, resulted from a gene duplication. The original phenotype was green, and the red resulted from it by a series of mutations. There are 15 amino acid differences between the two proteins, but only two of those are important for the difference in function of the two opsins. In fish, opsins have adapted to dim light in deep water by the *same* amino acid substitutions in eight different known instances.

Sixty-eight different species of weeds have acquired resistance to triazine herbicides. The *same* mutation is responsible for the resistance in all these species. It occurs in one of the subunits of the photosystem II complex. The herbicide binding site is the same as that of the essential cofactor plastoquinone. The interaction includes a hydrogen bond with a serine residue. In resistant weeds this serine has been replaced by glycine. Plastoquinone binds with lower affinity, but other interactions with the protein allow it to still function, whereas triazine binding is abolished. This is a striking example of parallel (independent) evolution by positive selection.

Mice colonized the beaches of Florida's Gulf coast about 3000 years ago. The beach mice have a light color coat, whose origin can be traced to a single amino acid mutation in the melanocortin-1-receptor. This mutation reduces the affinity of the receptor for the ligand (melanocortin), which loses the ability to stimulate formation of cyclic AMP, the second messenger leading to production of pigmentation. The light-color allele already existed in the mainland populations, but was subject to positive selection when the environment changed, in the beach mice populations.

Proteins in which interactions play a fundamental role are often in rapid evolution by positive selection. Protease inhibitors are a classical example. Bovine pancreatic trypsin inhibitor (BPTI) is a small protein that binds to the active site of trypsin (**Figure 3.35**) with a binding constant $K = 10^{13}$ M^{-1}, one of the largest known in proteins. Common binding constants are of the order of 10^6 M^{-1} to 10^9 M^{-1}. Essentially, BPTI binds to trypsin and never comes off. It inhibits trypsin because

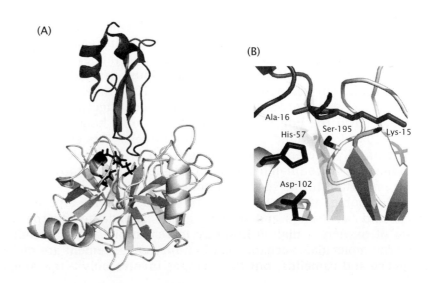

Figure 3.35 Structure of trypsin with its bound inhibitor (BPTI) (PDB 3OTJ). (A) Trypsin is shown in white, with the catalytic triad as stick models (black). BPTI is shown in dark gray. (B) The binding site. The catalytic triad of trypsin (His-57, Asp-102, and Ser-195) and the residues flanking the cleaved bond in BPTI (Lys-15 and Ala-16) are shown as stick models.

it binds to the active site, which is therefore permanently occupied. In fact, BPTI presents to trypsin a substrate-like bond between Lys-15 and Ala-16. This bond of BPTI is hydrolyzed but very slowly (months). Surprisingly, with the exception of the substrate-like bond itself, the areas of the BPTI sequence involved in binding to the protease are those that have changed most rapidly during evolution.

3.13 THE PROTEIN STRUCTURE ENCODES A MOLECULAR CLOCK

There May Exist a Molecular Evolutionary Clock

We have learned that different proteins, such as hemoglobin and cytochrome c, evolve at different rates. However, Figure 3.28 shows yet something else: it shows that homologous proteins appear to mutate at *constant rates*, independent of the organism that hosts them. That is, the rate of sequence change is constant in real time for each protein. This was first noticed by Zuckerkandl and Pauling (1965) when they compared sequences of hemoglobin chains from different mammals. They proposed that "there may thus exist a molecular evolutionary clock," which determines the rate of change of the protein sequence. A constant mutation rate in the DNA cannot be the reason for the constant mutation rate observed in the protein because otherwise all proteins from the same organism would evolve at the same rate. Thus, the evolutionary clock is determined by the protein molecule itself. The reason is that protein sequences change mainly through the accumulation of neutral mutations.

The molecular clock concept revolutionized the study of protein evolution. If clocks are constructed from the mutation rates of proteins that change very slowly, assuming a constant rate of change of these sequences over evolutionary time, we can extrapolate back in time and determine the point of occurrence of evolutionary events that are not accessible by the fossil record—that is, older than ~2.5 billion years ago. The concept of the molecular clock faces some difficulties, but for now, let us understand it better and explore the consequences of this amazing idea.

Molecular Clocks are Obtained by Matching Mutation Rates to Real Time

If there is a molecular clock, how do we set the time on it? We need some other clock that gives the correct time. That clock is the fossil record. Assuming that the mutation rate is constant along the branches of an evolutionary tree, one can calibrate the molecular clock for proteins using the fossil record. The intrinsic rate of mutation k (rate constant) is the number of accepted amino acid mutations (m) per residue (N) per unit time (t), given by Equation 3.2,

$$k = \frac{m/N}{t}.$$

Now, consider the evolutionary tree of the trypsin family shown in **Figure 3.36**. The length L of a tree branch is proportional to the fraction of accepted mutations (m/N). Note that L is not linearly proportional to the fractional identity (n_0/N); the relation is exponential

Figure 3.36 Evolutionary tree for trypsins. The numbers indicate the time in billion years.

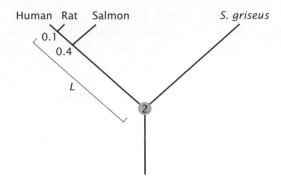

(shown in Equation 3.4). The length L corresponds to a time $t \approx 2$ billion years, because that is when the divergence between bacteria and eukaryotes began. To convert n_o/N (% identities) to m/N (accepted mutations), we can use Figure 3.15. The percent identity as a function of PAMs $\approx 100 \times m/N$. Alternatively, to a reasonable approximation, we can invert Equation 3.4 to obtain

$$\frac{m}{N} = -\ln\frac{n_o}{N}. \tag{3.14}$$

The sequences of trypsins from animals are $\approx 30\%$ identical to those of bacteria. In other words, the bacterial and animal trypsins have mutated 70% of their sequence relative to each other since their branch point ~ 2 billion years ago. Using Figure 3.15, we get PAMs ≈ 160 for 30% identity, which gives $m/N \approx 1.6$ (shown in Equation 3.14 yields $m/N = 1.2$). Thus, the rate of mutation in trypsins is 160 mutations, divided by 2 billion years, or 80 mutations per 100 residues per billion years; thus, $k = 8 \times 10^{-4}$ mutations per residue per million years (MY). This is the evolutionary clock rate for trypsin. Alternatively, we can calculate the characteristic time ($\tau = 1/k$) elapsed per mutation per residue (shown in Equation 3.3). Recall that after a time τ each residue has been mutated once on average (which does not mean that the whole sequence has changed). In the case of trypsin, $\tau \approx 1200$ MY. The characteristic times are $\tau = 120$ MY for fibrinopeptides, 580 MY for hemoglobin, and 2000 MY for cytochrome c. You can see in Figure 3.28 that those times correspond to $m/N = 1$.

Yet another way of thinking about the rate of sequence change is to ask how far back proteins can trace their ancestry. This time is called the *lookback time*. In a figurative sense, it is how far back a certain family can "look," or trace, its origin. Slow-changing proteins, such as cytochrome c, have long lookback times; they can be recognized by their similar sequences in very distantly related organisms. Fast-changing proteins, such as fibrinopeptides, have much shorter lookback times; matching two such sequences may be very difficult.

How accurate is the molecular clock? Over long periods of time, if a large number of sequence comparisons is used to set the clock, the molecular clock is fairly reliable. But over short periods it may deviate significantly from the long-time behavior. Protein evolutionary clocks reflect the evolution of *protein* sequences. The mutation rate at the DNA level varies among taxonomic groups because of differences in the accuracy of DNA replication, DNA repair, and levels of DNA damage. Because of all these factors, the clock *ticks erratically.*

Initially, it was thought that the molecular clock must depend on generation time. But it is now quite clear that it does not. The reasons have been much (and hotly) debated. But in the end, the molecular clock, with all its drawbacks—unclear influences, variations, erratic ticking, approximate nature—remains the most useful tool to assign time in molecular evolution.

3.14 ANCIENT PROTEINS, FROM EXTINCT ORGANISMS, CAN BE RESURRECTED IN THE LABORATORY

The effects of amino acid substitutions on protein structure and function are complex because they *depend on the context* in which they occur. Each residue interacts with many others in the protein structure. The consequences of amino acid replacements depend on changes in those interactions. Thus, a mutation engineered in two modern homologous proteins may have an certain effect on structure or function in one of them, but have none, or a different effect, in the other homolog. The interdependence of mutations is an example of what biologists call *epistasis*. It results from interactions between different residues in the protein structure. Epistasis means that the effect of a historical mutation in a modern protein may be different than in its ancestor. Therefore, epistasis complicates our understanding of how proteins evolve.

In 1963, Pauling and Zuckerkandl made a visionary proposal. They suggested that by comparing the sequences of contemporary homologous proteins it should be possible to *reconstruct* the sequence of the ancestral protein from which they evolved. They wrote that "it will in the future be possible to synthesize these presumed components of extinct organisms. Thus one will be able to study the physicochemical properties of these molecules and to make inferences about their functions."

This vision has now become reality. The sequences of the ancestral proteins have been reconstructed, and the actual proteins have been recreated in the laboratory. Reconstructing ancient genes allows the determination of the effects of each mutation in historical context. This is because not only is the ancestral sequence recovered from the dead, but sequence reconstruction also yields the *historical sequence* of mutations that occurred. The effect of each amino acid change can then be studied in the polypeptide sequence *in which it first occurred*.

Mutations Can be Permissive or Restrictive of Future Changes

There are two types of epistatic mutations. *Permissive* mutations do not alter protein structure and function on their own, and they often have minor effects on protein stability. In this sense, they are neutral mutations. But they allow other mutations to occur later, which, without the permissive mutations, would have deleterious effects, often on protein stability. These permissive mutations are not detected by mutagenesis on modern proteins. But ancient sequence reconstruction can trace those mutations backward in time, along the lineage that gave rise to the modern proteins.

Most mutations engineered in modern proteins lower the stability or leave it unaltered. It is rare to introduce mutations that increase stability. Similarly, many mutations that occur naturally in the DNA would decrease protein stability, and are therefore not accepted. This is a major obstacle to evolution of new functions. Permissive mutations render possible some pathways to protein evolution by increasing stability first, before functionally favorable mutations may subsequently decrease it (the protein "has no idea" that this is happening). Later, those sequences that have the permissive mutations will be able to undergo function-altering mutations; if advantageous or neutral, those mutations may be fixed in the population.

The other type of epistatic mutations are *restrictive mutations*. They may introduce local structural changes that are well tolerated by the protein at the historical time when they occur. But they prevent future changes because the combined mutations would lead to unfavorable residue interactions, such as steric hindrance or two adjacent like charges. Again, these mutations are neutral when they occur, but are essential later because they function as one-way road markers for evolution.

These concepts can be understood with the example depicted in **Figure 3.37**. Proteins B1, B2, C, and D are modern paralogs. Three gene duplications have occurred in their evolution. A1 is the common ancestor. A2 and A3 are the most recent common ancestors before each duplication. The shapes of the symbols in Figure 3.37 indicate different protein functions. The extant proteins C and D have specificities (circles) similar to the oldest ancestor (A1); B1 and B2 (squares) have another specificity. The function-switching mutations occurred between ancestors A2 and A3. They were preceded by permissive mutations, between ancestors A1 and A2, which allowed the functional switch to occur without the loss of protein stability. Further, restrictive mutations occurred between A3 and B1/B2. If engineered in the extant proteins B1 and B2, reversal of the functional mutations would fail to produce an active protein. Other restrictive mutations occurred between A2 and C, which prevent function switching to the B-type specificity.

Similarly, the functions of B1 and B2 cannot be switched to that of C because the B proteins accumulated restrictive mutations that prevent them from adopting the specificity of C (or D or A1). Conversely, C cannot be mutated to function like the B proteins, because restrictive mutations have accumulated in the historical path from A2 to C.

The permissive mutations that allowed the function switch between A2 and A3 are present in B1 and B2 but also in C, so they will not appear important for function if the sequences of the extant proteins are compared. Those seem to be neutral mutations. The restrictive mutations are not likely to be evident either, because they occurred in different residues in the paths to B1/B2 and C. Thus, mutagenesis of extant sequences is unlikely to reveal the functional importance of

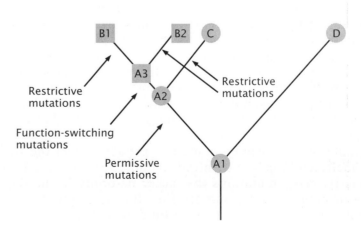

Figure 3.37 Hypothetical evolutionary tree illustrating the effects of epistasis on function-switching mutations in a set of paralogous proteins. The symbol shapes indicate different functions.

many residues in these proteins. However, ancient sequence reconstruction will, because the historical path of the mutations can be retraced. The function-switching mutagenesis can be performed in A2 and reversed in A3, identifying the residues involved and uncovering their true effects. As a result, the effects of permissive and restrictive mutations on function can be understood.

3.15 EVOLUTION OF COLOR VISION DEPENDS ON A FEW MUTATIONS IN MEMBRANE PROTEIN RECEPTORS

Ancient sequence reconstruction allows us to understand the connection between changes in sequence when they occurred and their effects on function, at the amino acid residue level. We will now study a few examples. The first are taken from the evolution of color vision.

G-proteins Transduce Receptor Binding Information to the Cell Interior

Most cellular receptors are integral membrane proteins. Signal transduction is usually initiated by binding of a *ligand*, a small molecule such as a hormone, to its receptor. These ligands are polar, water-soluble molecules, and therefore cannot cross the hydrophobic interior of the lipid bilayer. However, through binding to the receptor, the message that the ligand is present is conveyed to the cell interior.

G-protein coupled receptors (GPCRs), or seven transmembrane helix receptors, are the largest class of membrane receptors. Rhodopsin is one the best understood examples. These integral membrane receptors are associated with a peripheral GTP-binding protein, or G-protein (**Figure 3.38**A). G-proteins are heterotrimers of an α subunit (G_α) and a $\beta\gamma$ dimer ($G_{\beta\gamma}$). They are bound to the inner lipid monolayer of the cell membrane, in part by fatty acid chains covalently attached to the α and γ subunits. In the inactive state, the G-protein trimer has a molecule of GDP bound and is associated with the integral membrane receptor.

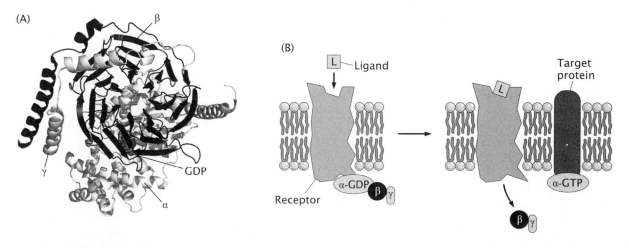

Figure 3.38 (A) G-protein heterotrimer (PDB 1GP2). The α subunit is shown in white, with a GDP molecule (gray) bound. The β subunit (black) has a β-propeller structure. The γ subunit (white) consists of two α-helices with a linker. (B) Schematic of the function of a membrane G-protein-coupled receptor. Activation by the ligand (L) induces a conformational change in the receptor, which alters its interaction with the G-protein. The G_α subunit then releases GDP, binds GTP, and dissociates from $G_{\beta\gamma}$. G_α now carries the signal further by associating with its target proteins.

Figure 3.39 Rhodopsin in a lipid bilayer. The chromophore retinal is shown in black (stick model) attached to Lys-296 (space-filling). The N- and C-termini are located in two different aqueous compartments, the intradiscal space and the cytosol, respectively (PDB 1U19).

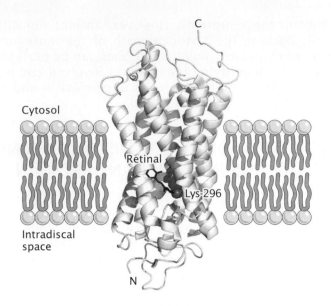

Upon ligand binding to its extracellular part, the receptor undergoes a conformational change, which is transmitted to the G-protein (see Figure 3.38B). The G-protein then releases GDP and binds GTP, and $G_{\beta\gamma}$ dissociates from G_α, which now becomes *activated*. G_α then interacts with other proteins, such as adenylate cyclase or phospholipase C, which further propagate the signal inside the cell. Eventually, the G-protein hydrolyzes GTP to GDP and becomes inactive, reassociating with $G_{\beta\gamma}$ and then with the receptor. The receptor itself becomes inactive when the ligand dissociates from its binding site (or by phosphorylation and interaction with the protein arrestin).

Color Vision is Mediated by Opsins

The sensation of vision is triggered in the brain when a photon of light reaches the retina of the eye and is absorbed by a visual pigment. Visual pigments are G-protein-coupled receptors that exist in the membranes of the retina cells. They consist of an integral membrane protein with seven transmembrane α-helices, called an *opsin*, and a *chromophore*, 11-*cis*-retinal, covalently attached through a Schiff base to the ϵ-amino group of a lysine residue (Lys-296) in the center of the protein. **Figure 3.39** shows the integral protein rhodopsin in a lipid bilayer.

What is special about opsins is that their ligand (11-*cis*-retinal) is *prebound*, but in an *inactive form*. When 11-*cis*-retinal absorbs a light photon, it isomerizes to all-*trans*-retinal, which is the *active ligand* (**Figure 3.40**). Upon activation, the opsin undergoes a conformational

11-*cis*-Retinal   All-*trans*-Retinal   Vitamin A (retinol)

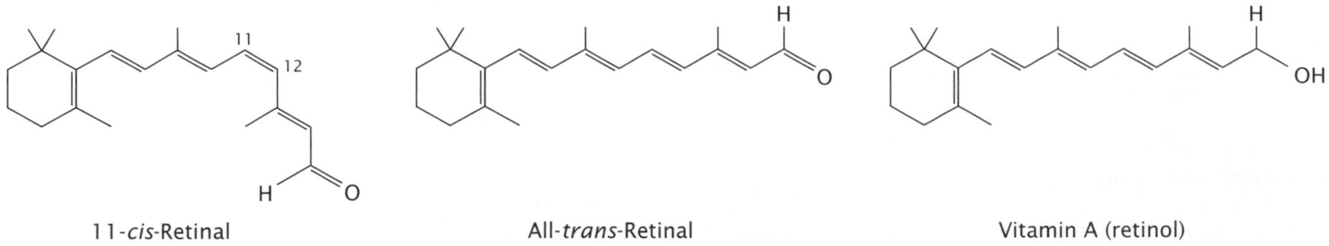

Figure 3.40 11-*cis*-Retinal, all-*trans*-retinal, and their precursor, vitamin A (retinol).

Table 3.5 Typical wavelengths of the various colors in the light spectrum

Color	Wavelength (λ)
Violet	410 nm
Blue	480 nm
Green	520 nm
Yellow	570 nm
Orange	610 nm
Red	680 nm

change, which is communicated to the cell interior through the interaction between the cytoplasmic side of the opsin and a G-protein called transducin. Transducin then releases GDP and binds GTP, thus triggering a cascade of events that culminates in the transmission of a signal to the brain by the optic nerve.

The retina of vertebrate eyes contains two types of cells, rods and cones. The visual pigments in the rod cells are called rhodopsins and are responsible for dim-light vision. The proteins (opsins) in the cone cells are called cone pigments or color visual pigments because they are responsible for the perception of color. Since the chromophore is the same in all visual pigments (except in a few cases where 11-*cis*-3,4-dehydroretinal is used), the wavelength of maximum absorption of light (λ_{max}) varies because of differences in the protein milieu. The visual pigments are named for their wavelengths of maximum absorbance. For reference, typical wavelengths of the colors in the light spectrum are listed in **Table 3.5**.

Color Vision Evolved Through Gene Duplications of an Ancient Rhodopsin

Humans have one kind of rhodopsin in rod cells and three kinds of opsins in cone cells. The rhodopsin is most sensitive to green light, with $\lambda_{max} = 497$ nm in humans. (Visual pigments are commonly designated by the letter P followed by the wavelength of maximum absorbance; for example, P497.) Rhodopsin senses dim-light intensity but does not provide color sensation in the brain. The color pigments come in three kinds. In humans, blue opsins (P414) have $\lambda_{max} = 414$ nm, green opsins (P530) have $\lambda_{max} = 530$ nm, and red opsins (P560) have $\lambda_{max} = 560$ nm. (The typical wavelength for red light is 680 nm; 560 nm actually corresponds to yellow, but the pigment is called red because it is the most *red-shifted* of them all. Recall that a red shift means that λ_{max} shifted to longer wavelength; a blue shift means that λ_{max} shifted to shorter wavelength.) The sequence similarity of human pigments and bovine rhodopsin are shown in **Table 3.6**.

Table 3.6 Percent amino acid identity among the sequences of bovine rhodopsin and human rhodopsin and color opsins[a]

Pigment	P497	P414	P530	P560[b]
Rhodopsin (bovine, P500)	93	45	44	42
Rhodopsin (human, P497)		47	45	44
Blue opsin (human, P414)			44	43
Green opsin (human, P530)				96

[a]Data from Yokoyama S [2000] *Prog Ret Eye Res* 19:385–419.
[b]P560, Red opsin (human).

Figure 3.41 Evolutionary tree of human opsins. Based on Yokoyama S [2000] *Prog Ret Eye Res* 19:385–419.

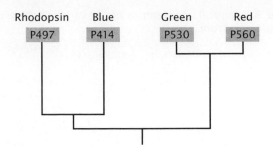

All opsins diverged after duplication of an ancient opsin gene. Human and bovine rhodopsins are more similar to each other than to any other opsin, which indicates that the rhodopsin gene existed in animals much before humans and cows diverged from each other in evolution, about 100 million years ago. However, cows do not have color vision. Thus, the gene duplication that gave rise to green and red opsins in humans appears to have occurred much more recently, about 30 million years ago. The evolutionary tree of human opsins is shown in **Figure 3.41**.

Trichromatic vision exists only in a few species of bony fishes, birds, reptiles, and primates. Most vertebrates lack either the green or the red pigment and are thus "color-blind" or dichromatic. Cavefish, on the other hand, have both green and red opsins but diverged from humans about 400 million years ago. Thus, the red–green gene duplication must have occurred *independently* in humans and fish after their divergence.

Dinosaurs were Probably Active at Night

Did terrestrial dinosaurs hunt at night? Reconstruction of the sequences of the visual pigments in extinct dinosaurs helped to answer this question. Dinosaurs descend from ancient archosaurs, which also gave rise to modern crocodiles (alligators) and birds (**Figure 3.42**). (Appearances can be deceptive: alligators and crocodiles look more like modern reptiles, but are evolutionarily closer to birds.)

The sequence of the visual pigment of ancient archosaurs was reconstructed from the rhodopsins of the alligator, pigeon, zebrafinch, and chick, which differ at most by 16% of amino acids, and from 26 outgroup sequences (pigments from vertebrates that are evolutionarily distant from the archosaurs). The ancestral rhodopsin expressed from this recreated sequence binds retinal, and when exposed to light, activates transducin similarly to bovine rhodopsin. This suggests that the ancient archosaurs were able to see well in dim light. Dinosaurs may have been night-active.

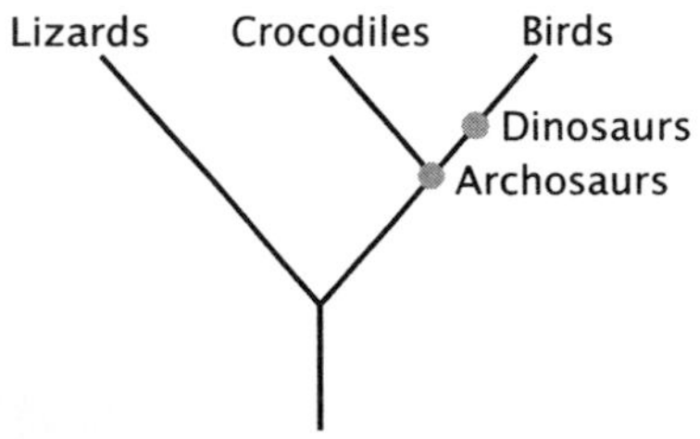

Figure 3.42 Evolutionary tree showing the divergence of modern reptiles, crocodiles (alligators), and birds.

Animals in Deep Sea Adapted to Blue Light Sensitivity by Mutations in their Rhodopsins

The coelacanth is a "living fossil," the only extant species of fishes that were common about 400 million years ago. Most fish rhodopsins (P510) absorb green light ($\lambda_{max}\sim$500–520 nm), but the coelacanths live in the Comoros archipelago in deep water, where only blue light reaches, with wavelengths narrowly centered at $\sim$480 nm. These fishes have adapted to dim light by using two pigments, one in rods and one in cones. The rod pigment (P485) is a rhodopsin with $\lambda_{max} = 485$; the cone pigment (P478) is rhodopsin-like as well, but has $\lambda_{max} = 478$ nm. Which amino acid mutations changed the color sensitivity from green to blue?

(A)

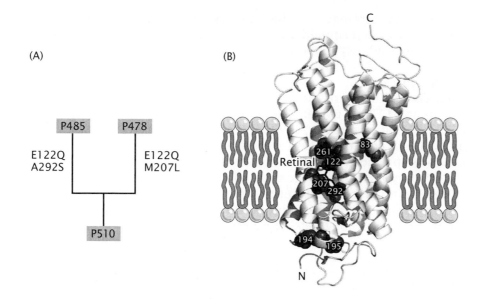

(B)

Figure 3.43 (A) Evolutionary tree of the visual pigments of the coelacanth. Based on Yokoyama S [2000] *Prog Ret Eye Res* 19:385–419. (B) Location of the seven critical residues (dark, space-filling) involved in adaptation to blue-light sensitivity in deep sea organisms, in the rhodopsin structure (PDB 1U19). (Based on Yokoyama S, Tada T, Zhang H & Britt L [2008] *Proc Natl Acad Sci USA* 105:13480–13485.)

To answer this question, the sequence of the ancestor of the modern coelacanth protein was reconstructed. It was found that a gene duplication gave rise to two branches, one leading to pigment P485, the other to P478 (**Figure 3.43**A). Many mutations occurred from the common ancestor rhodopsin to the two extant coelacanth pigments, but only two amino acid substitutions in each branch are responsible for the changes in color sensitivity. The ancestor rhodopsin absorbs light in the green ($\lambda_{max} = 510$). It contains a glutamate (Glu-122) where P485 and P478 have glutamine, alanine (Ala-292) where P485 has serine, and methionine (Met-207) where P478 has leucine. These residues are located in the center of the rhodopsin structure, near the retinal (see Figure 3.43B).

The simultaneous mutations Glu-122 to Gln (or E122Q for short) and Ala-292 to Ser (A292S) cause a blue shift of 25 nm in λ_{max}, from 510 to 485 nm. The blue shift to $\lambda_{max} = 478$ nm in P478 is also explained by two amino acid substitutions, one of which is the same (E122Q), but the other of which is different (M207L). However, the mutation E122Q occurred twice, *independently* in P485 and P478 (see Figure 3.43A). In fact, it is also present in rhodopsin-like pigments in some birds and reptiles; it occurred before the coelacanths diverged from the tetrapods (animals with four limbs). The mutation A292S, on the other hand, occurred independently in other evolutionarily distant animals (namely eels and dolphins). Thus, two mutations in the coelacanth rhodopsins are sufficient to revert to the green-sensitive protein. But this does *not* mean that if those mutations were performed in bovine rhodopsin, blue-light sensitivity would result. This is because many other *permissive* substitutions have occurred in the coelacanth proteins before the phenotype could change.

How about dim-light adaptation in other organisms that inhabit deep-sea waters? The evolutionary tree of the rhodopsins was reconstructed to include the coelacanth P485 pigment, the sequence of the earliest ancestor, and also those of more recent ancestors at the tree nodes (branch points). Then the specific effects of amino acid replacements along the historical sequence of mutations were recreated in the laboratory by introducing those mutations in the ancestors in which they occurred.

Using this strategy, it became evident that λ_{max} of modern rhodopsins is determined by only 15 changes at 12 of 350 residues (3% of the residues). Rhodopsin sequences differ by much more than 3%, but the other changes are neutral and do not affect function (light absorption). Four of the 15 mutations were *recurrent* in different organisms. Two of them are those found in the coelacanth (A292S occurred 9 times and E122Q, twice). The occurrence of parallel mutations, *independently* and frequently, in organisms that are distant evolutionarily but share a common feature—living in a deep-sea environment in this case—provides strong evidence that these changes constitute an evolutionary *adaptation*, the result of positive selection.

The chromophore 11-*cis*-retinal interacts with many different amino acid residues. The changes induced in λ_{max} when those amino acids mutate are often *not reversed* by reversing the mutation if it is engineered in a *modern protein*, for example bovine rhodopsin, or even in one of the modern pigments of deep-sea animals. To fully understand the effects of those replacements they need to be introduced in the direction and order in which they actually occurred in evolution. For example, the change A292S was found to decrease λ_{max} by ~10 nm. However, if introduced in one of the most recent ancestors, it has no effect on the phenotype, because for λ_{max} to change, two other mutations must occur as well.

It makes intuitive sense that the mutation A292S should affect λ_{max} because residue 292 is close to the retinal (see Figure 3.43B). However, residues 194 and 195, which also affect λ_{max}, are near the membrane interface, far from the chromophore. This example shows that the interactions between residues and cofactors may be direct or mediated by subtle changes in the protein conformation.

Blue (Violet) Opsins Evolved from UV Opsins

Most mammals and various birds cannot see ultraviolet (UV) light because their opsins absorb in the violet–blue region. UV or violet light sensitivity is determined by the λ_{max} of short-wavelength pigments (blue opsins). Was the ancestor of short-wavelength opsins sensitive to violet or to UV light? To answer this question, the sequence of the ancestor of bird and reptile opsins was reconstructed (**Figure 3.44**).

The ancestral pigment (P360) has $\lambda_{max} = 360$ nm, indicating that the ancestor of birds and reptiles (and humans) could see UV light. What amino acid mutations are responsible for the shift to blue sensitivity? There are nine amino acid differences between the bird ancestor (P393)

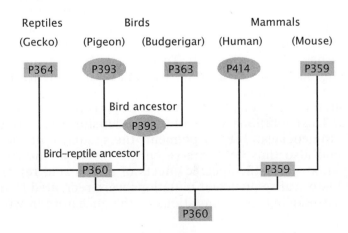

Figure 3.44 Evolutionary tree of the UV (rectangles) and violet-blue (ovals) pigments in reptiles, birds, and mammals. (Based on Shi Y & Yokoyama S [2003] *Proc Natl Acad Sci USA* 100:8308–8313, and Dean AM & Thornton JW [2007] *Nature Rev Genet* 8:675–687.)

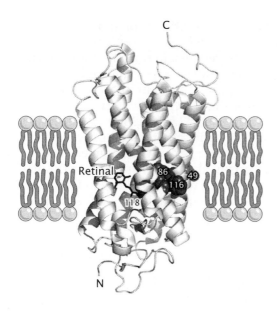

Figure 3.45 Location of the four critical residues involved in UV/blue-light sensitivity in the rhodopsin structure (PDB 1U19). (Based on Shi Y & Yokoyama S [2003] *Proc Natl Acad Sci USA* 100:8308–8313.)

and the bird-reptile ancestor (P360). Of those nine, four substitutions, introduced in the bird-reptile ancestor, were necessary and sufficient to produce a pigment with $\lambda_{max} = 393$ nm, the same phenotype as the bird ancestor and the modern pigeon opsins (**Figure 3.45**).

Mutations in Red Pigment Make it Green, but the Reverse Mutations in Green Pigment do not Make it Red

The human cone pigments red (P560) and green (P530) opsins differ in a number of amino acids (see Figure 3.41). Mutation of three of those residues (S180A, Y277F, and T285A; **Figure 3.46**) in the red pigment shifts λ_{max} to 530 nm, the phenotype of the green pigment. However, introducing the opposite changes in the modern green pigment does *not* produce the spectrum of the human red opsin ($\lambda_{max} = 560$ nm). This is because the changes in structure and function produced by a single amino acid mutation depend on the context in which they occur (epistasis). Amino acid replacements that occur during evolution alter the effects of a later mutation.

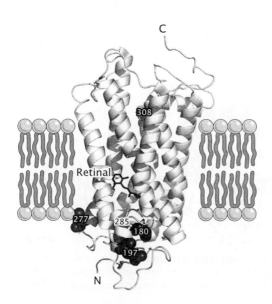

Figure 3.46 Location of the five critical residues involved in the red and green light sensitivity in rhodopsin (PDB 1U19). (Based on Yokoyama S, Yang H & Starmer WT [2008] *Genetics* 179:2037–2043.)

Thus, to understand the effects of individual mutations, we need to examine them on the reconstructed ancestor of the green and red paralogous opsins. By reconstructing the mutational history from the ancestor opsin to the modern green and red pigments in humans and other animals, it became apparent that five amino acid residues in the sequence control the spectra. The critical changes in the ancestor protein are S180A, H197Y, Y277F, T285A, and A308S (see Figure 3.46). Three of those are the same changes that convert red to green opsins. In fact, if those three changes are introduced in the ancestor protein, the human green opsin phenotype is obtained. But to obtain the red opsin from the green, the reverse of those changes and four more are necessary because of the interactions between those residues in the protein.

3.16 STEROIDS ACQUIRED NEW FUNCTIONS THROUGH THE EVOLUTION OF THEIR RECEPTORS

Steroid receptors are *not* membrane proteins. They are water-soluble proteins that reside in the cell nucleus, where they function as *transcription factors*. Their activators are steroid hormones. These lipidic molecules are soluble in the nonpolar interior of the bilayer and therefore can easily cross cell membranes to reach the receptors in the nucleus.

The estrogen receptor is shown in **Figure 3.47**. Steroid receptors have two domains: a hormone-binding domain and a DNA-binding domain. The estrogen-binding domain binds one estrogen molecule, shown in black in Figure 3.47A. The active form of the protein is a dimer of identical protomers. The DNA-binding domains of the dimer are shown as light and dark chains, one for each protomer, in Figure 3.47B. The domain has two "zinc-finger" motifs, in which zinc metal ions are coordinated by cysteine residues. When the steroid hormone binds to its receptor in the nucleus, the receptor dimerizes and binds to the DNA through the recognition helices, viewed from the top in Figure 3.47B. Once bound, the transcription factor activates the synthesis of mRNA that codes for the proteins that constitute the hormone response.

Figure 3.48 shows an overview of the metabolism of steroid hormones. The synthetic pathway begins with cholesterol. The first hormone produced is progesterone, which regulates reproductive cycle in females. At this point the pathway bifurcates. The first branch

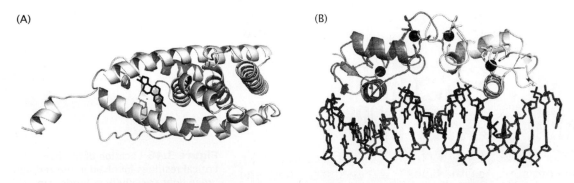

(A)

(B)

Figure 3.47 The estrogen receptor. (A) The estrogen-binding domain with estradiol bound, shown in black (PDB 1A52). (B) The DNA-binding domain. A domain from each monomer of the dimer is shown (light and dark molecules) interacting with DNA through the recognition helices (PDB 1HCQ). The zinc metal ions are shown as spheres, coordinated by cysteines (stick models).

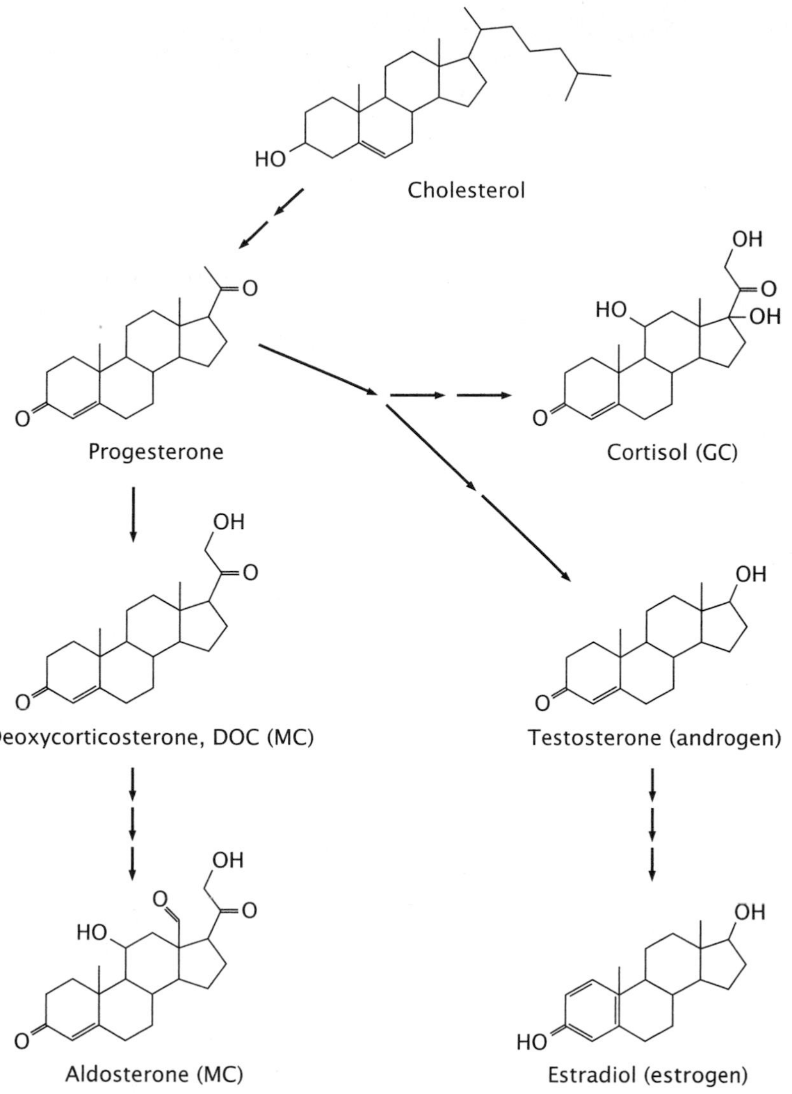

Figure 3.48 Metabolic pathway of the synthesis of steroid hormones from cholesterol. MC, mineralocorticoid; GC, glucocortioid.

yields the mineralocorticoids (MC): deoxycorticosterone (DOC) and aldosterone, which regulates reabsorption of inorganic ions (sodium, chloride, and bicarbonate). The second branch bifurcates further. One pathway yields cortisol, a glucocorticoid (GC), which regulates stress and inflammatory responses; the other pathway yields testosterone (androgen) and then estradiol (estrogen), which regulate sexual development in males and females, respectively.

We will study two examples of ancient sequence reconstruction that tell fascinating stories of the evolution of the steroid receptor family: the evolution of estrogen receptors (ER), and the evolution of glucocorticoid (GR) and mineralocorticoid receptors (MR) from a common corticoid receptor (CR).

Steroid Hormone Receptors Evolved Backward, from an Ancient Estrogen Receptor

To understand the evolution of function in the steroid receptors (SR), the sequence of the common ancestor (ancSR) was reconstructed. The phylogenetic tree was determined by aligning 73 receptor sequences (including 18 outgroup sequences). The sequence of ancSR, a protein that existed 600–1000 million years ago, was inferred by the maximum

Figure 3.49 Simplified evolutionary tree of the steroid hormone receptors. AR, androgen receptor; PR, progesterone receptor; MR, mineralocorticoid receptor; GR, glucocorticoid receptor; ER, estrogen receptor; ancCR, ancestor corticoid receptor; ancSR, ancestor steroid receptor. The steroid ligands and their target receptors are indicated (arrows). (Based on Thornton JW [2001] *Proc Natl Acad Sci USA* 98:5671–5676.)

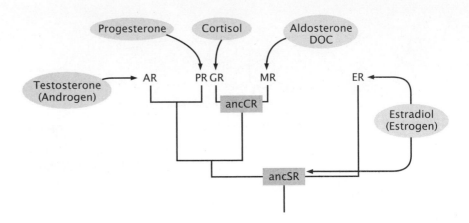

likelihood method. The modern receptors evolved from a series of gene duplications of this ancestral sequence. A simplified evolutionary tree of the steroid receptor family is shown in **Figure 3.49**.

Was ancSR an estrogen or a corticoid receptor? Or did it bind both hormones? Or neither? The sequence of ancSR is 71% identical to that of the estrogen receptor (ER), but much less similar to the other steroid receptors. Furthermore, the DNA-binding domain of ancSR is much more similar to that of the estrogen receptor than to the others. Therefore, it seems likely that ancSR was an estrogen receptor. This is very interesting: it says that the first ancestor steroid receptor was specific for estrogen, which is the *last* steroid hormone in the synthetic pathway (see Figure 3.48). The other steroid receptors evolved from ancSR (see Figure 3.49) after a series of gene duplications. They evolved new functions, such as binding of intermediate metabolites and triggering their own response in DNA transcription. Those steroid metabolites already existed before their receptors evolved, but had no signaling function.

Permissive Mutations were Essential for the Change that Produced Glucocorticoid Receptors

In humans, the modern glucocorticoid receptors (GR) bind cortisol, whereas mineralocorticoid receptors (MR) bind aldosterone. Aldosterone is a relatively new hormone, specific to tetrapods. What determines the different specificities of GR and MR? And how did mineralocorticoid specificity to aldosterone emerge? Sequence comparison of modern human GR and MR revealed that two residues are different at the steroid binding site: Ser-106 and Leu-111 occur in mineralocorticoid, whereas Pro-106 and Gln-111 occur in glucocorticoid receptors. These differences appear to determine specificity. However, the mutations of Ser-106 to Pro (S106P) and Leu-111 to Gln (L111Q) in MR, or their reverse in GR, produce inactive proteins.

To understand the problem, the sequences along the evolutionary tree of the corticoid receptors were reconstructed (**Figure 3.50**). This tree corresponds to the middle branch shown in Figure 3.49. The ancient corticoid receptor (ancCR) is the last common intermediate of GR and MR. A gene duplication of ancCR ~450 million years ago led to the two modern branches. After that, two more recent ancestors in the GR branch evolved, ancGR1 and ancGR2.

Consider the specificity of ancCR. When this ancient receptor was expressed *in vivo*, it was found to be activated by both aldosterone and cortisol. It was also activated by deoxycorticosterone (DOC), which

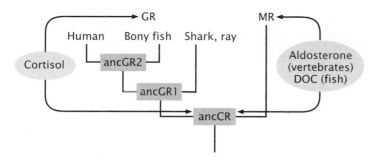

Figure 3.50 A simplified evolutionary tree of the corticoid receptors. MR, mineralocorticoid receptor; GR, glucocorticoid receptor; ancCR, ancestor corticoid receptor; ancGR1, ancestor GR with same specificity as ancCR; ancGR2, ancestor GR specific for cortisol. The steroid ligands and their targets are indicated (arrows). (Based on Dean AM & Thornton JW [2007] *Nature Rev Genet* 8:675–687, and Harms MJ & Thornton JW [2010] *Curr Opin Struct Biol* 20:360–366.)

is believed to be the original activator. Thus, ancCR appears to be a *generalist* protein; the modern receptors probably evolved by *becoming insensitive* to one or the other steroid. But why does ancCR bind aldosterone, which appeared *after* this receptor evolved? Aldosterone is structurally similar to deoxycorticosterone (DOC) (see Figure 3.48), and it binds to ancCR in the same way as DOC. Hence, ancCR binds aldosterone because of its structural similarity to DOC. DOC is the mineralocorticoid ligand in fish, but aldosterone is the ligand in extant vertebrates, which indicates that ancCR is similar to modern MR.

It is the modern GR that must have undergone the most changes, thus *losing affinity* for aldosterone. How did all this happen at the molecular level? The answer came from examining the binding affinities of ligands to the intermediate ancestors, ancGR1 and ancGR2. AncGR1 has a ligand specificity similar to modern MR; ancGR2 has a specificity similar to modern GR. Of all the changes that occurred between ancGR2 and modern GR, only five amino acid residues are conserved in GR that are different in MR. Engineering the mutations S106P and L111Q in ancGR1 switches the binding specificity in favor of cortisol. Thus, the double mutation has the predicted effect on the GR ancestor, though not on the modern MR.

The structure of the steroid-binding domain of the ancestor corticoid receptor (ancCR) is shown in **Figure 3.51**A; it is similar to that of the modern MR. The corresponding structure of the latest GR ancestor (ancGR2) is shown in Figure 3.51B; it is similar to that of the modern GR. Both structures are similar to that of the estrogen-binding domain (see Figure 3.47A), because these proteins are all homologous.

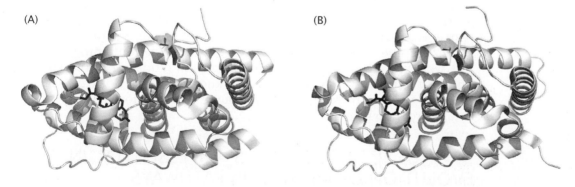

Figure 3.51 Steroid-binding domain of the corticoid receptors. (A) The ancestor (ancCR) of all corticoid receptors (MR and GR), which is MR-like (PDB 2Q1H). (B) The ancestor (ancGR2) of the bony vertebrate GR, which is GR-like (PDB 3GN8). Aldosterone is shown bound to ancCR (black) and a cortisol analog is shown bound to ancGR2.

Figure 3.52 Detailed structure of the binding site of the corticoid receptors with their ligands bound (black). (A) The ancestral corticoid receptor (ancCR) with aldosterone (PDB 2Q1H). (B) The latest ancestral glucocorticoid receptor (ancGR2) with a cortisol analog (PDB 3GN8). The main difference results from mutations S106P and L111Q. A hydrogen bond is made between the amide carbonyl of Gln-111 and the OH group at position 17 of cortisol.

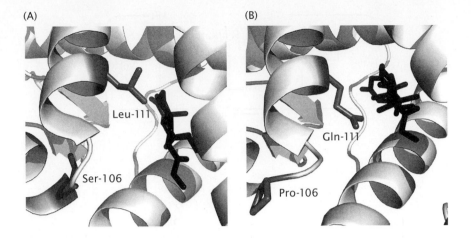

In ancGR2, the mutations S106P and L111Q reduce the equilibrium binding constant for aldosterone by a factor of $\sim 10^3$ compared to ancCR, but ancGR2 has a cortisol affinity similar to modern GR. The molecular mechanism of the change in affinity was understood by examining the effect of these mutations on the steroid binding site (**Figure 3.52**A). Pro-106 introduces a kink in the protein backbone, which reduces the interactions with all ligands, and therefore their binding affinity. The affinity of ancGR2 for cortisol is rescued by Gln-111 (see Figure 3.52B), through a hydrogen bond established between the amide carbonyl of the Gln-111 side chain and the OH group at position 17 of cortisol.

Why are nonfunctional proteins produced if the double mutation S106P and L111Q is introduced in modern MR, or if the reverse mutations are introduced in modern GR? The reason is that amino acid interactions limit or permit the manifestation of the effects of subsequent mutations (epistasis). In addition to the differences in residues 106 and 111, a few other replacements in modern GR are needed to abolish binding of aldosterone and DOC. These mutations, on their own, render the protein *less stable*. To accommodate them, a number of permissive mutations must take place first. Those mutations are by themselves neutral (or appear to be) because they have no significant effects on the protein function. But they are essential to permit future mutations that lower stability and alter binding specificity. The modern MR lacks those permissive mutations and therefore cannot be converted to a GR by the substitutions S106P and L111Q.

Conversely, restrictive mutations occurred in the evolutionary path to the modern GR that prevent the reversal mutations P106S and Q111L to restore the mineralocorticoid specificity. This is because those restrictive mutations produce unfavorable (steric) interactions or eliminate favorable ones that would occur if the helix containing residue 111 were placed in the position it occupies in modern MR and ancCR. This understanding was only possible knowing the historical order of the changes that took place in the course of evolution leading to the extant protein structures.

3.17 GENE DUPLICATIONS ARE ESSENTIAL IN THE EVOLUTION OF METABOLIC PATHWAYS

New enzymatic functions evolve from existing ones by sequence divergence following a gene duplication event. Horowitz (1945) proposed

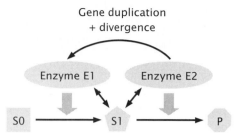

Figure 3.53 Hypothetical metabolic route, evolved by using enzyme E2 as a template for the new enzyme E1, which can catalyze the formation of substrate S1.

the hypothesis that metabolic pathways evolve backward. The last enzyme in the pathway would have appeared first. After a gene duplication, the preceding enzyme would have evolved, using the structure of the last enzyme in the pathway as a *template*. This concept arises from realizing that if two enzymes catalyze consecutive reactions in a metabolic route, the product of the first is the substrate of the next. That common metabolite binds to *both* enzymes.

Consider the metabolic pathway depicted in **Figure 3.53**. Substrate S1 must bind to both enzymes E1 and E2. Enzyme E1 could have evolved from a gene duplication of E2, by changing the active site in a way that still binds S1 but is capable also of catalyzing the reaction that converts S0 to S1. If those enzymes occurred in an organism that used to feed on substrate S1, and if because of an environmental change S1 were no longer available, those organisms in the population that could produce S1 from S0 (because they had enzyme E1) would have a decisive advantage, and would eventually dominate the population. In this way, the metabolic pathway leading to product P would grow backward by one step. Although the Horowitz hypothesis is not a general mechanism of evolution of metabolic pathways, elements of this idea are pertinent to the evolution of some metabolic routes, such as the tryptophan and histidine synthetic pathways, and, as we have just learned, in the evolution of steroid receptors.

A more plausible hypothesis is that specific enzymes evolved from generalist ones, which are more promiscuous and act on a number of substrates, albeit with lesser specificity and efficiency. Generalist enzymes probably dominated at one point, and were replaced by specific enzymes as the regulation of metabolic pathways evolved. Generalist enzymes are less prone to regulation, since they catalyze a variety of reactions. However, we would be wrong to think that the majority of extant proteins are specific. They are not. For example, in *E. coli* most reactions in the overall metabolism (~65%) are catalyzed by generalist enzymes, even though they account for only 37% of the enzymes in the bacterium. However, most of the metabolic flux in the cell goes through specialist enzymes. An analogy is the traffic between two cities. Most of the connections between two cities are through backroads, but most of the traffic between those two cities flows through the major highways. The fluxes through specialist enzymes are more sensitive to changes in environment and nutrients. Enzymes that are more sensitive to metabolic regulation are usually more specific. It seems that specialist enzymes evolved, following gene duplication events, which allow for the more precise regulation of metabolic pathways.

The shortcomings of Horowitz's hypothesis notwithstanding, some of the best examples of evolution of metabolic routes appear to have arisen on the basis of the reuse of previous functional enzyme *templates*. We will examine three cases. In the first, substrate *binding specificity* of the template is conserved as the new function evolves;

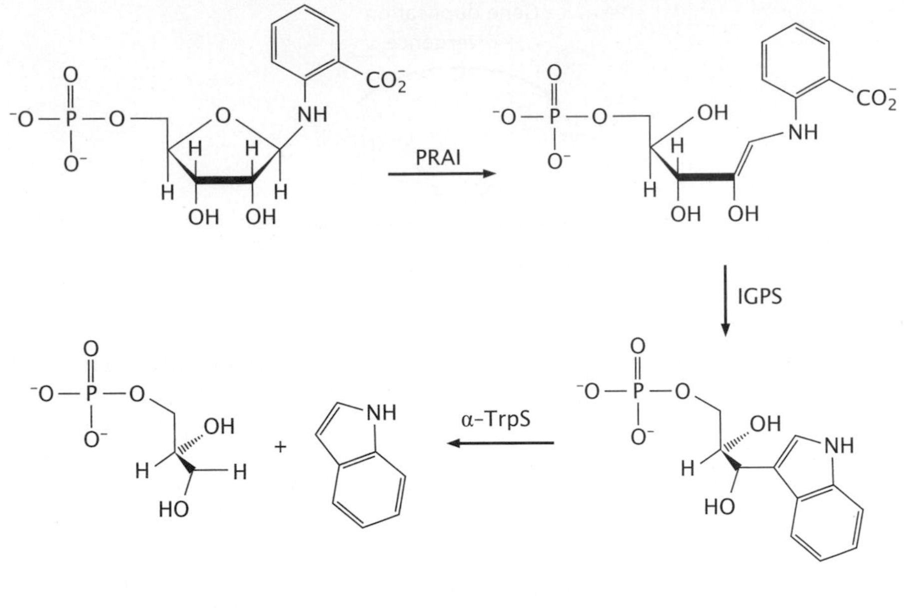

Figure 3.54 Reactions catalyzed by PRAI, IGPS, and α-TrpS. (Based on Wise EL & Rayment I [2004] *Acc Chem Res* 37:149–158.)

in the second, the chemical *reaction mechanism* is conserved; and in the third, the *architecture of the active site* is conserved.

Substrate Binding Specificity can be the Template for Enzyme Evolution

The substrate binding specificity is the template for the development of new functions in three enzymes that catalyze consecutive reactions in tryptophan biosynthesis: phosphoribosyl anthranilate isomerase (PRAI), indole glycerol phosphate synthase (IGPS), and the tryptophan synthase α subunit (α-TrpS). These are homologous enzymes that bind common substrates but catalyze different reactions (**Figure 3.54**).

PRAI, IGPS, and α-TrpS are classical TIM barrels, also called $(\beta/\alpha)_8$ barrels (**Figure 3.55**). The proteins are homologous despite the low sequence similarity (15–20%). The active sites of these enzymes are located in the same place in the structure, on one end of the central β-barrel. The proteins bind their common substrates in the same position and essentially in the same manner. PRAI and IGPS both bind 1-(*o*-carboxyphenylamino)-1-deoxyribulose-5-phosphate; IGPS and α-TrpS both bind indoleglycerol-3-phosphate (IGP) (see Figure 3.54). However, the mechanisms of the reactions catalyzed by these enzymes,

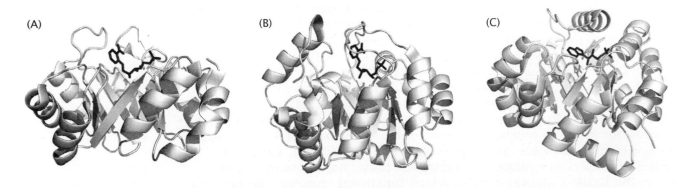

Figure 3.55 Structures of (A) phosphoribosyl anthranilate isomerase, PRAI (PDB 1LBM); (B) indole glycerol phosphate synthase, IGPS (PDB 1LBF); and (C) tryptophan synthase α subunit, α-TrpS (PDB 1QOQ). The bound substrates are shown in black.

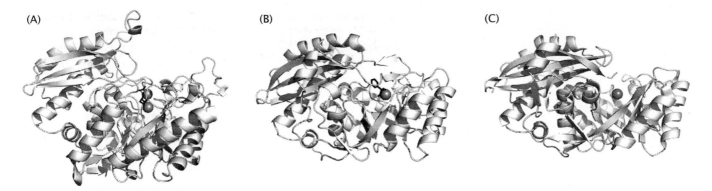

Figure 3.56 Structures of (A) enolase (PDB 1ONE), (B) mandelate racemase (MR, PDB 1MDR), and (C) muconate lactonizing enzyme (MLE, PDB 1MUC). Bound substrates are shown as stick models. Divalent cations (cofactors) are shown as spheres.

and the amino acid residues involved in the catalysis at the active sites, are completely different.

A Conserved Reaction Mechanism Can be the Template for Evolution

In our second case, the mechanism of the reaction catalyzed is the common feature. Homologous enzymes of the enolase family, enolase, mandelate racemase (MR), and muconate lactonizing enzyme (MLE), share a common mechanism but catalyze reactions on different substrates. These enzymes share only ~25% sequence identity, but are clearly homologous. Their three-dimensional structures consist of a TIM barrel domain and another α/β domain (**Figure 3.56**).

The common mechanism of the reaction and the specific reactions catalyzed by these three enzymes are shown in **Figure 3.57**.

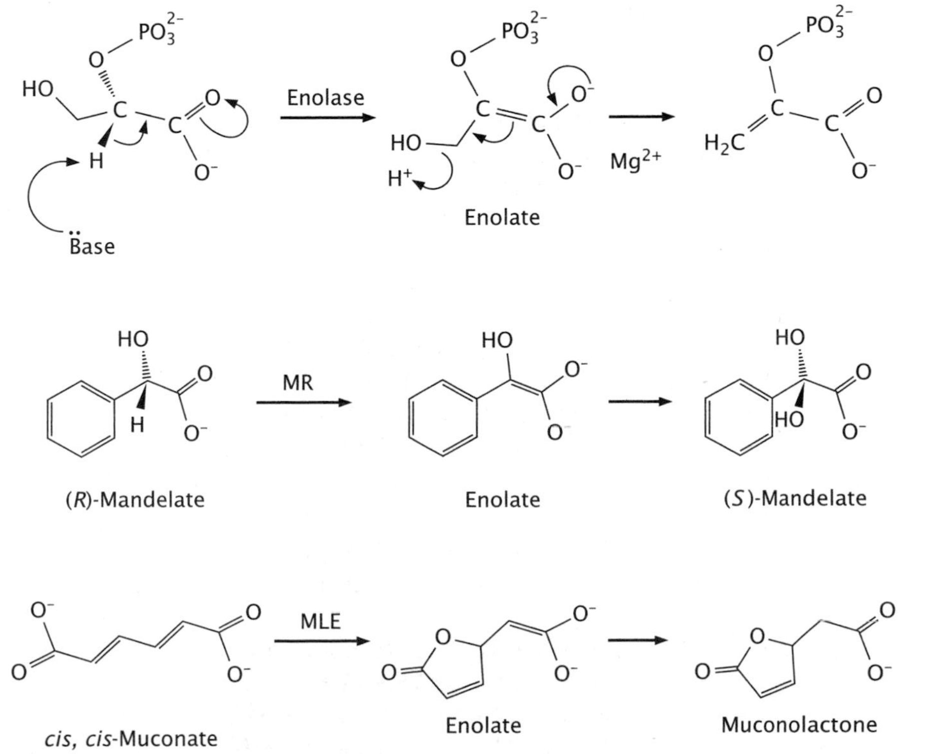

Figure 3.57 Mechanism of the enolase reaction and the reactions catalyzed by enzymes of the enolase family (enolase, ML, and MLE). (Based on Wise EL & Rayment I [2004] *Acc Chem Res* 37:149–158.)

The mechanism involves acid–base catalysis and an enolate intermediate, but the active-site residues are different in the three enzymes. A divalent cation, usually Mg^{2+}, stabilizes the negatively charged enolate and is essential in the mechanism of all enzymes of this family.

The Architecture of the Active Site Can be the Template for Evolution

In our third case, neither the substrate-binding site nor the reaction mechanism is conserved, but the overall architecture of the active site is conserved. It serves as the template, or scaffold, for the evolution of new functions. We consider the homologous enzymes orotidine monophosphate decarboxylase (OMPDC), 3-keto-L-gulonate-6-phosphase decarboxylase (KGPDC), D-ribulose-5-phosphate 3-epimerase (RPE), and 3-hexulose-6-phosphate synthase (HPS). Their structures are again TIM barrels (**Figure 3.58**).

These enzymes catalyze different reactions, on different substrates, and with different mechanisms. OMPDC and KGPDC share the architecture and even the identities of several residues in the active site. The same Asp-X-Lys-X-X-Asp motif occurs in the active site of both enzymes, but the conserved residues have different catalytic functions. HPS also has that same motif in its active site, and its mechanism shares some similarities with that of KGPDC, but not with OMPDC. RPE has a different mechanism and does not contain the same catalytic motif.

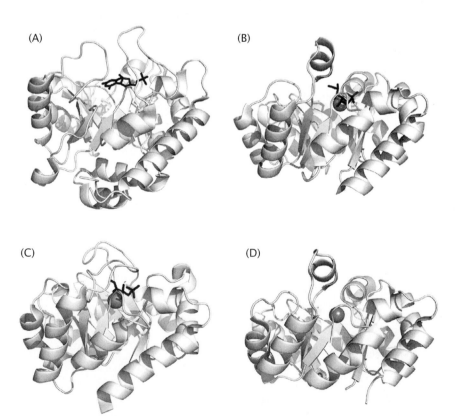

Figure 3.58 Structures of enzymes (A) orotidine monophosphate decarboxylase, OMPDC (PDB 1DQX); (B) 3-keto-L-gulonate-6-phosphase decarboxylase, KGPDC (PDB 1Q6O); (C) D-ribulose-5-phosphate 3-epimerase, RPE (PDB 3OVQ); and (D) 3-hexulose-6-phosphate synthase, HPS (PDB 3AJX). Bound metal ions are shown as spheres; substrates, in black.

3.18 SUMMARY

Evolution has shaped the structures of the proteins that exist today. Whereas many structural and aspects of protein function and stability can be understood without recourse to evolution, a deeper understanding is achieved by examining structure and function in an evolutionary perspective. In particular, the interplay of amino acid mutations as they historically occurred is necessary to understand the mutual dependence of the effect of amino acid residues on protein structure and function.

Proteins that diverge by evolution from a common ancestor are called homologous. Homologous proteins have similar structures. Their sequences diverge exponentially in time, by the accumulation of mutations. There are two kinds of homologous proteins: orthologs and paralogs. Orthologous proteins are products of the same gene in different organisms. They diverge because of speciation, as mutations independently accumulate in the organisms that carry those proteins. Paralogous proteins diverge following gene duplications. The duplicated gene can mutate without affecting the viability of the organism and evolves independently of the original gene. Often paralogous proteins evolve new functions. Protein homology can be established by sequence comparison. Sequences that are unrelated by evolution are as similar as random sequences. Sequence similarity in excess of what is expected by chance indicates homology.

Mutations in proteins arise from mutations in the DNA. However, it is the protein structure, not the genetic code, that determines which mutations are accepted and their frequency. Protein structures are surprisingly tolerant to amino acid substitutions, especially on their surface. The most conserved regions are typically the active sites of enzymes, the functional centers of other proteins that transport oxygen or electrons, and the areas of interaction with other molecules. Most mutations that occur in proteins are neutral to structure and function. Therefore, neutral mutations determine the bulk of the evolution of protein sequences. Because the rate of mutation of a protein is a consequence of its structure, the structure encodes a molecular clock characteristic of that protein. Molecular clocks are the most powerful tool to assign time in molecular evolution. However, important functional changes result from positive selection. Often, this is the result of adaptation to new environments. Those changes often occur independently in the same proteins in different organisms subject to similar environmental constraints, such as deep seas or high altitudes.

Common patterns occur in the evolution of protein receptors and enzymes, where an ancestral protein works as a template for the development of a new function by the accumulation of mutations. Thus, modern steroid receptors evolved backward, from an ancient estrogen receptor, as the ability to bind intermediates of the estrogen synthetic pathway developed in proteins that resulted from gene duplications. Similarly, new enzymatic functions have developed from existing templates, such as in the tryptophan and histidine synthetic pathways, by using a common substrate binding, a common reaction mechanism, or a common active-site architecture.

3.19 PROBLEMS

3.1. The TIM barrel fold of triose phosphate isomerase (A) in the figure below (PDB 1MO0) is similar to that of many other proteins with different functions, such as thiamine phosphate synthase (B) in the figure (PDB 1G4S). Are these proteins homologous?

(A)

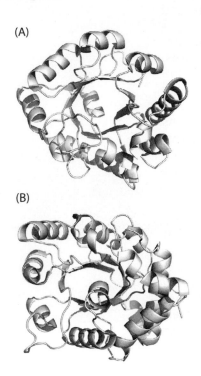

(B)

(a) Obtain the sequences of two proteins from the PDB web site (use the PDB numbers to locate the protein in the database and then download the sequence in FASTA format).
(b) Use BLAST to align the two sequences. Use the BLAST web site at NIH (http://blast.ncbi.nlm.nih.gov), choose Protein Blast (blastp), and check the box "Align two or more sequences." Use a BLOSUM scoring matrix (you will probably have to experiment with the version), but use the default gap penalties for the matrix you choose, at least as a start.
(c) Find the best alignment. What is the value of the bit score and the E-value (or expect)? Do you think these proteins are homologous? Discuss your results.

3.2. The figure below shows the structures of (A) α-lactalbumin (PDB 1ALC) and (B) lysozyme (PDB 1VDQ). The structures of these proteins are almost identical, but their functions are completely different. The function of lysozyme is the degradation of polysaccharides in bacterial walls; it protects the egg from bacterial infection. α-Lactalbumin regulates lactose synthesis by binding to galactosyl transferase. When complexed with α-lactalbumin, galactosyl transferase synthesizes lactose. The only apparent functional similarity between α-lactalbumin and lysozyme is that both proteins participate in reactions that involve saccharides. Are these two proteins homologous? To answer, follow the same steps as in the previous problem.

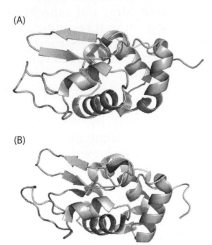

(A)

(B)

3.3. Below is a table of *percent differences* of five orthologous proteins from the *A family*, from five different organisms. Construct an evolutionary tree from these data. Is this protein evolutionary tree expected to match the organismal phylogenetic tree? Briefly explain.

	A2	A3	A4	A5
A1	33	30	5	56
A2		24	32	58
A3			30	60
A4				54

3.4. Below is a table of *percent identities* of five homologous proteins from the *C family*, from five different organisms. Construct an evolutionary tree from these data.

	C2	C3	C4	C5
C1	79	45	46	82
C2		44	42	93
C3			80	41
C4				44

3.5. The following is a table of *percent identities* of five homologous proteins from the *T family*, from five different organisms.

	T2	T3	T4	T5
T1	**42**	40	43	44
T2		**83**	57	**59**
T3			**70**	68
T4				98

The *correct* evolutionary tree for the T family is shown below.

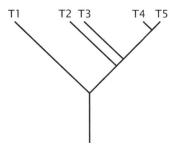

By comparing the table and tree, however, you will find that the table contains *one error*. Which one of the bold entries in the table is wrong? Briefly explain.

3.6. Read the article by Baum et al. (2005). Take the Tree-Thinking Quizzes I and II at (http://www.sciencemag.org/content/suppl/2005/11/07/310.5750.979.DC1).

3.7. Assess whether the pairs of proteins PRAI/IGPS and IGPS/α-TrpS of the tryptophan biosynthesis pathway are homologous using the procedure learned in this chapter.

Determine the bit-score and the E-value with BLAST. Determine also the percent identity using the Needleman–Wunsch algorithm (use the program EMBOSS Needle at the EMBO web site, http://www.ebi.ac.uk/Tools/psa/emboss_needle/). Discuss your results.

3.8. Now compare enzymes across families, using the same tests, to see if they are homologous. The catalytic domains of those enzymes have similar folds $((\beta/\alpha)_8$ barrel), but are they homologous? Compare
(a) α-TrpS and enolase;
(b) α-TrpS and OMPDC.

3.20 FURTHER READING

General

Creighton TE (1993) Proteins. Structure and Molecular Properties, 2nd ed. Freeman.

Dietrich MR (1998) Paradox and persuasion: Negotiating the place of molecular evolution with evolutionary biology. *J Hist Biol* 31:85–111.

Dobzhansky T (1973) Nothing in biology makes sense except in light of evolution. *Am Biol Teach* 35:125–129.

Doolittle RF (1987) Of URFS and ORFS. A Primer on How to Analyze Derived Amino Acid Sequences. University Science Books.

Kuriyan J, Konforti B & Wemmer D (2013) The Molecules of Life: Physical and Chemical Principles. Garland Science.

Orengo CA & Thornton JM (2005) Protein families and their evolution—A structural perspective. *Annu Rev Biochem* 74:867–900.

DNA Structure

Watson JD & Crick FHC (1953) A structure for deoxyribose nucleic acid. *Nature* 171:737–738.

Bertrand K, Korn L, Lee F et al (1975) New features of the regulation of the tryptophan operon. *Science* 189:22–26.

Rogozin IB, Carmel L, Csuros M & Koonin EV (2012) Origin and evolution of spliceosomal introns. *Biol Direct* 7:11.

How Protein Sequences Change in the Course of Evolution

Dayhoff MO (1978) Atlas of Protein Sequence and Structure, vol. 5, suppl. 3. National Biomedical Research Foundation.

Dickerson RE (1971) The structure of cytochrome c and the rates of molecular evolution. *J Mol Evol* 1:26–45.

Zuckerkandl E & Pauling L (1965) Evolutionary divergence and convergence of proteins. In Evolving Genes and Proteins (Bryson V & Vogel HJ eds), pp 97–166. Academic Press.

Protein Sequence Comparison

Karlin S & Altschul SF (1990) Methods for assessing the statistical significance of molecular sequence features by using general scoring schemes. *Proc Natl Acad Sci USA* 87:2264–2268.

Altschul SF, Madden TL, Schäffer AA et al (1997) Gapped BLAST and PSI-BLAST: A new generation of protein database search programs. *Nucl Acids Res* 25:3389–3402.

Doolittle RF, Feng DF, Johnson MS & McClure MA (1986) Relationship of human protein sequences to those of other organisms. *Cold Spring Harb Symp Quant Biol* 51:447–455.

Henikoff S & Henikoff JG (2000) Amino acid substitution matrices. *Adv Prot Chem* 54:73–97.

Lipman DJ & Pearson WR (1985) Rapid and sensitive protein similarity searches. *Science* 227:1435–1441.

Pearson WR (2001) Protein Sequence Comparison and Protein Evolution. University of Virginia.

Pearson WR & Sierk ML (2005) The limits of protein sequence comparison? *Curr Opin Struct Biol* 15:254–260.

Pearson WR (2013) An introduction to sequence similarity ("homology") searching. *Curr Protoc Bioinform* 42:3.1.1–3.1.8.

Sievers F, Wilm A, Dineen D et al (2011) Fast, scalable generation of high-quality protein multiple sequence alignments using Clustal Omega. *Mol Sys Biol* 7:539.

Evolutionary Trees and Reconstruction of Ancestral Sequences

Doolittle WF & Brown JR (1994) Tempo, mode, the progenote, and the universal root. *Proc Natl Acad Sci USA* 91:6721–6728.

Baum DA, DeWitt Smith SD & Donovan SS (2005) The tree-thinking challenge. *Science* 310:979–980.

Yang Z, Kumar S & Nei M (1995) A new method of inference of ancestral nucleotide and amino acid sequences. *Genetics* 141:1641–1650.

Zhang J & Nei M (1997) Accuracies of ancestral amino acid sequences inferred by the parsimony, likelihood, and distance methods. *J Mol Evol* 44:S139–S146.

Molecular Clocks and the Neutral Theory of Evolution

Kimura M (1968) Evolutionary rate at the molecular level. *Nature* 217:624–626.

King JL & Jukes TH (1969) Non-Darwinian evolution. *Science* 164:788–798.

Kumar S (2005) Molecular clocks: Four decades of evolution. *Nature Rev Genet* 6:654–662.

Nei M (2005) Selectionism and neutralism in molecular evolution. *Mol Biol Evol* 22:2318–2342.

Nei M, Suzuki Y & Nozawa M (2010) The neutral theory of molecular evolution in the genomic era. *Annu Rev Genom Human Genet* 11:265–289.

Takahata N (2007) Molecular clock: An anti-neo-Darwinian legacy. *Genetics* 176:1–6.

Molecular Adaptation

Creighton TE & Darby NJ (1989) Functional evolutionary divergence of proteolytic enzymes and their inhibitors. *Trends Biochem Sci* 14:319–324.

Domingues VS, Poh Y-P, Peterson BK et al (2012) Evidence of adaptation from ancestral variation in young populations of beach mice. *Evolution* 66:3209–3223.

Perutz MF (1983) Species adaptation in a protein molecule. *Mol Biol Evol* 1:1–28.

Storz JF & Moriyama H (2008) Mechanisms of hemoglobin adaptation to high altitude hypoxia. *High Alt Med Biol* 9:148–157.

Reconstruction of Ancestral Proteins

Chang BS & Donoghue MJ (2000) Recreating ancestral proteins. *TREE* 15:109–114.

Chang BS, Johnsson K, Kazmi MA et al (2002) Recreating a functional ancestral archosaur visual pigment. *Mol Biol Evol* 19:1483–1489.

Dean AM & Thornton JW (2007) Mechanistic approaches to the study of evolution: The functional synthesis. *Nature Rev Genet* 8:675–687.

Harms MJ & Thornton JW (2010) Analyzing protein structure and function using ancestral gene reconstruction. *Curr Opin Struct Biol* 20:360–366.

Harms MJ & Thornton JW (2013) Evolutionary biochemistry: Revealing the historical and physical causes of protein properties. *Nature Rev Genet* 14:559–571.

Ortlund EA, Bridgham JT, Redinbo MR & Thornton JW (2007) Crystal structure of an ancient protein: Evolution by conformational epistasis. *Science* 317:1544–1548.

Pauling L & Zuckerkandl E (1963) Chemical paleogenetics. Molecular "resaturation studies" of extinct forms of life. *Acta Chem Scand* 17:S9–S16.

Shi Y & Yokoyama S (2003) Molecular analysis of the evolutionary significance of ultraviolet vision in vertebrates. *Proc Natl Acad Sci USA* 100:8308–8313.

Thornton JW (2001) Evolution of vertebrate steroid receptors from an ancestral estrogen receptor by ligand exploitation and serial genome expansions. *Proc Natl Acad Sci USA* 98:5671–5676.

Thornton JW (2004) Resurrecting ancient genes: Experimental analysis of extinct molecules. *Nature Rev Genet* 5:366–375.

Yokoyama R & Yokoyama S (1990) Convergent evolution of the red- and green-like visual pigment genes in fish, Astyanax fasciatus, and human. *Proc Natl Acad Sci USA* 87:9315–9318.

Yokoyama S (2000) Molecular evolution of vertebrate visual pigments. *Prog Ret Eye Res* 19:385–419.

Yokoyama S, Tada T, Zhang H & Britt L (2008) Elucidation of phenotypic adaptations: Molecular analyses of dim-light vision proteins in vertebrates. *Proc Natl Acad Sci USA* 105:13480–13485.

Yokoyama S, Yang H & Starmer WT (2008) Molecular basis of spectral tuning in the red-and green-sensitive (M/LWS) pigments in vertebrates. *Genetics* 179:2037–2043.

Evolution of Metabolic Pathways

Gertl JA & Babbitt PC (2001) Divergent evolution of enzymatic function: Mechanistically diverse superfamilies and functionally distinct suprafamilies. *Annu Rev Biochem* 70:209–246.

Horowitz NH (1945) On the evolution of biochemical syntheses. *Proc Natl Acad Sci USA* 31:153–157.

Nam H, Lewis NE, Lerman JA et al (2012) Network context and selection in the evolution to enzyme specificity. *Science* 337:1101–1104.

Wise EL & Rayment I (2004) Understanding the importance of protein structure in nature's routes for divergent evolution in TIM barrel enzymes. *Acc Chem Res* 37:149–158.

Web Sites for Evolution

EMBL-EBI. http://www.ebi.ac.uk

UniProt. http://www.uniprot.org

ExPASy. http://www.expasy.org

Protein Stability

4.1 WHY DO PROTEINS FOLD?

Proteins fold in water into compact globular structures. These structures are not amorphous globules. Locally, the polypeptide chain adopts certain regular conformations, α-helices and β-sheets, which we call secondary structure. These secondary structure elements are organized according to certain folding motifs, forming the three-dimensional or tertiary structure of the protein. Why do regular secondary structures form? What interactions keep the protein folded? Hydrogen bonds between the backbone peptide groups (–CONH–) stabilize helices and sheets. However, if the protein were to unfold, those groups could just as well establish hydrogen bonds with water molecules. Are hydrogen bonds within the protein more favorable? Perhaps. But most hydrophilic amino acid sequences do not form secondary structures in water. They remain unfolded, establishing hydrogen bonds with water. To understand why a protein folds we need to understand the interactions that stabilize its structure.

There are several types of interactions that contribute to protein stability: electrostatic interactions (charge–charge, charge–dipole, and dipole–dipole), which occur between charged amino acid residues (Lys, Arg, Asp, Glu, His), between charges and dipoles, and between two dipoles (including protein and water dipoles); van der Waals interactions (London dispersion forces and van der Waals repulsions), which are especially important in the densely packed protein interior; and, most important, hydrogen bonds, which occur between amide C=O and N—H peptide groups, side chains, or with water. All these types of interactions occur in most physical–chemical systems. However, other interactions are important for protein stability that are not so common in simpler chemical systems: disulfide bonds, which are covalent bonds between two cysteine residues, and most important, the *hydrophobic effect*, which represents the tendency of nonpolar groups to be excluded from contact with water. In almost all of these interactions, the role of water is fundamental. Finally, the *conformational entropy* of the polypeptide chain is the major driving force for protein unfolding.

The hydrophobic effect probably provides the major contribution to protein stability. Yet, it contributes almost nothing to the formation of secondary structure. Most polypeptides have little propensity to form α-helices or β-sheets in water. The main reason secondary structure forms in globular, water-soluble proteins is that there is *no water* in the hydrophobic interior of the protein. There, the protein must

We (the undivided divinity that operates within us) have dreamed the world. We have dreamed it resistant, mysterious, visible, ubiquitous in space and firm in time; but we have allowed in its architecture tenuous and eternal interstices of unreason to know it is false.

Jorge Luis Borges

satisfy its hydrogen bonding needs within itself. Adopting the conformations of α-helices and β-sheets guarantees the largest possible number of hydrogen bonds between the amide N—H and C=O groups of the polypeptide backbone. Thus, secondary structure forms, to a large extent, as consequence of the *confinement* of the polypeptide chain to a space *without water*.

Protein denaturation and the interactions that are involved in protein stability are two intertwined subjects. We will learn that there is not a single type of interaction that, alone, controls the equilibrium between folded and unfolded protein. Rather, the position of the equilibrium is determined by the difference between many sets of interactions, each of which is small in itself. Among those, the hydrophobic effect is dominant. But it does not determine the secondary structure. Thus, the beautiful three-dimensional structures of proteins are the result of small and hard-to-predict contributions from many diverse microscopic interactions. We will spend most of this chapter studying those various interactions and their contributions to protein stability. But first, we will examine the thermodynamics of protein denaturation.

4.2 PROTEIN DENATURATION IS DESCRIBED BY EQUILIBRIUM THERMODYNAMICS

The problem of protein stability has interested biochemists for a long time. In his classical article in 1959, Kauzmann writes, "The changes that take place in protein molecules during denaturation constitute one of the most interesting and complex classes of reactions that can be found either in nature or in the laboratory. These reactions are important because of the information they can provide about the more intimate details of protein structure and function. They are also significant because they challenge the chemist with a difficult area for the application of chemical principles."

Protein Denaturation is Reversible

We begin by briefly summarizing what we learned in Chapter 1 on the thermodynamics of protein folding. For most small proteins, there are only two well-populated states, native (N) and denatured (D). The *thermodynamic hypothesis* of protein folding, proposed by Anfinsen, states that, under physiological conditions, the native structure corresponds to the Gibbs energy minimum of the system. The minimum is defined with respect to the extent of the unfolding reaction. An equilibrium exists between the native and denatured states (**Figure 4.1**), which is shifted to the native (left), under physiological conditions (and

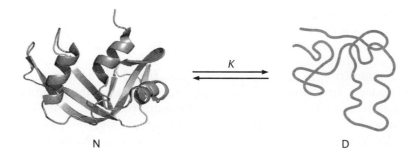

Figure 4.1 The equilibrium between the folded (N) and the unfolded state (D) of a protein is governed by the equilibrium constant K. (Left structure: PDB 7RSA.)

N   D

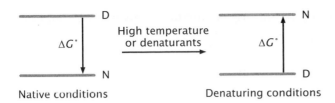

Figure 4.2 Relative value of the Gibbs energies between native (N) and denatured (D) states of the protein, under native (left) or denaturing (right) conditions.

about room temperature). The existence of this *equilibrium* means that protein denaturation is *reversible*.

It is good to develop the habit of visualizing this type of equilibrium using a simple sketch like the one shown in **Figure 4.2**. On the left is the change in Gibbs energy under native, physiological conditions: the native state is the most favorable. Upon increasing the temperature beyond the melting temperature (T_m) of the protein, or by addition of denaturants, such as urea or guanidinium chloride, the denatured state becomes the most favorable: the order of the Gibbs energies switches.

The equilibrium constant for denaturation is given by

$$K = \frac{[D]}{[N]},\qquad(4.1)$$

and the standard Gibbs energy change upon denaturation is

$$\Delta G^o = -RT \ln K.\qquad(4.2)$$

Under physiological conditions, the equilibrium $N \rightleftharpoons D$ is shifted toward the native state, and $\Delta G^o > 0$ for denaturation (see Figure 4.2, left). But at temperatures higher than T_m or under high concentrations of urea, the equilibrium lies toward the denatured state and $\Delta G^o < 0$ for denaturation (see Figure 4.2, right). In the case of ribonuclease A at pH 5, the melting temperature is $T_m \approx 61°C$, the enthalpy change upon unfolding is $\Delta H^o \approx 110$ kcal/mol, and the entropy change is $\Delta S^o \approx 0.33$ kcal/mol/K. The Gibbs energy change is

$$\Delta G^o = \Delta H^o - T\Delta S^o.\qquad(4.3)$$

At room temperature $\Delta G^o \approx 10$ kcal/mol (12 kcal/mol assuming ΔH^o and ΔS^o are independent of temperature, and 9.4 kcal/mol if their temperature dependence is taken into account). This is a small difference, equivalent to the energy of only a couple of hydrogen bonds. It means that the protein is only *marginally stable*.

Equation 4.3 shows that ΔG^o is the result of a *balance* between the enthalpy and the entropy changes, and that this balance is controlled by the temperature. As T increases, the *weight* of the entropy term, $T\Delta S^o$, increases. When $T = T_m$, the two terms exactly balance each other: $\Delta G^o = 0$ and the equilibrium constant $K = 1$. Above T_m the entropy term is larger in absolute value, and most protein molecules in the ensemble are unfolded. However, even under denaturing conditions, the equilibrium constant is finite. This means that there are some folded protein molecules in the ensemble even at high temperatures. Their three-dimensional structure is that of the native protein.

In Chapter 1 we studied this equilibrium using the partition function. We assigned a *statistical weight* to each state, native and denatured. Those statistical weights were relative probabilities: *relative* to a reference state. The partition function is the sum of those relative probabilities. The native state was chosen as the reference and assigned a statistical weight of 1; the denatured state was assigned a statistical weight of K, the equilibrium constant, because we must have

$$\frac{[D]}{[N]} = \frac{p_D}{p_N} = K, \tag{4.4}$$

where p_N and p_D are the probabilities of states N and D. Thus, the partition function is simply

$$Q = 1 + K. \tag{4.5}$$

Q is also the normalizing factor to convert statistical weights to absolute probabilities (p_N and p_D), which must vary between 0 and 1. The probability of each state is the statistical weight divided by Q. Those probabilities are identical to the fractions f_N and f_D of the native and denatured state in the protein population,

$$p_N = f_N = \frac{1}{1 + K} \tag{4.6}$$

and

$$p_D = f_D = \frac{K}{1 + K}. \tag{4.7}$$

The Enthalpy Change Determines the Variation of the Equilibrium Constant with Temperature

We can learn a lot about a reaction by studying how it varies with temperature. Heat denatures proteins because the temperature changes the balance between the enthalpy and the entropy contributions to the Gibbs energy of protein unfolding. As the temperature increases, the relative weight of the enthalpy decreases and the weight of the entropy increases. But this is not all. In this section we will collect a number of important relations that describe the effect of temperature on thermodynamic functions. These relations are necessary to understand protein stability, and we will make frequent use of them from now on.

We begin with Equation 4.3, which defines the Gibbs energy change,

$$\Delta G^o = \Delta H^o - T \Delta S^o.$$

If we differentiate the Gibbs energy with respect to the temperature, we obtain

$$\frac{d\Delta G^o}{dT} = -\Delta S^o. \tag{4.8}$$

Alternatively, if we divide ΔG^o by T in Equation 4.3,

$$\frac{\Delta G^o}{T} = \frac{\Delta H^o}{T} - \Delta S^o, \tag{4.9}$$

and then differentiate with respect to temperature, recalling the calculus rule

$$\frac{d}{dT}\left(\frac{1}{T}\right) = -\frac{1}{T^2}, \tag{4.10}$$

we obtain the important Gibbs–Helmholtz equation

$$\frac{d(\Delta G^o/T)}{dT} = -\frac{\Delta H^o}{T^2}. \tag{4.11}$$

Another very important relation in thermodynamics is the van't Hoff equation. It expresses the temperature dependence of an equilibrium constant. The equilibrium constant is related to ΔG^o by Equation 4.2, which we can rewrite as

$$R \ln K = -\Delta G^o / T. \qquad (4.12)$$

If we now substitute this relation into the Gibbs-Helmholtz equation (shown in Equation 4.11), we obtain the van't Hoff equation,

$$\frac{d \ln K}{dT} = \frac{\Delta H^o}{RT^2}. \qquad (4.13)$$

Equation 4.13 shows that what controls the variation of the equilibrium constant with temperature is the enthalpy difference, ΔH^o, between the denatured and the native state of the protein (**Figure 4.3**),

$$\Delta H^o = H^D - H^N. \qquad (4.14)$$

In Equation 4.13, since RT^2 is always positive, the sign of ΔH^o determines whether K increases or decreases with temperature: if $\Delta H^o > 0$, K increases with T; if $\Delta H^o < 0$, K decreases as T increases. For the denaturation reaction $N \rightleftharpoons D$, ΔH^o is positive. This means that the enthalpy of the native state is lower than that of the denatured state (see Figure 4.3). The enthalpy (energy) favors the native state. As the temperature increases, however, the relative weight of the favorable enthalpy decreases and the advantage of the folded state relative to the unfolded state decreases.

Recall that ΔH^o is the *heat exchanged* with the surroundings at constant pressure. For protein unfolding, $\Delta H^o > 0$. The reaction is endothermic: the protein *absorbs heat* when it unfolds. Le Châtelier's principle tells us that the reaction is more favorable at higher temperatures, when molecules in the solution (solvent and solutes) have on average more kinetic energy to donate. Therefore, K increases with T (shown in Equation 4.13). The converse is true in cases when $\Delta H^o < 0$.

Thermal Denaturation of Proteins can be Studied by Calorimetry

Ultimately, it is the heat capacity C_p that controls the temperature dependence of the other thermodynamic functions at constant pressure. In general, the temperature dependence of the enthalpy is given by

$$\frac{dH}{dT} = C_p, \qquad (4.15)$$

and the temperature dependence of the entropy is given by

$$\frac{dS}{dT} = \frac{C_p}{T}. \qquad (4.16)$$

In protein thermal denaturation, or unfolding, the heat capacity itself varies with temperature. To emphasize this dependence, we refer to it as the *heat capacity function* (of temperature) and write it as $C_p(T)$. The heat capacity function can be measured experimentally by differential scanning calorimetry (DSC). This is the most informative method to study thermal unfolding of proteins. In a DSC experiment, the temperature is *scanned* (increased or decreased) and the heat absorbed (or released) by the sample is measured. The heat absorbed per degree is

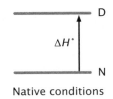

Figure 4.3 The enthalpy always increases in a transition from the native (N) to the denatured (D) state of a protein.

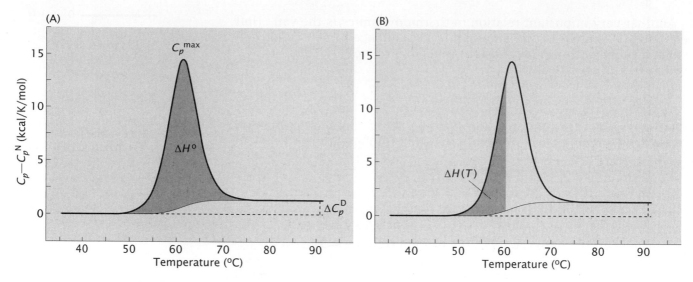

Figure 4.4 The heat capacity of ribonuclease A in a DSC experiment. (A) The shaded area under the curve is the total enthalpy change of denaturation (ΔH°). The heat capacity of the native state, C_p^{N} (horizontal dashed line at zero), is subtracted from the curve. The DSC curve was calculated at pH 5, with $\Delta H^{\circ} = 111$ kcal/mol, $T_m = 61.3^{\circ}$C, and $\Delta C_p^{D} = 1.3$ kcal/K/mol. (B) The heat $\Delta H(T)$ absorbed up to a temperature T ($T = 60^{\circ}$C in the figure) is the shaded area under the curve until that temperature. Data from Privalov PL [1997] *J Chem Thermodynam* 29:447–474, and Catanzano F, Graziano G, Cafaro V et al. [1997] *Biochemistry* 36:14403–14408.

the heat capacity. The thick solid line in **Figure 4.4**A is a DSC curve for ribonuclease A.

Let us take a moment to understand the DSC curve. Essentially, it consists of a peak that sits on a *baseline* represented by the S-like curve in Figure 4.4. We call this baseline C_p^{base}. At temperatures to the left side of the peak, essentially all proteins are folded; to the right side of the peak, essentially all proteins are unfolded; in the peak region—the *transition region*—both folded and unfolded proteins exist in solution. Outside the transition region, little heat is absorbed. When the protein unfolds, it absorbs heat, giving rise to the peak in the DSC curve. The heat absorbed is the enthalpy change ΔH°, which is the difference between the enthalpy of the unfolded and the folded state of the protein (see Figure 4.3). ΔH° is the *shaded area* under the DSC curve in Figure 4.4A. The maximum value of the heat capacity (C_p^{max}) is reached at T_m. In addition, there is a difference between the heat capacity of the unfolded and the folded states of the protein, denoted by ΔC_p^{D} (see Figure 4.4A). The denatured protein has a higher intrinsic heat capacity. This difference is why the baseline increases about the transition region.

We measure the difference between the heat capacity function $C_p(T)$ and the intrinsic heat capacity of the native state $C_p^{N}(T)$, at each temperature,

$$C_p(T) - C_p^{N}(T). \qquad (4.17)$$

In Figure 4.4, we set $C_p^{N}(T) = 0$ (dashed line) because the native state is our reference. (However, in reality the absolute C_p^{N} is not zero.) At high temperatures, after the transition, $C_p(T)$ follows the intrinsic heat capacity of the denatured state $C_p^{D}(T)$. C_p^{D} is larger than C_p^{N}, which is extrapolated to high temperatures (dashed line). In an actual experiment, $C_p^{N}(T)$ and $C_p^{D}(T)$ both increase very slightly, but in parallel, with temperature; they are shown flat in Figure 4.4 for the sake of simplicity.

Since we set $C_p^N(T) = 0$ for the reference, we write simply $C_p(T)$ instead of $C_p(T) - C_p^N(T)$ in much of this discussion.

To gain a deeper understanding of the relation between the thermodynamic parameters (ΔH^o, C_p, T_m) and protein denaturation, we will calculate the DSC curve of Figure 4.4. We concentrate here on the most important aspects; details are given in Appendix A. The DSC curve can easily be calculated under the *two-state approximation*, which assumes that only the native and denatured states exist. This approximation is essentially valid for most small proteins. The heat capacity function $C_p(T)$ consists of two parts: the peak, which we call the excess heat capacity, $\Delta C_p(T)$, and the baseline, $C_p^{base}(T)$,

$$C_p(T) = C_p^{base}(T) + \Delta C_p(T). \tag{4.18}$$

We now calculate these two parts separately. Let us begin with the baseline $C_p^{base}(T)$, the thin S-like curve in Figure 4.4. This curve is the heat capacity of the proteins, folded and unfolded, that exist in solution as a function of temperature. C_p^{base} is simply an *average* of the intrinsic heat capacities of the folded and unfolded states,

$$C_p^{base} = f_N C_p^N + f_D C_p^D, \tag{4.19}$$

weighted by their respective fractions, f_N and f_D, which can be obtained from the partition function (shown in Equations 4.6 and 4.7). There is a "bump" in the baseline about T_m because folded proteins are converted to unfolded proteins and $C_p^D > C_p^N$. Since $f_N = 1 - f_D$ and we set $C_p^N(T) = 0$, we can simplify Equation 4.19 to

$$C_p^{base}(T) = f_D(T)\Delta C_p^D, \tag{4.20}$$

where ΔC_p^D is the heat capacity change upon unfolding, the difference between heat capacities of the denatured and native states,

$$\Delta C_p^D = C_p^D - C_p^N. \tag{4.21}$$

Note that $C_p^{base}(T)$ changes with temperature because $f_D(T)$ changes with temperature (see Appendix A). ΔC_p^D is constant to a good approximation.

Now, let us calculate $\Delta C_p(T)$, the part of the heat capacity function $C_p(T)$ that gives rise to the peak in Figure 4.4. The peak is due to the heat absorbed when the protein unfolds. ΔH^o is the *total* area under the $\Delta C_p(T)$ curve (shaded in Figure 4.4A). $\Delta H(T)$ is the area under the $\Delta C_p(T)$ curve *up to a certain temperature* T (shaded in Figure 4.4B; $T = 60^oC$ in this example). $\Delta H(T)$ increases as we increase the temperature, until it coincides with ΔH^o at the end of the scan. At intermediate temperatures, $\Delta H(T)$ is a *fraction* of ΔH^o; it is the fraction $f_D(T)$ of denatured proteins at temperature T,

$$\Delta H(T) = f_D(T)\Delta H^o. \tag{4.22}$$

Thus, we see that $\Delta H(T)$ also changes with temperature because $f_D(T)$ changes with temperature.

Mathematically, the area $\Delta H(T)$ under the heat capacity curve is the integral of $C_p(T)$. Conversely, the derivative of $\Delta H(T)$ is $C_p(T)$. Thus, we can write the version of Equation 4.15 for protein unfolding as

$$\Delta C_p(T) = \frac{d\Delta H(T)}{dT}. \tag{4.23}$$

If we now substitute the expression for $\Delta H(T)$ from Equation 4.22 into Equation 4.23 and calculate the derivative (see Appendix A), we obtain

$$\Delta C_p(T) = \frac{(\Delta H^\circ)^2}{RT^2} \frac{K}{(1+K)^2}. \tag{4.24}$$

The equilibrium constant for denaturation, K, itself depends on temperature. If we invert Equation 4.3, we obtain

$$K = e^{-\Delta G^\circ / RT}. \tag{4.25}$$

At the melting temperature (T_m), $\Delta G^\circ = 0$. Therefore, from Equation 4.3, $\Delta S^\circ = \Delta H^\circ / T_m$. Substitution of this result in Equation 4.25 then yields

$$K = \exp\left(-\frac{\Delta H^\circ}{R}\left(\frac{1}{T} - \frac{1}{T_m}\right)\right). \tag{4.26}$$

Finally, we substitute Equation 4.26 for K in Equation 4.24, to obtain $\Delta C_p(T)$. Add the baseline, $C_p^{\text{base}}(T)$, and we obtain the complete heat capacity curve $C_p(T)$ shown in Figure 4.4.

For ribonuclease A (124 amino acid residues, MW$\approx$ 13,700), $\Delta C_p^{\text{max}} \approx$ 14 kcal/mol/K (pH 5) and $\Delta C_p^{\text{D}} \approx 1.3$ kcal/mol/K. Other small proteins have similar values, which increase in proportion to their molecular weights. For example, myoglobin (153 residues, MW $\approx$ 17,900), has $\Delta C_p^{\text{D}} \approx 2.7$ kcal/mol/K; cytochrome c (104 residues, MW $\approx$ 12,400), $\Delta C_p^{\text{D}} \approx 1.7$ kcal/mol/K; α-chymotrypsin (241 residues, MW $\approx$ 25,200), $\Delta C_p^{\text{D}} \approx 3.0$ kcal/mol/K; and lysozyme (129 residues, MW $\approx$ 14,300), $\Delta C_p^{\text{D}} \approx 1.4$ kcal/mol/K.

The Heat Capacity Measures the Width of the Energy Distribution

Why does the heat capacity reach a maximum $(\Delta C_p^{\text{max}})$ about T_m? At temperatures much below or much above T_m, essentially only one state exists, either N or D; the heat capacity observed is that of the predominant state. We call this the "within-states" heat capacity. Its value in the denatured state is larger than in the native state by ΔC_p^{D}. Close to T_m, however, appreciable amounts of both native and denatured proteins exist. Now, if heat is provided to the system, proteins absorb that heat and denature; but the temperature does not increase. Conversely, when proteins fold, they release heat (ΔH°), preventing the temperature from decreasing. Thus, about the T_m the protein solution has the largest capacity *to resist* a change in temperature, by absorbing or releasing heat as they unfold or refold. Because it arises from shuffling heat back and forth *between* the two states, we call this the "between-states" heat capacity. At T_m the fractions of folded and unfolded proteins are equal (one-half each): neither is small. Thus, the system has always a *buffer* of folded and unfolded proteins to absorb or release heat. This is the same reason why pH buffers are strongest close to the pK: that is when large fractions of both the acidic and the basic form of the ionizable group exist in solution.

Mathematically, the heat capacity is maximal at T_m because the fraction $K/(1+K)^2$ in Equation 4.24 is maximal at T_m. This is easier to see if we express the equilibrium constant of denaturation as

$$K = \frac{p_D}{p_N} = \frac{f_D}{1 - f_D} \tag{4.27}$$

to obtain

$$\frac{K}{(1+K)^2} = f_D(1 - f_D), \qquad (4.28)$$

which has its maximum value (1/4) when $f_D = 1/2$ (at T_m, $p_N = p_D$ or $f_D = 1/2$). Note that $f_D(1 - f_D)$ becomes zero when either the proteins are all folded ($f_D = 0$) or all unfolded ($f_D = 1$).

To interpret the excess heat capacity, we have to consider the enthalpy (energy) distributions of the protein (**Figure 4.5**). The heat capacity at constant pressure (volume) is a measure of the *width of the distribution* of enthalpy (energy). The within-states heat capacity corresponds to changes *within* the enthalpy distribution of each state.

In Figure 4.5, the peak on the left represents the probability distribution of the enthalpy of the native state (centered at zero enthalpy because we have chosen the native state as the reference). This distribution is *narrow* because the native conformation is well defined. The conformations belonging to the native state vary only negligibly from the mean, and the same is true of their enthalpies. In general, the *variance* of the distribution of enthalpy $(\sigma_H)^2$, where σ_H is the *standard deviation* of the distribution, is proportional to the heat capacity,

$$C_p = \frac{(\sigma_H)^2}{RT^2}. \qquad (4.29)$$

In the native state, C_p^N is small because σ_H^N is small. Thus, the width of the distribution is small (horizontal line in the left peak in Figure 4.5). The peak on the right, centered at higher enthalpy (ΔH^o relative to the native state), represents the probability distribution of the enthalpy in the denatured state. The distribution is broad because there are many conformations that differ significantly in enthalpy. In the denatured state, C_p^D is larger because σ_H^D is larger. Indeed, if σ_H^N is small, we can write

$$\Delta C_p^D = C_p^D - C_p^N \approx C_p^D. \qquad (4.30)$$

Thus, the width of the enthalpy distribution in the denatured state is essentially a measure of ΔC_p^D. Note, however, that ΔC_p^D is the change of heat capacity of the system, not just that of the protein. The folded and unfolded conformations interact differently with water, and the configurations of the water molecules around the protein contribute greatly to ΔC_p^D.

In the transition region, the probabilities of the native and denatured states of the protein are *both* appreciable. This means that the protein

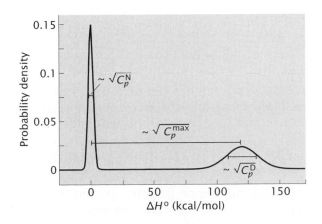

Figure 4.5 Enthalpy distributions of the native (left peak) and denatured (right peak) states of the protein. The reference is the native state; therefore H^N is set to zero. The x-axis shows the enthalpy difference (ΔH^o) relative to the native state. The horizontal lines indicate the width of the distributions. In the transition region between the two states, the distribution is bimodal (two peaks). The width of the bimodal distribution is the long horizontal line between the two peaks.

can adopt either the folded conformation or one of the many conformations belonging to the unfolded state. The distribution around T_m contains both peaks: it is bimodal (see Figure 4.5). The protein can transition *from one peak to the other*. The effective width of the bimodal distribution is the distance between the two peaks, indicated by the long horizontal line in Figure 4.5. This is why the heat capacity is maximal at the T_m. The "between-states" heat capacity corresponds to jumps from one peak to other, with absorption or release of heat.

Proteins Have a Temperature of Maximum Stability

The DSC experiment provides the values of several thermodynamic parameters at once: T_m, ΔH^o, and ΔC_p^D. In addition, knowing T_m and ΔH^o, the entropy change can be calculated from

$$\Delta S^o = \frac{\Delta H^o}{T_m},\tag{4.31}$$

and the Gibbs energy change, from Equation 4.3

$$\Delta G^o = \Delta H^o - T\Delta S^o.$$

Moreover, knowing the heat capacity change of denaturation (ΔC_p^D), we can calculate the Gibbs energy of denaturation as a function of temperature. To do so, we begin by writing the forms of Equations 4.15 and 4.16 applicable to the unfolding transition,

$$\frac{d\Delta H^o}{dT} = \Delta C_p^D,\tag{4.32}$$

and

$$\frac{d\Delta S^o}{dT} = \frac{\Delta C_p^D}{T}.\tag{4.33}$$

Differentiating Equation 4.8 once more with respect to T and using Equation 4.16, we can see that ΔC_p^D completely determines the temperature dependence of the Gibbs energy of denaturation,

$$\frac{d^2\Delta G^o}{dT^2} = -\frac{d\Delta S^o}{dT} = -\frac{\Delta C_p^D}{T}.\tag{4.34}$$

To obtain the temperature dependence of ΔG^o, we integrate Equations 4.32 and 4.33. The result is

$$\Delta H^o(T) = \Delta H^o(T_m) + (T - T_m)\Delta C_p^D\tag{4.35}$$

and

$$\Delta S^o(T) = \Delta S^o(T_m) + \Delta C_p^D \ln\frac{T}{T_m},\tag{4.36}$$

where we chose T_m as a reference temperature at which both ΔH^o and ΔS^o are known. If we substitute these results into Equation 4.3 for the Gibbs energy, we obtain the temperature dependence of ΔG^o,

$$\Delta G^o(T) = \Delta H^o\left(1 - \frac{T}{T_m}\right) + (T - T_m)\Delta C_p^D - \Delta C_p^D\, T \ln\frac{T}{T_m}.\tag{4.37}$$

Further, the enthalpy difference between folded and unfolded states (ΔH^o) really depends on the temperature at which the transition occurs

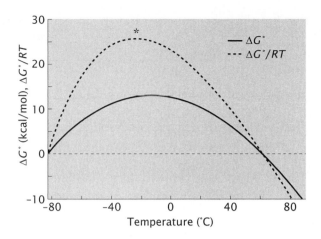

Figure 4.6 The Gibbs energy of denaturation, $\Delta G^0(T)$ (solid line) and $\Delta G^0/RT$ (dashed line) for ribonuclease A as a function of temperature, at pH 5. The maximum in $\Delta G^0/RT$ (*) corresponds to the temperature of maximum stability because that is when the fraction of folded protein is largest. (Calculated from Equation 4.37 with $\Delta H^0 = 111$ kcal/mol, $\Delta C_p^D = 1.3$ kcal/mol/K, and $T_m = 61.3°C$.)

(T_m), as given by Equation 4.32. To change T_m, the experiment can be performed at a different pH or in the presence of a denaturant, such as urea. However, over the temperature range of the unfolding transition, $\Delta H^0 \approx$ constant.

The Gibbs energy of denaturation as a function of temperature is shown in **Figure 4.6** for ribonuclease A. Three aspects are worth noticing. First, the curve of $\Delta G^0(T)$ crosses zero at a high temperature (T_m). At this point *heat denaturation* occurs. Second, the curve crosses zero *also* at a low temperature (T_c). At this point *cold denaturation* occurs. The temperature T_c is not attainable for ribonuclease because water freezes before that, but T_c is above 0°C for many proteins. Third, there is a maximum in $\Delta G^0(T)$, which means the protein has a *temperature of maximum stability*. (The curve is convex up because the second derivative of $\Delta G^0(T)$ with respect to T is negative ($\Delta C_p^D > 0$ in Equation 4.34). The function $\Delta G^0(T)$ is just like the function $-x^2$, which has a similar shape: it is also convex up because its second derivative with respect to x is negative.) The temperature of maximum stability is actually defined by the maximum in $\Delta G^0/RT$ (dashed line in Figure 4.6). This function reaches a maximum when the equilibrium constant for *unfolding* reaches a minimum. It is then that the fraction of folded protein is the largest.

We have described thermal unfolding of proteins, and the changes in the thermodynamic functions (ΔG^0, ΔH^0, ΔS^0, and ΔC_p) that accompany this large change in protein conformation. However, we do not yet know why those changes arise. To understand that, we need to examine the interactions that stabilize the native structure, and find out what causes the protein to unfold when the temperature increases or a denaturant is added to the solution.

4.3 THE ENTROPY OF THE UNFOLDED POLYPEPTIDE IS LARGE BECAUSE OF ITS HUGE NUMBER OF CONFORMATIONS

Several types of interactions contribute to stabilize the native state of a protein. On the other plate of the balance, favoring the unfolded state, there is only the entropy that arises from the vast number of *conformations* of the unfolded polypeptide chain. The *conformational entropy* is given by Boltzmann's formula,

$$S_{conf} = k \ln W,\qquad(4.38)$$

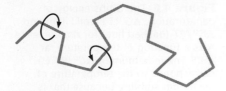

Figure 4.7 The unfolded chain has much more freedom of rotation around the C_α—N and C_α—CO bonds (ϕ and ψ angles).

where k is the Boltzmann constant and W is the total number of microstates, which in this case are the different conformations of the chain.

The native conformation of a protein is unique and well defined. To a first approximation, we can consider that the native protein has essentially one state. Thus, $W_N = 1$ and the conformational entropy is zero for the native state. When the protein unfolds, the number of possible conformations increases dramatically. The conformational entropy of the denatured state is much larger because of all the rotations that become possible around the C_α—N and C_α—CO bonds (ϕ and ψ angles) of the polypeptide backbone (**Figure 4.7**), which were fixed in the native structure. The increased number of allowed conformations of the side chains also contributes to the difference, but to a much lesser extent.

How many conformations does a 100-residue polypeptide have? Along the backbone, there are two covalent bonds per residue about which rotation is possible. As a rough estimate, the number of backbone conformations around each C_α is $\omega_D \approx 9$. As in the case of butane, which has one *anti* and two *gauche* conformations, we may consider that there are three allowed angles for rotation about the N_i—C_i^α bond (ϕ angle), and another three for rotation about the C_i^α—C_i' bond (ψ angle), making a total of $3 \times 3 = 9$ allowed angular regions about each α carbon. A more accurate estimate is $\omega_D = 7$. (Several estimates from experiment and theory yield values of $\omega_D/\omega_N = 3$ to 14, most of them about 7–10. From the change in entropy of protein unfolding, an estimate of $\omega_D/\omega_N \approx 6$ is obtained. In an interesting experiment, a comparison was made between the unfolded state and a completely stretched protein obtained by pulling on its ends and measuring the work (Gibbs energy change) by atomic force microscopy. Assuming that the fully stretched protein has only one conformation ($W = 1$, as in the native state), one obtains $\omega_D = 9.6$. Computational studies using a representative set of unfolded conformations for the denatured state yield $\omega_D/\omega_N = 3$ to 7. These studies compare the regions of the Ramachandran diagram (see Figures 2.40 and 2.41) accessible to the native and unfolded states to obtain ω_D/ω_N.)

Let us see what these values correspond to in terms of Gibbs energy. Assuming one conformation for each residue in the native state ($\omega_N = 1$), Equation 4.38 (with $R = N_A k$, where N_A is Avogadro's number) yields

$$S_{conf}^N = R\ln 1 = 0. \qquad (4.39)$$

Assuming $\omega_D = 7$ per residue in the denatured state, we have

$$S_{conf}^D = R\ln 7 = 3.9 \text{ cal/K/mol-residue} \qquad (4.40)$$

per mole of residues. Hence, for the protein denaturation reaction, per residue we have

$$\Delta S_{conf} = S_{conf}^D - S_{conf}^N = 3.9 \text{ cal/K/mol-residue.} \qquad (4.41)$$

Therefore, at room temperature, the conformational Gibbs energy change is

$$\Delta G_{conf} = -T\Delta S_{conf} = -1.16 \text{ kcal/mol-residue.} \qquad (4.42)$$

For a small globular protein with 100 residues, $W_D = (\omega_D)^{100}$. Assuming $\omega_D = 7$ possible conformations around each α carbon, there are

$7^{100} \approx 3 \times 10^{84}$ conformations in the unfolded state! This is an enormous number. The corresponding conformational part of Gibbs energy change upon unfolding is $\Delta G_{conf} = -116$ kcal/mol. This very large Gibbs energy favors the denatured state. It is especially significant if we recall that the total ΔG for protein unfolding is only ≈ 10 kcal/mol (marginal stability). All the other interactions together must balance, and just slightly exceed ΔG_{conf} with the opposite sign, if the protein is to fold to its native conformation.

The conformational entropy itself (the number of conformations of the unfolded state) does *not* depend on temperature. However, as the temperature increases, the *weight* of $T\Delta S_{conf}$ increases in proportion. A temperature is eventually reached where $T\Delta S_{conf}$ balances all the favorable interactions combined. That point is the melting temperature of the protein. Above T_m the denatured state is favored.

4.4 ELECTROSTATIC INTERACTIONS IN PROTEINS ARE WEAK

A number of different types of interactions unite against the configurational entropy to render the protein just marginally stable under physiological conditions. We begin with electrostatic interactions. Electrostatic interactions are long-ranged; that is, they are felt over large distances. An interaction energy is considered long-ranged if it decays as the third power of the distance (r) or slower. The energy of charge–charge interactions decays as $1/r$, that of charge–dipole interactions as $1/r^2$, and that of fixed dipole–dipole interactions as $1/r^3$. For comparison, the energy of van der Waals attractive interactions between induced dipoles decays as $1/r^6$, which is short-ranged.

Electrostatic Interactions are Described by the Coulomb Potential

The interaction between two charges is described by the Coulomb potential. The energy of two point charges Q_1 and Q_2 separated by a distance r is

$$E = \frac{Q_1 Q_2}{4\pi\epsilon_0 D r},\tag{4.43}$$

where D is the dielectric constant of the medium ($D = 1$ for the vacuum) and ϵ_0 is the permitivity of vacuum, which has the value 8.854×10^{-12} $C^2J^{-1}m^{-1}$. The charge of a proton is 1.60×10^{-19} coulomb (C). If the signs of the two charges are equal, this energy is a positive number, which means that the energy increases as the distance between the charges decreases—the charges repel each other. If the signs are opposite, the energy is negative, which means that the energy decreases as the distance decreases—the charges attract each other.

The interaction between a charge and a dipole is the dot product ($\cdot$) of the dipole moment ($\vec{\mu}$) and the electric field ($\vec{V}$) caused by the charge at the dipole position,

$$E = -\vec{V} \cdot \vec{\mu}\tag{4.44}$$

The magnitude of a dipole moment is the product of the dipole length (ℓ) by the charge separated ($\pm\delta Q$),

$$\mu = \ell(\delta Q).\tag{4.45}$$

Figure 4.8 Interaction between a charge ⊕ and a dipole (arrow) at an angle θ. The dipole is defined by its length ℓ and the charge separated, $+\delta Q$ and $-\delta Q$.

Figure 4.9 Definition of angles between two dipoles.

The direction of the dipole is from its negative to its positive end. Thus, the interaction energy between a charge and a dipole can be written as

$$E = \frac{Q\mu}{4\pi\epsilon_0 D r^2}\cos\theta, \qquad (4.46)$$

where θ is the angle between the dipole and the line linking the dipole to the charge Q (**Figure 4.8**). Equation 4.46 shows that the energy decays with the square of the distance $(1/r^2)$. Note that the energy is positive (unfavorable interaction) if a positive charge is closer to the + end of dipole (the arrowhead), and negative (favorable interaction) if a positive charge is closer to the − end of the dipole (origin of the dipole vector).

For two interacting dipoles, the energy is given by

$$E = \frac{\mu_1\mu_2}{4\pi\epsilon_0 D r^3}(\cos\theta_{12} - 3\cos\theta_1\cos\theta_2), \qquad (4.47)$$

where μ_1 and μ_2 are the magnitudes of the dipole moments, θ_{12} is the angle between the two dipoles, and θ_1 and θ_2 are the angles between each dipole and the line that links the two dipoles (their centers, more precisely) (**Figure 4.9**). The interaction energy decays as $1/r^3$ (shown in Equation 4.47).

The Work Done Against the Electrostatic Forces is the Gibbs Energy Change

Consider two positive charges, Q_1 at point a, and Q_2 at the origin (**Figure 4.10**). The *work* done against the electrostatic forces $\vec{F}$ in bringing the charge Q_1 from point a to point b,

$$W = -\int \vec{F}\cdot d\vec{r}, \qquad (4.48)$$

at constant temperature and pressure, is the Gibbs energy change of the system. The integral is calculated along the *path* of the charge moved, and $d\vec{r}$ is a unit vector in the direction of the displacement. If the two charges have the same sign, the force and the displacement have opposite directions; therefore, the dot product $\vec{F}\cdot d\vec{r} < 0$. The total work is the sum of the contribution of $\vec{F}\cdot d\vec{r}$ at each point along the path and is therefore positive because of the minus sign in front of the integral in Equation 4.48. Thus, the work done in moving one like charge toward the other, against their repulsive force, is positive. We can also write Equation 4.48 as

$$W = -\int_a^b F\,dr \qquad (4.49)$$

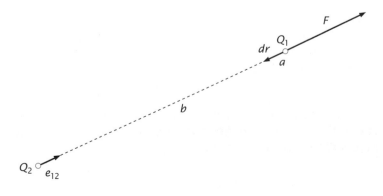

Figure 4.10 The force on charge Q_1 due to charge Q_2, when Q_1 is moved from point a to b. Both charges are assumed positive (or both negative); the force points away from the line that connects them. The work in bringing two like charges closer is positive.

where a and b are the initial and final distances ($b < a$) from Q_2 on the line.

In Figure 4.10, the electrostatic force $\vec{F}$ acting on Q_1 is the result of the charge Q_2. The force has the value

$$\vec{F} = \frac{Q_1 Q_2}{4\pi\epsilon_0 D r^2} \vec{e_{12}},\qquad(4.50)$$

where $\vec{e_{12}}$ is the unit vector pointing from the location of Q_2 to the location of Q_1.

The work done at constant pressure in bringing a charge from infinity to a distance a of another is the electrostatic component of the Gibbs energy change. Assuming point charges, the result is

$$\begin{aligned}
\Delta G^0 &= -\int_\infty^a \frac{Q_1 Q_2}{4\pi\epsilon_0 D r^2}\, dr \\
&= \frac{Q_1 Q_2}{4\pi\epsilon_0 D}\left(\frac{1}{a} - \frac{1}{\infty}\right) \\
&= \frac{Q_1 Q_2}{4\pi\epsilon_0 D a}.
\end{aligned}\qquad(4.51)$$

Substituting in this equation the numerical values of the constants, and expressing the length in angstroms (1 Å $=10^{-10}$m), energy in kilocalories per mole (1 cal = 4.184 J), and the charges in elementary units (q_1 and $q_2 = -1, +1$) instead of coulombs (C), we can write Equation 4.51 in a form more useful for practical purposes,

$$\Delta G^0 = \frac{332\, q_1 q_2}{D a},\qquad(4.52)$$

where ΔG^0 is given in kcal/mol and a is in Å. If the two charges have opposite signs, ΔG^0 is negative, which means that the interaction is favorable, as expected.

Another useful quantity is the Bjerrum length (ℓ_B), which is the distance at which the electrostatic energy between two unit charges in vacuum is equal to the thermal energy (RT per mole or kT per molecule).

$$\ell_B = \frac{e^2}{4\pi\epsilon_0 kT} = 560\ \text{Å}\qquad(4.53)$$

at room temperature, where e is the unit charge. Then, to calculate the Gibbs energy change corresponding to moving a charge from a to b at room temperature, we can simply use

$$\frac{\Delta G^0}{RT} = \frac{\ell_B q_1 q_2}{D}\left(\frac{1}{a} - \frac{1}{b}\right).\qquad(4.54)$$

Water Weakens Electrostatic Interactions

Let us now see what happens in water. In classical electrostatics, changing the medium means changing the dielectric constant (D). The dielectric constant is a measure of the polarizability of the medium: it represents the ability of a solvent to respond to, or counteract, the effects of electrostatic fields from charges and dipoles by rearranging or reorienting its molecules. At room temperature, the dielectric constant of water is $D_w \approx 80$, which means that charges and dipoles sense each other *much less* in water than in a vacuum (where $D = 1$). In a liquid hydrocarbon, $D \approx 2$. What the value of D is in the interior or on the surface of a protein is a much-debated issue. If the nonpolar interior of a protein were assumed to resemble a liquid hydrocarbon, $D \approx 2\text{--}4$ would be appropriate. But the very definition of a dielectric constant inside a protein is complicated.

The dielectric constant in classical continuum electrostatics is a macroscopic property. This constant, however, is inappropriate to describe events at the molecular level, such as local rearrangements in a protein. Values of $D = 2\text{--}4$ underestimate the effective dielectric constant in a protein. Calculations of electrostatics using the Poisson–Boltzmann equation require an effective or apparent dielectric constant $D \approx 6\text{--}20$ to obtain realistic results.

Using experimentally measured shifts in the pK_a of amino acids away from their normal values, we can estimate that $D \approx 10\text{--}15$ in the protein interior. This, however, is not the true dielectric constant of the protein nonpolar interior; it includes contributions from large rearrangements of the polypeptide chain and repositionings of water molecules that occur as a response to charge burial. On average, the Gibbs energy of destabilization that arises from placing a charged group inside the protein, calculated from pK_a shift values, is ≈ 5 kcal/mol. This value is small but comparable to the total ΔG^o of denaturation of the protein. However, it is much smaller than what would be predicted by using $D = 2\text{--}4$ for the protein interior, which yields electrostatic Gibbs energies ≈ 20 kcal/mol. Thus, $D \approx 10$ is a better estimate of the dielectric in the protein interior if we use Equations 4.51 or 4.52.

At close range, in systems such as ionic crystals, electrostatic interactions are among the strongest noncovalent interactions. In proteins, however, they are much weaker because of the larger effective dielectric constants; but they can change the balance between the folded and unfolded states of a protein.

Formation of Ion Pairs in Water is Driven by Entropy

The dielectric constant of water tells us a lot about the origins of ionic interactions in proteins in aqueous solutions. When we think of charge–charge interactions in a protein, we typically have in mind the formation of ion pairs involving the carboxylate group from the side chains of aspartate or glutamate, the ϵ-amino group of lysine, or the guanidinium group of arginine. In a vacuum, or in a low dielectric, the attraction between positive and negative charges is driven by energy. However, in water or on the protein surface, formation of ion pairs is mainly driven by the entropy of water. To see this, let us begin by considering the temperature dependence of the electrostatic Gibbs energy. If we know how ΔG^o depends on temperature, we can calculate the enthalpy and entropy changes from Equations 4.8 and 4.11,

$$\frac{d(\Delta G^o/T)}{dT} = -\frac{\Delta H^o}{T^2}$$

$$\frac{d\Delta G^o}{dT} = -\Delta S^o,$$

where ΔG^o is given by Equation 4.51. Now differentiate Equation 4.51 with respect to T, assuming that only D varies with temperature. Using the chain rule and recalling that for any function $f(T)$ ($f = D$ in this case)

$$\frac{d}{dT}\frac{1}{f} = -\frac{1}{f^2}\frac{df}{dT}, \tag{4.55}$$

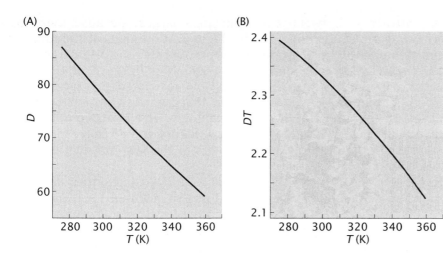

Figure 4.11 Temperature dependence of (A) the dielectric constant of water, D, and (B) the product the dielectric constant by the temperature, DT. (Data from [2008] CRC Handbook of Chemistry and Physics, 88th ed. CRC Press.)

we obtain the entropy change as

$$\Delta S^{o} = \frac{Q_1 Q_2}{(4\pi\epsilon_0)D^2 a}\frac{dD}{dT} = \frac{\Delta G^{o}}{D}\frac{dD}{dT}, \tag{4.56}$$

and the enthalpy change as,

$$\Delta H^{o} = \frac{Q_1 Q_2}{(4\pi\epsilon_0)D^2 a}\frac{d(DT)}{dT} = \frac{\Delta G^{o}}{D}\frac{d(DT)}{dT}. \tag{4.57}$$

Figure 4.11 shows how D and the product DT vary with temperature in water: they both *decrease*. Thus, the factors dD/dT and $d(DT)/dT$ in Equations 4.56 and 4.57 are both negative, which means that ΔS^{o} and ΔH^{o} have the *opposite sign* of ΔG^{o}. If ion pair formation is favorable ($\Delta G^{o} < 0$) then $\Delta H^{o} > 0$, which means that the enthalpy (energy) actually *opposes* ion pair formation. It is the entropy change ΔS^{o}, which is also positive, that *favors* the ion pair. Thus, ion pair formation *in water* is driven by entropy.

Ion Pairs in Proteins Make a Small Contribution to Protein Stability

Consider acetic acid, which has $pK_a = 4.8$ at room temperature. The standard Gibbs energy change for *dissociation* of H^+ from acetic acid is ($\ln K_a = 2.303 \log K_a$)

$$\Delta G^{o}_{diss} = -RT \ln K_a \tag{4.58}$$

$$= 2.303\, RT\, pK_a \tag{4.59}$$

$$= +6.5 \text{ kcal/mol.} \tag{4.60}$$

Thus, for *association* of the positive proton with the negative acetate, the Gibbs energy change is $\Delta G^{o} = -6.5$ kcal/mol. In this case, $\Delta H^{o} = +0.4$ kcal/mol, close to zero but slightly *opposing* association; and $\Delta S^{o} = +23$ cal/K/mol, *favoring* association. Why is the enthalpy (energy) change for bringing two opposite charges together positive? Where does the positive entropy change come from when the acetate ion and the proton associate in water? Well, it comes *from water*.

When an anion is placed in water, the water molecules orient themselves around it with the positive end of their dipole moments toward the anion, as shown in **Figure 4.12**. The converse happens for a cation in water. This charge–dipole interaction is very strong, almost as strong

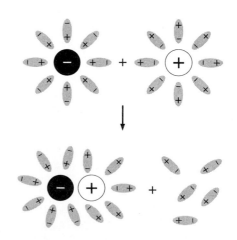

Figure 4.12 Formation of an ion pair in water, with concomitant release of water molecules from the hydration shells. The ions are represented by the circles (+, −) and the water molecules, by the elliptical dipoles.

as the attraction between two opposite charges. Because there are many water molecules involved in the hydration shell of the ion, the summed energy of these interactions compensates the original interaction between cation and anion. This is why $\Delta H^o \approx 0$. However, the water molecules in the hydration shell are now restricted in space, which represents an entropic cost. If the two charges, positive and negative, are brought together, the water molecules of the hydration shell are released (see Figure 4.12). The extra *freedom* gained corresponds to an increase in the possible arrangements of the system—that is, to an increase in entropy.

On the surface of proteins, charged amino acid residues are common. What is the contribution of an ion pair, formed for example by the ε-amino group (NH_3^+) of a lysine and the β-carboxylic group (COO^-) of an aspartate, to the stability of a protein? Using Equation 4.52, and assuming a distance between the positive and negative ion of about 4 Å, the electrostatic contribution of this interaction is only −1 kcal/mol. Nevertheless, if there were five of these interactions on the protein surface, 5 kcal/mol is significant compared to the total ΔG^o of denaturation. In addition, NH_3^+ and COO^- groups interact electrostatically, but are often also hydrogen bonded to each other. A hydrogen-bonded ion pair is called a *salt bridge*. Its contribution to protein stability is greater.

Burying Charges in the Nonpolar Interior of a Protein has Large Energetic Costs

In the nonpolar interior of a protein, the favorable electrostatic attraction between two unlike charges would be much larger, because the dielectric constant is much smaller. However, this favorable contribution is obliterated by the larger energetic cost of placing charges in a nonpolar environment. The energy of an ion with charge Q and radius a inside a medium of dielectric constant D is

$$E \approx \frac{1}{2}\frac{Q^2}{4\pi\epsilon_0 Da}.$$

(4.61)

For this reason, charged residues are almost never found inside proteins, with the very few exceptions being those that have functional significance. Examples are charges in the active sites of enzymes and in the transmembrane regions of integral membrane protein, such as mitochondrial ATPase, cytochrome c oxidase, and bacteriorhodopsin (**Figure 4.13**). Several charged residues are buried in the protein nonpolar interior or in contact with the lipid bilayer core. In bacteriorhodopsin, several water molecules are also found in the protein nonpolar interior, making hydrogen bonds with the charged residues (see Figure 4.13).

It is possible to engineer mutations in proteins, changing a nonpolar interior residue, such as leucine, to a charged one, such as arginine. Those changes have several consequences. The protein still folds but is less stable. The pK_a values of ionizable residues inside the protein often become very different from those in water. We say they are *anomalous* pK_a values. Acidic residues (Asp or Glu) undergo pK_a shifts to larger values, typically from the normal $pK_a = 4$ to ≈ 8 (but can reach 10). With such high pK_a values, the carboxylic acid groups are protonated at pH 7, and therefore uncharged. Buried lysines, on the other hand, undergo pK_a shifts to smaller values, typically from the normal

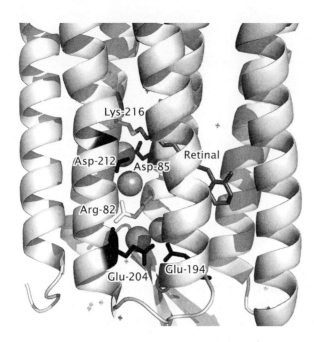

Figure 4.13 The nonpolar, transmembrane region of the integral protein bacteriorhodopsin includes several buried charged residues. Residues Asp-85, Asp-212, Glu-194, and Glu-204 are shown in black; Arg-82 is shown in white. The chromophore retinal (dark gray) is attached to Lys-216 (light gray). Water molecules, also buried inside the protein, are shown as gray spheres (PDB 1C3W).

$pK_a = 10$ to ≈ 7 (but can reach 5). They become deprotonated, and therefore neutral, at pH 7. Thus, the pK_a shifts to render the side chain *neutral inside the protein*, because the energy of a charge in a nonpolar environment is very large.

Arginine residues, however, are always charged, even when buried. The normal $pK_a = 12$ of arginine is large because its side chain has high affinity for the proton. Deprotonation, to achieve neutrality, would be too costly energetically. What happens then is that the protein interior rearranges to accommodate the charge. In some cases, another residue with the opposite charge (Asp or Glu) moves closer to the buried arginine to make a salt bridge, or water becomes associated with this arginine in the protein interior, making hydrogen bonds (see Figure 4.13).

4.5 VAN DER WAALS INTERACTIONS ARE WEAK BUT THEIR TOTAL CONTRIBUTION IN THE PROTEIN CORE IS SUBSTANTIAL

When we speak of van der Waals interactions, we usually have in mind London dispersion forces. These forces are always attractive, and result from induced dipole–induced dipole interactions. However, van der Waals interactions comprise both London dispersion forces and hard-core repulsions. We can write the van der Waals energy as a Lennard–Jones potential,

$$E = \frac{A}{r^{12}} - \frac{B}{r^6} \tag{4.62}$$

where A and B are positive constants, and r is the distance between atoms or molecules. **Figure 4.14** shows a plot of the van der Waals energy as a function of distance between two methane molecules.

Van der Waals Interactions are Short-ranged

At very close distance, when the electronic clouds begin to overlap (van der Waals contact), all molecules repel each other. This is the van der

Figure 4.14 Distance dependence of the van der Waals energy of methane (CH_4). The two gray circles represent two methane molecules, one at the origin and the other at the distance of lowest energy. Calculated from Equation 4.62.

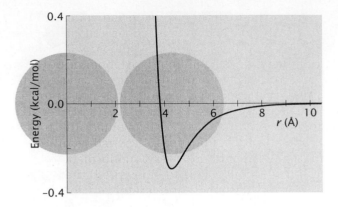

Waals repulsion energy, which decays with distance as $1/r^{12}$ (the first term in Equation 4.62). At slightly greater distances, sufficient to avoid electronic overlap, the interaction between two molecules is attractive. This is the result of London dispersion forces, which are usually pictured as resulting from attraction between instantaneous dipoles and induced dipoles. This interaction is strongest at very short distances, just at van der Waals contact, which corresponds to the energy minimum in Figure 4.14. The energy of the London dispersion interaction is negative (favorable) and decays with a $1/r^6$ distance dependence (the second term in Equation 4.62).

A striking illustration of the strength of van der Waals interactions at close contact comes from biology. Geckos adhere to walls, not because of suction cups, but because the base of their feet contains about 500,000 keratinous hairs (proteins). The van der Waals interactions between these fibers and the surface holds the gecko on vertical walls (**Figure 4.15**).

Van der Waals Interactions Stabilize the Nonpolar Core of Proteins

Because of their strong distance dependence, London dispersion forces are very sensitive to packing. In a loosely packed hydrocarbon medium, such as in a monolayer of a detergent at the air/water interface, the average interaction energy of a methylene group ($-CH_2-$) is about 0.4 kcal/mol, which is less than the thermal energy at room temperature (RT). However, in a densely packed crystalline hydrocarbon, the enthalpy of sublimation per $-CH_2-$ group is about 2 kcal/mol, which is comparable to the van der Waals contribution to the enthalpy of sublimation of ice (3.6 kcal/mol).

We can expect, then, that while London dispersion forces will not provide energy to bring large molecules together or allow for distant parts of a protein to sense each other, they will play an important role—perhaps even a predominant role—in the densely packed, nonpolar interior of a globular protein. The interactions between nonpolar side chains inside the protein core are not identical to those in a liquid hydrocarbon or solid hydrocarbon. Nevertheless, van der Waals interactions between nonpolar side chains are essential for protein stability. This is demonstrated by mutations of large amino acids to small amino acids.

Mutations Val → Ala and Leu → Ala were introduced in the nonpolar core of several proteins. Because a large side chain is replaced by a small one, a cavity is created inside the protein nonpolar core. The

Figure 4.15 The Moorish gecko (*Tarantola mauritanica*) adheres to vertical walls through van der Waals interactions. (Adapted from Almeida NF, Almeida PF, Gonçalves H et al. [2001] Anfíbios e Répteis de Portugal. With the permission of FAFAS.)

mutant proteins have fewer interactions between hydrophobic areas and water in the denatured state, but the protein core in the native structure is less well packed. Those two effects largely compensate each other, resulting in little change in ΔS^o of denaturation. However, the mutant proteins are less stable, because ΔH^o of unfolding is smaller. Loss of van der Waals interactions in the native state raises the enthalpy of the native state and results in a smaller ΔH^o of denaturation (**Figure 4.16**). Thus, it appears that large-to-small amino acid mutations impair van der Waals interactions in the protein core.

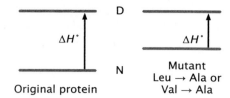

Figure 4.16 Change in ΔH^o of unfolding by large-to-small amino acid mutations.

4.6 HYDROGEN BONDS ARE ESSENTIAL FOR PROTEIN STABILITY

The role of hydrogen bonds in protein stability and structure has been recognized since the work of Pauling. Hydrogen bonds provide the major stabilization for the regular secondary structures of proteins—namely α-helices and β-sheets. The energy of a hydrogen bond is ≈ 5 kcal/mol. This is of the same order of magnitude as the Gibbs energy for protein denaturation, which means that leaving one hydrogen bond unmade could amount to denaturing a protein. In fact, protein structures show that all possible hydrogen bonds are established in the native state, either within the protein or with water. However, hydrogen bonds probably do not provide a major energetic contribution to protein stability. In fact, whether their overall contribution is even favorable has been long debated. But before we consider that, we need to learn what a hydrogen bond is.

A Hydrogen Bond is an Attraction between Two Electronegative Atoms Mediated by an H Atom

A hydrogen bond X–H$\cdots$Y is an attractive interaction between a group X–H, in which a covalent bond exists between the electronegative atom X and the H atom, and another electronegative atom or group Y (**Figure 4.17**A). The electropositive H atom inserts between the two electronegative atoms X and Y, bringing them together. The X–H group is the hydrogen-bond donor; the Y group is the acceptor. The X and Y atoms are typically oxygen, nitrogen, or fluorine. A hydrogen bond has a large electrostatic component, which consists of the interaction between the positively polarized end of the X–H dipole (the H atom) and the negatively polarized end of the dipole ending at the electronegative Y atom. There is also a covalent component arising from charge transfer between donor and acceptor, and a dispersion component. Bonds with more covalent character have been called "low-barrier" hydrogen bonds.

In proteins, oxygen atoms are the most common acceptors of hydrogen bonds. Note that the nitrogen atom of an amide bond is not a hydrogen-bond acceptor because of resonance delocalization of its lone electron pair throughout the amide group (see Figure 2.30). The most common hydrogen-bond donors are the N–H of amides and amines, and the O–H groups of alcohols (Ser, Thr). Histidine residues also participate in hydrogen bonding through the nitrogen atoms of their imidazole rings, both as donors (N–H) and as acceptors (N:). The most important hydrogen bonds, however, are those between the C=O and N–H of the backbone peptide groups, because they are the reason for the existence of α-helices and β-sheets.

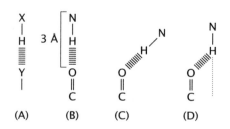

Figure 4.17 (A) Hydrogen bond between X–H (donor) and Y (acceptor). The hydrogen bond length is the distance between X and Y. (B–D) Hydrogen bonds between N–H (donor) and C=O (acceptor) in three different orientations. The hydrogen bond length is 3.0 Å in this example.

The energy (enthalpy) of a hydrogen bond varies with its length and orientation. The length of a hydrogen bond is defined as the distance between the centers of the two heavy atoms (see Figure 4.17B), such as N and O in $N-H\cdots O=C$ (not to the hydrogen atom). The bonds are stronger when the hydrogen bond lies in the same direction as the N–H covalent bond (see Figure 4.17B), but the dipoles do not have to be exactly aligned (see Figure 4.17C). The bonds are weaker if the hydrogen bond is not aligned with the N–H covalent bond (see Figure 4.17D). In model calculations, the strongest hydrogen bonds are achieved if the covalent bond of the donor is aligned with the orbital that contains the lone pairs on the acceptor side, but in proteins this effect is significantly blurred by local interactions determined by the protein structure. In protein crystal structures, hydrogen atoms are often not visible, because the electron density around the hydrogen nucleus is too weak; their positions are inferred from those of the heavy atoms to which they are covalently bonded and hydrogen bonded. Typical hydrogen bond lengths are about 3.0 Å, but they vary between 2.8–3.6 Å. A 2.8-Å hydrogen bond is strong, whereas a 3.6-Å one is weak.

Hydrogen bonds largely determine the properties of water. In ice, there are four hydrogen bonds per water molecule, with an average length of 2.76 Å. The heat of sublimation of ice is $\Delta H^o_{subl} = 11.6$ kcal/mol; subtracting the van der Waals contribution of 3.6 kcal/mol leaves 8.0 kcal/mol of hydrogen-bonding energy. Since there are two bonds per molecule, this means that each hydrogen bond contributes an enthalpy $\Delta H^o = 4$ kcal/mol. However, the heat required to melt ice is much smaller, ($\Delta H^o_m = 1.4$ kcal/mol), because *in liquid water most hydrogen bonds are maintained.* In fact, only $\approx 10\%$ of the bonds present in ice are broken in water. Liquid water is still extensively hydrogen bonded. Water can be viewed as continuously changing, flickering clusters of hydrogen-bonded molecules. The average lifetime of a hydrogen bond in water is $\sim 10^{-12}$ seconds.

The Net Contribution of Hydrogen Bonds to Protein Stability is Probably Small but Favorable

Hydrogen bonds are essential for protein structure, but how much they contribute to the stability of the native state is still a controversial question 60 years after Schellman's (1955) classic article. The strength of hydrogen bonds depends not only on the donor–acceptor pair and on their distance, but also on their local environment. The different polarizabilities (dielectric constants) of water and the protein interior affect the energy (enthalpy) of the hydrogen bond. To complicate matters, hydrogen bonds can be established between protein groups and water. If hydrogen bonds contribute to protein stability, there must be an advantage in making them between protein groups rather than with water.

To understand the problem, we need to consider the thermodynamics of formation of hydrogen bonds between groups within the protein, between those groups and water, and among water molecules. For groups exposed to water, there is *competition* by the water molecules to form hydrogen bonds. If we knew the energies of hydrogen bonds in water and inside the protein, we might think that we could simply count their numbers, and it would then be just a matter of bookkeeping to calculate the *net* enthalpy of hydrogen bond formation (ΔH_{hbond}). However, the energies of hydrogen bonds in water and in the protein are *not* additive. The hydrogen bond strength determined with model compounds (amides) in water provides misleading estimates for the hydrogen bonds involving peptide groups (CONH), because many factors involved in the hydration and desolvation of those groups are not additive. In addition, hydrogen bonds seem to form cooperatively in a helix: the energy of a hydrogen bond increases (ΔH_{hbond} becomes more negative) and the bond length decreases from the termini to the middle of the helix.

Nevertheless, to obtain a rough estimate of the changes involved, consider the thermodynamic cycle of **Figure 4.18**. We begin on the top left, with the C=O and H—N groups of a peptide group in water; they are not hydrogen bonded to each other but to water. The left arrow represents transfer of the equivalent of a peptide group to the protein interior, *without* making a hydrogen bond. The bottom arrow is the formation of a C=O$\cdots$H—N hydrogen bond. Alternatively, the C=O$\cdots$H—N hydrogen bond can be made in water (top arrow) and the hydrogen-bonded group can be transferred to the protein interior (right arrow).

The hydrogen-bonding changes that accompany protein folding correspond to going from the top left of Figure 4.18 (peptide group hydrogen bonded to water) to the bottom right (hydrogen-bonded peptide group in the protein). Let us estimate ΔG^0 for each step. Calorimetric experiments of alanine-based α-helical polypeptides show that the formation of a hydrogen bond C=O$\cdots$H—N between peptide groups has a favorable $\Delta H_{hbond}^{w} \approx -1$ kcal/mol in water (superscript w). This value is smaller than the enthalpy of a hydrogen bond ($\Delta H^0 \approx -4$ kcal/mol), because ΔH_{hbond}^{w} is the *difference* between the enthalpy of water–water and peptide–peptide bonds on one hand, and water–peptide bonds on the other. If the entropic contribution favoring the unfolded state stems mainly from the decrease in the number of conformations of the polypeptide when the bond is made, we obtain $T\Delta S_{hbond}^{w} \approx -1$ kcal/mol per residue. Then we must conclude that $\Delta G_{hbond}^{w} \approx 0$. In fact, ΔG_{hbond}^{w} is probably slightly negative, but not more than ≈ -0.3 kcal/mol, at room temperature for residues exposed to water.

In the protein interior (superscript n), probably $\Delta H_{hbond}^{n} \approx -4$ kcal/mol. If, as in water, the entropic contribution is $T\Delta S_{hbond}^{n} \approx -1$ kcal/mol, then $\Delta G_{hbond}^{n} \approx -3$ kcal/mol. However, in the conformationally restricted space of the protein interior, it is possible that $T\Delta S_{hbond}^{n} \approx 0$ for many hydrogen bonds. In other words, a large part of the entropic penalty involved in bringing the two hydrogen-bond partners together may have already been paid when the protein collapsed to a globular structure. Therefore, it is likely that ΔG_{hbond}^{n} is favorable by -3 to -4 kcal/mol in the protein interior.

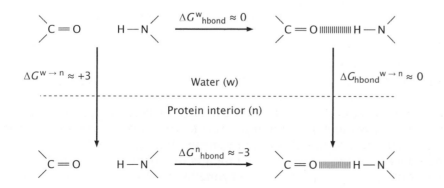

Figure 4.18 Thermodynamic cycle for the formation of hydrogen bonds in water and inside a protein. The values refer to ΔG^0 in kcal/mol.

However, the Gibbs energy of transfer of a non-hydrogen-bonded peptide group from water to the protein interior ($\Delta G^{w \to n}$) is certainly very unfavorable. Its value should be intermediate between transfer from water to octanol ($\Delta G^{w \to oct} \approx +2$ kcal/mol) and transfer to an alkane ($\Delta G^{w \to alk} \approx +7.5$ kcal/mol). Thus, probably $\Delta G^{w \to n} \approx +3$ to $+4$ kcal/mol.

In Figure 4.18, as in any thermodynamic cycle, the total ΔG must be the same independently of the path between the initial and final states. This means that $\Delta G^{w \to n} + \Delta G^n_{hbond} = \Delta G^w_{hbond} + \Delta G^{w \to n}_{hbond}$. If we use $\Delta G^{w \to n} \approx +3$, $\Delta G^n_{hbond} \approx -3$, and $\Delta G^w_{hbond} \approx 0$ kcal/mol, then the transfer of a hydrogen-bonded peptide group from water to the protein interior must have $\Delta G^{w \to n}_{hbond} \approx 0$ (the vertical step on the right in Figure 4.18). The uncertainties in these estimates, however, are probably of about ± 1 kcal/mol. Thus, the overall transfer of C=O and H—N from water (hydrogen bonded to water) to the protein interior, making a C=O$\cdots$H—N hydrogen bond, could contribute ± 1 kcal/mol to ΔG^o of protein folding.

Hydrogen bonds between peptide groups stabilize α-helices and β-sheets. They also make a large favorable enthalpy contribution to protein stability. If a hydrogen bond (buried or water-exposed) contributes on average $\Delta H_{hbond} \approx -1$ to -2 kcal/mol, and considering that $\approx 70\%$ of the peptide groups participate in intramolecular hydrogen bonds, then in a 100-residue protein the total contribution of hydrogen bonds to the enthalpy of folding is ~ -100 kcal/mol, similar to the hydrophobic effect, and 10 times larger than the total ΔG^o of folding. The contribution of hydrogen bonds to the Gibbs energy, however, is much less clear. The total ΔG, from the top left to the bottom right of the thermodynamic cycle (see Figure 4.18) is probably favorable but less than -1 kcal/mol in absolute value. However, even if this value is small, the protein *cannot* fold without the formation of peptide hydrogen bonds in its interior, where water does not exist as an alternative bonding partner. The energetic cost would be too high because burying a non-hydrogen-bonded peptide group corresponds to $\Delta G^{w \to n} \approx +3$ kcal/mol. Therefore, hydrogen bonds can tip the balance of protein folding in one direction or the other depending on environmental conditions—namely in the presence or absence of denaturants and protective osmolytes, discussed at the end of this chapter.

4.7 THE HYDROPHOBIC EFFECT IS THE REASON WHY PROTEINS FOLD TO GLOBULAR STRUCTURES IN WATER

Proteins fold to compact globular structures in water to hide their nonpolar residues from contact with water. This "dislike" or *phobia* of nonpolar compounds for water is called the *hydrophobic effect*. The hydrophobic effect is not an *interaction* in the proper sense, but an *effect* that is a consequence of the hydrogen-bonding properties of water. The hydrophobic effect acts on the *unfolded protein*; it is in this state that the nonpolar side chains interact with water. The hydrophobic effect does not apply only to protein stability. Indeed, it is one of the most important concepts in biochemistry, essential in the formation of membranes, protein–protein and enzyme–substrate interactions, binding of molecules to membranes, and drug binding to DNA and proteins. The concept of the hydrophobic effect is one of the most familiar and, to a point, intuitively simple. Everyone knows that oil and water don't mix. When you begin to study chemistry, you learn the general rule that "like dissolves like." It makes intuitive sense: the "repulsion" between water and oil fits in this concept. However, as you learn more about the hydrophobic effect, you will see that this apparent simplicity is illusory.

The Chemical Potential of Transfer Measures the Difference in Interactions of a Molecule with its Surroundings in Two Solvents

To quantify the hydrophobic effect, we need to determine the Gibbs energy of transfer of a hydrocarbon from a nonpolar solvent to water. The Gibbs energy of transfer of a hydrocarbon A is the difference between the chemical potential of A in water ($\mu_{A,w}$) and in the nonpolar solvent ($\mu_{A,n}$). The chemical potential of a pure substance is its Gibbs energy per mole. In a solution, the chemical potential of a component A is defined by how the Gibbs energy of the system changes with an increase in the number of moles of component A (n_A),

$$\mu_A = \left(\frac{dG}{dn_A}\right)_{P,T,n_i} \tag{4.63}$$

at constant pressure (P), temperature (T), and numbers of moles of all other components (n_i), as indicated by the subscripts in Equation 4.63. In Chapter 1, we wrote the chemical potential of a component A in a very dilute solution as

$$\mu_A = \mu_A^o + RT\ln[A], \tag{4.64}$$

where μ_A^o is the standard chemical potential and [A] is the concentration of A. The *standard* Gibbs energy of transfer between two solutions (ΔG_A^o) is then given by

$$\Delta\mu_A^o = \mu_{A,w}^o - \mu_{A,n}^o = -RT\ln K_{eq}, \tag{4.65}$$

where K_{eq} is the equilibrium constant, or partition coefficient, for the "reaction" of transferring a molecule of hydrocarbon A from the nonpolar solvent to water,

$$K_{eq} = \left(\frac{[A_w]}{[A_n]}\right)_{eq}. \tag{4.66}$$

Thus, we can *measure* $\Delta\mu_A^o$ by determining the concentrations [A_w] and [A_n] at equilibrium, provided both solutions are very dilute.

Consider now the thought experiment illustrated in **Figure 4.19**. A hydrocarbon A is transferred from a *fixed position* in a nonpolar solvent (n) to a *fixed position* in water (w). This transfer, between two fixed positions in the two solvents, is called the Ben-Naim *standard transfer process*.

Why do we impose the restraint that the transfer occur between fixed positions in the two solvents? The reason is that, if we do not, the accompanying Gibbs energy change includes differences in the translational entropy of the hydrocarbon molecule in the two solutions or phases, which have nothing to do with *interactions* between hydrocarbons and water. This is especially important if we consider transfers between the gas phase and a liquid solvent (which we will consider shortly). If we transfer the hydrocarbon from a fixed position in the gas phase to a fixed position in water, we obtain the Gibbs energy of

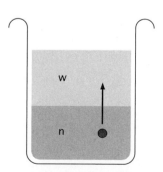

Figure 4.19 Transfer of a solute from a nonpolar solvent (lower layer) to water (upper layer).

Table 4.1 Thermodynamic data for transfer of hydrocarbons, from a nonpolar hydrocarbon liquid to water (n → w), from the gas phase to water (g → w), and from the gas to nonpolar liquid (g → n), at 25°C (Ben-Naim standard transfer process, indicated by *).[a] The values of ΔG^*, ΔH^*, and $T\Delta S^*$ are in kcal/mol. The values of ΔC_p (in cal/K/mol) are taken directly from the sources. ΔC_p for propane, isobutane, and neopentane conform to the Ben-Naim standard transfer; the value for butane (in parentheses) probably has a large error.

	n → w			g → w	g → n	n → w
Solute	ΔG^*	ΔH^*	$T\Delta S^*$	ΔG^*	ΔG^*	ΔC_p
Propane	+4.0	−1.3	−5.3	+2.0	−2.1	+74
Butane	+4.8	−0.7	−5.5	+2.1	−2.7	(+63)
Isobutane	+4.9	−0.6	−5.5	+2.3	−2.5	+81
Pentane	+5.7	−0.5	−6.2	+2.3	−3.5	+96
Neopentane	+5.4	−0.4	−5.8	+2.5	−2.9	+87
Benzene	+3.7	+0.6	−3.1	−0.8	−4.5	+54

[a]The values represent a combination of data from Jorgensen WL, Gao J & Ravimohan C [1985] *J Phys Chem* 89:3470–3473; Lee B [1991] *Biopolymers* 31:993–1008; Makhatadze, GI & Privalov PL [1994] *Biophys Chem* 50:285–291; Ben-Naim A & Marcus Y [1984] *J Phys Chem* 81:2016–2027; Murphy KP [1994] *Biophys Chem* 51:311–326; Gill SJ, Nichols NF & Wadsö I [1976] *J Chem Thermodyn* 8:445–452; and Sturtevant JM [1977] *Proc Natl Acad Sci USA* 74:2236–2240.

hydration. In general, if the solvent is not water, transfer from the gas is called the Gibbs energy of *solvation.*

The Gibbs energy change per mole (or per molecule) that accompanies the transfer of a solute A between fixed positions in two solvents or phases is called the Ben-Naim chemical potential of transfer. We designate it by $\Delta\mu_A^*$ or ΔG_A^*. $\Delta\mu_A^*$ measures what we want: the difference in interactions of the molecule with its surroundings in the two solvents. To measure $\Delta\mu_A^*$, we need to relate it to the concentrations of hydrocarbon A in water and in the nonpolar solvent in equilibrium. In a very dilute solution, the solute A is indeed surrounded only by solvent, either water or nonpolar solvent, and the Ben-Naim chemical potential of transfer equals the traditional standard Gibbs energy of transfer,

$$\Delta\mu_A^* = \Delta\mu_A^o, \tag{4.67}$$

provided that the concentrations [A_w] and [A_n] are expressed in a molar concentration scale, or in a concentration scale proportional to the number of molecules per unit volume of the solution (number density). Otherwise, some contributions to the Gibbs energy of transfer that depend on the concentration of the solvent do not cancel in the transfer process.

The "dislike" between water and hydrophobic (nonpolar) compounds is manifested by the *positive* Gibbs energy of transfer from hydrocarbon to water (n → w), listed in the first column of **Table 4.1**. Based on the values for the linear alkanes, we can infer that transferring a methylene group (−CH_2−) from a nonpolar solvent to water corresponds to $\Delta G^* \approx +1$ kcal/mol.

The Immiscibility of Hydrocarbons in Water at Room Temperature is Caused by an Unfavorable Entropy

Imagine we remove a molecule of the solute hydrocarbon from the nonpolar solvent. To do so, we must break the London dispersion forces that the solute established with the solvent. Next, we want to dissolve our hydrocarbon in water. This process can be viewed as consisting of two steps. First, we create a cavity in water to place the hydrocarbon; in doing so, we must break hydrogen bonds between water molecules. Second, we place the hydrocarbon molecule in that cavity; this process creates new interactions between the hydrocarbon and water (London forces), but *no new hydrogen bonds*. Since the new interactions (between the hydrocarbon and water) are weaker than the original ones (between water molecules), the solvation process is expected to have a positive enthalpy of transfer ($\Delta H^* > 0$).

However, when ΔH^* is measured from the dependence of ΔG^* on temperature given by the Gibbs–Helmholtz equation (shown in Equation 4.11),

$$\frac{d(\Delta G^*/T)}{dT} = -\frac{\Delta H^*}{T^2},$$

or directly by calorimetry, it is found that $\Delta H^* \approx 0$ or even negative at room temperature (see Table 4.1). This means that ΔH^* does *not* favor separation of the hydrocarbon and water. Indeed, close to room temperature, because heat is released upon hydrocarbon transfer to water ($\Delta H^* < 0$), the equilibrium constant for transfer to water *decreases* with increasing temperature, as given by the van't Hoff equation (4.13),

$$\frac{d \ln K_{eq}}{dT} = \frac{\Delta H^*}{RT^2}.$$

Thus, heating will not improve the solubility of hydrocarbons in water.

The interpretation of the hydrophobic effect emerges from these observations. At room temperature, the enthalpy of transfer of a hydrocarbon to water is negative or close to zero, thus slightly favoring dissolution. It is the *entropy* term, $T\Delta S^*$, which is negative and much larger than ΔH^* in absolute value, that determines the low solubility of a hydrocarbon in water. Water forms a hydration shell around a hydrocarbon, stabilized by favorable van der Waals (dispersion) interactions. It thus appears to become more *ordered*, or *structured*, than bulk water: hence the unfavorable entropy change. This hydration shell, however, is dynamic, not a rigid cage.

Placing a hydrocarbon in water replaces hydrogen bonds between water molecules by relatively weaker van der Waals interactions between water and hydrocarbon. Therefore, it was expected that ΔH^* would be positive—but it is not. One way to understand the observed $\Delta H^* \leq 0$ for hydrocarbon transfer from a nonpolar solvent to water is if the hydrogen bonds between water molecules in the hydration shell are stronger. Support for this idea comes from the positive values of the heat capacity of transfer (ΔC_p) from a nonpolar solvent to water (see Table 4.1). ΔC_p reflects mainly the change in the *heat capacity of water*, which *increases* as a hydrocarbon molecule is introduced in it. That is, when a hydrocarbon is placed in water, water appears to acquire additional modes of storing energy—in the hydrogen bonds of the hydration shell. As heat is furnished to the system, it is used to "melt" the hydration shells rather than to increase the temperature of the solution. The hydrophobic effect thus appears as a consequence of the hydrogen-bonding structure of water.

The Essence of the Hydrophobic Effect Arises from Transfer of a Hydrocarbon from Gas to Water

To understand the hydrophobic effect in depth, however, we need to break up the transfer process into two steps (**Figure 4.20**). Transfer

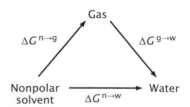

Figure 4.20 Decomposition of the transfer process of a hydrocarbon from a nonpolar solvent to water (n → w) into a transfer from nonpolar solvent to gas (n → g) and from gas to water (g → w).

Table 4.2 Gibbs energy of transfer (ΔG^* in kcal/mol, Ben-Naim standard process) of hydrocarbons at 25°C, from a nonpolar hydrocarbon liquid to water (n → w), from the gas phase to water (g → w), and from the gas to nonpolar liquid (g → n); and the two separate components of the transfer from gas to solvent: the Gibbs energy of cavity formation (ΔG_c) and the energy of interaction of the solute with the solvent (E_a)[a].

	n → w	g → w			g → n		
Solute	ΔG^*	ΔG^*	ΔG_c	E_a	ΔG^*	ΔG_c	E_a
Propane	+4.0	+2.0	+8.6	−6.6	−2.1	+4.3	−6.4
Butane	+4.8	+2.1	+10.4	−8.3	−2.7		
Isobutane	+4.9	+2.3	+10.1	−7.8	−2.5	+5.9	−8.4
Pentane	+5.7	+2.3	+12.3	−10.0	−3.5		
Neopentane	+5.4	+2.5	+11.1	−8.6	−2.9	+6.6	−9.5

[a]Data from the same sources as in Table 4.1.

of a hydrocarbon from a nonpolar solvent to water (n → w) can be separated into transfer from hydrocarbon to gas (n → g) and from gas to water (g → w). Table **Table 4.2** lists the Gibbs energies for the various components of the transfer process at 25°C.

The Gibbs energy of transfer of a hydrocarbon from a nonpolar solvent to water ($\Delta G^{n \to w}$) is the sum of the values for the two processes: from nonpolar solvent to gas and from gas to water,

$$\Delta G^{n \to w} = \Delta G^{n \to g} + \Delta G^{g \to w}. \qquad (4.68)$$

Since $\Delta G^{n \to g} = -\Delta G^{g \to n}$, we can write

$$\Delta G^{n \to w} = -\Delta G^{g \to n} + \Delta G^{g \to w}. \qquad (4.69)$$

So, there are two parts to $\Delta G^{n \to w}$. If we look at the values of ΔG^* in Table 4.2, we see that the two processes contribute almost equally to the total $\Delta G^{n \to w}$. Transfer from gas to water or to nonpolar solvent have about the same magnitude, and because they have opposite signs, they add up when combined according to Equation 4.69.

Now, we can think of the transfer of a hydrocarbon from the gas to a liquid (to water or to a nonpolar solvent) as consisting of two steps: first we create a cavity in the liquid and then we place the solute hydrocarbon in that cavity (**Figure 4.21**). Thus, the Gibbs energy of transfer from gas to liquid consists of two terms,

$$\Delta G^* = \Delta G_c + E_a. \qquad (4.70)$$

The first term, ΔG_c, is the Gibbs energy of cavity formation, which is the *work* required to *open the cavity* in the solvent. It involves only solvent–solvent interactions, and represents the *reorganization* of the solvent to accommodate the solute. In water, it may involve breaking some hydrogen bonds (but not many). The second term, E_a, is the *interaction energy* between the solute and solvent, which is released when the solute hydrocarbon is placed in the cavity. The term E_a involves only van der Waals dispersion interactions. Those interactions are favorable ($E_a < 0$). They occur between the hydrocarbon solute and the water molecules in the *hydration shell*.

We see from Table 4.2 that E_a is about the same in nonpolar solvents and water. Those terms constitute the largest contributions to ΔH^* of the transfer of alkanes from a pure hydrocarbon to water. That they are similar in both phases explains why $\Delta H^* \approx 0$ (see Table 4.1). We also see from Table 4.2 that the main difference in the two transfer processes is in the Gibbs energy of creating the cavity. This *cavity work*

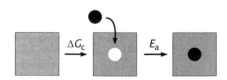

Figure 4.21 Decomposition of the transfer of a hydrocarbon (black sphere) from a gas to a solvent, represented as the opening of a cavity (white) in the solvent followed by inserting the solute in the cavity.

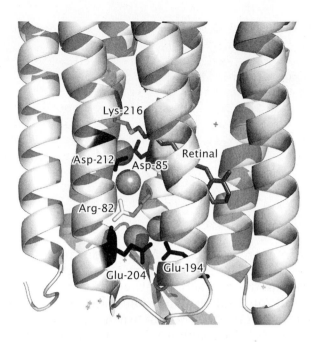

Figure 4.13 The nonpolar, transmembrane region of the integral protein bacteriorhodopsin includes several buried charged residues. Residues Asp-85, Asp-212, Glu-194, and Glu-204 are shown in black; Arg-82 is shown in white. The chromophore retinal (dark gray) is attached to Lys-216 (light gray). Water molecules, also buried inside the protein, are shown as gray spheres (PDB 1C3W).

$pK_a = 10$ to ≈ 7 (but can reach 5). They become deprotonated, and therefore neutral, at pH 7. Thus, the pK_a shifts to render the side chain *neutral inside the protein*, because the energy of a charge in a nonpolar environment is very large.

Arginine residues, however, are always charged, even when buried. The normal $pK_a = 12$ of arginine is large because its side chain has high affinity for the proton. Deprotonation, to achieve neutrality, would be too costly energetically. What happens then is that the protein interior rearranges to accommodate the charge. In some cases, another residue with the opposite charge (Asp or Glu) moves closer to the buried arginine to make a salt bridge, or water becomes associated with this arginine in the protein interior, making hydrogen bonds (see Figure 4.13).

4.5 VAN DER WAALS INTERACTIONS ARE WEAK BUT THEIR TOTAL CONTRIBUTION IN THE PROTEIN CORE IS SUBSTANTIAL

When we speak of van der Waals interactions, we usually have in mind London dispersion forces. These forces are always attractive, and result from induced dipole–induced dipole interactions. However, van der Waals interactions comprise both London dispersion forces and hard-core repulsions. We can write the van der Waals energy as a Lennard–Jones potential,

$$E = \frac{A}{r^{12}} - \frac{B}{r^6} \tag{4.62}$$

where A and B are positive constants, and r is the distance between atoms or molecules. **Figure 4.14** shows a plot of the van der Waals energy as a function of distance between two methane molecules.

Van der Waals Interactions are Short-ranged

At very close distance, when the electronic clouds begin to overlap (van der Waals contact), all molecules repel each other. This is the van der

Figure 4.14 Distance dependence of the van der Waals energy of methane (CH_4). The two gray circles represent two methane molecules, one at the origin and the other at the distance of lowest energy. Calculated from Equation 4.62.

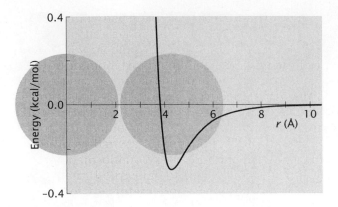

Waals repulsion energy, which decays with distance as $1/r^{12}$ (the first term in Equation 4.62). At slightly greater distances, sufficient to avoid electronic overlap, the interaction between two molecules is attractive. This is the result of London dispersion forces, which are usually pictured as resulting from attraction between instantaneous dipoles and induced dipoles. This interaction is strongest at very short distances, just at van der Waals contact, which corresponds to the energy minimum in Figure 4.14. The energy of the London dispersion interaction is negative (favorable) and decays with a $1/r^6$ distance dependence (the second term in Equation 4.62).

A striking illustration of the strength of van der Waals interactions at close contact comes from biology. Geckos adhere to walls, not because of suction cups, but because the base of their feet contains about 500,000 keratinous hairs (proteins). The van der Waals interactions between these fibers and the surface holds the gecko on vertical walls (**Figure 4.15**).

Van der Waals Interactions Stabilize the Nonpolar Core of Proteins

Because of their strong distance dependence, London dispersion forces are very sensitive to packing. In a loosely packed hydrocarbon medium, such as in a monolayer of a detergent at the air/water interface, the average interaction energy of a methylene group ($-CH_2-$) is about 0.4 kcal/mol, which is less than the thermal energy at room temperature (RT). However, in a densely packed crystalline hydrocarbon, the enthalpy of sublimation per $-CH_2-$ group is about 2 kcal/mol, which is comparable to the van der Waals contribution to the enthalpy of sublimation of ice (3.6 kcal/mol).

We can expect, then, that while London dispersion forces will not provide energy to bring large molecules together or allow for distant parts of a protein to sense each other, they will play an important role—perhaps even a predominant role—in the densely packed, nonpolar interior of a globular protein. The interactions between nonpolar side chains inside the protein core are not identical to those in a liquid hydrocarbon or solid hydrocarbon. Nevertheless, van der Waals interactions between nonpolar side chains are essential for protein stability. This is demonstrated by mutations of large amino acids to small amino acids.

Mutations Val → Ala and Leu → Ala were introduced in the nonpolar core of several proteins. Because a large side chain is replaced by a small one, a cavity is created inside the protein nonpolar core. The

Figure 4.15 The Moorish gecko (*Tarantola mauritanica*) adheres to vertical walls through van der Waals interactions. (Adapted from Almeida NF, Almeida PF, Gonçalves H et al. [2001] Anfíbios e Répteis de Portugal. With the permission of FAFAS.)

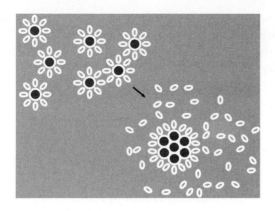

Figure 4.22 Interpretation of the hydrophobic effect: hydrocarbons (black) in dilute aqueous solutions are surrounded by a hydration shell of water molecules. As the hydrocarbon molecules associate to separate from water, the water molecules of the hydration shell are released, and the entropy increases.

(ΔG_c) is much larger in water than in a liquid alkane. In both cases $\Delta G_c > 0$, but in water it is larger than E_a in absolute value, and in the nonpolar solvent it is smaller. It is this difference in the cavity work that determines the hydrophobic effect.

The Gibbs energy of cavity formation in water arises exclusively from the reorganization of water molecules, which results in a decrease in the entropy of water at room temperature. The number of *configurations* (W) available to pure water is therefore reduced in the presence of the hydrocarbon solute, resulting in a decrease in entropy, according to the Boltzmann equation ($S = k \ln W$). This decrease in configurations occurs because of the interference of the hydrocarbon solute with the many ways of establishing hydrogen bonds among water molecules. Conversely, when the hydrocarbon molecules come together and segregate in water, the water molecules in the hydration shell are released and water regains more configurations (**Figure 4.22**).

The Heat Capacity of Water Increases when a Hydrocarbon is Placed in it

The heat capacity change (ΔC_p) of the system when a hydrocarbon is transferred from a nonpolar liquid (hydrocarbon) to water is always *positive* (see Table 4.1). The reason $\Delta C_p > 0$ is that the heat capacity *of water increases* significantly when a hydrocarbon is placed in it. If we now compare ΔC_p of transfer of a hydrocarbon from a nonpolar solvent to water (n → w) and from gas to water (g → w), we notice that both values are large and similar (**Table 4.3**). Thus, most of the ΔC_p for

Table 4.3 Heat capacity change for the transfer of hydrocarbons from the liquid hydrocarbon to water (n → w) and from the gas to water (g → w)[a].

Solute	ΔC_p	
	n → w	g → w
Propane	+74	+79
Butane	(+63)	+92
Isobutane	+81	
Pentane	+96	
Neopentane	+87	
Benzene	+54	+70

[a]Values of ΔC_p (cal/K/mol) at 25°C (see Table 4.1 for sources and notes).

$n \to w$ transfer originates from the $g \to w$ step (see Figure 4.20); most of it comes from the formation of the cavity in water.

To understand how a large positive ΔC_p can arise, consider the breaking of a hydrogen bond in water as an equilibrium reaction,

$$\text{Hbond (intact)} \rightleftharpoons \text{Hbond (broken)}.$$

Now let us define the equilibrium constant K_b for hydrogen-bond breaking in *bulk water* by (subscript b for b̲ulk),

$$K_b = \frac{[\text{broken Hbonds}]}{[\text{intact Hbonds}]} = e^{-\Delta G_b/RT}, \tag{4.71}$$

where the corresponding Gibbs energy for breaking a hydrogen bond consists of the enthalpic and entropic components,

$$\Delta G_b = \Delta H_b - T\Delta S_b. \tag{4.72}$$

Knowing the equilibrium constant, we can calculate the fraction of bonds broken in bulk water, using the partition function as we did before,

$$f_b = \frac{K_b}{1 + K_b}. \tag{4.73}$$

A similar set of equations can be written for the *hydration shell* (subscript h for h̲ydration) around a hydrocarbon in water. Then, by exactly the same procedure used to calculate the excess heat capacity function of protein unfolding (shown in Equations 4.22–4.24 and Appendix A), we obtain the heat capacity change caused by transfer of a hydrocarbon from a nonpolar solvent to water

$$\Delta C_p = n_h(C_{p,h} - C_{p,b})$$
$$= \frac{1}{RT^2}\left((\Delta H_h)^2 \frac{K_h}{(1 + K_h)^2} - (\Delta H_b)^2 \frac{K_b}{(1 + K_b)^2}\right). \tag{4.74}$$

Here n_h is the number of hydrogen bonds in the hydration shell, $C_{p,h}$ is the heat capacity of water in the hydration shell, and $C_{p,b}$ is the heat capacity of bulk water. We have assumed that ΔH_b does not vary appreciably with temperature.

Now, few hydrogen bonds are broken in the hydration shell or in the bulk, with or without hydrocarbon. Therefore, K_b and $K_h \ll 1$, and the denominators $(1 + K_h)^2$ and $(1 + K_b)^2$ do not change much with K_h or K_b. We can now see from Equation 4.74 that $\Delta C_p > 0$ can arise in two situations. If $\Delta H_h = \Delta H_b$, we need $K_h > K_b$; this means that the number of broken hydrogen bonds in the hydration shell is larger than in bulk water. If $\Delta H_h > \Delta H_b$ (you need more energy to break bonds in the hydration shell), then we can have $K_h = K_b$. A combination of these two extremes is also possible.

The Heat Capacity Change Determines how the Enthalpy and the Entropy of Transfer Vary with Temperature

The heat capacity change determines how the enthalpy and the entropy changes vary with temperature (shown in Equations 4.35 and 4.36),

$$\Delta H^*(T) = \Delta H^*(T_0) + (T - T_0)\Delta C_p \tag{4.75}$$

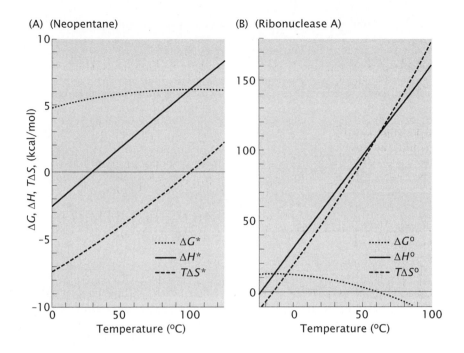

Figure 4.23 Temperature dependence of the thermodynamic functions for (A) transfer of neopentane from liquid neopentane to water and (B) the denaturation of ribonuclease A (pH 5). The curves for neopentane are calculated from the parameters in Table 4.1; those for ribonuclease, from the parameters indicated in Figure 4.6.

and

$$\Delta S^*(T) = \Delta S^*(T'_0) + \Delta C_p \ln \frac{T}{T'_0} \qquad (4.76)$$

where T_0 and T'_0 are reference temperatures, at which ΔH^* and ΔS^* are known. **Figure 4.23**A shows the variation of ΔH^* and ΔS^* with temperature for neopentane transfer from the pure hydrocarbon to water. Since $\Delta C_p > 0$ for hydrocarbon transfer to water and T is always a positive quantity, both ΔH^* and ΔS^* increase with temperature.

Close to room temperature (25°C), $\Delta H^* \approx 0$ and $\Delta S^* \ll 0$, but as the temperature increases, ΔH^* becomes positive and ΔS^* becomes less negative (smaller in absolute value). Eventually, a temperature is reached at which $\Delta S^* = 0$ ($T \approx 100°$C). Beyond this temperature, ΔS^* becomes positive. Over the entire temperature range, the Gibbs energy remains unfavorable ($\Delta G^* > 0$) for hydrocarbon transfer to water. Moreover, it remains almost constant, because of compensating effects of ΔH^* and ΔS^* in ΔG^* ($=\Delta H^* - T\Delta S^*$). The most positive value of ΔG^* of transfer is actually obtained when $\Delta S^* = 0$. (Why?)

Comparison of the temperature dependence of the thermodynamic functions of transfer of alkanes from a nonpolar solvent to water with that of protein denaturation demonstrates strong similarities. Figure 4.23 shows the corresponding graphs for neopentane (A) and ribonuclease A (B). The graphs are similar because the hydrophobic effect plays a major role in both cases. The figures differ in the ordinate scale because the protein is much larger. More important, the plots also differ in that the line for $T\Delta S$ (dashed) has a positive vertical shift in the plot for ribonuclease A compared to neopentane. This is because the entropy change in protein denaturation contains the large contribution from the conformational entropy increase, which is absent in neopentane.

The contribution of the hydrophobic effect to the heat capacity change upon protein denaturation is $\Delta C_p^D/n \approx 14$ cal/K/mol, where n is the number of residues. In a 100-residue protein, $\Delta C_p^D = 1.4$ kcal/K/mol. Since the entropy of transfer of hydrocarbon to water becomes smaller in absolute value as the temperature increases (see

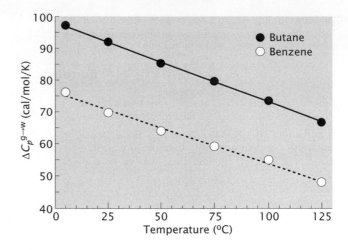

Figure 4.24 The temperature dependence of the heat capacity for the transfer of butane and benzene from gas to water. (Data from Makhatadze GI & Privalov PL [1994] *Biophys Chem* 50:285–291.)

Figure 4.23), we might be tempted to conclude that the hydrophobic effect disappears at high temperatures. However, this is not so. ΔC_p of transfer to water is positive and it determines the change of both ΔS^* and ΔH^* with temperature. The heat capacity change itself decreases with temperature (**Figure 4.24**). However, unlike the entropy of transfer, ΔC_p does *not* become zero at any attainable temperature. ΔC_p is positive close to room temperature, and it is still positive at high temperatures, even when $\Delta S^* = 0$, and at temperatures above that. Thus, the special properties of water do not disappear even if $\Delta S^* = 0$. Indeed, the large positive heat capacity change associated with the transfer of a hydrocarbon to water is the very *signature of the hydrophobic effect*.

The hydrophobic effect on the heat capacity arises mainly from the reorganization of the hydrogen-bonding structure of water to create a cavity where the nonpolar solute can be placed. The different organization of the hydration shell changes the distribution of hydrogen bonds and therefore the enthalpy distribution, which is measured by the heat capacity.

The Hydrophobic Effect Makes the Largest Contribution to Protein Stability

In summary, the hydrophobic effect is essential to protein stability. Transfer of the nonpolar amino acid side chains from the protein hydrophobic core, in the folded state, to water, in the unfolded state, makes the largest contribution to the Gibbs energy of protein unfolding at 25°C. **Table 4.4** lists the Gibbs energies of transfer of amino acid side chains from three nonpolar solvents (octanol, cyclohexane, and the interior of a lipid bilayer) to water.

These *hydrophobicity scales* are in qualitative agreement with each other. Quantitatively, however, they depend on the nonpolar solvent used. Which solvent is most appropriate to mimic the nonpolar interior of a protein is a matter of debate.

From studies of protein unfolding, we know that the hydrophobic effect contribution to protein stability arising from the burial of nonpolar amino acid side chains is ≈ 1.4 kcal/mol per residue. In a small protein with 100 residues, this amounts to $\Delta G \approx 140$ kcal/mol. This is a very large number, 10 times larger than the entire ΔG^o of protein unfolding (~ 10 kcal/mol) and similar to the entire ΔH^o of protein

Table 4.4 Hydrophobicity scales: Gibbs energy of transfer of amino acid side chains and a non-hydrogen-bonded peptide group (CONH) from a nonpolar medium to water.

Amino Acid	Gibbs energy of transfer (kcal/mol) nonpolar solvent → water		
	$\Delta G^{oct \to w \, a}$	$\Delta G^{alk \to w \, b}$	$\Delta G^{bil \to w \, c}$
Arg(+)	−0.7	−15	−2.1
Lys(+)	−1.7	−5.6	−3.8
Asp(−)	−2.5	−8.7	—
Asp(0)	+0.7	—	−1.4
Glu(−)	−2.5	−6.8	—
Glu(0)	+1.0	—	−0.1
Asn	+0.3	−6.6	−1.9
Gln	+0.4	−5.5	−1.4
His(+)	−1.2	—	−3.2
His(0)	+1.0	−4.7	—
Ser	+0.7	−3.4	−0.3
Thr	+0.9	−2.6	−0.2
Gly	0	+0.9	−0.2
Ala	+0.7	+1.8	+1.6
Pro	+1.0	—	+3.1
Cys	+1.2	+1.3	+1.1
Met	+1.8	+2.4	+2.3
Val	+1.6	+4.0	+2.3
Ile	+2.3	+4.9	+3.1
Leu	+2.4	+4.9	+3.3
Phe	+2.9	+3.0	+3.8
Tyr	+1.9	−0.1	+2.7
Trp	+3.2	+2.3	+2.0
CONH	−2.0	$−7.8^{d}$	—

$^{a}\Delta G^{oct \to w}$ is the Wimley–White scale for transfer from octanol → water (at pH 1 or 9). Data from Wimley WC, Creamer TP & White SH [1996] *Biochemistry* 35:5109–5124.
$^{b}\Delta G^{alk \to w}$ is the Radzicka–Wolfenden scale for transfer from cyclohexane → water at pH 7 (values listed under the predominant form at pH 7). Data from Radzicka A & Wolfenden R [1988] *Biochemistry* 27:1664–1670.
$^{c}\Delta G^{bil \to w}$ is the Moon–Fleming scale for transfer from a lipid bilayer → water at pH 3.8. Data from Moon CP & Fleming KG [2011] *Proc Natl Acad Sci USA* 108:10174–10177.
^{d}Value from Ben-Tal N, Sitkoff D, Topol IA, et al. [1997] *J Phys Chem B* 101:450–457.

unfolding. The hydrophobic effect probably contributes more than 60% of the total Gibbs energy of protein stability.

4.8 DISULFIDE BONDS DESTABILIZE THE UNFOLDED STATE

Disulfide bonds are common in small secreted proteins, where they are essential for protein stability. Examples are the hormone insulin, bovine pancreatic trypsin inhibitor (BPTI), and phospholipase A2, which is a major component of the venom of vipers, such as the cottonmouth. The cellular exterior is an oxidizing environment (because of the presence of oxygen) and thus favors the formation of disulfide bonds (S–S). The cellular interior is a reducing environment, which favors the reduced, thiol state (SH).

Figure 4.25 The primary effect of disulfide bonds on protein denaturation is to decrease the entropy, and thus increase the Gibbs energy, of the unfolded state.

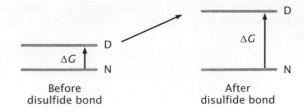

Before disulfide bond

After disulfide bond

Disulfide bonds increase protein stability, but they do so in a way that is not intuitive. At first sight, they appear to "hold together" the tertiary structure of the protein. However, the interactions present in the native structure do not change (except for the S–S covalent bond itself) when the disulfide bond is reduced to two SH groups. All other interactions remain in place. This seems to indicate that the S–S bond has little effect on the Gibbs energy of the *folded* state (**Figure 4.25**).

What happens in the unfolded state? If there are no disulfide bonds, when the protein unfolds it gains a very large amount of conformational entropy (ΔS_{conf}). Consequently, the Gibbs energy decreases by $-T\Delta S_{conf}$. However, if the protein contains disulfide bonds, they *restrict* the possible conformations of the *unfolded* state. Therefore, the *conformational entropy* of the unfolded state is *smaller* in the presence of disulfide bonds. Thus, disulfide bonds increase the Gibbs energy of the unfolded state, rather than decrease the Gibbs energy of the folded state (see Figure 4.25). The consequence is that ΔG of denaturation increases: the native state is stabilized *relative* to the denatured state.

4.9 COSOLVENTS CHANGE PROTEIN STABILITY AND SOLUBILITY

The stability of proteins is not only influenced by temperature and pH, but also by the presence of other solutes, or *cosolvents*, such as salts, denaturants, sugars, polyols, detergents, and other organic solvents, such as methanol, ethanol, or trifluoroethanol. Understanding the effects of cosolvents on protein stability and solubility is important for several reasons. First, the mechanisms by which they affect protein stability teach us valuable lessons about the interactions that stabilize the native structures of proteins. They also allow us to understand how protein folding and stability are controlled in nature. Finally, the subject has practical utility in the purification and refolding of proteins.

The effects of cosolvents on protein stability and solubility have been known for a long time. Hofmeister (1888) ordered several salts in a series according to their capacity to precipitate proteins from egg white. In fact, one method to fractionate proteins is based on differential precipitation upon addition of high concentrations (1–4 M) of ammonium sulfate, $(NH_4)_2SO_4$. This process is known as *salting out*. Salts that precipitate proteins also stabilize them; the proteins retain their three-dimensional structure in the precipitate. The opposite process, *salting in*, occurs when the solubility of the protein is increased by adding salts. It takes place when low concentrations of salts are added, which increase the protein solubility in water by screening charges on the protein surface. A familiar example is the dissolution of egg white (mainly albumin and avidin) in water: it dissolves much better if you add a little salt.

Denaturants and Protective Osmolytes Compensate the Effects of Each Other

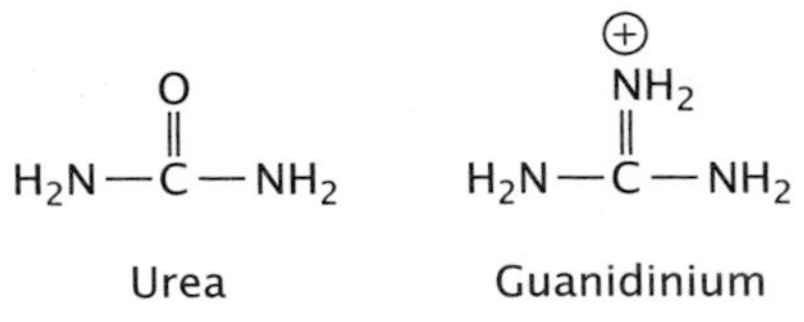

A common method to stabilize proteins and to preserve enzyme activity is to store them in aqueous solution with high concentrations of sucrose (1 M) or glycerol (10%). On the other hand, high concentrations (4–8 M) of denaturants, such as guanidinium chloride and urea (**Figure 4.26**), cause proteins to unfold. They also increase the solubility of denatured proteins and nonpolar compounds in water.

Organisms protect themselves from osmotic shock and freezing by using mixtures of denaturants and protective osmolytes (**Figure 4.27**). Fish and amphibians accumulate urea to increase their cellular osmotic pressure, but compensate the denaturing effect of urea on their proteins by also accumulating a protective osmolyte in an exact molar ratio. The osmolytes used by these organisms are zwitterionic molecules, such as betaine, trimethylamine oxide (TMAO), and sarcosine, or neutral compounds, such as sugars, polyols, and amino acids (glycine). Urea destabilizes proteins: the T_m of the protein decreases with increasing urea concentration. However, if urea and TMAO are added in a fixed 2:1 ratio, even at high concentrations, the T_m of the protein does not change. These seemingly unrelated observations can be understood in a simple manner by considering how cosolvents interact with the *protein surface*.

Figure 4.26 Structures of urea and the guanidinium ion.

Proteins are Unfolded by Denaturants

In a typical experiment in protein denaturation, the concentration of urea is increased in the range of 0–8 M, and the fraction of denatured protein (f_D) is measured by monitoring a suitable physical property that is proportional to the amount of folded or unfolded protein, such as a circular dichroism spectrum (see Figure 2.49) or the fluorescence intensity of a tryptophan residue in the protein (see Figure 2.21).

Figure 4.28A shows a denaturation curve obtained in this type of experiment. At low urea concentration, the proteins are folded; the dashed line on the initial region of the plot represents the dependence of the intrinsic fluorescence F_N of the folded protein on urea concentration. At high urea concentration, the proteins are unfolded; the dashed line in the final region of the plot represents the dependence of the intrinsic fluorescence F_D of the unfolded state on urea concentration. In the transition region, both folded and unfolded proteins are present; the fluorescence intensity (F_i) is a weighted average of the fluorescences, F_N and F_D, of the two states, weighted by their fractions, f_N

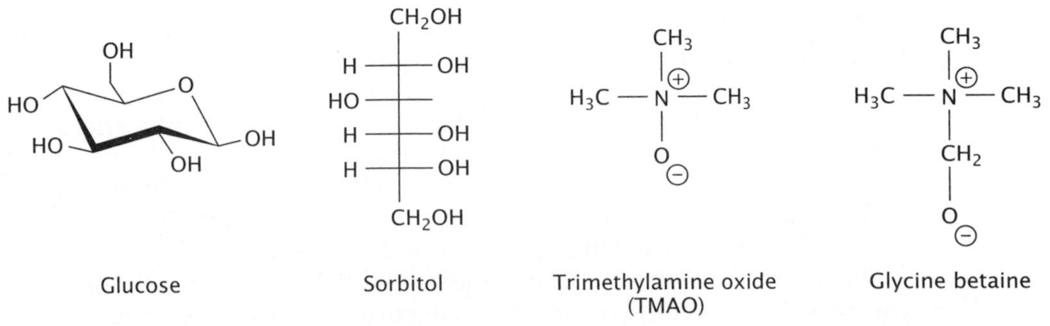

Figure 4.27 Chemical structures of some protective osmolytes.

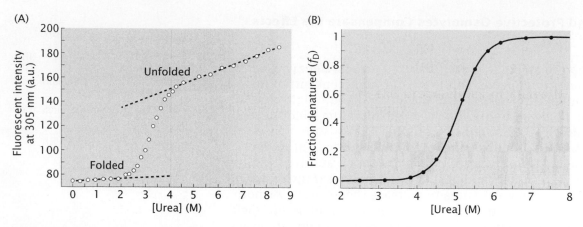

Figure 4.28 Denaturation of ribonuclease A by urea. (A) Experimental fluorescence intensity data of ribonuclease A at pH 3.5 (open circles). The dashed lines show the fluorescence dependence of the folded and unfolded states on the urea concentration. (B) Normalized fraction of unfolded protein (f_D) as a function of the urea concentration, calculated for ribonuclease A at room temperature, pH 5. (A, Data from Pace CN, Laurents DV & Thomson JA [1990] *Biochemistry* 29:2564–2572.)

and f_D,

$$F_i = f_N F_N + f_D F_D. \tag{4.77}$$

The data points in Figure 4.28B can be obtained from similar data at pH 5. The shape of this curve can be understood using the partition function. Given the equilibrium constant for unfolding, $K = [D]/[N]$, the probabilities of the two states can be determined. The native state is the reference, with statistical weight 1, and the denatured state has statistical weight K. The partition function is

$$Q = 1 + K \tag{4.78}$$

and probabilities of each state, or their fractions, are

$$f_N = \frac{1}{1 + K} \tag{4.79}$$

$$f_D = \frac{K}{1 + K}. \tag{4.80}$$

Figure 4.28 shows that f_D, and therefore K, varies with urea concentration. In the transition region, which is the steeper part of the curve, in this case for [urea] = 4–6 M, we can calculate the equilibrium constant by

$$K = \frac{f_D}{f_N}, \tag{4.81}$$

where $f_N = 1 - f_D$. (Outside the transition region, the protein is almost all folded or all unfolded ($f_D \approx 0$ or $f_N \approx 0$); therefore, the error in calculating K from Equation 4.81 is very large.)

If we calculate K as a function of urea concentration through Equation 4.81 from the data in Figure 4.28, we obtain the Gibbs energy of denaturation through $\Delta G^o = -RT \ln K$. The plot of ΔG^o as a function of urea concentration is shown in **Figure 4.29**A. You can obtain similar data for addition of a protective osmolyte, such as TMAO, and obtain ΔG^o in the same way, which is shown in Figure 4.29B.

The plots in Figure 4.29 are linear; the Gibbs energy of *unfolding* varies with cosolvent concentration according to the equations

$$\Delta G^o = \Delta G^o_{water} + m[urea] \tag{4.82}$$

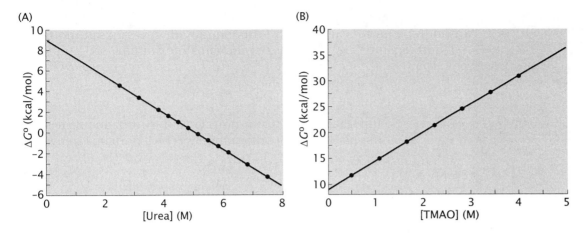

Figure 4.29 The ΔG° of unfolding of ribonuclease A as a function of the concentration of (A) urea, a denaturant, and (B) TMAO, a stabilizing osmolyte.

or

$$\Delta G^{\circ} = \Delta G^{\circ}_{\text{water}} + m[\text{TMAO}], \qquad (4.83)$$

where $\Delta G^{\circ}_{\text{water}}$ is the Gibbs energy of denaturation in water, and the slope is called the *m*-value. The *m*-value is negative for urea and other denaturants, and is positive for TMAO and other protective osmolytes or stabilizers.

In closing, note that we can analyze the curve in Figure 4.28B directly by using Equation 4.80 for the fraction f_{D} of denatured protein. The equilibrium constant K in the expression for f_{D} is given by Equation 4.25. Using Equation 4.82 for ΔG°, we can write it explicitly as a function of urea concentration,

$$K = e^{-(\Delta G^{\circ}_{\text{water}} + m[\text{urea}])/RT}. \qquad (4.84)$$

Cosolvents Affect Proteins by Preferential Association or Exclusion from the Protein Surface

Now we know how *to measure* the effect of cosolvents on protein stability. Our next task is to understand its physical basis. How do cosolvents interact with proteins? What is the relation of the *m*-value to the thermodynamics of protein–cosolvent interactions? An important observation is that the effect of cosolvents on protein stability and solubility is only noticeable at very *high* concentrations of the cosolvent. This indicates that cosolvent interactions with the protein are weak and *nonspecific*; they do not involve binding to a specific site on the protein. The key to understanding cosolvent effects is the difference between the affinity of the protein *surface* toward *water* and toward the *cosolvent*.

Cosolvents that accumulate at the protein surface cause denaturation. They are said to show *preferential binding* to the protein (**Figure 4.30**A). Binding is not specific, but the concentration of denaturant around the protein surface is larger than in the bulk solution. Cosolvents that are excluded from the protein surface *stabilize* the protein and cause protein precipitation (see Figure 4.30B). This is called *preferential hydration* or cosolvent *exclusion*: as the cosolvent is excluded from the surface, the concentration of *water* is larger at the surface than in the bulk solution.

Figure 4.30 Preferential binding (A) and preferential hydration (B) of a protein (gray ellipse) by cosolvent molecules (black) in an aqueous solution. (Based on Timasheff SN & Arakawa T [1989] In Protein Structure: A Practical Approach [TE Creighton ed] pp 331–345. IRL Press.)

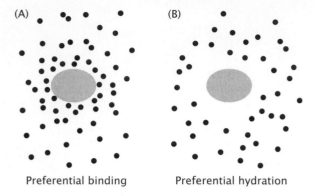

Preferential binding   Preferential hydration

Cosolvents Change the Gibbs Energy Difference between Protein States

To understand the effect of cosolvents on protein stability and solubility we need to consider three states of the protein: soluble (native, monomeric), precipitated (native), and denatured (**Figure 4.31**). We also need to know two pieces of information: the difference in the *affinity* of the *protein surface* for water and for the cosolvent, and the *changes* in protein surface *area* between the soluble (folded), the precipitated, and the unfolded states of the protein.

Consider first the effect of cosolvents on protein *stability*. Here the relevant states of the protein are the folded and the unfolded conformations. Let us examine protein denaturation by urea and protein stabilization by TMAO. The cosolvent affects mainly the Gibbs energy of the *unfolded* state, because the interaction occurs with the protein surface (with protein groups exposed to the solvent) and the unfolded state has a *larger surface*; it is extended, whereas the folded state is globular. **Figure 4.32** shows a diagram of the Gibbs energy of the protein states in the absence and in the presence of cosolvents. Urea decreases the Gibbs energy of the *denatured* state because it interacts favorably with its surface. TMAO *increases* the Gibbs energy of the denatured state because it interacts less favorably with the surface than water does.

Consider next the effect of cosolvents on *solubility*. Here the relevant states are the soluble protein and the precipitated protein. Both states are *folded*. Some salts, such as $(NH_4)_2SO_4$, promote protein precipitation because their unfavorable interaction with the protein surface renders it a *high-energy interface*. Minimizing this energy means minimizing the surface: the protein precipitates in reponse to the addition of $(NH_4)_2SO_4$ (see Figure 4.31). The interaction affects mainly the Gibbs

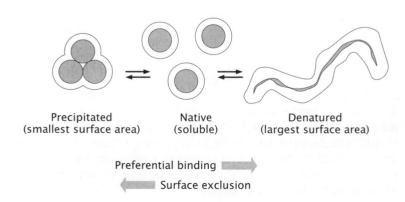

Precipitated   Native   Denatured
(smallest surface area)   (soluble)   (largest surface area)

Preferential binding ▶
◀ Surface exclusion

Figure 4.31 Effect of precipitation and denaturation on the size of the protein–solution interface.

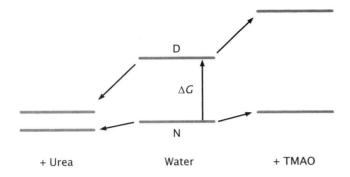

energy of the soluble, native form of the protein (**Figure 4.33**), because the precipitate has a smaller surface area.

In summary, we must always make a *comparison*. On protein *stability*, we must compare the effect of cosolvent on the Gibbs energy of the native and denatured states (see Figure 4.32). If the cosolvent increases ΔG of denaturation, it stabilizes the protein; if the cosolvent decreases ΔG of denaturation, it destabilizes the protein. On protein *solubility*, we must compare the effect of cosolvent on the Gibbs energy of the soluble and precipitated states (see Figure 4.33). In protein denaturation, the cosolvent interacts mainly with the unfolded state, because it has a larger surface area than the folded state. In protein precipitation, the cosolvent interacts mainly with the soluble protein, because it has a larger surface area than the precipitate.

Preferential Binding and Exclusion are Defined in Precise Thermodynamic Terms

We have now understood the effects of cosolvents qualitatively. We are ready to learn how to express them precisely and how to relate preferential binding or exclusion to the *m*-value obtained from experiment. Three parameters are important in understanding the cosolvent effects on protein stability and solubility: the *Gibbs energy of transfer* of the protein from water to a cosolvent solution ($\Delta\mu_p$), the *preferential interaction parameter* (μ'), and the *preferential binding parameter* (Γ).

The Gibbs energy of transfer $\Delta\mu_p$ of the protein is the difference between the chemical potential of the protein (p) in water (μ_p^w) and in the cosolvent solution (μ_p^{cs}). It measures the preference of the protein for residing in a cosolvent solution in comparison with pure water,

$$\Delta\mu_p = \mu_p^{cs} - \mu_p^w. \tag{4.85}$$

The chemical potential of the protein is given by

$$\mu_p = \mu_p^0 + RT \ln a_p, \tag{4.86}$$

where the protein *activity* $a_p = \gamma_p[p]$ is the product of the concentration [p] by the activity coefficient (γ_p), and μ_p^0 is the standard chemical potential of the protein. A similar equation applies to the cosolvent (cs), or to any chemical component in general. Note that the chemical potential of a component always increases with its own concentration (if all else is kept constant).

The preferential interaction parameter μ' is a *differential* Gibbs energy of interaction of the protein with the cosolvent, relative to the interaction with water. It measures the effect of the concentration of cosolvent on the chemical potential of the protein, or the effect of protein concentration on the chemical potential of cosolvent,

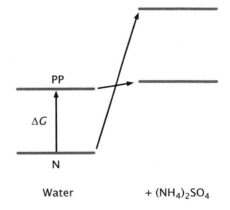

Figure 4.33 The effect of ammonium sulfate on the solubility of the native protein occurs mainly on the Gibbs energy of the soluble native protein (N), not the precipitate (PP).

which are identical,

$$\mu' = \left(\frac{d\mu_p}{d[cs]}\right)_{[p]} = \left(\frac{d\mu_{cs}}{d[p]}\right)_{[cs]}. \tag{4.87}$$

In Equation 4.87 the subscripts indicate the variables that are kept constant. Temperature (T) and pressure (P) are also constant, but are not explicitly indicated in the subscript to simplify the notation. The two differential chemical potentials have the same value: they are thermodynamically *linked*. We also say there is a *linkage* relationship between them. In words: if the cosolvent stabilizes the protein in solution, then the protein stabilizes the cosolvent in solution to the same extent (at constant water concentration). Or simply put, likes and dislikes are mutual. You can think of μ' as a Gibbs energy change due to the interaction. If the interaction is favorable, $\mu' < 0$. If the interaction is unfavorable, $\mu' > 0$. We use a *prime* in μ' to remind ourselves that it is a derivative of a chemical potential.

Consider a cosolvent solution in water. To this solution, you add a small amount of protein $d[p]$. The *preferential binding parameter* Γ is defined by

$$\Gamma = \left(\frac{d[cs]}{d[p]}\right)_{\mu_{cs}}. \tag{4.88}$$

Γ is the amount of cosolvent $d[cs]$ that you would have to add to the solution (or remove, if it has a negative sign) to restore the equilibrium that existed *before* protein was added. That is, you are trying to maintain constant the chemical potential of the cosolvent (μ_{cs}), or to maintain its bulk concentration, when $d[p]$ protein is added.

Now, there is a simple relation between Γ and μ'. From Equation 4.88 we can write,

$$\Gamma = -\frac{(d\mu_{cs}/d[p])_{[cs]}}{(d\mu_{cs}/d[cs])_{[p]}} \tag{4.89}$$

$$= -\frac{\mu'}{(d\mu_{cs}/d[cs])_{[p]}} \tag{4.90}$$

$$\approx -\frac{\mu'[cs]}{RT}. \tag{4.91}$$

To derive this relation, begin with Equation 4.88 and use the formula from calculus

$$\left(\frac{dx}{dy}\right)_z \left(\frac{dy}{dz}\right)_x \left(\frac{dz}{dx}\right)_y = -1, \tag{4.92}$$

and also

$$\left(\frac{dy}{dz}\right)_x = \frac{1}{(dz/dy)_x}, \tag{4.93}$$

setting $x = [cs]$, $y = [p]$, and $z = \mu_{cs}$. Then write the equivalent of Equation 4.86 for the cosolvent using the concentration instead of the activity (this is why Equation 4.91 is approximate),

$$\mu_{cs} = \mu_{cs}^o + RT \ln[cs], \tag{4.94}$$

and differentiate it with respect to $[cs]$ to obtain

$$\frac{d\mu_{cs}}{d[cs]} = \frac{RT}{[cs]}. \tag{4.95}$$

Substitution of Equation 4.95 in 4.90 yields Equation 4.91.

These equations appear complicated but have a simple interpretation. Suppose that a cosolvent, TMAO for example, is *excluded* from the protein surface. Then, if we begin with a solution of TMAO in water and add protein, the chemical potential of TMAO increases (its bulk concentration increases), because it is "repelled" by the protein to the bulk. The preferential interaction parameter is positive, $\mu' = (d\mu_{cs}/d[p]) > 0$ (the interaction is unfavorable). To restore the chemical potential of TMAO, we must *remove* it from the bulk solution; thus, $\Gamma = (d[cs]/d[p])_{\mu_{cs}} < 0$. "Negative binding" means *depletion* of the cosolvent from the protein surface. Note that the factor $(d\mu_{cs}/d[cs])_{[p]}$ in the denominator of Equation 4.91 is always positive, because the chemical potential of a component increases with its own concentration.

Conversely, if you add protein to a solution of urea in water, urea becomes concentrated around the protein, because the protein prefers urea to water. Thus, urea is depleted from the bulk solution. To maintain the chemical potential of urea constant, its bulk concentration needs to be *increased*. Therefore, $\Gamma = (d[cs]/d[p])_{\mu_{cs}} > 0$ for urea, meaning "positive binding." Note that it is the bulk concentration that sets the chemical potential: the bulk is the *reservoir* of cosolvent.

To study cosolvent effects on protein *stability*, we need to consider the changes in μ' and Γ between the native and the denatured state as a protein unfolds. The difference between the preferential interaction parameter in protein unfolding is the change in μ',

$$\Delta\mu' = \mu'^D - \mu'^N. \tag{4.96}$$

The difference between the preferential binding parameter is the change in Γ that accompanies protein denaturation,

$$\Delta\Gamma = \Gamma^D - \Gamma^N. \tag{4.97}$$

The relation of the *m*-value to $\Delta\mu'$ and $\Delta\Gamma$ can now be derived. The *m*-value is defined by its relation to the change of ΔG^o of denaturation with solvent concentration,

$$m = \frac{d\Delta G^o}{d[cs]} \tag{4.98}$$

$$= \frac{d\mu_p^D}{d[cs]} - \frac{d\mu_p^N}{d[cs]}$$

$$= \mu'^D - \mu'^N$$

$$= \Delta\mu'. \tag{4.99}$$

So, you see, the *m*-value is just the change $\Delta\mu'$ in the preferential interaction parameter. Now, from Equation 4.91 we know that $\Delta\mu'$ and $\Delta\Gamma$ are related by

$$\Delta\Gamma \approx -\frac{\Delta\mu'[cs]}{RT}. \tag{4.100}$$

Therefore, we can also relate the experimental *m*-value to $\Delta\Gamma$ by

$$m \approx -\frac{RT\Delta\Gamma}{[cs]}. \tag{4.101}$$

In summary, the *m*-value obtained experimentally is a direct measure of preferential binding or exclusion. Urea accumulates more around the unfolded than around the folded protein. Therefore, $\Gamma^D > \Gamma^N$, which means that $\Delta\Gamma > 0$ and $m < 0$. Conversely, TMAO and other protective osmolytes are excluded from the unfolded protein surface more than from the folded structure. Hence, $\Gamma^D < \Gamma^N$, which means that $\Delta\Gamma < 0$ and $m > 0$.

Urea and Guanidinium Bind Preferentially to Peptide Groups (CONH) and to a Lesser Extent to Nonpolar Groups

We have understood the effects of cosolvents from a thermodynamic point of view, in terms of the extent of the protein surfaces exposed in each state (folded, unfolded, precipitated) and the preferential binding or exclusion of the cosolvents from those surfaces. Now we need to find the molecular mechanism of this preference or exclusion.

Let us begin with denaturants. Urea and guanidinium chloride (GuHCl) are the most common protein denaturants and solubilizers (see Figure 4.26). Urea can donate hydrogen bonds, with its N—H groups, and can accept them with its C=O group. These are the same groups (CONH) that establish the hydrogen bonds in the polypeptide backbone. A hydrogen bond between N—H of urea and the peptide C=O group is especially strong. Urea also has a favorable interaction with the aliphatic carbon atoms of the amino acid side chains, which are exposed in the denatured state. Similarly, GuH^+ interacts favorably with the peptide C=O group, even more so than urea.

Protective Osmolytes are Preferentially Excluded from Peptide Groups (CONH)

Protective osmolytes or stabilizers, such as TMAO and glycine betaine (see Figure 4.27), are accumulated by living organisms to *compensate* for the effects of urea. What is the molecular mechanism of their preferential exclusion, relative to water, from the protein surface? The oxygen atom of the C=O groups of the peptide backbone is the main contributor to the backbone surface; therefore, interactions with it determine to a large extent the ability of cosolvents to compete with water. Neither TMAO nor betaine contain any hydrogen bond donors. Therefore, they cannot compete with water for the interaction with the C=O oxygen, and are excluded from the protein surface, which is thus preferentially hydrated.

Sugars and Polyols Increase the Surface Tension of Water

Sugars, such as glucose and sucrose, and polyols, such as glycerol and sorbitol, are also natural protective osmolytes (see Figure 4.27). Sugars and polyols stabilize proteins by increasing the surface tension of water. They have a high concentration of hydroxyl groups and fit very well in the hydrogen-bonding network of water. Moreover, the interaction between the peptide group (CONH) and the sugars or polyols is less favorable than the interactions of each of them separately with water. The overall effect is the exclusion of these compounds from protein surfaces (preferential hydration) (see Figure 4.30) and an increase in protein stability.

For example, sorbitol stabilizes ribonuclease A (**Figure 4.34**). The Gibbs energies of both native and denatured states increase in the presence of sorbitol; but because the denatured state has a larger surface area, it is *more destabilized* than the native state, resulting in overall stabilization of the folded protein.

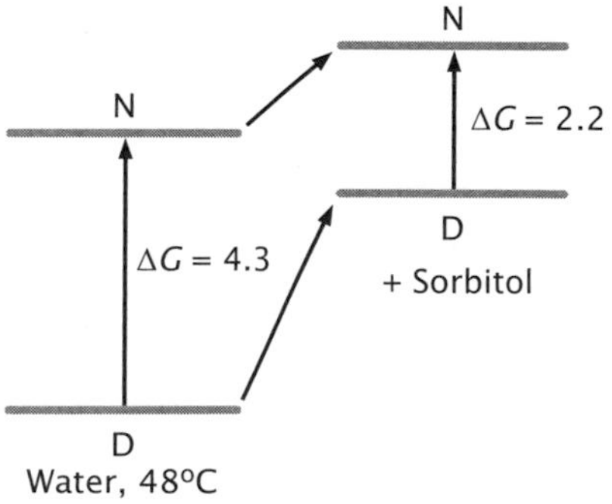

Figure 4.34 Effect of sorbitol on the stability of ribonuclease A at 48°C. ΔG in kcal/mol. Note that at this temperature the denatured state is more favorable. (Based on Timasheff SN [1993] *Annu Rev Biophys Biomol Struct* 22:67–97.)

The Effects of Hofmeister Ions Arise from Different Interactions with Peptide Groups (CONH) and Nonpolar Groups

At low concentrations, ions cause *salting in* through screening of Coulomb potentials from the protein in water. Now we want to understand the effect of Hofmeister ions at *high concentration*, typically >1 M. The Hofmeister series for anions and cations are the following:

Salting out   *Salting in*
$$Na^+ > K^+ > NH_4^+ > Li^+ > Mg^{2+} > Ca^{2+} > GuH^+$$
$$SO_4^{2-} > HPO_4^{2-} > F^- > CH_3COO^- > Cl^- > NO_3^- > Br^- > I^- > ClO_4^- > SCN^-$$

Minor changes in the rank order occur depending on the specific process examined. Ions on the left side of the series cause *salting out*, which means they *decrease* protein solubility in water. Ions on the right side of the series cause *salting in*, meaning that they *increase* protein solubility. In a salt, the effect of cations and anions is additive, with the anion usually having a greater effect.

Ions that *decrease* the solubility of the native state *increase the surface tension* of water. Contact between the cosolvent solution and any surface becomes more unfavorable than contact between pure water and that surface. This leads to cosolvent exclusion and preferential hydration of the protein surface. The protein–solution interface becomes a high Gibbs energy region, and the system responds to decrease its Gibbs energy by decreasing the amount of interface: the proteins precipitate (see Figure 4.31). High concentrations of salts from the left side of the Hofmeister series, such as Na_2SO_4 or $(NH_4)_2SO_4$, decrease protein solubility and are commonly used to precipitate proteins.

Hofmeister ions also affect protein stability, as shown for example by their effect on the T_m of ribonuclease A in **Figure 4.35**. For example, salts such as Na_2SO_4 or $(NH_4)_2SO_4$ (from the left side of the series) increase protein stability, because they are preferentially excluded from the protein surface, and the denatured protein has a larger surface area than the native structure. Guanidinium (GuH^+), on the other hand, is a common denaturant. However, GuH^+ salts can stabilize the native state of proteins depending on the counterion used. For example, $(NH_4)_2SO_4$ is a strong protein stabilizer and precipitant, whereas $GuHSCN$ is a strong denaturant and solubilizing agent, but $GuHSO_4$ is a mildly stabilizing agent.

The molecular mechanism of the Hofmeister ions is still debated, after more than 100 years of research. Understanding it involves the same principles as in the accumulation of urea and the exclusion of protective osmolytes from the protein surface. Namely, the effect of the ions in protein solubilization and stabilization depends on whether they interact favorably or unfavorably *relative to water* with the different components of the protein surface. When a protein unfolds, the *additional* solvent-exposed area is about 2/3 nonpolar and 1/3 polar (peptide groups and polar side chains). The effect of Hofmeister ions on *stability* depends on their preferential accumulation or exclusion from contact with the newly exposed groups. Their effect on *solubility* depends on the interaction with the surface of the native protein, which is mostly polar, but the interaction with the peptide group is complex.

Figure 4.36 summarizes our current understanding on Hofmeister ion effects in the form of a two-dimensional plot. The *x*-axis represents the interaction with backbone peptide groups (CONH); the

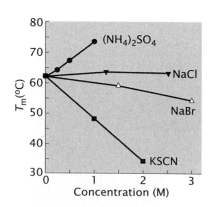

Figure 4.35 Effect of Hofmeister salts on the melting temperature (T_m) of ribonuclease A. (Data from von Hippel PH & Wong K-Y [1965] *J Biol Chem* 240:3909–3923.)

Figure 4.36 Two-dimensional graph of Hofmeister ion effects. The *x*-axis represents the interaction with backbone peptide groups (CONH). The *y*-axis represents the interaction with nonpolar groups. The solid diagonal line represents the order of the Hofmeister ions from salting out (bottom left) to salting in (top right). The position on the Hofmeister line is obtained by projecting the ion position onto this line, as indicated by the dashed line. The gray line represents the "null point," where the effects of binding and exclusion cancel each other, so those ions are "neutral" with regard to salting in or out. (Qualitative, based mainly on Pegram LM & Record MT [2008] *J Chem Phys* 112: 9428–9436, supplemented by Record MT, Guinn, E, Pegram L & Capp M [2013] *Faraday Discuss* 160:9–44, and Rembert KB, Paterová J, Heyda J, et al. [2012] *J Am Chem Soc* 134:10039–10046.)

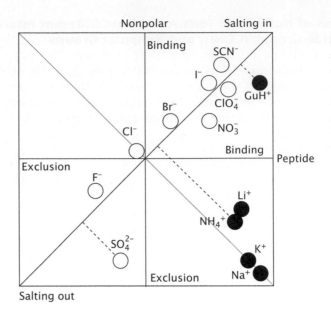

y-axis represents the interaction with nonpolar groups. The separation in those two axes is not simple because different calculations weigh differently the components of the amide groups (C,N,O), and it is difficult to separate those components experimentally. However, the projection of each point in Figure 4.36 onto the solid diagonal line gives a measure of the Hofmeister effect of each ion.

Cations preferentially accumulate around backbone peptide groups, probably because of attraction to the partial negative charge of the C=O oxygen (see Figure 2.30), but are strongly excluded from nonpolar groups. Conversely, anions are repelled from the C=O oxygen. However, large, "soft" anions (ions in which the negative charge resides on atoms of the third and fourth rows of the periodic table), such as I^- and ^-SCN, interact favorably with the partial positive charge on the amide nitrogen atoms (see Figure 2.30) of the polypeptide backbone and with the α-carbon. This softens their exclusion or even attracts them to the backbone peptide groups.

Most cations, such as Na^+, K^+, and NH_4^+, are excluded from nonpolar groups of the protein, which compensates from their favorable interaction with the amide oxygen, rendering them essentially neutral toward protein precipitation and denaturation (see Figure 4.36). The exception is GuH^+, which is why it causes salting in. The interactions of anions with nonpolar groups of the protein vary. For example, SO_4^{2-} and F^- are excluded, whereas ClO_4^- and I^- are accumulated around nonpolar groups. Thus, the order of anions and cations in the Hofmeister series depends on a combination of their interactions with the amide C=O and N—H groups, but also with the α-carbon, and with side chain nonpolar groups.

Polyethylene Glycol has Mixed Effects

Polyethylene glycol (PEG), such as PEG 600 or PEG 1000, are polymers of ethylene glycol units, $HO—(CH_2CH_2O)_n—H$, with variable average number of units *n*. (The numbers 600 or 1000 refer to the average molecular weight). PEG 1000 destabilizes chymotrypsinogen at 20°C by increasing the Gibbs energy of the native state and decreasing the Gibbs energy of the denatured state of the protein (**Figure 4.37**). This occurs because PEG 1000 is much smaller than the protein and is therefore *sterically excluded* from contact with the native protein surface. This is an entirely entropic effect that arises because many more arrangements (configurations) of the molecules in solution are possible if the sizes of two components are similar. On this basis, PEG tends to cause preferential hydration (see Figure 4.30). However, because of the nonpolar character of its ethylene groups ($-CH_2CH_2-$), PEG interacts favorably with hydrophobic side chains exposed in the unfolded state, which tends to cause protein denaturation. PEG is used to precipitate and crystallize proteins: if the temperature is low enough that the protein is kept from denaturing, the preferential hydration effect of PEG leads to precipitation.

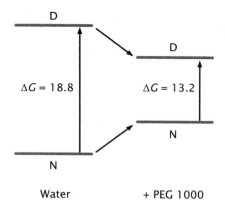

Figure 4.37 Effect of PEG 1000 on the stability of chymotrypsinogen. ΔG in kcal/mol. (Based on Timasheff SN [1993] *Annu Rev Biophys Biomol Struct* 22:67–97.)

4.10 MARGINAL STABILITY IS SUFFICIENT FOR FUNCTION

We end this chapter by returning to the topic of protein *marginal stability*. Under physiological conditions, the native state is only *slightly* more favorable than the unfolded state. We have seen that $\Delta G \approx 10$ kcal/mol for the denaturation of most small proteins, which is of the same order of magnitude as some of the interactions between protein groups. For example, the energy of a hydrogen bond is ~5 kcal/mol, and the enthalpy difference between a hydrogen bond made within the protein or between protein and water is ~1 kcal/mol. Breaking or exchanging a few of those interactions is sufficient to tip the Gibbs energy balance between folded and unfolded protein. Why are proteins marginally stable? Is there a functional advantage to marginal stability?

The traditional answer given by biochemists is that proteins need to be marginally stable to function. With a small Gibbs energy difference between the native and denatured conformations, the protein is able to unfold and refold reversibly. The idea is that a certain flexibility is built into the structure if the differences in interactions are small. Most mutations destabilize proteins, but it is possible to introduce amino acid replacements that increase stability. Nature appears to select for marginal stability, which seems to indicate it is important for function. The corollary is that hyperstable proteins should not function well.

However, directed evolution experiments in the laboratory have shown otherwise. In these experiments, protein sequences were subject to random mutations and the mutant proteins were selected for function. Then the selected proteins were mutated again, and so on. After several iterations, hyperstable proteins were obtained that were also hyperfunctional. Thus, nature *could* select for hyperstability *but does not*. Why? The answer appears to reside in a combination of statistical thermodynamics and protein evolution.

First, marginally stable proteins function well. There is no reason to select *against* them. Very stable proteins also function well, so there is no reason to select against them either. As long as the proteins are stable enough to function, marginal or hyperstability are *neutral traits* in evolution. It seems that proteins are only marginally stable and not hyperstable because there are *many more sequences* that fold to marginally stable than to hyperstable proteins. Indeed, the number of sequences decreases exponentially as stability increases. Therefore, it is much *more likely* that proteins will be marginally stable, because evolution randomly accepts sequences that are equally functional.

The "driving forces" at play here are analogous to those in protein denaturation. If the native state were only minimally favored by interactions (or not at all) relative to the unfolded state, then all proteins would be unfolded because there are many more conformations belonging to the unfolded state. Similarly, there are many more sequences that fold to marginally stable proteins than to hyperstable proteins. In this sense, we may say that the marginally stable "state" has a much greater "entropy" than the hyperstable "state," and their "energies" (meaning function) are about the same. To what extent this argument regarding the marginal stability of proteins can be extended to the predominance of weak interactions in biomolecules and biological systems is an important and unresolved question.

4.11 SUMMARY

The native or folded conformation of a protein is the thermodynamically stable state under physiological conditions. At high temperatures or in the presence of denaturants, the protein unfolds. However, protein denaturation is reversible. An equilibrium exists between the folded and the unfolded state. Under native conditions, the protein folds again, because the native conformation corresponds to the Gibbs energy minimum.

Proteins fold to globular structures to hide their hydrophobic residues from contact with water. The hydrophobic effect provides the main contribution to protein stability. Ultimately, this effect is mainly a consequence of the hydrogen-bonding structure of water and the changes water undergoes when it comes in contact with nonpolar compounds. The regular α-helical and β-sheet secondary structures form in globular proteins to satisfy the peptide hydrogen bonds. This is necessary because there are no water molecules to serve as hydrogen-bonding partners in the protein hydrophobic interior. Transfer of a non-hydrogen-bonded peptide group (CONH) from water to the protein nonpolar interior would cost several kilocalories per mole. A few unmade hydrogen bonds thus amount to the entire Gibbs energy difference between the folded and the unfolded state. Therefore, all hydrogen bonds inside the protein must be established.

When the protein unfolds, the conformational entropy of the polypeptide chain increases enormously. This is the major driving force for protein thermal denaturation. The entropy change comes multiplied by the temperature in the Gibbs energy difference between folded and unfolded states. Therefore, that term increases with temperature, eventually becoming larger than the enthalpy change. At that point the unfolded state is favored.

Denaturants such as urea and guanidinium chloride cause protein unfolding by preferentially binding to the protein surface, especially by hydrogen bonding to peptide groups. The protein conformational equilibrium is shifted to the unfolded state, because its surface area is larger and the peptide groups are more exposed to the solvent. Conversely, protective osmolytes such as TMAO or betaine stabilize the native state of the protein through preferential hydration. They are excluded from the protein surface in favor of water mainly because of their inability to serve as hydrogen-bonding donors.

4.12 PROBLEMS

4.1. The Gibbs energy associated with the repulsive interactions of a pair of protonated lysine side chains of a protein that protrude into the water, at a distance of about 2 Å from each other, is about +2 kcal/mol at room temperature. If this pair of lysines were still protonated and at the same distance, but were buried in the hydrophobic interior of the protein, what would the Gibbs energy be (approximately)?

4.2. The dielectric constant (D) of water is shown in Figure 4.11A as a function of temperature. Recalling that the entropy of formation of an ion pair depends on the dielectric constant according to

$$\Delta S^{\circ} = \frac{\Delta G^{\circ}}{D} \frac{dD}{dT},$$

calculate the relation between the entropy and the Gibbs energy of formation of an ion pair in water at room temperature. Obtain the result from the graph.

4.3. Suppose you have two dipoles in a protein, as sketched below in (A). One of these dipoles is an N—H bond and the other is a C=O bond.

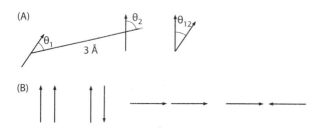

Read the following carefully. The dipoles moments of the N—H and C=O are 1.63 and 2.3 D (debye), respectively. Convert the values to the appropriate units. Recall that $\mu = Qd$, where Q is the charge separated and d is the distance between the charges (bond length in this case), so that the SI units are coulomb × meter (C · m), not debye.
(a) Consider the four possible orientations of the dipoles shown in (B). The distance between the centers of the two dipoles is always assumed to be 3 Å. First, calculate just the *orientational* part of the energy—that is, the part that has to do with the angles. At this initial stage, set all the remaining quantities equal to 1. Qualitatively, this will give you the correct result in terms of relative energies of the four orientations shown in (B), which will allow you to identify any possible mistakes you might make later. The factor in front of the orientational part of the energy is just a constant, the same for all cases in the calculation (but different for water and the protein interior).
(b) Now calculate the energy of the interaction for the four possible orientations of the dipoles shown in (B). Do the calculations first assuming the dipoles are in water and, second, assuming they are in the interior of a protein (use a dielectric constant $D = 10$ for the protein interior). Define θ between −180 and +180 degrees. List your results in a table, with entries for the different cases calculated.
(c) Rank the four possible orientations of the two dipoles in order of most to least favorable.

(d) What does your answer tell you regarding the formation of hydrogen bonds between N—H and C=O in proteins?
(e) What does your answer tell you about the strength of a hydrogen bond in water and in the protein interior?

4.4. In Problem 14 of Chapter 2, using the kinemage c2Motifs.kin, you measured the φ and ψ angles for the residues of an α-helix, and calculated the mean and standard deviation. Now, estimate the area allowed for the α-helix in the Ramachandran diagram by calculating the area of a square defined by $A_{\alpha} = (3\Delta\phi) \times (3\Delta\psi)$, where the Δs are the standard deviations you obtained in Problem 14 of Chapter 2. We are using three standard deviations to make sure all allowed conformations are included. Next, estimate the total area allowed in the Ramachandran diagram using the normal limits for an amino acid other than glycine or proline. Call this area A_n. The ratio A_{α}/A_n gives a measure of the restriction imposed on the central residue for forcing it to be in an α-helical conformation. The entropy loss it experiences because of this restriction is $\Delta S_{conf} \approx R \ln A_{\alpha}/A_n$. Calculate the value of ΔS_{conf}, and the corresponding Gibbs energy change, for placing a residue in an α-helix at room temperature.

4.5. When we studied Ramachandran diagrams, we learned how Gly and Pro affect the φ,ψ angles available to each residue and therefore the conformations that polypeptides can adopt. Now suppose that for a certain protein you have made two types of mutants: one with mutations Gly → Ala and another with mutations Pro → Ala. You found out, by determining the tertiary structure by NMR, that the mutations are compatible with the tertiary structure of the folded protein.
(a) Sketch a Gibbs energy diagram that shows how the mutations Gly → Ala qualitatively affect the Gibbs energy of the native and denatured states of the protein, and their difference.
(b) Do the same for the Pro → Ala mutations.
(c) What happens to the stability of the protein in the two cases?

4.6. Close to its melting temperature, lysozyme undergoes reversible thermal unfolding at pH 7 with $\Delta H^{\circ} = 130$ kcal/mol and $\Delta S^{\circ} = 373$ cal/K/mol. Close to room temperature, $\Delta H^{\circ} = 60$ kcal/mol and $\Delta S^{\circ} = 155$ cal/K/mol.
(a) What is the melting temperature of lysozyme (in °C)?
(b) What is the Gibbs energy change for unfolding of lysozyme at 25°C?
(c) What percentage of lysozyme molecules are unfolded, on average, at 25°C? (Use the partition function.)

4.7. Using the data from the previous problem and $\Delta C_p^D \approx 1.4$ kcal/mol/K, calculate the heat capacity function of lysozyme in a DSC experiment. This is analogous to the curve for ribonuclease plotted in Figure 4.4.

4.8. Figure 4.5 shows the distribution of enthalpy of ribonuclease around the T_m. How would the two peaks in the distribution change as you move away from the transition region?

4.9. Hydrogen bonds between protein groups and water that are established both in the native and in the denatured states do not contribute to protein stability. Why?

4.10. The ΔG° of unfolding of ribonuclease A (RNase) at room temperature is about 12 kcal/mol. RNase has four disulfide bonds. If those bonds are reduced (S—S to SH), the protein becomes much less stable, adopting the folded state only about 1% of the time (or 1% of the proteins are folded at any moment in time). Supposing that each disulfide bond contributes the same amount of Gibbs energy to the stability of RNase, what is the Gibbs energy contributed by each disulfide bond?

4.11. Using the information in Table 4.1, recalling that $\Delta C_p = d\Delta H/dT$ (shown in Equation 4.32), and assuming that the heat capacity change (ΔC_p) does not vary with temperature, calculate the enthalpy change for transfer of benzene to water at 90°C.

4.12. The following values were obtained for the thermodynamics of dissolving a certain unknown compound in water at room temperature: $\Delta H = -0.4$ kcal/mol, $\Delta S = -24$ cal/K/mol, and $\Delta C_p = +2$ cal/K/mol. Is it likely that the compound is ionic or hydrophobic? Explain.

4.13. The major force leading to protein folding is the hydrophobic effect. This "force," however, is not specific and entails no special orientation of groups. Rather, it would be expected to lead to an amorphous aggregation of nonpolar groups in water. Nevertheless, protein structures are highly organized and regular. Briefly explain this *apparent* contradiction.

4.14. The figure below shows the effects of four cosolvents on the stability of a protein. Which one works by preferential binding?

4.15. If ammonium sulfate $((NH_4)_2SO_4)$ increases the Gibbs energy of the native state (which it does), why doesn't it denature proteins? As the Gibbs energy of the native state increases, it should become higher than that of the denatured state. Why doesn't this happen? What is wrong with this argument?

4.13 FURTHER READING

General

Creighton TE (1993) Proteins. Structure and Molecular Properties, 2nd ed. Freeman.

Cantor CR & Schimmel PR (1980) Biophysical Chemistry. Freeman.

Kyte J (2006) Structure in Protein Chemistry, 2nd ed. Garland.

Feynman RP, Leighton RB & Sands M (1977) The Feynman Lectures on Physics, vol. 2, 2nd ed. Addison Wesley.

Thermodynamics of Protein Folding

Anfinsen CB (1973) Principles that govern the folding of protein chains. *Science* 181:223–230.

Baldwin RL (2007) Energetics of protein folding. *J Mol Biol* 371:283–301.

Brandts JF (1964) The thermodynamics of proteins denaturation. I. The denaturation of chymotrypsinogen. *J Am Chem Soc* 86:4291–4301.

Brandts JF (1964) The thermodynamics of proteins denaturation. II. A model of reversible denaturation and interpretations regarding the stability of chymotrypsinogen. *J Am Chem Soc* 86:4302–4314.

Dill KA (1990) Dominant forces in protein folding. *Biochemistry* 29:7133–7155.

Privalov PL (1979) Stability of proteins. Small globular proteins. *Adv Protein Chem* 33:167–241.

Privalov PL (1997) Thermodynamics of protein folding. *J Chem Thermodynam* 29:447–474.

Pace CN, Shirley BA, McNutt M & Gajiwala K (1996) Forces contributing to the conformational stability of proteins. *FASEB J* 10:75–83.

Pace CN, Scholtz JM & Grimsley GR (2014) Forces stabilizing proteins. *FEBS Lett* 588:2177–2184.

Schellman JF (1955) The stability of hydrogen-bonded peptide structures in aqueous solution. *C R Trav Lab Carlsberg Ser Chim* 29:230–259.

Conformational Entropy

Baxa MC, Haddadian EJ, Jha AK, Freed KF & Sosnick TR (2012) Context and force field dependence of the loss of protein backbone entropy upon folding using realistic denatured and native state ensembles. *J Am Chem Soc* 134:15929–15936.

D'Aquino JA, Gomez J, Hilser VJ et al (1996) The magnitude of the backbone conformational entropy change in protein folding. *Proteins* 25:143–156.

Karplus M, Ichiye T & Petitt BM (1987) Configurational entropy of native proteins. *Biophys J* 52:1083–1085.

Thompson JB, Hansma HG, Hansma PK & Plaxco KW (2002) The backbone conformational entropy of protein folding: Experimental measures from atomic force microscopy. *J Mol Biol* 322:645–652.

Zhang C, Cornette JL & Delisi C (1997) Consistency in structural energetics of protein folding and peptide recognition. *Protein Sci* 6:1057–1064.

Electrostatic Interactions

Harris MJ, Schlessman JL, Sue GR & García-Moreno E (2011) Arginine residues at internal positions in a protein are always charged. *Proc Natl Acad Sci USA* 108:18954–18959.

Isom DG, Castañeda CA, Cannon BR, Velu PD & García-Moreno E (2010) Charges in the hydrophobic interior of proteins. *Proc Natl Acad Sci USA* 107:16096–16100.

Isom DG, Castañeda CA, Cannon BR & García-Moreno E (2011) Large shifts in pK_a values of lysine residues buried inside a protein. *Proc Natl Acad Sci USA* 108:5260–5265.

Kukic P, Farrell D, McIntosh LP et al (2013) Protein dielectric constants determined from NMR chemical shift perturbations. *J Am Chem Soc* 135:16968–16976.

Parsegian A (1969) Energy of an ion crossing a low dielectric membrane: Solutions to four relevant electrostatic problems. *Nature* 221:844–846.

Takano K, Tsuchimori K, Yamagata Y & Yutani K (2000) Contribution of salt bridges near the surface of a protein to the conformational stability. *Biochemistry* 39:12375–12381.

Van der Waals Interactions

Autumn K, Liang YA, Hsieh ST et al (2000) Adhesive force of a single gecko foot-hair. *Nature* 405:681–685.

Baase WA, Liu L, Tronrud DE & Matthews BW (2010) Lessons from the lysozyme of phase T4. *Protein Sci* 19:31–641.

Hill TL (1980) An Introduction to Statistical Thermodynamics. Dover, Appendix IV.

Loladze VV, Ermolenko DN & Makhatadze GI (2002) Thermodynamic consequences of burial of polar and non-polar amino acid residues in the protein interior. *J Mol Biol* 320:343–357.

Salem L (1962) Attractive forces between long saturated chains at short distances. *J Chem Phys* 37:2100–2113.

Hydrogen Bonds

Bolen DW & Rose GD (2008) Structure and energetics of the hydrogen-bonded backbone in protein folding. *Annu Rev Biochem* 77:339–362.

Desiraju GR (2011) A bond by another name. *Angew Chem Int Ed* 50:52–59.

Lopez MM, Chin D-H, Baldwin RL & Makhatadze GI (2002) The enthalpy of the alanine peptide helix measured by isothermal titration calorimetry using metal-binding to induce helix formation. *Proc Natl Acad Sci USA* 99:1298–1302.

Nisius L & Grzesiek S (2012) Key stabilizing elements of protein structure identified through pressure and temperature perturbation of its hydrogen bond network. *Nature Chem* 4:711–717.

Myers JK & Pace CN (1996) Hydrogen bonding stabilizes globular proteins. *Biophys J* 71:2033–2039.

Perrin CL & Nielson J 1997. "Strong" hydrogen bonds in chemistry and biology. *Annu Rev Phys Chem* 48:511–544.

Scholtz JM, Marqusee S, Baldwin RL et al (1991) Calorimetric determination of the enthalpy change for the α-helix to coil transition of an alanine peptide in water. *Proc Natl Acad Sci USA* 88:2854–2858.

Wang J & Purisima E (1996) Analysis of thermodynamic determinants in helix propensities of non-polar amino acids through a novel free energy calculation. *J Am Chem Soc* 118:995–1001.

Hydrophobic Effect

Baldwin RL (1996) Temperature dependence of the hydrophobic interaction in protein folding. *Proc Natl Acad Sci USA* 83:8069–8072.

Baldwin RL (2012) Gas–liquid transfer data used to analyze hydrophobic hydration and find the nature of the Kauzmann-Tanford hydrophobic factor. *Proc Natl Acad Sci USA* 109:7310–7313.

Baldwin RL (2013) The new view of hydrophobic free energy. *FEBS Lett* 587:1062–1066.

Baldwin RL (2014) Dynamic hydration shell restores Kauzmann's 1959 explanation of how the hydrophobic factor drives protein folding. *Proc Natl Acad Sci USA* 111:13052–13056.

Ben-Naim A (1978) Standard thermodynamics of transfer. Uses and misuses. *J Phys Chem* 82:792–803.

Ben-Naim A & Marcus Y (1984) Solvation thermodynamics of nonionic solutes. *J Phys Chem* 81:2016–2027.

Chandler D (2005) Interfaces and the driving force of hydrophobic assembly. *Nature* 437:640–647.

Jorgensen WL, Gao J & Ravimohan C (1985) Monte Carlo simulations of alkanes in water: Hydration numbers and the hydrophobic effect. *J Phys Chem* 89:3470–3473.

Kauzmann W (1959) Some factors in the interpretation of protein denaturation. *Adv Protein Chem* 14:1–63.

Lee B (1991) Solvent reorganization contribution to the transfer thermodynamics of small nonpolar molecules. *Biopolymers* 31:993–1008.

Makhatadze GI & Privalov PL (1994) Energetics of interactions of aromatic hydrocarbons with water. *Biophys Chem* 50:285–291.

Muller N (1990) Search for a realistic view of hydrophobic effects. *Acc Chem Res* 23:23–38.

Privalov PL & Gill SJ (1988). Stability of protein structure and hydrophobic interaction. *Adv Protein Chem* 39:191–234.

Tanford C (1980) The Hydrophobic Effect, 2nd ed. Krieger.

White SH & Wimley WC (1999) Membrane protein folding and stability: Physical principles. *Annu Rev Biophys Biomol Struct* 28:319–365.

Wu J & Prausnitz JM (2008) Pairwise-additive hydrophobic effect for alkanes in water. *Proc Natl Acad Sci USA* 105:9512–9515.

Cosolvent Effects on Protein Stability and Solubility

Baldwin RL (1996) How Hofmeister ion interactions affect protein stability. *Biophys J* 71:2056–2063.

Canchi DR & García AE (2013) Cosolvent effects on protein stability. *Annu Rev Phys Chem* 64:273–293.

Guinn EJ, Pegram LM, Capp M et al (2011) Quantifying why urea is a protein denaturant whereas glycine betaine is a protein stabilizer. *Proc Natl Acad Sci USA* 108:16932–16937.

Pegram LM & Record MT (2008) Thermodynamic origin of Hofmeister ion effects. *J Chem Phys* 112:9428–9436.

Record MT, Guinn E, Pegram L & Capp M (2013) Introductory lecture: Interpreting and predicting Hofmeister salt ion and solute effects on biopolymer and model processes using the solute partitioning model. *Faraday Discuss* 160:9–44.

Rembert KB, Paterová J, Heyda J et al (2012) Molecular mechanisms of ion-specific effects on proteins. *J Am Chem Soc* 134:10039–10046.

Scholtz JM, Grimsley GR & Pace N (2009) Solvent denaturation of proteins and interpretations of the m-value. *Methods Enzymol* 466:546–565.

Timasheff SN (1993) The control of protein stability and association by weak interactions with water: How do solvents affect these processes? *Annu Rev Biophys Biomol Struct* 22:67–97.

Timasheff SN (2002) Protein-solvent preferential interactions, protein hydration, and the modulation of biochemical reactions by solvent components. *Proc Natl Acad Sci USA* 99:9721–9726.

Yancey PH, Clark ME, Hand SC et al (1982) Living with water stress: Evolution of osmolytes systems. *Science* 217:1214–1222.

Marginal Stability and Evolution

Goldstein RA (2008) The structure of protein evolution and the evolution of protein structure. *Curr Opin Struct Biol* 18:170–177.

Harms MJ & Thornton JW (2013) Evolutionary biochemistry: Revealing the historical and physical causes of protein properties. *Nature Rev Genet* 14:559–571.

Taverna DM & Goldestein RA (2002) Why are proteins marginally stable? *Proteins* 46:105–109.

Protein Folding

5.1 PROTEIN FOLDING—WHAT'S THE QUESTION?

The title of this section is borrowed from an article by Lattman and Rose (1993), which frames the protein folding problem as two questions. The first—what are the determinants of *protein stability*?—is a thermodynamic question. This was the topic of Chapter 4. We studied the physical interactions, both within the protein and between the protein and its surrounding aqueous environment, that determine folding. We learned how the thermodynamics relate to the structure, but we left out a few aspects that we now consider: (a) the foldability of sequences, and (b) the cooperativity of folding.

The second question—what are the determinants of *protein conformation*?—is a structural one. Ribonuclease and lysozyme, for example, are two proteins with very different structures but similar sizes (**Figure 5.1**). In principle, one protein *could* adopt the fold of the other, but it never does. If a protein is mutated, it may become less stable and unfold at lower temperatures, but if it folds it always folds to the native structure. The sequence codes for the structure. But how? How is the three-dimensional native structure specified by the amino acid sequence? If you know the sequence, can you predict the structure? In general, the answer is no. But in many cases it is possible to predict the structure with remarkable accuracy.

There is yet a third question on protein folding—what is the *folding mechanism*? How does the protein find its native structure? There are only two thermodynamically stable states, folded (native) and unfolded

Paracelsus remained alone. Before extinguishing the lamp and sitting down in the tired armchair, he turned the tenuous handful of ash in the cup of his hand and said a word in low voice. The rose reappeared.

Jorge Luis Borges

(A) (B)

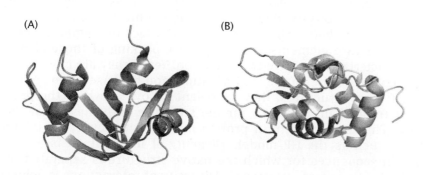

Figure 5.1 The structures of (A) ribonuclease (PDB 7RSA) and (B) lysozyme (PDB 1VDQ).

(denatured). The native state is a well-defined, compact conformation, with regular structure, and often a high degree of symmetry. The unfolded state is a very large *ensemble* of more or less extended conformations, with only local regularity. Kinetically, however, there must exist *intermediates*; that is, conformations that the protein samples on its way from the unfolded to the folded state. We want to know the *structural pathway* that the protein takes from one state to the other: the folding mechanism.

5.2 PROTEINS FOLD TO UNIQUE STRUCTURES

A native protein sequence folds to a unique three-dimensional structure. This structure is remarkably resistant to the effects of mutations. In the course of evolution, the sequence changed mainly because of *neutral mutations* (see Chapter 3), which essentially do not affect the function or the structure. In the laboratory, we can introduce mutations that affect the structure. Usually, then, the protein loses stability, sometimes to the point that it does not fold properly. In the vast majority of cases, however, the protein does not fold to some *other* structure. That sequence just does not fold to anything. Thus, some sequences fold whereas others do not. What makes a sequence *foldable* to a unique structure?

A Random Sequence Does not Fold to a Unique Structure

Theoretical models have helped us to understand why some sequences fold and others do not. Random sequences of different kinds of amino acids do not fold to a unique structure. Instead, they adopt many alternative conformations, none of which resembles that of a native protein. In addition, the simpler the amino acid alphabet (the different amino acid types), the more difficult it is for the polymer to fold to a unique structure. For example, the simplest protein model is a sequence of hydrophobic (H) and polar (P) residues in a chain, the so-called HP model. It is possible to choose HP sequences that have a "folded" conformation in which the H residues are inside and the P residues are in contact with water. Usually, however, other conformations with similar free (Gibbs) energies are possible, and the folded state is also an *ensemble*, albeit of fewer conformations than the unfolded state.

For a sequence to fold to a unique structure, a *large energy gap* must exist between the native fold and all other conformations, including other compact globular structures. These kinds of conformations, which have been observed in protein experiments and simulations, are called *molten globules*. Though globular, they lack the regularity of secondary structure elements and the precise packing of those elements that characterize the native conformation. Yet, if they are close enough in free energy to the native conformation, and because there are many more of these structures, they represent decoys in which the protein can get trapped in its search for the structure with the lowest Gibbs energy. This is a problem in protein models with few types of interactions, such as the HP model. *Diversity* of interactions is essential to obtain sequences for which the native structure is separated from decoys by a large free energy gap. Diversity of interactions is achieved through diversity of amino acid types. Whether exactly 20 amino acids is the minimum number needed is open to debate, but that seems to be the answer of evolution.

Foldable sequences must contain a statistically representative number of amino acid types that is larger than the number of conformations per residue. The greater the number of conformations per residue, the more difficult it is to fold, because the number of conformations of the entire protein increases. Restrictions imposed on the number of possible conformations, such as the requirement of hydrogen bonds, consequently improve folding.

Proteins Fold Cooperatively from the Denatured to the Native Conformation

Most proteins have only two thermodynamic states, native and denatured. What this means is that folding intermediates are *not significantly populated* at equilibrium. Intermediates exist in the folding pathway, but in the middle of the transition, under equilibrium conditions, they account only for ~5% of the molecules or fewer. This is true for the majority of single-domain proteins. *Thermodynamically*, they are *two-state* systems. Even for many proteins with multiple domains, the two-state approximation holds. In a protein with two domains, for example, one domain is often more stable, but the interactions between domains (because they are favorable only in the native state) stabilize the less-stable domain. The consequence is that both domains often denature cooperatively and simultaneously. There are, however, many proteins, particularly with multiple domains, for which the two-state approximation is not valid. Further, it is possible to create conditions that specifically stabilize certain intermediates.

Under nonequilibrium conditions, when we study the *kinetics* of protein folding, intermediates can be stable enough to accumulate and become a significant fraction of the protein population. Those intermediates are important to understand the folding pathways and have been the subject of extensive research. Still, many small single-domain proteins fold *kinetically* as two-state, without any significant intermediates. We will concentrate most on these proteins in this chapter because it is the nature of the *transition state* that provides the greatest insight into the essence of protein folding.

The reason that only the folded and unfolded conformations exist at equilibrium is that the intermediates have a much higher free energy. This is why folding is *cooperative*. Once the protein begins to fold (or unfold) its free energy increases. The protein then has "to decide" whether it wants to go back, thereby decreasing its free energy, or to continue to fold—but in this case it must go all the way to the end, not linger in the high free energy intermediate states. The essence of cooperativity is to render intermediates *unfavorable*: in the limit, folding becomes two-state.

What structural features of the native conformation enhance folding cooperativity? *Local contacts* in the native state (between residues close to each other in the sequence) improve the folding *speed*. However, structures relying heavily on local contacts are not very cooperative. *Distant contacts* (between residues far away from each other in the sequence) enhance folding *cooperativity* and favor the existence of a single native state. To better understand the concept of cooperativity, we turn now to a simpler system than a globular protein: a helical polypeptide.

5.3 THE COIL–HELIX TRANSITION IS A SIMPLE CASE OF COOPERATIVE FOLDING

A completely unstructured polypeptide, which we call a *random coil*, folds cooperatively to an α-helix. The helix–coil transition can be monitored experimentally by circular dichroism (CD), because the CD spectrum of the helix and the coil are very different (see Figure 2.49). Typically, the CD signal (ellipticity) difference is measured at $\lambda = 222$ nm and converted to a *fractional helicity* (θ_h), as shown in the example of **Figure 5.2**.

Most α-Helices are Unstable in Water

Consider a polypeptide chain in water. Let us assume that each residue can exist only in two states: helical (h), when the dihedral angles ϕ and ψ are those of an α-helix, and coil (c), when the dihedral angles have any other values. There are many more angles belonging to the coil state. This means there are many *microstates* in the random coil, whereas there is only one in the α-helix. Therefore, the random coil is favored by conformational entropy. The helix must be favored by enthalpy if it is to form at all. This favorable enthalpy arises from the establishment of hydrogen bonds between the peptide groups of residues at positions i and $i + 4$ along the chain.

In Chapter 4 we discussed the changes in entropy and enthalpy associated with hydrogen-bond formation that accompany the acquisition of secondary structure. Here we will study their role in the cooperativity of this process. We will use lower case letters for thermodynamic functions *per residue*: Δs, Δh, and Δg, for entropy, enthalpy, and Gibbs energy changes. Upper case letters will be used for the entire polypeptide chain. The favorable Δh in helix formation arises mainly from hydrogen bonds between residues i and $i + 4$. Its value depends somewhat on the amino acid type. Some estimates are as follows. For most amino acids without β-branches (Ala, Leu, Phe, probably Lys), $\Delta h \approx -1$ kcal/mol. β-Branched and some other polar amino acids (Val, Ile, Thr, Gln, Asn, Ser) have somewhat smaller values, with $\Delta h \approx -0.6$ kcal/mol. Glycine has an even lower value, $\Delta h \approx -0.4$ kcal/mol. However, there is considerable uncertainty in these values.

The entropy decreases when a residue becomes part of a helix because the number of conformational states per residue (ω) is larger in the coil than in the helix. When the first hydrogen bond is made, between residue **1** (i) and **5** ($i + 4$), the ϕ and ψ angles of the residues

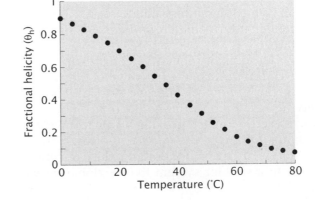

Figure 5.2 Helix–coil transition in a 50-residue, alanine-rich peptide (Ac–Y(AEAAKA)$_8$F–NH$_2$). The data were measured by CD (ellipticity at 222 nm) and converted to fractional helicity (θ_h). (Data from Scholtz JM, Marqusee S, Baldwin RL et al. [1991] *Proc Natl Acad Sci USA* 88:2854–2858.)

1 and **5** are not constrained, but those of the *three intervening residues* (residues **2**, **3**, and **4**) become fixed at the α-helical values (**Figure 5.3**).

There are on average about 7–9 conformational states accessible to each residue in a random coil ($\omega_c \approx 7 - 9$), but there is only one in the helix ($\omega_h = 1$). According to the Boltzmann formula (shown in Equation 1.15), the conformational entropy is

$$s_{conf} = R\ln \omega, \tag{5.1}$$

where R is the gas constant ($R = 1.987$ cal/K/mol; the gas constant is the same as the Boltzmann constant, except that it is expressed per mole instead of per molecule). Thus, the entropy change in a residue switch from coil $\rightarrow$ helix is

$$\Delta s_{conf} = R\ln \omega_h - R\ln \omega_c$$
$$= -R\ln \omega_c \tag{5.2}$$

(since $\ln \omega_h = \ln 1 = 0$). With $\omega_c = 7$ we obtain $\Delta s_{conf} \approx -3.9$ cal/K/mol. This corresponds to a change in the conformational Gibbs energy of $\Delta g_{conf} = -T\Delta s_{conf}$ per residue; at about room temperature, $\Delta g_{conf} \approx +1.1$ kcal/mol.

When the first hydrogen bond in an α-helix is established, three residues are fixed; consequently, $\Delta G_{conf} = 3 \times (-T\Delta s_{conf}) \approx +3.3$ kcal/mol around room temperature. Let us assume that the enthalpic contribution is $\Delta h = -1.3$ kcal/mol for one hydrogen bond. Then, forming the first bond is very unfavorable, with $\Delta G = +2.0$ kcal/mol. This is a five-residue helix because the bond was formed between residues 1 and 5 ($i + 4$). Adding subsequent residues to the helix, however, results in $\Delta g = -0.2$ kcal/mol per residue (-1.3 kcal/mol from Δh plus $+1.1$ kcal/mol from $-T\Delta s_{conf}$). Therefore, when 10 more residues have been added, to form a 15-residue helix, we reach a total $\Delta G = 0$; beyond that, the helix is favorable. In globular proteins, the average length of an α-helix is ≈ 15 residues. Although the diameter of the native structure constrains the helical size, the average length coincides with the length at which helices begin to be marginally stable.

In summary, most sequences have a low helical content in water because, on average, for an individual residue $\Delta g = \Delta h - T\Delta s \approx 0$ around room temperature. However, sequences rich in residues with high helical propensity, such as Ala, Leu, and Lys, have slightly more favorable values of Δg and form marginally stable helices (see Figure 5.2).

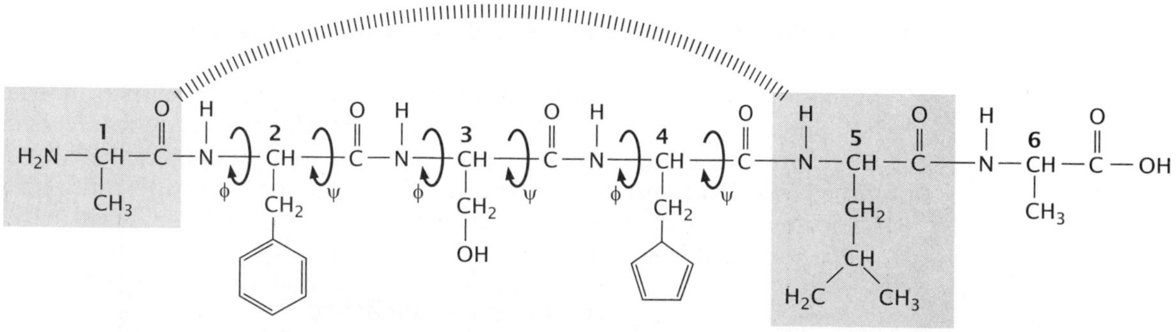

Figure 5.3 Formation of the first hydrogen bond (dotted line) in an α-helix, between residues **1** (*i*) and **5** (*i* + 4) (boxed), fixes the ϕ and ψ angles of the three intervening residues (**2**, **3**, **4**) at their α-helical values.

α-Helix Formation is Cooperative Because it is Easier to Propagate than to Initiate the Helix

The coil → helix transition is cooperative. To be cooperative means that the occurrence of an event makes it easier for other events of the same kind to occur. Initiation of the helix is difficult because making the first hydrogen bond is accompanied by a large loss of possible conformations, and thus of entropy. Once the helix is initiated, however, adding more residues to it becomes progressively easier. We would like to develop a deeper understanding of the concept of cooperativity. To achieve this, we will study the coil–helix transition using the partition function approach.

We need to begin by defining two quantities. The equilibrium constant (K) for switching a residue from the coil to the helical state *if the preceding residue is helical* is

$$K = e^{-\Delta g/RT}. \tag{5.3}$$

The equilibrium constant for converting a residue in the coil state to the helical state *at the beginning of a helix* is

$$\sigma K = e^{-(\Delta h - 3T\Delta s_{conf})/RT}. \tag{5.4}$$

One hydrogen bond is made, giving rise to the term Δh, and three pairs of ϕ, ψ angles are fixed, giving rise to the term $3T\Delta s_{conf}$ in Equation 5.4. We can also write

$$\sigma K = e^{-\Delta g/RT + 2\Delta s_{conf}/R}. \tag{5.5}$$

This equation defines the *cooperativity parameter* σ as the factor by which K is modified (multiplied) when a residue switch initiates the helix. From Equations 5.3 and 5.5, we see that

$$\sigma = e^{2\Delta s_{conf}/R}.$$

For most amino acids $\sigma \sim 0.01$ to 0.1 and $K \sim 1$ around room temperature.

To understand cooperativity in the coil → helix transition, we will study several models of the transition. We begin with a model that is not cooperative at all, the model of *independent residues*, and then move to one that shows maximal cooperativity, the *all-or-none* model. Neither is realistic. But they will prepare us to study two other models that are increasingly more realistic: the *zipper* model and the Zimm–Bragg model.

The Model of Independent Residues is not Cooperative

Suppose for now that it is as easy to initiate as to elongate the helix. That would mean $\sigma = 1$. What would be the consequences? Each residue has two states, coil (c) or helix (h). In this model, each residue behaves *independently* of all other residues. The polypeptide chain has N residues, each of which can adopt either state. For example with $N = 20$, we could have a conformation specified by

cccchchhhcccchhchchcc

Experimentally, we can measure the fraction of helical residues using CD. If there are n helical residues in a total of N, the fraction of helical residues (fractional helicity) is

$$\theta_h = \frac{n}{N}. \tag{5.6}$$

Now let us write the partition function for the polypeptide chain. We begin by writing the partition function for *one residue*, which we call *q*. We assign a *statistical weight* to each state of the amino acid residue (coil or helix). The statistical weight is simply the *relative probability* of that state—relative to a *reference state*. Here we will choose the coil as our reference state. The residue conversion from coil to helix,

$$c \rightleftharpoons h,$$

is controlled by the equilibrium constant

$$K = \frac{p_h}{p_c}, \tag{5.7}$$

which is the ratio of the probabilities of helix (p_h) to coil (p_c). (These probabilities can be absolute or relative, as long as they are referenced to the same state.) Since we chose the coil as reference, its relative probability is just 1. With the same reference, the helical state is K *times more probable* (shown in Equation 5.7), so the relative probability of the helix is K. The partition function for one residue is the sum of the two relative probabilities,

$$q = 1 + K. \tag{5.8}$$

This partition function is exactly the same as for an entire protein, because each residue is a *two-state* system: it is either coil or helix, just like the protein is either folded or unfolded.

The partition function of the entire polypeptide chain (Q) is very easy to obtain in this case because the residues are independent. The first residue can be h or c, the second can be h or c, and so on. To write Q we need to know the probabilities of all possible conformations of the chain. These conformations vary in two ways: they vary in the *different numbers* of helical (n) and coil residues ($N - n$) that the chain contains; and for given $N - n$ coil and n helical residues present, the conformations vary also in the *different ways of arranging* the h and c residues along the chain. We need to take all this into account. It seems complicated but you will see it is actually simple.

Let us begin with the example of the conformation specified above, and write the statistical weights of each residue underneath it:

```
c  c  c  c  h  c  h  h  h  c  c  c  h  h  h  c  h  c  h  c  c
↓  ↓  ↓  ↓  ↓  ↓  ↓  ↓  ↓  ↓  ↓  ↓  ↓  ↓  ↓  ↓  ↓  ↓  ↓  ↓  ↓
1  1  1  1  K  1  K  K  K  1  1  1  K  K  1  K  1  K  1  1
```

What is the contribution of *this conformation* to the partition function? We have 8 h's and 12 c's. Each h contributes a factor of K and each c contributes a factor of 1 to Q. Statistical weights are relative probabilities; therefore, they combine like standard (absolute) probabilities. The probability of two *independent events* happening together is just the *product* of the individual probabilities. Here the *events* are being helix or coil, and they are independent because the states of any residues are independent of each other. Thus, the statement that the residue 1 is coil *and* 2 is coil *and* 3 is coil *and* 4 is coil *and* 5 is helix, and so on, is equivalent to saying that the relative probability of this conformation is

$$1 \times 1 \times 1 \times 1 \times K \times \cdots = 1^{12} K^8 = K^8. \tag{5.9}$$

Perhaps it is easier to see this by calculating the Gibbs energy of the conformation we wrote, which is just the sum of the Gibbs energies of each residue (because they are independent). The Gibbs energy change for the residue switch c ⇌ h is Δg. The coil is the reference, so the c residues contribute zero to the Gibbs energy of the chain. The eight helical residues contribute $8\Delta g$ altogether. Thus, for this conformation, the Gibbs energy is $\Delta G = 8\Delta g$. The equilibrium constant for the transition from all-coil to this particular conformation is

$$e^{-\Delta G/RT} = e^{-8\Delta g/RT}$$
$$= K^8, \tag{5.10}$$

which is the result we had found.

This conformation makes a contribution of K^8 to the partition function of the polypeptide. But any other different *arrangement* that has 8 h's and 12 c's also contributes K^8 to Q. We would like to lump all these terms together because Q would be easier to write. Then, altogether, the conformations with 8 h's and 12 c's contribute $\Omega_{20,8}K^8$ to Q, where $\Omega_{20,8}$ is the number of conformations with 8 h's in 20 residues. How many are there? The number of such conformations is the number of possible *different arrangements* of 8 h's and 12 c's in the chain. To see how many there are, you can start by calculating the number of possible *permutations* (or exchanges) of all the h's and c's in the chain. That number is the *factorial* of N (written as $N!$), which is given by

$$N! = N \times (N-1) \times (N-2) \times (N-3)\ldots \times 2 \times 1. \tag{5.11}$$

(To see how this result arises, imagine you have a bag with 8 h's and 12 c's and you want to place them on each of the N positions in the chain. You can place the first residue—coil or helix—at *any* of the N positions; this brings in the factor of N. Then you can place the second at any of the remaining $N-1$ positions; this brings in the factor of $N-1$. You can place the third at any of the remaining $N-2$ positions, which brings in the factor of $N-2$, etc., until you get to the last residue, and there is only one place for it, which brings in the factor of 1.)

Now note that $N!$ includes arrangements that are identical: if you permute two h's or two c's, you still have the same arrangement. Therefore, to obtain the number of *different* arrangements, we need to divide $N!$ by the numbers of permutations that produce identical arrangements. Those numbers are the permutations of all the h's, which is $n!$, and the permutations of all the c's, which is $(N-n)!$ Therefore, the number of *different* arrangements of n h's and $(N-n)$ c's in a total of N sites is

$$\Omega_{N,n} = \frac{N!}{n!(N-n)!}. \tag{5.12}$$

In particular, the statistical weight of the term that corresponds to $n = 8$ helical residues in a chain of $N = 20$ is

$$\frac{20!}{8!12!}K^8.$$

To obtain the partition function, we need all possible states, from $n = 0$ (all coil) to $n = N$ (all helix). The partition function is the sum of the statistical weights for all possible states of the system. Each term is of the form

$$\frac{N!}{n!(N-n)!}K^n. \tag{5.13}$$

To obtain Q we just need to sum over all possible values of n. It turns out that this sum can be written in closed form because this is just the binomial formula you learned in calculus:

$$Q = \sum_{n=0}^{N} \frac{N!}{n!(N-n)!} K^n \qquad (5.14)$$

$$= (1+K)^N \qquad (5.15)$$

$$= q^N. \qquad (5.16)$$

So, you see, the partition function for the model of independent residues is very simple. Now this is very interesting: the total Q is just the partition function for one residue ($q = 1 + K$) raised to the power of the total number of residues, N. This is because the residues are independent. Partition functions behave like probabilities. The partition function of two independent events is the product of the individual partition functions. Thus, $Q = (1+K)(1+K)\ldots(1+K)$, N times, which is $(1+K)^N$.

Now we would like to know what is the average number of helical residues in the chain. For one isolated residue with two states, just divide each term (1 or K) by their sum q to obtain the absolute probabilities, or the fractions (θ), of each state. Hence,

$$\theta_h = \frac{K}{1+K} \qquad (5.17)$$

$$\theta_c = \frac{1}{1+K}. \qquad (5.18)$$

What about the fraction of helical residues in the entire polypeptide chain? Previously, we have learned a recipe to calculate θ: you obtain these probabilities by taking the derivative of the natural logarithm of the partition function with respect to the natural logarithm of the equilibrium constant. (shown in Equation 5.19 is derived in Appendix B.)

$$\theta_h = \frac{1}{N} \frac{d \ln Q}{d \ln K} \qquad (5.19)$$

$$= \frac{1}{N} \frac{K}{Q} \frac{dQ}{dK}. \qquad (5.20)$$

If you apply Equation 5.19 to $Q = (1+K)^N$ (Equation 5.15), you obtain

$$\theta_h = \frac{K}{1+K}. \qquad (5.21)$$

This is exactly the same as for *one isolated residue*—as it should be— because the *residues are independent*. It makes no difference whether they are connected in a chain or separated. The probabilities of their conformations do not change. (An isolated residue cannot hydrogen bond to anything, but we are trying to understand the concept of independent units.)

The plot of θ_h as a function of the value of the equilibrium constant K for the switch of a residue from coil to helix in the model of independent residues is shown by the solid line in **Figure 5.4**. Mathematically, this is called a hyperbolic function. The fractional helicity increases with K until it reaches a plateau.

Figure 5.4 The fraction of helix per residue as a function of the equilibrium constant K for the conversion from coil to helix, in the model of independent residues (solid line) and in the all-or-none model (dashed line) for $N = 20$.

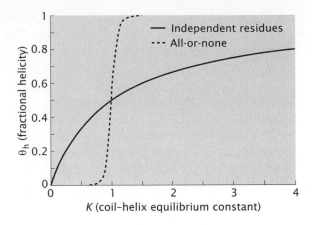

How do we vary K? The simplest way is by varying the temperature. In Chapter 4 we encountered this question regarding the transition between native and denatured proteins. The equilibrium constant is related to the Gibbs energy difference (Δg) between coil and helix,

$$K = e^{-\Delta g/RT}. \tag{5.22}$$

Using

$$\Delta g = \Delta h - T\Delta s \tag{5.23}$$

and because $\Delta g = 0$ at the melting temperature ($T = T_m$),

$$0 = \Delta h - T_m\Delta s,$$

we have

$$\Delta s = \frac{\Delta h}{T_m}.$$

Now substituting this result into Equation 5.23, the Gibbs energy change for switching a residue from coil to helix can be written as

$$\Delta g = \Delta h \left(1 - \frac{T}{T_m}\right). \tag{5.24}$$

Finally, using Equation 5.24 in Equation 5.22, we obtain the equilibrium constant as a function of temperature

$$K = \exp\left[-\frac{\Delta h}{R}\left(\frac{1}{T} - \frac{1}{T_m}\right)\right]. \tag{5.25}$$

The All-or-none Model has Maximal Cooperativity

The model of independent units is not cooperative, but we needed it to lay the ground work for the next steps. Now we move to the other extreme. In the all-or-none model the chain can only exist as a random coil or as a helix. From the point of view of polypeptide *chain populations*, maximal cooperativity means that only *two states of the chain* are appreciably present at equilibrium: either all residues are coil or they are all helix. There are no intermediate conformations. From the point of view of an *individual polypeptide molecule*, maximal cooperativity means that once the coil → helix transition begins, it goes to completion (or back to all-coil). The transition does not stall in the middle.

The partition function of the chain is especially simple in the all-or-none case: it has only two terms, all-coil and all-helix. Again we choose the coil state as the reference for each residue. In the coil state of the chain, all residues are coil; each residue contributes a statistical weight of 1 to Q. The probability of the all-coil state is just the product of the statistical weights for each residue, resulting in a total statistical weight of $1^N = 1$. In the helix state, all residues are helix; each contributes K, so we get K^N. The residues are not independent now; their state depends on the state of *all other residues*. The partition function of the chain is therefore

$$Q = 1 + K^N. \tag{5.26}$$

The fraction of *chains* that are helical is now the same as the fraction of helical residues because the transition is all-or-none. This fraction is

$$\theta_h = \frac{1}{N} \frac{d \ln Q}{d \ln K}$$

$$= \frac{K^N}{1 + K^N}. \tag{5.27}$$

Again the result is very simple. This should be a hint that instead of calculating anything, we could simply have reasoned that the probability of each state is just its statistical weight divided by the sum of all statistical weights, which is the partition function. Since the entire chain has only two states,

$$\theta_h = \frac{K^N}{1 + K^N} \tag{5.28}$$

$$\theta_c = \frac{1}{1 + K^N}. \tag{5.29}$$

The plot of the fractional helicity for the all-or-none model with $N = 20$ residues is shown by the dashed line in Figure 5.4. Compare it with the model of independent residues (dotted line). As $N \to \infty$, the all-or-none transition becomes infinitely sharp: *a step function*.

The Zipper Model is Suitable for Small Polypeptide Chains

Now we begin to study some more realistic models for the coil $\to$ helix transition. We have examined the two extreme cases, so we know that the next models will fall in between. First we consider the zipper model. The concept of this model is that once the helix is initiated, it continues to elongate and then stops. There are *no breaks* in the middle of the helix. The polypeptide forms *one continuous helix*, possibly with coil segments at each end, which we call *terminal disorder*. (The analogy with a clothes zipper is that it is difficult to connect the initial pieces, but once this is done, it zips up easily.)

We have already learned all the concepts we need to write the partition function. As previously, the statistical weight for the all-coil state is 1. For a chain containing a helix, we could have the following conformation, with the statistical weights contributed by each residue indicated below.

$$
\begin{array}{cccccccccccccccccccc}
c & c & c & c & c & c & h & h & h & h & h & h & h & h & h & h & c & c & c & c \\
\downarrow & \downarrow & \downarrow & \downarrow & \downarrow & \downarrow & \downarrow & \downarrow & \downarrow & \downarrow & \downarrow & \downarrow & \downarrow & \downarrow & \downarrow & \downarrow & \downarrow & \downarrow & \downarrow & \downarrow \\
1 & 1 & 1 & 1 & 1 & 1 & \sigma K & K & K & K & K & K & K & K & K & K & 1 & 1 & 1 & 1
\end{array}
$$

Each c contributes 1 to the corresponding term in Q, and each h residue contributes K, *except that the first* h *contributes* σK. The factor σ represents the difficulty of initiating a helix ($\sigma \leq 1$). The particular conformation above contributes σK^{10} to Q. In general, a sequence with n helical residues contributes σK^n. However, there are many sequences that contribute σK^n because the first residue of the helix can occur at any of $N - n + 1$ positions. To see why, suppose we push all the h residues to the end:

$$c\quad c\quad c\quad c\quad c\quad c\quad c\quad c\quad c\quad c\quad h\quad h\quad h\quad h\quad h\quad h\quad h\quad h\quad h\quad h$$

This conformation would also contribute σK^n to Q. To lump together all conformations with n helical residues, we need to find out how many there are. The number of h's is n, and the number of c's is $N - n$. But now notice: the first h can be placed at *any* of the first $N - n + 1$ positions. To obtain all possible conformations, we just need to let n vary from 1 to N. Thus, the partition function for the zipper model is

$$Q = 1 + \sum_{n=1}^{N}(N - n + 1)\,\sigma K^n. \tag{5.30}$$

If $n = 0$, we have the all-coil conformation, which has a statistical weight of 1. Using the series summation from calculus, Q can be written as

$$Q = 1 + \frac{\sigma K^2}{(K-1)^2}\left(K^N + \frac{N}{K} - N - 1\right). \tag{5.31}$$

The average fraction of helix per residue (θ_h) in the zipper model can be calculated using Equation 5.19. We omit the mathematical expression because it is not illuminating. Instead, we plot θ_h as a function of K, in **Figure 5.5**, for several values of σ ($N = 20$ residues). The *sharpness* of the coil–helix transition increases as σ *decreases*.

The Zimm–Bragg Model Allows any Number of Helices to Exist in a Chain

The limitation of the zipper model is that it only allows one helix to exist in each polypeptide chain. This is acceptable for small chains, because the cost of initiating a helix is large, so it is unlikely for initiation to happen twice (or more) with a little coil interruption. However, in long polypeptide chains, this unfavorable situation is compensated

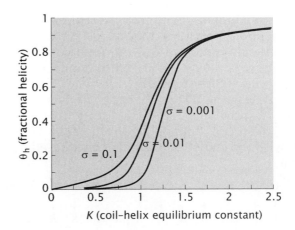

Figure 5.5 The helical fraction per residue as a function of the equilibrium constant K for the conversion from coil to helix in the zipper model with $N = 20$ for various values of σ.

by the favorable entropy arising from a *large number of ways of inter-rupting the helix*. There is only one way of producing a helix of *n* residues once the first residue is fixed; but if you can interrupt an *n*-residue helix at any point, there are $n - 1$ ways of producing it. For large helices, the interruption is favorable because there are many places to break the helix. The model of Zimm and Bragg removes the single-helix restriction.

The derivation of the partition function for the Zimm–Bragg model is more complicated (see for example Cantor and Schimmel [1980]). We give only the result here. For long chains,

$$Q = q^N \tag{5.32}$$

where

$$q = \frac{1 + K + \sqrt{(1 - K)^2 + 4\sigma K}}{2}. \tag{5.33}$$

The average fractional helicity is obtained using Equation 5.19,

$$\theta_h = \frac{1}{N} \frac{d \ln Q}{d \ln K}$$

$$= \frac{d \ln q}{d \ln K}$$

$$= \frac{K}{2q} \left(1 + \frac{K - 1 + 2\sigma}{\sqrt{(1 - K)^2 + 4\sigma K}} \right). \tag{5.34}$$

The plot of the fractional helicity (θ_h) in the Zimm–Bragg model is shown by the solid lines in **Figure 5.6** for $N = 20$ residues and several values of σ. As in the zipper model, the *sharpness of the coil–helix transition increases as σ decreases*. The transition becomes more cooperative as σ decreases, because a small σ means an *unfavorable interface* between a coil and a helical residue. If this interface becomes extremely unfavorable, the polypeptide must avoid it, which means it needs to "decide" if it is going to complete the transition and become all helix, or go back to the all-coil state. The result is that the transition approaches the all-or-none case as σ decreases.

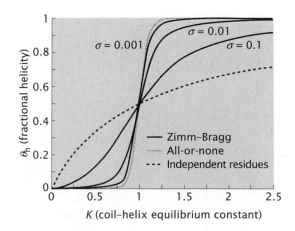

Figure 5.6 The helical fraction per residue as a function of the equilibrium constant K in the Zimm–Bragg model with $N = 20$ for several values of σ (solid lines). The plots for the all-or-none model and the model of independent residues are shown for comparison.

You can think of Equation 5.19 as "interrogating" the partition function with respect to the value of the [helix]/[coil] equilibrium constant, K. If we rewrite it as

$$\frac{d \ln Q}{d \ln K} = N\theta_h, \tag{5.35}$$

the right-hand side of Equation 5.35 is the average number of helical residues. The larger the K, the more favorable is the helical state, and the larger is $N\theta_h$. Similarly, we can find the average number $\langle j \rangle$ of helical *stretches* in a chain by "interrogating" the partition function with respect to σ,

$$\langle j \rangle = \frac{d \ln Q}{d \ln \sigma} \tag{5.36}$$

$$= \frac{N\sigma K^2}{q\sqrt{(1-K)^2 + 4\sigma K}}. \tag{5.37}$$

We use σ now instead of K because σ tells us about how favorable the *interface* is between a coil and a helical residue. As σ decreases, the coil–helix interface becomes less favorable, and consequently, there are fewer helical stretches: for very small σ, it is more favorable to have one long helical stretch than many small ones. For example, if $N = 20$, $K = 1$, and $\sigma = 0.01$, then $\langle j \rangle = 1$; we have only one helical stretch per chain, on average.

How well does the Zimm–Bragg model describe the coil–helix transition of a real polypeptide? Most isolated α-helices are unstable in water. However, they can be designed to have greater stability by using amino acid types with high helical propensity, such as alanine. The helix can also be slightly stabilized by electrostatic interactions or salt bridges between side chains of residues at positions i and $i+3$ or $i+4$ along the chain. **Figure 5.7** shows the coil–helix transition for a 50-residue alanine-rich peptide. The data points are the same as in Figure 5.2, but they are plotted as a function of the equilibrium constant K instead of temperature. The line is a fit of the fractional helicity in the Zimm–Bragg model (shown in Equations 5.33–5.34) to the data with one adjustable parameter ($\sigma = 0.012$).

Figure 5.7 Helix–coil transition in a 50-residue, alanine-rich peptide, with the sequence Ac–Y(AEAAKA)$_8$F–NH$_2$ (the N-terminus is acetylated (Ac = CH$_3$C=O) and the C-terminus is amidated). Same data as in Figure 5.2, measured by CD as a function of temperature in the original. The line is a fit of the Zimm–Bragg model (shown in Equations 5.33–5.34) to the data, with $\sigma = 0.012$ ($\Delta s_{conf} = -4.4$ cal/K/mol, or $\omega_c = 9$). The data were converted to fractional helical content (θ_h) as a function of the equilibrium constant K for switching a residue from coil to helix (shown in Equation 5.25), using $T_m = 307.7$ K (midpoint of the transition), and $\Delta h = -1.35$ kcal/mol for the formation of a hydrogen bond that accompanies the residue switch to the helical state. (Data from Scholtz JM, Marqusee S, Baldwin RL et al. [1991] *Proc Natl Acad Sci USA* 88:2854–2858.)

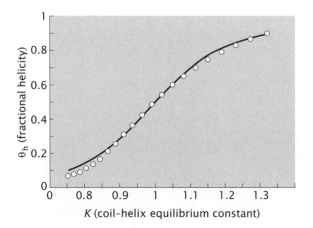

Lessons from the Coil–Helix Transition

We have learned several lessons from the coil–helix transition. Let us summarize them. A polypeptide chain has a multitude of conformations in the unfolded state, which result mainly from rotations around the ϕ, ψ angles about the N–C_α and C_α–CO bonds of each peptide residue. When the chain folds to a helix, that conformational freedom is lost, and there is essentially only one conformation per residue. Hence, the entropy of the polypeptide chain drops enormously when it folds to a helix. This is also true for protein folding, except that the native structure does not consist only of helices.

It is difficult to initiate the helix because the initial "gain"—the energy *released* upon formation of a hydrogen bond—does not compensate the initial loss in conformational entropy. Helix initiation is associated with an unfavorable Gibbs energy change. This gives rise to the cooperativity factor σ that *decreases* the likelihood of helix initiation ($\sigma \leq 1$).

If, however unlikely, this initial interaction (hydrogen bond) is established, then subsequent interactions are established much more easily. The helix "zips up," each new residue added to the growing helix increasingly stabilizing its structure. Hence the cooperativity of the coil–helix transition: once initiated, it tends to go to completion—or go back. Intermediate conformations, containing little helix, are very sparsely populated, because they "paid" all the costs of the initiation but have drawn no "benefit." The Gibbs energy of these intermediates is high and therefore their probability is low: their population is small. The initial hydrogen bond is a *nucleation point*, around which more structure can form.

The more cooperative a transition, the more the system approaches two-state behavior, with virtually nothing else in between. Kinetically, however, there must exist intermediates. Kinetic intermediates are characterized by having large Gibbs energy barriers on either side ($>RT$). How important they are depends on the fraction of folding pathways that go through them.

5.4 SEQUENCE SPECIFIES STRUCTURE, BUT HOW?

Since the work of Anfinsen, we know that the sequence contains all the information necessary to specify the native structure. But how? What is the *folding code*? This is the second question of protein folding: Can we predict the protein structure from the sequence? As we will see in this and the next section, we can in some cases, but this is a very difficult problem.

Amino Acids have Preferences for Certain Secondary Structures

The initial attempts to answer this question started with the prediction of secondary structure. The hope was that, once the secondary structure was predicted, it would be possible to "somehow" fold it into the protein tertiary structure. Chou and Fasman (1978) used the structures of a representative set of proteins to calculate the frequency of occurrence of each amino acid in each type of secondary structure. On this basis, they defined conformational preference parameters, P_α, P_β, and P_t, which represent the propensity of each amino acid to occur in α-helices, β-strands, and turns (t) (**Table 5.1**).

Table 5.1 Chou and Fasman (1978) secondary structure preference parameters of amino acids for α-helices, β-strands, and turns (t)[a]

Amino acid	Propensity		
	P_α	P_β	P_t
Ala	**1.41**[b]	0.52	1.01
Val	0.90	**1.87**	0.41
Leu	**1.34**	1.14	0.57
Ile	1.09	**1.67**	0.47
Met	**1.30**	1.14	0.52
Cys	0.66	**1.40**	0.54
Asp	0.99	0.39	**1.24**
Glu	**1.59**	0.52	1.01
Asn	0.76	0.48	**1.34**
Gln	**1.27**	0.98	0.84
Lys	**1.23**	0.69	1.07
Arg	**1.21**	0.72	0.82
His	**1.05**	0.80	0.81
Gly	0.43	0.58	**1.77**
Pro	0.34	0.31	**1.32**
Ser	0.57	0.96	**1.22**
Thr	0.76	0.17	**0.90**
Phe	1.16	**1.33**	0.59
Tyr	0.74	**1.45**	0.76
Trp	1.02	**1.35**	0.65

[a]Data from Williams RW, Chang A, Juretić D & Loughran S [1987] *Biochim Biophys Acta* 916:200–204.
[b]Values in bold indicate the preferred structure.

Those propensities give the probability of an amino acid to occur in a given type of secondary structure relative to the average probability of all amino acids to occur in that structure. For example, $P_\alpha \approx 1$ for Asp indicates that aspartic acid occurs in α-helices as often as the average amino acid, and $P_\alpha \approx 1.6$ for Glu indicates that glutamic acid occurs in α-helices 1.6 times more often than the average amino acid.

However, the hope of predicting secondary structure on the basis of the simple statistics of occurrence of amino acids was much too optimistic. Equally unsuccessful was the idea of predicting the tertiary structure from the secondary structure. Elements distributed over the entire sequence determine the fold, and a small number of interactions can change it dramatically, as will become evident in a moment. However, the different amino acids have clearly different secondary structural preferences. For example, Ala, Leu, Glu, and Lys favor α-helices over β-strands; β-branched amino acids, such as Ile and Val, and aromatic amino acids, Trp, Phe, and especially Tyr, favor β-strands; and Gly, Asn, and Pro favor turns over α or β structures.

The Paracelsus Challenge: Can we Switch Folds?

The logical continuation at this point would be to describe the improvements that have been made over the Chou–Fasman scheme until the current methods of protein structure prediction. We will get to those methods. But first, to convey the full complexity of the challenge before us, we will take a dramatically different approach. We will tell a story.

It is a fascinating tale of challenge, failure, and success in modern biochemical research. The story, however, ends with a curse. The curse is still with us. But let us start at the beginning.

In 1994, Rose and Creamer made a bold prediction: that by the end of the 1990s the problem of predicting protein structure from the sequence would be solved. As if to dare destiny, they quoted Niels Bohr: "Prediction is very uncertain, especially about the future." Rose and Creamer proposed a challenge. The motivation goes back to the structures of lysozyme and ribonuclease (see Figure 5.1). Suppose these proteins had exactly the same size. Then we would know for sure that it would be possible to convert the structure of lysozyme into that of ribonuclease, by mutating all its amino acids if need be. So the problem of converting one protein fold into another actually has a solution (unlike the alchemical proposition of converting one metal into another). But wait. Do we need to convert all amino acids in the sequence to achieve the conversion of one fold to the other? Probably not. Probably there is a "tipping point" somewhere in the middle. How much of a sequence do we need to mutate before the fold changes?

Rose and Creamer (1994) proposed this problem: Can you take two sequences of proteins with very different structures but similar sizes and convert one fold (native conformation) to the other by mutating no more than half of the sequence? To stimulate work on this field, they established the Paracelsus[1] challenge, with an award of 1000 dollars, to be given to the first scientist to achieve that conversion.

It is Easier to Convert β-Sheet to α-Helix than α-Helix to β-Sheet

Two research groups took up the challenge at about the same time. Historically those who succeeded were the first, but we will tell first the story of those who failed. Yuan and Clarke (1998) began with two proteins: the phage 434 Cro protein (a DNA-binding protein) and the B1 domain of protein G (G_B), a streptococcal cell-surface protein that binds immunoglobulin G (IgG). They mutated residues that were different in the two proteins (not more than 50%) to try to convert the structure of Cro into that of protein G (**Figure 5.8**).

The designed protein, named "crotein G," was 50% identical to the Cro protein but also 62% identical to protein G_B. However, crotein G did not fold cooperatively; its transition was not two-state. Protein aggregation was a problem, which is a sign that the protein did not fold properly. As we have learned, not all sequences fold. Crotein G folds to some structure whose CD spectrum resembles that of protein G_B, but the protein does not adopt a unique native structure. Yuan and Clarke (1998) recognized they had failed to meet the challenge.

Dalal et al. (1997) made a clever choice. Because it is easier to convert a β-sheet to an α-helix than the other way around, they started with the B1 domain of protein G (α/β structure) and tried to convert it to an α-helical protein, the Rop protein (**Figure 5.9**).

The strategy of Dalal et al. (1997) was also different. Instead of trying to mutate the residues of protein G_B to the corresponding amino acids

---

[1] Paracelsus (1493–1541) was a German physician born in the village of Einsiedeln, in what is now Switzerland. His name often appears connected to alchemy, and the search for the ability to transform a substance into another, which is why this challenge bears his name. What we know for certain, however, is that Paracelsus was interested in the use of chemistry in medicine, which is why he is often considered the first pharmaceutical chemist (The Encyclopaedia Britannica, 11th ed, vol 20, pp 749–750, 1911. Cambridge University Press.)

Figure 5.8 Conversion of the fold of 434 cro protein (PDB 2CRO) to that of the B1 domain of G protein (PDB 2GB1) proposed by Yuan and Clarke (1998).

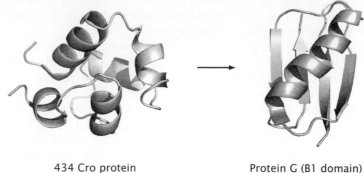

434 Cro protein   Protein G (B1 domain)

in Rop, they sought to identify residues that were key determinants of each fold and removed or incorporated those amino acids as needed to achieve the structural change. Residues in protein G_B with high helix propensity were retained; those with a high propensity for β-strand were replaced in regions that needed to be helical; and hydrophobic residues were incorporated at the **a** and **d** positions of the heptad repeat of a coiled coil (see Chapter 2), because the Rop protein forms a head-to-tail dimer, which is stabilized by those hydrophobic contacts, resulting in a four-helix bundle.

The CD spectrum of the new protein, named *Janus*, clearly showed that it was helical. The NMR spectrum indicated that the protein was folded, not aggregated, adopting a unique structure. Thus, Dalal et al. (1997) succeeded in converting the fold of protein G_B to that of Rop, changing only 50% of the residues. Rose (1997) acknowledged the Paracelsus challenge had been met, and admitted, "Frankly, we thought it could not be done." (He also wished they had bet only a T-shirt instead of $1000.)

Mutation of a Single Residue Switches the Entire Fold

We will now jump 10 years in our story. Clearly two proteins can be 50% identical and fold to different structures. Can this quest be taken further? How many residues is it really necessary to change to convert one structure to another? In 2007, Bryan, Orban, and colleagues took the Paracelsus challenge to its limit. They started with the A and B domains of protein G: G_A, which has an all-α fold, and G_B, which has an

Figure 5.9 Conversion of the fold of protein G_B (PDB 2GB1) to that of the Rop protein (PDB 1ROP). Rop is a homodimer, but it is the monomer that has the length of protein G_B. Proposed by Dalal et al. (1997).

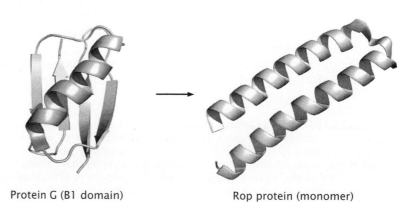

Protein G (B1 domain)   Rop protein (monomer)

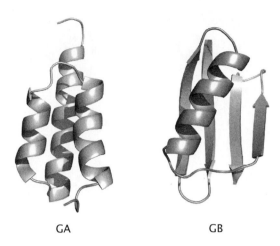

Figure 5.10 Structures of the all-α domain, G_A (PDB 2FS1), and the α/β domain, G_B of protein G (PDB 1PGA).

GA   GB

α/β fold (**Figure 5.10**). They mutated an increasing number of amino acids of G_A to those of G_B, and vice versa. In this manner, in successive steps, the proteins were made 77%, 88%, 91%, and 95% identical in sequence (called G_A77/G_B77 through G_A95/G_B95). However, each mutant still folded correctly, to their respective all-α or α/β native structures.

The G_A and G_B domains of protein G have 56 amino acids each. In the case of G_A95 and G_B95 (95% identity), *only three amino acids are different* between the two sequences (**Figure 5.11**). Yet the folds are still different (**Figure 5.12**). Critical to the fold is the *hydrophobic core* of the two proteins. In G_A95, the hydrophobic core consists of residues A12, A16, L20, I30, I33, A36, V42, and I49. Leu-45 packs against Ile-33 and Ile-49. In G_B95, the core is formed by Y3, L5, L7, A16, A20, A26, F30, A34, W43, Y45, F53, and V54. Tyr-45 is part of the core of G_B, making a strong contact with Phe-52 and a hydrogen bond to Asp-47. Thus, most of the *core* positions are *not common* to the two folds. Residues buried in G_A95 are at least partially exposed to water in G_B95 and vice versa.

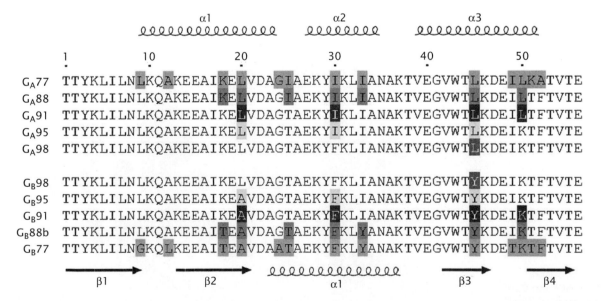

Figure 5.11 Sequences of the mutants of protein G. Top, G_A (all-α domain). Bottom, G_B (α/β domain). The residues mutated are highlighted and the secondary structure is indicated. (Adapted from Alexander PA, He Y, Chen Y, et al. [2009] *Proc Natl Acad Sci USA* 106:21149–21154. Copyright [2009] With permission from National Academy of Sciences.)

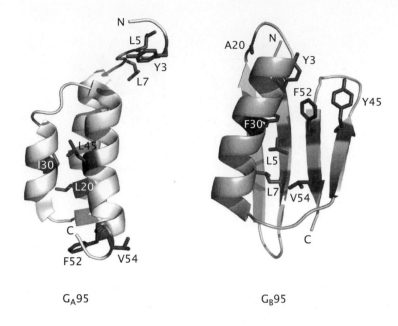

Figure 5.12 Structures of the G$_A$95 (PDB 2KDL) and G$_B$95 (PDB 2KDM). The residues that form the hydrophobic core in either conformation are shown as stick models. (Based on Alexander PA, He Y, Chen Y et al. [2009] *Proc Natl Acad Sci USA* 106:21149–21154.)

Finally, two of the three different residues were made identical in both proteins. The two new proteins, G$_A$98 and G$_B$98, now differ only by *one amino acid* residue, but each still folds to its native conformation. That one amino acid, Leu-45 in G$_A$98 and Tyr-45 in G$_B$98, holds the key to the overall structure *in these mutants* (**Figure 5.13**). The mutation L45Y in G$_A$98 eliminates the few favorable hydrophobic contacts that Leu-45 had (with Ile-33 and Ile-49), thus destabilizing the G$_A$ fold. Conversely, this mutation stabilizes the G$_B$98 fold by creating favorable interactions between the aromatic rings of Tyr-45 and Phe-52, both hydrophobic core residues in G$_B$98.

However, we must be careful. It is not the effect of *just this one side chain* that causes the protein to change fold. Important contacts are

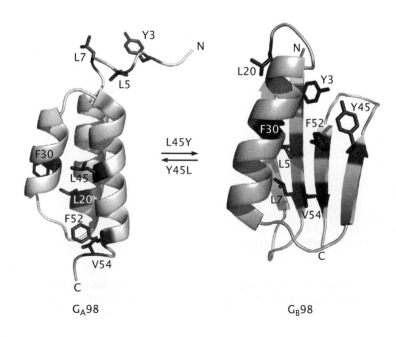

Figure 5.13 Structural switch between G$_A$98 (PDB 2LHC) and G$_B$98 (PDB 2LHD) with the Leu-45 ↔ Tyr-45 mutation. (Based on Alexander PA, He Y, Chen, Y et al. [2009] *Proc Natl Acad Sci USA* 106:21149–21154.)

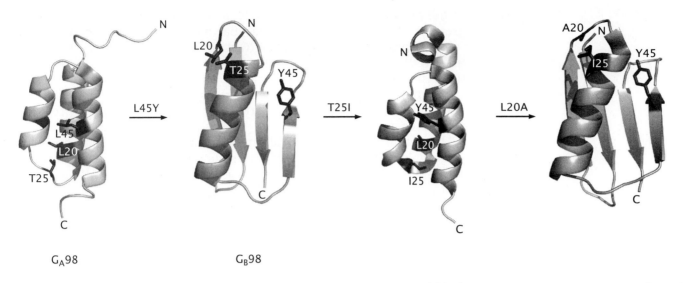

Figure 5.14 Successive switches between the all-α fold of $G_A 98$ and the α/β fold of $G_B 98$ (PDB 2LHC, 2LHD, 2LHG, and 2LHE). The residues mutated are shown as sticks. (Based on He Y, Chen Y, Alexander PA et al. [2012] *Structure* 20:283–291.)

provided by the protein termini in G_B, but not in G_A, in which those termini are unstructured. Mutation of that one amino acid tips the balance between the two possible folded conformations, but only in the *context* of all the other mutations that were previously made to render the sequences of G_A and G_B 98% identical. If performed in $G_A 91$ or $G_A 95$, the mutation L45Y does *not* convert the protein to the fold of G_B, but merely renders the G_A structure less stable, causing it to unfold. Similarly, if performed in G_B, the mutation Y45L does not cause a change to the fold of G_A, but destabilizes G_B.

We have encountered this scenario before, when we studied the concept of *epistasis* in protein evolution. Evolutionary biochemists call epistasis the dependence of the functional or structural effect of a mutation on the *state* (that is, the amino acid types) of the residues at *other* positions in the sequence. We called *permissive mutations*, those prior changes in sequence that allowed the later occurrence of a function-switching mutation. The effect of mutations depends on context because it depends on the interactions the mutated residues establish with other parts of the protein. In fact, the mutation L45Y is *not unique* in its fold-switching ability. If Thr-25 is mutated to Ile in $G_B 98$ (T25I), the fold reverts to that of G_A. And if now Leu-20 is mutated to Ala (L20A), the fold changes yet again to that of G_B (**Figure 5.14**).

The mutational experiments on the A and B domains of protein G raise many questions. Can we ever hope to simulate folding of a protein, or even predict its structure, if a single amino acid difference determines which of two folds a protein adopts? This is the "Paracelsus curse" to computational and theoretical biochemists. The challenge is now much greater. At present, the effect of such minimal mutations on conformation cannot be predicted.

5.5 COMPUTATIONAL METHODS CAN PREDICT PROTEIN STRUCTURES AND SIMULATE FOLDING

You might think the prospect of providing an answer to our second question in protein folding—if I give you the sequence, can you tell me the structure?—is hopeless. Yet, remarkable progress has occurred in the prediction of protein structures from the sequence. One of the greatest incentives to reach this goal has been another challenge: a folding competition called CASP (Critical Assessment of protein Structure Prediction), first held in 1994. In CASP, research groups from more than 100 laboratories all over the world compete to propose the best structures for proteins whose structures are known but have not yet been published. In the end, the structure is revealed and the proposed structures are assessed.

The assumption of structure prediction methods, which is based on Anfinsen's work, is that the native structure is the conformation that corresponds to the global Gibbs energy minimum. For all we know, this assumption is correct, with very rare exceptions. The problem, then, is to find the native structure in a reasonable amount of time. There are two broad types of approaches to predict protein structure: physics-based and knowledge-based. Some methods combine both.

Energy Functions Define the Interactions between Residues

In physics-based methods, the idea is to arrive at the native structure simply from a knowledge of the interactions within the protein (between residues) and between the protein and water. This requires, first, a potential energy function, or *force field*, which defines the forces between the various atoms, and second, a search method that finds the lowest Gibbs energy conformation, subject to those forces. In some physics-based methods, we begin with the unfolded protein and try to find the native structure. In this case, prediction of the structure amounts to simulating the entire folding process. In other cases, the process begins by generating a number of possible conformations of the protein, and then searching those conformations to find native-like structures, and discard false free energy minima.

Physics-based potential energy functions are highly detailed. They attempt to reproduce the physical interactions of all atoms in the protein molecule and with the solvent. These interactions include the energetic contributions of bond lengths and angles, torsional angles, van der Waals interactions, and electrostatic interactions. The most common force fields are AMBER, CHARMM, GROMOS, and OPLS. They have been used extensively in combination with molecular dynamics simulations.

There are, however, highly simplified models with much simpler interactions that are nevertheless very useful in understanding protein folding. One is the one-dimensional *Ising model*. It was originally developed to describe transitions in magnetic systems under a field. A one-dimensional magnet is represented as a chain of spins, each of which can exist in one of two states: spin up (↑), oriented parallel to the external magnetic field, and spin down (↓), antiparallel to the field. For example:

$$\uparrow\uparrow\uparrow\uparrow\uparrow\uparrow\downarrow\downarrow\downarrow\uparrow\uparrow\downarrow\downarrow\downarrow\uparrow$$

Further, in the Ising model, there is an unfavorable interaction between adjacent opposite spins (↑↓), and favorable interactions between parallel spins (↑↑ and ↓↓).

Now, this may seem strange to you: What do magnetic spins have to do with proteins? Well, we have encountered this kind of model already, in the coil–helix transition. The Zimm–Bragg model is identical to the Ising model: the polypeptide is represented as a one-dimensional chain of residues, each residue has two states, helix and coil, and there is an unfavorable interaction when a coil and helix residue are adjacent. In an Ising-like model of a protein, the chain is a sequence of beads, one for each residue. Each residue can exist in two states, native and coil, and transitions are allowed to occur between those states. In protein folding, unlike in the coil–helix transition, interactions occur in three-dimensional space between residues that are not necessarily nearest neighbors in the sequence. To simulate protein folding, we need to specify the interactions between all possible pairs of residue states.

In the simplest Ising-like protein models, the interactions are specified by the Gō potential. In the Gō model, we assign a *favorable interaction* to a contact that is established between two residues that are in *contact in the native structure*. Each native contact contributes an attractive energy of −1 and each nonnative contact contributes a repulsive energy of +1. (Recall that the lower the energy, the more favorable the conformation, so attractive contacts lower the energy whereas repulsive contacts increase it.) We need to know what the native contacts are to begin with, so we need to know the structure from X-ray crystallography or NMR spectroscopy. But once we know it, we can use the Gō model to find out *how* the protein folds. The Gō model is not necessarily an Ising-like model, where residues are structureless beads. Often, a representation of the protein is used that includes all atoms (except hydrogens, usually). This type of model has proven especially informative in protein folding simulations.

Conformational Search Methods are Used to Find the Native Structure

To find the lowest free energy state, which corresponds to the native structure, we need to search the protein conformational space. There are two main methods to achieve this goal: molecular dynamics (MD) and Monte Carlo simulations. The two methods are very different in concept and approach, but either one can be combined with a potential energy function to predict protein structure and simulate folding.

Monte Carlo simulations change the protein conformation by choosing trial moves and then deciding whether to accept or reject the move. The trial moves are *randomly* generated by rotations around protein single bonds. (It is to the role of *chance* in the selection of events (moves), which is a feature of the games played in the Monte Carlo casino in Monaco, that the method owes its name.) In general, the trial moves are not physically realistic (unlike in MD simulations) in the sense that they would not occur like that in a real protein. This, however, does not matter. The energy before (E_i) and after (E_f) the move is calculated, and the move is accepted or rejected in accordance with the Metropolis criterion: if the final energy is lower than the initial, $\Delta E = E_f - E_i < 0$, the move is accepted. (We can also use the Gibbs energy instead of the energy, but we must exclude the conformational entropy contribution, which is automatically incorporated in Monte Carlo simulations, because states with more conformations appear more often *by chance*.) If $\Delta E > 0$, we calculate the probability of the transition by $P_t = e^{-\Delta E/RT}$ and compare it with a random number between 0 and 1 generated by the computer (note that if $\Delta E > 0$, P_t varies between 0 and 1, as a proper probability). If $P_t >$ random number, the move is accepted; if not, it is rejected. These steps are repeated until equilibrium is reached. The Metropolis Monte Carlo method generates an ensemble of conformations that is Boltzmann distributed and should yield the folded state if that corresponds to the minimum in free energy.

You may think, justifiably, that the simple definition of interactions of the Gō model would never work to simulate protein folding. After all, there are many and complex interactions in proteins, which are captured by the detailed physics-based force fields to a good measure but not by the Gō model. The amazing thing is that the Gō model works. This has a profound implication: *native contacts determine the fold*. However, it is the diversity of interactions, afforded by the 20 different amino acids, that allows the evolution of structures with sufficiently specific native contacts to determine the fold. Minimalist sequences, with few types of amino acids, do not fold to unique structures because native contacts are not sufficiently favored over nonnative ones. Folding of small proteins has been simulated by the Metropolis Monte Carlo method with the Gō potential, using all-atom protein representations, from the completely unfolded state to the native conformation (**Figure 5.15**).

In MD simulations, Newton's equations of motion are solved for all atoms of the protein. At each time step, the position and the velocity of each atom is calculated under the effect of the force field. A new conformation of the protein results from the motion of its atoms. With a physics-based force field, the calculated moves are physically realistic, in the sense that they could happen in a real protein on the same timescale. The problem with MD simulations is that the simulations take a very long time. This is because so many atoms are included in the calculations, detailed interaction energies are used, and the time increment between moves must be small (to allow the calculation of derivatives of the position and velocity of each atom).

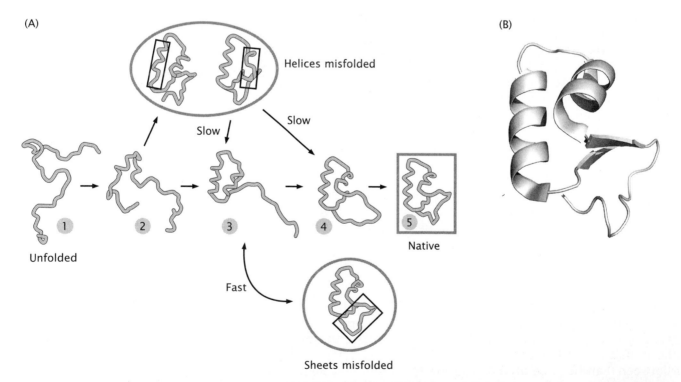

Figure 5.15 (A) Monte Carlo simulations of folding of the small protein crambin using the Gō model. Several conformations along the folding path are shown, including a few misfolds. (B) A high-resolution structure of crambin, determined by neutron diffraction is shown for comparison (PDB 4FC1). (A, Adapted from Shimada J, Kussell EL & Shakhnovich EI [2001] *J Mol Biol* 308:79–95. With permission from Elsevier.)

Full MD simulations of protein folding starting from the unfolded state have only been performed for small proteins, because the amount of computational time required is prohibitive. This situation, however, is likely to change in the near future. Moreover, starting from the native state close to T_m, one can unfold the protein and watch what conformations it acquires. Assuming that the protein traces back its pathway along the reaction coordinate from the folded to the unfolded state, it should reach the transition state shortly after it begins to unfold. This is the *principle of microscopic reversibility* applied to protein folding. If the MD simulations are performed close to the T_m of the protein, unfolding and refolding events are observed.

Folding of small proteins can be simulated in supercomputers. One of the most striking demonstrations of the progress in our ability to reproduce protein folding in a computer is the simulation of the folding of 12 small proteins with very different native structures by molecular dynamics in a specially designed supercomputer named "Anton." The proteins studied have lengths of 10–80 amino acid residues, and span a diversity of folds, including a simple hairpin, all-α-helix, all-β-sheet, and α/β structures. In the simulations, those proteins were observed to fold from a completely disordered state to a folded conformation that closely resembles the native structure determined experimentally by X-ray crystallography or NMR (**Figure 5.16**).

The accuracy of a protein model obtained by simulation or prediction is assessed by calculating the root mean square deviation (RMSD) of the centers of the α carbons (C_α, see Figure 2.4) in the simulated and experimental structure,

$$C_\alpha\text{-RMSD} = \sqrt{\frac{\sum_{n=1}^{N}(r_{\alpha,n}^{\text{sim}} - r_{\alpha,n}^{\text{exp}})^2}{N}}, \qquad (5.38)$$

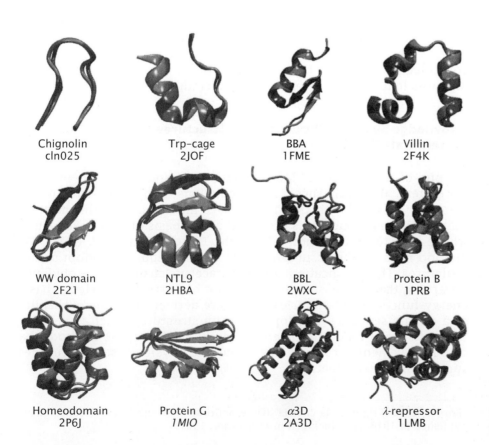

Chignolin
cln025

Trp–cage
2JOF

BBA
1FME

Villin
2F4K

WW domain
2F21

NTL9
2HBA

BBL
2WXC

Protein B
1PRB

Homeodomain
2P6J

Protein G
1MIO

α3D
2A3D

λ-repressor
1LMB

Figure 5.16 Comparison of the native structures of 12 small proteins obtained experimentally (light gray) with their folded conformations in MD simulations (dark gray). The common name and the PDB entry of each protein are indicated. (Adapted from Lindorff-Larsen K, Piana S, Dror RO & Shaw DE [2011] *Science* 334:517–520. With permission from American Association for the Advancement of Science.)

where N is the number of residues and r_α are the positions of the α carbons. Most of the simulated structures in Figure 5.16 have C_α-RMSD < 2 Å. Thus, molecular dynamics can simulate the correct native structure of proteins.

Now be careful: you might think that this implies that the force field used must be very accurate. Well, it does *not*. All it means is that the energy of the native conformation is much lower than the energy of any other conformation, including incorrectly folded structures. Recall that Gō models are also able to simulate folding to the native structure, although the Gō potential is all but accurate. However, molecular dynamics is the most physically realistic way to simulate the actual movements of a real molecule.

What Model is Best Depends on the Question you Ask

What is the best model for protein folding simulations? Is there a place for simple models, such as the Ising or the Gō models in protein folding? If detailed MD simulations are so much more realistic, shouldn't we use them always, at least if practical? Perhaps, but the answer to your question depends on the question you want to answer. For example, the mechanisms observed in the simulation of the folding of villin subdomain (see Figure 5.16) using an Ising model are remarkably similar to those obtained by the most detailed MD simulations. The assumption, in the Ising model, that the native structure grows from nucleation in a few regions of the protein is borne out by the simulations. Ising models have the advantage of using only very few parameters, which allows for an understanding of the most fundamental aspects of the process.

On the topic of exactitude in science, the Argentinian writer (of fantastic literature) Jorge Luis Borges tells a story[2] about an empire where the science of cartography was especially developed. Maps of increasing size and detail were produced, only to be deemed unsatisfactory. Until finally, the cartographers built a map the size of the empire that coincided with it point by point. Eventually, however, this enormous map was found to be useless and was abandoned.

Knowledge-based Methods Predict Structures Based on the Protein Data Bank

In spite of the success of MD simulations of small proteins in supercomputers, the most successful predictions of protein structure are generally obtained with knowledge-based methods. What this means is that the protein structure is predicted on the basis of similar sequences whose structures are known from the Protein Data Bank (PDB). The method rests on the assumption that similar sequences adopt similar structures. The ramifications of the Paracelsus challenge are obviously a warning that this assumption may not be correct. Knowledge-based energy functions are empirical. They are derived from the probability of a certain sequence to adopt a certain structure, which is given by the statistics of its occurrence in the PDB. There are also methods that combine elements of knowledge-based and physics-based approaches, such as the *Rosetta* method.

---

[2]Borges JL, "Del rigor en la ciencia" (On exactitude in science), in *Museo*, originally published in 1946, later included in *El Hacedor*, 1961.

The simplest knowledge-based method is *template* or *homology* modeling. We simply search (using PSI-BLAST, for example) for sequences that are closely related to that of the protein whose structure we want to predict. When such similar proteins exist in the PDB, predicting the structure of our protein just amounts to modeling the structure of the target sequence on a known structure. The difficulty arises if no sufficiently similar sequence exists whose structure is known.

If no closely related protein exists in the PDB, we can resort to the method of *fragment assembly*. The initial part of the strategy is to find the structures of sequences that are similar to *short segments* of our target sequence. Then, those structured segments are combined to produce the tertiary structure. The Rosetta method incorporates this approach.

In Rosetta, there are two phases in the prediction of protein structure. The first phase consists in dividing the target sequence into short stretches and folding them by fragment assembly. This is achieved by using a library of peptide fragments (typically nonamers) with similar sequences whose structures (or better, distributions of structures) are known from the PDB. At this point the chain is represented in a coarse grained manner by collapsing several atoms in beads, and the protein is folded to a low-resolution model by assembling those structured fragments through the establishment of nonlocal contacts. In the second phase, the low-resolution models are replaced by all-atom representations and the structures are refined by Metropolis Monte Carlo simulation using a physics-based energy function. In Rosetta, this energy function includes van der Waals interactions, solvation free energy, and hydrogen bonds. The method has achieved remarkable success in predicting protein structures (**Figure 5.17**). The agreement between the predicted and experimental structures is demonstrated by a C_α-RMSD <1.5 Å for five of the proteins shown.

We are now beginning to be able to predict and simulate protein folding from the sequence. Given the sequence, a discriminating force field,

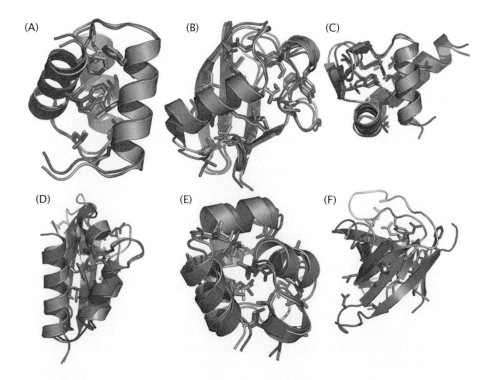

(A) (B) (C)

(D) (E) (F)

Figure 5.17 Comparison of the native structures of six small proteins obtained experimentally (light gray) with their folded conformations predicted with Rosetta (dark gray). (A) Hox-B1, (B) ubiquitin, (C) RecA, (D) KH domain of Nova-2, (E) 434 repressor, (F) Fyn tyrosine kinase. The structures (A–E) are well predicted, but in (F) the prediction erroneously replaced a large loop by an extra β-hairpin. (Adapted from Bradley P, Misura KMS & Baker D [2005] *Science* 309:1868–1871. With permission from American Association for the Advancement of Science.)

and sufficient computer power, MD simulations of folding of most proteins should be possible—in principle. For now, this can only be done for small proteins. But structure prediction with methods like Rosetta requires less computer power. However, somehow we still hear the nagging question: Do you really understand how it folds? We want to *see* how the protein folds, but also *understand* how it gets there: we need to learn about the *folding mechanism*.

5.6 THE KINETIC MECHANISM DESCRIBES HOW THE PROTEIN FOLDS

In this and the next section we consider our third question: What is the *mechanism* of protein folding? How does the protein find its native structure? Does it randomly search all possible conformations until it finds the most favorable? Certainly not, because this would take longer than the age of the universe—and yet the protein finds it. This apparent contradiction became known as the *Levinthal paradox*. Consider a protein with $N = 100$ residues, and assume that each residue can exist in $\omega = 7$ conformations by rotations about its ϕ and ψ angles. The number of conformations of the protein in the denatured state is $W \approx \omega^N$. For a 100-residue protein, this corresponds to $7^{100} = 3 \times 10^{84}$ conformations. If the protein takes $\sim 10^{-13}$ seconds (which is about the time for a rotation) to change conformation and try a new one, then it takes $\sim 3 \times 10^{71}$ seconds to find the native structure, or 10^{64} years! The universe is only $\sim 10^{10}$ years old (about 14 billion years). However, small proteins fold in microseconds to milliseconds.

Protein folding does *not* proceed by a random search over all conformations. There are two alternatives. One is that the folded state is not necessarily the overall Gibbs energy minimum, but a local minimum, which is found because the sequence specifies a *pathway* to reach the native conformation *quickly*. This was the solution suggested by Levinthal.

The other alternative, which is favored by most protein chemists, is that the native state is the global Gibbs energy minimum. Under native conditions, proteins always find the *same* folded conformation no matter from what point they begin the search. This is a compelling argument that the native fold corresponds to the global Gibbs energy minimum. *How*, then, does the protein find it? There must be ways to make the search much faster. Much has been learned in this regard from experiments and simulations. We will study this problem in the remaining sections of this chapter. We will try to understand what we mean by mechanism, what are the kinetics of protein folding, what is the transition state, and how we learn about its structure.

A Mechanism is a Description of the Steps from Reactants to Products Along a Reaction Coordinate

Probably when you studied organic chemistry, you learned how to describe a reaction mechanism. Typically you used curved arrows to indicate the movement of electrons (bonds or lone pairs). For example, for an S_N2 nucleophilic substitution reaction between the hydroxide ion and ethyl chloride, you wrote something like the mechanism shown in **Figure 5.18**. The species marked with the symbol ‡ is the transition state.

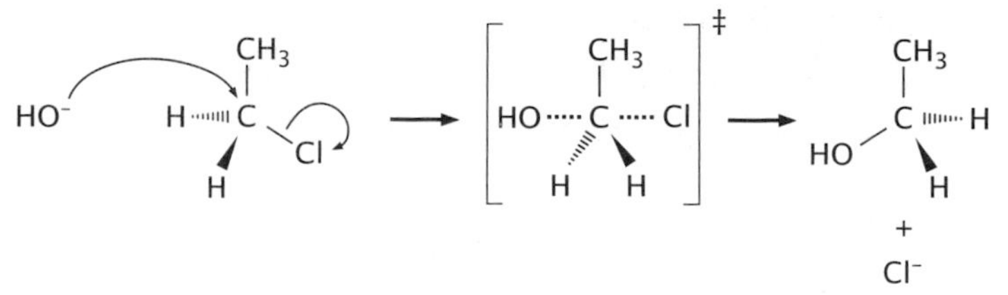

Figure 5.18 Mechanism of an S_N2 nucleophilic substitution.

Other times there were intermediates, as in the case of the nucleophilic acyl substitution reaction in **Figure 5.19**, between the hydroxide ion and acetyl chloride, where a tetrahedral intermediate is formed.

In those cases things were simple. The reaction started with the reactants; we knew what they looked like. Then the reaction proceeded to a transition state, whose structure we did not know so well because it is so short-lived and thus difficult to characterize, but we could nevertheless make some educated guesses about what it probably looks like—because we knew which bonds were made or broken going into and out of the transition state. Then, we knew what the product looked like; we knew which bonds had been formed or broken. The *reaction coordinate* represented the formation and breaking of a few covalent bonds.

In the protein folding problem, nothing is simple. We know fairly well the structure of the product, the native state, from NMR or X-ray crystallography. Beyond that, however, the problem is very complicated. The reactants are *not* well defined. There is not *one* unfolded structure. What we call the unfolded state is an *ensemble* of a huge number of conformations, usually extended, with variable degrees of randomness, but by no means completely lacking of secondary structure. The pathway to the native fold begins at *each one* of those conformations and is not—cannot be—the same for all. The transition state, also, is *not one* structure, but an ensemble, albeit much smaller than that of the unfolded state. We know it is a compact structure with a high degree of organization, with native-like topology and contacts and a variable amount of secondary structure.

Now, what is the reaction coordinate in protein folding? An unfolded conformation reaches the folded structure by rotations around single bonds, the most important of which are those defined by the ϕ and ψ dihedral angles. This, however, constitutes a multidimensional space (with 2^N dimensions, where N is the number of residues, and the base 2 corresponds to the two angles ϕ and ψ, if we ignore side-chain rotations). Other "global coordinates" have been tried to describe folding, such as the volume of the molecule (compactness) or the number of correct (native) contacts. Here, we will define the reaction coordinate somewhat loosely as the set of most likely paths leading from the unfolded ensemble to the folded structure, through the transition state ensemble (**Figure 5.20**).

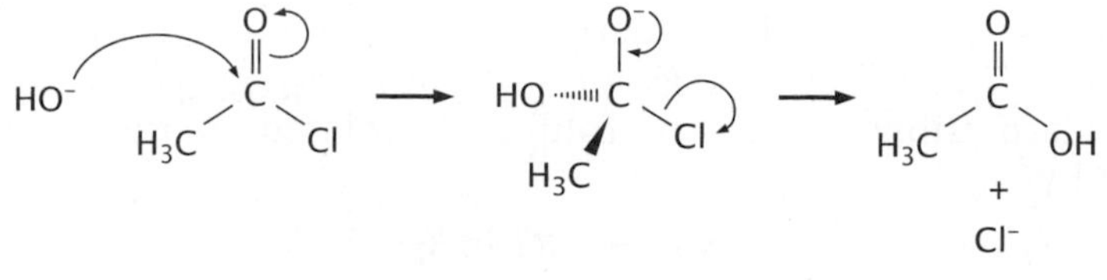

Figure 5.19 Mechanism of a nucleophilic acyl substitution.

Figure 5.20 Gibbs energy as a function of the reaction coordinate for two-state protein folding.

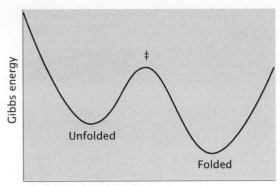

"Reaction coordinate"

The Kinetics Describe the Rates of the Steps Along the Folding Pathway

We turn now to the kinetics of protein folding. This entails listing the states the protein may assume from the unfolded to the folded conformation, including any intermediates that are significantly populated along the folding pathway. It also includes a determination of the rates at which one state is formed from the previous one. We will concentrate on protein folding reactions that follow *two-state* kinetics. This is true for most small, single-domain proteins and is a good approximation for many others. However, there are many cases where kinetic intermediates occur that are sufficiently long-lived so that they cannot be ignored in the kinetics, even if only two thermodynamic states exist under equilibrium conditions. If intermediates occur, the kinetics are more complicated. The approach we will learn can be easily extended to include intermediates, but here we will limit ourselves to providing a few guidelines for those cases.

Rate Constants Measure Activation Barriers

The kinetics of folding are characterized by rate constants. A large rate constant tells us that the reaction is intrinsically fast. Let us see how the kinetics are related to the thermodynamics. Consider the reversible folding reaction, from the denatured (D) to the native (N) conformation of the protein. The rate of this reaction is determined by the *molecular rate constants* of folding (k_f) and unfolding (k_u),

$$D \underset{k_u}{\overset{k_f}{\rightleftharpoons}} N. \tag{5.39}$$

The equilibrium constant for folding (K_F) is the ratio of the folding to the unfolding rate constants,

$$K_F = \frac{k_f}{k_u}. \tag{5.40}$$

The Gibbs energy of *folding* (ΔG_f^o) is the change for the reaction written in the direction from the denatured (D) to the native (N) protein. In Chapter 4 we wrote this reaction in the direction of *denaturation* (unfolding). The Gibbs energy of unfolding is simply $\Delta G_u^o = -\Delta G_f^o$. The standard Gibbs energy of folding is related to the equilibrium constant by

$$\Delta G_f^o = -RT \ln K_F. \tag{5.41}$$

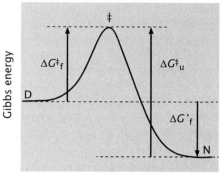

Figure 5.21 Relation between the equilibrium Gibbs energy of folding (ΔG_f^0), and the Gibbs activation energy of folding ($\Delta G_f^{\ddagger}$) and unfolding ($\Delta G_u^{\ddagger}$).

Note that the *standard state* usually means that the concentration is 1 M, but in an ideal state in which the protein behaves as if it were in an infinitely dilute solution, so that no interactions between protein molecules occur. Here we can be a little more relaxed, because the protein concentration is actually *not* important. The folding reaction is unaffected by protein concentration, as long as the proteins do not interact. In practice, this means that the protein concentration needs be sufficiently low, but a particular concentration need not be specified. The standard state also means that the pressure is 1 atm (see Chapter 1 for the definition).

Figure 5.21 shows a simplified diagram of the Gibbs energy for the folding reaction. The Gibbs activation energy of folding ($\Delta G_f^{\ddagger}$) is the barrier in the folding direction, D → N; it is the difference between the Gibbs energy of the transition state ($G^{\ddagger}$) and that of the the unfolded state. The Gibbs activation energy of unfolding ($\Delta G_u^{\ddagger}$) is the barrier in the unfolding direction, N → D.

The Gibbs activation energies are related to the rate constants of folding (k_f) and unfolding (k_u) by

$$k_f = k_0 e^{-\Delta G_f^{\ddagger}/RT} \tag{5.42}$$

$$k_u = k_0 e^{-\Delta G_u^{\ddagger}/RT}. \tag{5.43}$$

The pre-exponential factor k_0, which is related to the frequency of barrier crossing, must be the same in both directions because, as shown in Figure 5.21,

$$\Delta G_f^0 = \Delta G_f^{\ddagger} - \Delta G_u^{\ddagger}. \tag{5.44}$$

Thus,

$$K_F = \frac{k_f}{k_u}$$

$$= e^{-\Delta G_f^0/RT} \tag{5.45}$$

$$= e^{-(\Delta G_f^{\ddagger} - \Delta G_u^{\ddagger})/RT}, \tag{5.46}$$

so the factor k_0 cancels out in k_f/k_u.

Two-state Folding Kinetics are Exponential in Time

Protein folding kinetics can be measured in a *perturbation* experiment by suddenly changing the conditions of the system. This can be achieved, for example, by a *temperature-jump* (T-jump), or by a *concentration-jump* in which, for example, pH or a denaturant concentration suddenly changes. Under the new conditions, the state of the protein evolves in time until a new equilibrium is reached. For example in stopped-flow, two solutions are rapidly injected into a small cell (~50 μL) and are mixed in ~1 ms. Thus, we can suddenly change the pH or the concentration of a denaturant, from initial conditions that favor the unfolded state to final ones that favor the folded state, or vice versa. If the fluorescence emission of the protein is different between the native and denatured states, we can follow the reaction by monitoring the fluorescence as a function of time.

An example of a stopped-flow fluorescence experiment is shown in **Figure 5.22**. A protein is initially at a pH where it is mostly unfolded, and then it is mixed very rapidly with a buffer at a pH that favors the folded state. In the new buffer, the protein folds with an apparent rate constant k that describes the approach to equilibrium. The reciprocal of this rate constant is the apparent *characteristic time* of folding, τ,

$$\tau = 1/k \tag{5.47}$$

which is indicated in Figure 5.22. The time τ characterizes the kinetics.

We can also define the *half-time* $t_{1/2}$ as the time elapsed when the fraction of folding reaches one-half. In a two-state transition, this means that half of the proteins are folded at $t_{1/2}$ ($\theta = 1/2$). The characteristic time τ is related to the half-time by $t_{1/2} = \tau \ln 2$ (we will see why shortly). We can use either τ or $t_{1/2}$ to characterize the curve. However, τ, or equivalently the apparent rate constant k (shown in Equation 5.47), is a more fundamental quantity, because it is directly related to the molecular rate constants k_f and k_u.

Consider again the interconversion between the unfolded (D) and folded (N) states of the protein (shown in Equation 5.39).

$$D \underset{k_u}{\overset{k_f}{\rightleftharpoons}} N.$$

Suppose that, at time $t = 0$, we begin with a certain concentration [D(0)] of unfolded and [N(0)] of folded protein. We want to know how the concentrations [D(t)] and [N(t)] vary as a function of the time (t) after the reaction started. To do that, we must write and solve the *rate equations* for this reversible reaction. The amount of denatured state D decreases in time in proportion to the amount that folds to N. The corresponding rate is $-k_f[D]$, where the minus sign indicates a *decrease* in the concentration of D. The amount of D increases in time in proportion to the

Figure 5.22 Protein folding kinetics measured by the fluorescence change as a function of time following a pH-jump in a stopped-flow instrument (hypothetical experiment). The line is a fit of Equation 5.57, $F = A\left(1 - e^{-t/\tau}\right)$, to the data. (Data from the author's laboratory, adapted from a different stopped-flow fluorescence experiment.)

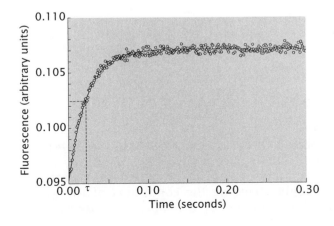

amount of N that unfolds to D. The corresponding rate is $+k_u[N]$, where the plus sign indicates an *increase* in the concentration of D. Similarly, [N] decreases in time in proportion to the amount that unfolds to D; the rate is $-k_u[N]$. Finally, [N] increases in time in proportion to the amount of D that folds to N; the rate is $+k_f[D]$. Thus, the concentrations change with time as given by the following rate equations:

$$\frac{d[D]}{dt} = -k_f[D] + k_u[N] \tag{5.48}$$

$$\frac{d[N]}{dt} = +k_f[D] - k_u[N]. \tag{5.49}$$

This is a system of *coupled* differential equations, with *initial conditions* $[D] = [D(0)]$ and $[N] = [N(0)]$. The procedure to solve it is described in Appendix C. We will write the results in terms of the fractions of unfolded (θ_D) and folded (θ_N) states, instead of molar concentrations,

$$\theta_D = \frac{[D]}{[D] + [N]} \tag{5.50}$$

$$\theta_N = \frac{[N]}{[D] + [N]}. \tag{5.51}$$

The solution of Equations 5.48 and 5.49 can then be written as

$$\theta_D(t) = \frac{1}{1 + K_F} + \left(\theta_D(0) - \frac{1}{1 + K_F} \right) e^{-kt} \tag{5.52}$$

$$\theta_N(t) = \frac{K_F}{1 + K_F} - \left(\theta_D(0) - \frac{1}{1 + K_F} \right) e^{-kt}, \tag{5.53}$$

where $k = k_f + k_u$ is the apparent rate constant and $K_F = k_f/k_u$ is the equilibrium constant for folding.

It is worthwhile to pause for a moment and note some important features of Equations 5.52–5.53. First, the fractions of both denatured and native protein change in time as a *single exponential* function, characterized by the *same* apparent rate constant k. Further, k is the *sum* of the two molecular rate constants for folding and unfolding: $k = k_f + k_u$. The reason for the sum is that, the larger *both* rate constants, the faster the equilibrium is reached. The time dependence of θ_D and θ_N is determined *only* by the factor e^{-kt}.

Second, when the *net* reaction stops, the interconversion $D \rightleftharpoons N$ continues. If we set the time $t = \infty$ in Equations 5.52 and 5.53, thus making the exponential factors zero ($e^{-\infty} = 0$), we obtain the equilibrium values of the fractions of denatured and native states,

$$\theta_D(t) = \frac{1}{1 + K_F} \tag{5.54}$$

$$\theta_N(t) = \frac{K_F}{1 + K_F}. \tag{5.55}$$

Those fractions are the same as we found in Chapters 1 and 4 (shown in Equations 1.75 and 4.5) from the terms in the partition function, except now we are using the unfolded state as the reference, so we would write the partition function as $Q = 1 + K_F$.

Third, to gain a better understanding of the shape of these exponential functions, and to see how $\theta_N(t)$ describes the data in Figure 5.22, let us simplify Equations 5.52 and 5.53 to the case in which all the protein is denatured in the beginning of the experiment ($\theta_D(0) = 1$ and $\theta_N(0) = 0$). Then we obtain

$$\theta_D(t) = \frac{K_F}{1 + K_F}\, e^{-kt} + \frac{1}{1 + K_F} \tag{5.56}$$

$$\theta_N(t) = \frac{K_F}{1 + K_F}\left(1 - e^{-kt}\right). \tag{5.57}$$

Figure 5.23 shows a plot of $\theta_D(t)$ and $\theta_N(t)$ as a function of time. The unfolded fraction (see Figure 5.23A) follows an exponential decay (e^{-kt}) from $\theta_D(0) = 1$ to the final, equilibrium value $\theta_D(\infty) = 1/(1 + K_F)$. The folded fraction (see Figure 5.23B) follows an exponential rise ($1 - e^{-kt}$) from $\theta_N(0) = 0$ to the final equilibrium value $\theta_N(\infty) = K_F/(1 + K_F)$. Both curves have a total *amplitude* change of $K_F/(1 + K_F)$ in the ordinate axis (*y*-axis), from the beginning to the horizontal dashed line. This is a decrease in the case of the unfolded fraction and an increase in the case of the folded fraction, but the magnitude of the *change* is the same because one state is converted to the other.

What amplitude changes (in θ) do τ and $t_{1/2}$ correspond to? If we set $t = \tau = 1/k$ in Equation 5.57, we obtain

$$\theta_N(\tau) = \frac{K_F}{1 + K_F}\left(1 - \frac{1}{e}\right), \tag{5.58}$$

where $e = 2.71828\ldots$ is the base of the natural logarithm (Euler's number). Thus, at $t = \tau$ the folded fraction θ_N has increased by $1 - 1/e \approx 0.63$ of its total amplitude change (see Figure 5.23B). The total amplitude change is $K_F/(1 + K_F)$, which is ≈ 0.91 in this example; thus, the amplitude that corresponds to τ is $\theta_N(\tau) = 0.63 \times 0.91 = 0.57$. At $t = t_{1/2}$, θ_N has increased by $1/2$ of its total amplitude change. Similarly, at $t = \tau$ the unfolded fraction θ_D has decreased by $1 - 1/e \approx 0.63$ of its total amplitude change (see Figure 5.23A). In this example, the total amplitude change is again ≈ 0.91 but in the *opposite* direction; the amplitude that corresponds to τ is $\theta_D(\tau) = 1 - 0.63 \times 0.91 = 0.43$. At $t = t_{1/2}$, θ_D has decreased by $1/2$ of its total amplitude change.

From Figure 5.23 you also see that $t_{1/2}$ is a little *shorter* than τ. We can now derive the exact relation between them. Since $k = 1/\tau$, we can write the exponential factor in Equation 5.57 as $1 - e^{-t/\tau}$. This term becomes $1/2$ when $t = t_{1/2}$. Therefore,

$$1 - e^{-t_{1/2}/\tau} = \frac{1}{2} \tag{5.59}$$

Figure 5.23 Folding kinetics calculated with $k_f = 10$ s^{-1} and $k_u = 1$ s^{-1}. (A) Unfolded fraction. (B) Folded fraction. The horizontal lines indicate the final values ($t = \infty$), which are $\theta_D(\infty) = 1/(1 + K_F)$ in A, and $\theta_N(\infty) = K_F/(1 + K_F)$ in B. The vertical lines (read on the time axis) indicate $t_{1/2}$ and τ. The amplitude of the change from the beginning to $t_{1/2}$ is 0.50 of the total amplitude change. The amplitude of the change from the beginning to τ is $1 - 1/e \approx 0.63$ of the total amplitude change.

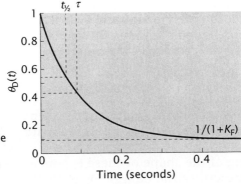

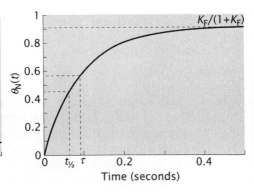

or

$$e^{-t_{1/2}/\tau} = \frac{1}{2}. \qquad (5.60)$$

Taking the natural logarithm of both sides of this equation, we obtain the relation

$$t_{1/2} = \tau \ln 2. \qquad (5.61)$$

Since $\ln 2 \approx 0.69$, $t_{1/2} < \tau$.

In summary, in the reaction $D \rightleftharpoons N$, both the folded and the unfolded fractions change exponentially in time, with an apparent rate constant k that is the sum of the molecular rate constants of folding and unfolding, k_f and k_u. Because $k = k_f + k_u$, *the larger* of the two molecular rate constants dominates the overall rate of the reaction. The characteristic time of the reaction is $\tau = 1/k$, which is also called the *relaxation time*. The *extent* of the reaction is determined by the equilibrium constant for folding. Namely, the final values of the folded and unfolded fractions, $K_F/(1 + K_F)$ and $1/(1 + K_F)$, respectively, are determined by K_F. Ultimately, however, they too are determined by the molecular rate constants, because $K_F = k_f/k_u$.

Sequential Multistep Reactions give Rise to Multiexponential Kinetics

The reaction we studied is particularly simple. We have only two states connected by an equilibrium, $D \rightleftharpoons N$. The kinetics of this reaction are described by a single exponential function of time, which we can write in simplified form as

$$\theta_D \sim e^{-kt} \qquad (5.62)$$

$$\theta_N \sim 1 - e^{-kt}, \qquad (5.63)$$

where $k = k_f + k_u$. But what happens if we have a kinetic intermediate? What happens if the steps are not reversible? We will not solve all of these problems in detail, but we will provide guidelines for each case.

One-step irreversible reaction:

$$D \xrightarrow{k_f} N. \qquad (5.64)$$

This case yields single exponentials for D and N, with an apparent rate constant identical to the folding rate constant $k = k_f$.

Kinetic intermediate: an intermediate (I) affects the kinetics if the Gibbs energy barriers on either side of I are significantly larger than the ambient kinetic energy, RT (**Figure 5.24**).

In this case we have two steps. To make things simpler, assume the two steps are irreversible, and all protein is denatured initially.

$$D \xrightarrow{k_1} I \xrightarrow{k_2} N. \qquad (5.65)$$

Formation of the final product (N) is described by a sum of two exponentials in time, because both steps affect the rate of appearance of N.

$$\theta_N = 1 - \frac{1}{k_2 - k_1} \left(k_2 e^{-k_1 t} - k_1 e^{-k_2 t} \right). \qquad (5.66)$$

Figure 5.24 Gibbs energy as a function of the reaction coordinate in the case an intermediate (I) exists.

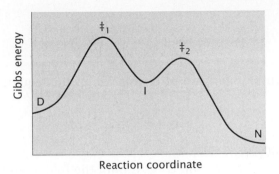

The reactant (D) disappears as a one-exponential function of time, because its concentration depends only on k_1,

$$\theta_D = e^{-k_1 t}. \tag{5.67}$$

Since there is no flux back to D, how much I or N is formed does not affect the amount of D. The concentration of the intermediate I is affected by *both* rate constants, because it appears from D with rate constant k_1 and disappears to N with rate constant k_2. Therefore, its time course is described by two exponentials,

$$\theta_I = 1 - \frac{k_1}{k_2 - k_1} \left(k_2 e^{-k_1 t} - k_1 e^{-k_2 t} \right). \tag{5.68}$$

Before we move to the next case, let us pause for a moment. Look carefully at Equation 5.66 for the time dependence of the native state. What happens if you interchange k_1 and k_2? Nothing happens. The equation remains exactly the same. (This is *not* true of Equation 5.68 for the intermediate.) What this means is that by observing the product alone, you cannot tell which rate constant corresponds to each step. This may be puzzling, but it is true. To assign the rate constants you need to observe the intermediate or the reactant as well.

Figure 5.25 shows the time change of the fractions of D, I, and N in the scheme of Equation 5.65, with $k_1 = 10$, $k_2 = 1\,s^{-1}$ (see Figure 5.25A), and with $k_1 = 1$, $k_2 = 10\,s^{-1}$ (see Figure 5.25B). The curve for N (the product) is exactly the same in both panels. The intermediate I *accumulates* if it forms fast and disappears slowly (see Figure 5.25A), but it does *not* accumulate if it forms slowly and disappears fast (see Figure 5.25B). The denatured state decays fast if k_1 is large (see Figure 5.25A) but decays slowly if k_1 is small (see Figure 5.25B). Figure 5.25 also shows that the shape of the curve for N is not a simple exponential but has a small *lag time* in the beginning. That is because before N is formed, I has to be formed from D.

Figure 5.25 Folding kinetics with an intermediate, D → I → N (shown in Equation 5.65). (A) $k_1 = 10$ and $k_2 = 1\,s^{-1}$; (B) $k_1 = 1$ and $k_2 = 10\,s^{-1}$.

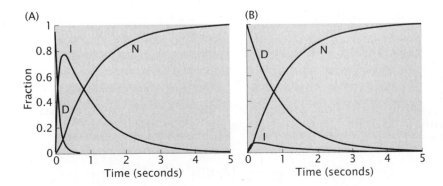

Kinetic intermediate with reversible reactions: if the reactions are reversible,

$$D \underset{k_{u1}}{\overset{k_{f1}}{\rightleftharpoons}} I \underset{k_{u2}}{\overset{k_{f2}}{\rightleftharpoons}} N, \qquad (5.69)$$

the equations are more complicated, but some points are worth noting. The kinetics of each of the three species are described by two exponential functions of time. Now even D decays as a two-exponential function, because I communicates with D through the back reaction with rate constant k_{u1}. In general, the two exponentials have apparent rate constants that are complicated functions of the four molecular rate constants (k_{f1}, k_{u1}, k_{f2}, and k_{u2}).

Several intermediates: What happens if we have three sequential steps?

$$D \;\rightleftharpoons\; I_1 \;\rightleftharpoons\; I_2 \;\rightleftharpoons\; N. \qquad (5.70)$$

By now you might guess that with three steps the kinetics will contain three exponentials, and so on. You can solve these rate equations with the method described in Appendix C, preferably with the help of a computer program that does symbolic mathematics.

Small Proteins Fold in Microseconds to Milliseconds

How fast *do* proteins fold? Many small proteins with only one domain and ≤ 100 residues fold in a two-state manner, with a relaxation time $\tau \sim 1\,\mu s$ to $\sim 10\,ms$. If folding is not too fast ($\tau > 1\,ms$), the kinetics can be measured by rapid mixing methods, such as stopped-flow fluorescence or circular dichroism, following a jump in pH or denaturant concentration. Alternatively, the rates can be measured by analysis of NMR line shapes. For the fastest reactions, however, faster perturbation methods are needed. In a T-jump experiment, a laser is used to raise the temperature of the protein solution almost instantaneously by $\sim 10^\circ C$, and relaxation to the new equilibrium is observed. We learned in Chapter 4 that proteins can be denatured by heat or by cold. If we perform a T-jump from the cold-denatured state, the protein refolds; if we do a T-jump from the folded state, the protein unfolds. Finally, single-molecule techniques are now available, some of which are described below. In all these methods, the data are often measured in the presence of a denaturant, such as urea or guanidinium hydrochloride (GuHCl), which slows down the folding reaction, and are then extrapolated to pure water.

For example, the B1 domain of protein G folds in $\tau \approx 10\,ms$, and the cold shock protein CspB folds in $\tau \approx 1\,ms$ in water. A truncated form of the N-terminal domain of the phase λ repressor protein (residues 6–85 of 102) was estimated to fold in $\tau \approx 300\,\mu s$ in water. But some small proteins fold faster, with $\tau \approx 10$–$100\,\mu s$, and superfast folders have relaxation times $\tau \approx 1$–$5\,\mu s$, as in the case of the villin subdomain. MD simulations yield results consistent with these timescales. Namely, the small proteins studied by Lindorff-Larsen et al. (see Section 5.5) folded in $\tau \sim 1$–$65\,\mu s$ in the MD simulations, in reasonable agreement with experimental folding kinetics. Usually, the larger the protein, the longer the folding time.

How fast *can* proteins fold? Is there a *speed limit* for protein folding? It turns out there is. Even if there were no activation energy barriers

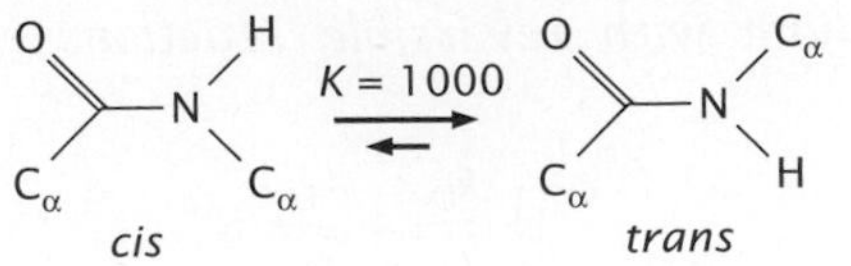

Figure 5.26 The *cis–trans* equilibrium of a general peptide bond.

for folding, proteins could never fold faster than the time it takes for the chain segments to move in space through rotations until the native structure is found. This conformational search is a process of diffusion of chain segments and is not instantaneous. Secondary structure elements form very fast: α-helices and loops fold in ~10–100 ns, and β-hairpins fold in ~100 ns to 1 μs. The speed limit for small proteins corresponds to a time $\tau_0 \approx 1$ μs. For each protein, this limit would be the time needed to fold without an activation barrier. If you set $\Delta G^{\ddagger} = 0$ in Equation 5.42, you see that the speed limit is just k_0, which is the reciprocal of τ_0.

Peptidyl–proline Bond Isomerization is Slow

When we studied protein structure in Chapter 2, you learned that almost all peptide bonds are in the *trans* configuration. In the unfolded state, where essentially no other factors influence the peptide bond configuration, the equilibrium favors *trans* over *cis* by 1000:1 (**Figure 5.26**).

However, when a proline residue follows the peptide bond, the equilibrium constant is only [*trans*]/[*cis*] = 4:1 in the unfolded state (**Figure 5.27**). In the folded state, the proline *trans/cis* ratio depends on the protein. The peptide bond configuration is determined by the balance of the contributions to the Gibbs energy from the intrinsic propensity, which favors *trans*, and from all the other interactions with the rest of the protein in the native state, which may favor the *cis* configuration for that particular bond. Whatever the bond configuration, it is *fixed* in the *folded* structure: it is either *cis* or *trans* in all molecules of a given protein in the native state.

Suppose now that a protein contains only one proline and the peptide bond preceding this proline is *cis* in the native state. After the protein unfolds and the *cis* ⇌ *trans* equilibrium is established, only 20% of the protein molecules have *cis*-proline, whereas the other 80% have *trans*-proline (4:1 ratio). Now, when the protein refolds, 20% of the molecules have the "correct" proline peptide bond isomer and fold fast, but in the other 80% that bond needs to revert to *cis*-proline. The problem is that *trans–cis* isomerization is slow. It becomes *rate-limiting* for those protein molecules that have switched to the *trans*-proline bond in the denatured state.

$$D_{trans} \;\overset{slow}{\rightleftharpoons}\; D_{cis} \;\overset{fast}{\rightleftharpoons}\; N. \tag{5.71}$$

If you monitor the folding of this protein, you obtain kinetics that look like the curve in **Figure 5.28**. Unlike the curve for two-state kinetics that we have seen before (see Figure 5.22), this one has *two phases*: a

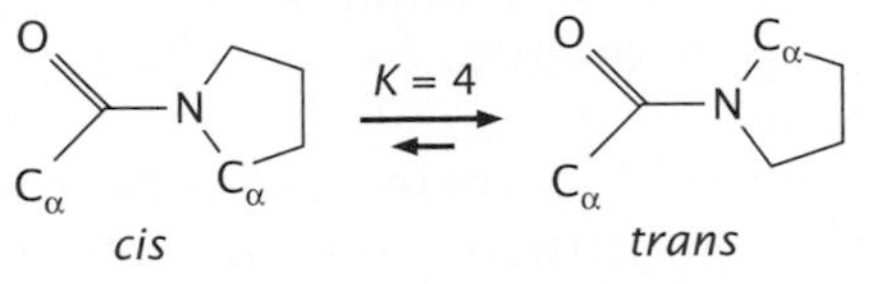

Figure 5.27 The *cis–trans* equilibrium of a peptide bond preceding a proline residue.

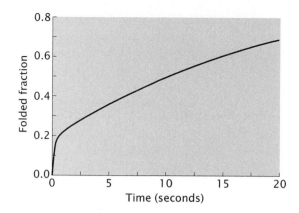

Figure 5.28 Folding kinetics of a protein with one proline, which is *cis* in the native state. The relaxation time of the *trans* ⇌ *cis* isomerization step is τ = 20 seconds in this curve.

fast phase and a slow phase. The fast phase ends at ~0.5 seconds in this example, at an amplitude of 0.2 of the folded fraction, because it corresponds to the 20% of proteins that had the *cis*-proline bond and folded fast. The slow phase is limited by the rate of *trans* ⇌ *cis* isomerization, which has a relaxation time τ ≈ 10–50 seconds, and must happen before the protein can fold.

In proteins with *cis*-prolines in the native state, their mutation to alanine often results in the disappearance of the slow phase. This is strong evidence that slow proline isomerization is the reason for the slow phase in many folding kinetics. In practice, however, the situation is a bit more complicated. Some proteins can fold to the native structure even with the "incorrect" proline isomer, which then isomerizes to the *cis* configuration in the folded state. In fact, in that case, folding enhances the rate of proline *trans* ⇌ *cis* conversion. Further, not all biphasic kinetics are due to proline isomerization.

Cooperative Protein Folding is Observed by Single-molecule Experiments

Now, let us ask a *different* question about kinetics: How long does an *individual protein* molecule take to fold? You may say, "Wait a minute, haven't we just answered that question?" Actually, no, we have not. The kinds of folding times that we have been discussing are *relaxation times* or *lifetimes*. You can think of the lifetime τ_u of the unfolded state as the time an unfolded protein molecule takes to "make up its mind" about whether to fold. The time at which it "decides" to fold is *stochastic*: it is unpredictable for each individual molecule. We can only say that, *on average*, unfolded proteins take a time τ_u to fold—more precisely, to commit to fold.

The mean waiting time in the unfolded state, or lifetime τ_u, is the reciprocal of the folding rate constant,

$$\tau_u = \frac{1}{k_f}, \tag{5.72}$$

because that is also the rate at which a *population* of unfolded proteins would disappear (if they could not unfold again). Similarly, the mean waiting time in the folded state, or lifetime τ_f, is the reciprocal of the unfolding rate constant,

$$\tau_f = \frac{1}{k_u}, \tag{5.73}$$

Figure 5.29 Transition path (tp) from the unfolded to the folded state, across the Gibbs activation energy barrier.

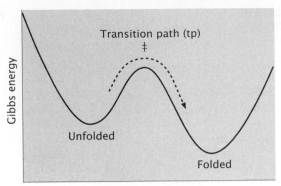

"Reaction coordinate"

because that is the rate at which a *population* of folded proteins would disappear (if they could not fold back). Also, the relaxation time (τ) of a protein population is

$$\tau = \frac{1}{k_f + k_u}. \tag{5.74}$$

None of these times, however, is the time that *an individual* protein molecule actually takes to fold *once it initiates the transition* from the unfolded to the folded state. This time is much, much shorter than τ or τ_u. It is the time a protein takes to cross the Gibbs energy barrier between the unfolded and folded states. It is called the *transition path time*, τ_{tp} (tp, for transition path). The transition path from the unfolded to the folded state, over the Gibbs activation energy barrier, is shown in **Figure 5.29**.

Can we measure τ_{tp}? Yes, but we need to perform experiments on single protein molecules, not on an ensemble of proteins. This can be done using the technique of fluorescence (Förster) resonance energy transfer (FRET). In single-molecule FRET, a protein is covalently modified by attaching two fluorophores to positions in the protein sequence that are *close to each other* when the protein is folded (**Figure 5.30**), but far apart when the protein is unfolded. In this way, FRET can occur between the two fluorophores in the folded state, but not in the unfolded state.

We learned about fluorescence in Chapter 2. When a fluorophore absorbs light, one of its electrons is excited to a higher energy level (see Figure 2.20). When the electron returns to its ground state, the fluorophore releases the extra energy by emitting a light photon with that energy. In FRET, instead of releasing the energy by emission of a photon, the excited fluorophore transfers that energy (by resonance, not emission) to another fluorophore nearby (see Figure 5.30). Two conditions are necessary for FRET to occur: First, the emission wavelength of the excited fluorophore (donor) must overlap with the absorption wavelength of the neighboring fluorophore (acceptor); and second, the donor and the acceptor must be close to each other, typically within ~ 20–50 Å. When the protein is folded, the two fluorophores are close together, and the FRET efficiency is high; but when the protein unfolds, the fluorophores move apart and FRET drops. Therefore, in the folded state, a large amount of photons are emitted by the acceptor fluorophore; but in the unfolded state, most photons are emitted directly by the donor (because FRET does not occur).

To measure single-molecule fluorescence with a microscope, the proteins need to be attached to a solid support (see Figure 5.30), so that we can focus on one molecule at a time without it drifting from the field

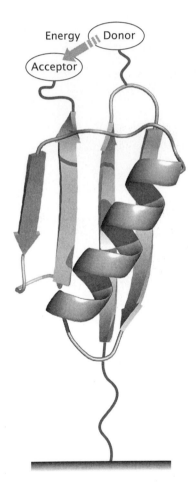

Figure 5.30 A molecule of protein G B1 (PDB 1PGA) covalently labeled with two fluorescent molecular probes, which are close together only when the protein is folded. The donor fluorophore transfers its excitation energy to the acceptor fluorophore, which then emits a photon of a longer wavelength than the those emitted by the donor. (Based on Chung HS, Louis JM & Eaton WA [2009] *Proc Natl Acad Sci USA* 106:11837–11844.)

of view. The photons directly emitted by the donor (unfolded state) and by the acceptor (folded state) have different colors (wavelengths). Therefore, folded and unfolded proteins are distinguishable, and the state of a single protein can be followed in real time as it undergoes reversible folded ⇌ unfolded transitions under equilibrium conditions. **Figure 5.31** shows the result of this experiment on protein G_B. The high FRET efficiency (top level) corresponds to the folded state (F) and the low FRET (bottom level) to the unfolded state (U).

Two important observations can be made. First, the folding transition in protein G_B is *two-state*; no intermediate FRET levels are observed (see Figure 5.31). Second, the transition path time, shown by the vertical, abrupt changes in FRET, is much shorter than the waiting time from either the folded or the unfolded states (horizontal stretches in Figure 5.31). When the experiment is performed with a finer time resolution, it is found that $\tau_{tp} \leq 10$ μs. The transition path time seems to be determined by the internal friction of the protein molecule, as if the interactions within the protein hinder its conformational change. Finally, the Gibbs activation energy in protein folding is very small. Both ensemble and single-molecule experiments yield values of $\Delta G^{\ddagger} \approx 1$–3 kcal/mol close to the T_m of most proteins.

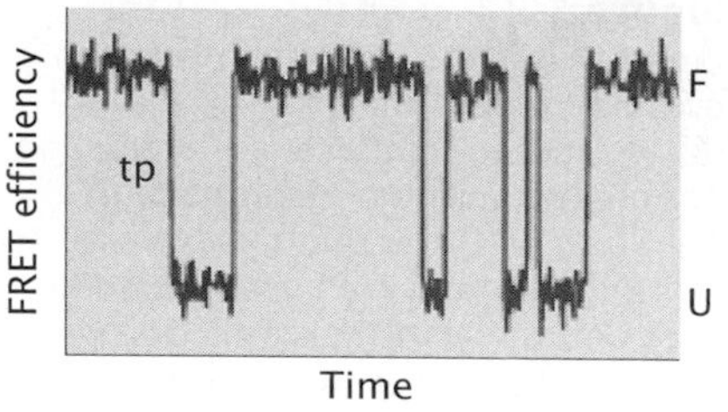

Figure 5.31 Single-molecule trajectory, measured by FRET, between the folded (top level) and unfolded (bottom level) states. The transition path (tp) corresponds to the abrupt vertical jumps from one state to the other. The waiting times in each state correspond to the horizontal regions in FRET. (Adapted from Chung HS, Louis JM & Eaton WA [2009] *Proc Natl Acad Sci USA* 106:11837–11844. With permission from National Academy of Sciences.)

5.7 THE STRUCTURE OF THE TRANSITION STATE CAN BE PROBED BY EXPERIMENT AND SIMULATION

What is left? We have learned about the kinetics of protein folding and how they relate to the Gibbs energy of the activation barrier for folding. The top of this activation barrier is the transition state. Probing the structure of the transition state is a notoriously difficult task, because of its very *transient* nature. No standard technique to determine native protein structure can be used for this purpose. However, through a combination of experiments and computer simulations, we can learn about the structure of the transition state and the mechanism of protein folding.

ϕ-Value Analysis Compares Amino Acid Contacts in the Native and Transition States

The most common experimental approach that yields information on the structure of the transition state is ϕ-value analysis, a method developed by Alan Fersht. Consider the diagram of **Figure 5.32**. In ϕ-value analysis we compare the Gibbs energy of the transition state and the equilibrium Gibbs energy of folding for a mutant and the wild-type protein. Here we use the denatured state as reference. Suppose you mutate an amino acid residue in a protein. The mutation is chosen to be conservative, in the sense that the mutant has the same native structure as the wild-type protein, and its stability does not change much. Ideally, the chosen amino acid has interactions with other residues in the native state (*native interactions*) but not in the denatured state. If a native interaction is present in the transition state, the mutation will affect the transition state in the *same* way as it affects the native state (see Figure 5.32A). If not, the mutation will affect the native state but *not* the transition state (see Figure 5.32B).

Assuming that there are no specific interactions between residues in the denatured state, it makes sense to choose D as the reference and set

Figure 5.32 φ-Analysis. (A) The mutation affects a native interaction present in the transition state (φ = 1). (B) The mutation affects an interaction not present in the transition state (φ = 0). The Gibbs energy levels for the wild-type protein are shown as solid lines and those of the mutant protein as dashed lines. The change in the equilibrium Gibbs energy of folding ($\Delta\Delta G_f^o$) upon mutation and the change in the Gibbs energy of the transition for folding ($\Delta\Delta G_f^\ddagger$) are indicated. (Based on Fersht A [1999] Structure and Mechanism in Protein Science. Freeman.)

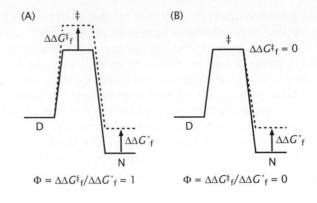

its Gibbs energy to zero (see Figure 5.32). The fundamental question we ask is this: Which native interactions are present in the transition state? To answer, we calculate the φ-value. First, we determine the *changes* in the equilibrium ($\Delta\Delta G_f^o$) and the activation ($\Delta\Delta G_f^\ddagger$) Gibbs energies of folding upon mutation,

$$\Delta\Delta G_f^o = \Delta G_{f,\text{mutant}}^o - \Delta G_{f,\text{wild type}}^o \tag{5.75}$$

$$\Delta\Delta G_f^\ddagger = \Delta G_{f,\text{mutant}}^\ddagger - \Delta G_{f,\text{wild type}}^\ddagger. \tag{5.76}$$

The φ-value is the Gibbs energy change in the transition state relative to the change in the native state,

$$\phi = \frac{\Delta\Delta G_f^\ddagger}{\Delta\Delta G_f^o}. \tag{5.77}$$

If φ = 1, the interactions of the mutated residue change in the transition state as much as in the native state (see Figure 5.32A). Therefore, those native interactions must be present in the transition state, which means that the region of the mutation is folded in the transition state. If φ = 0, the interactions of the mutated residue change in the native state but not in the transition state (see Figure 5.32B). Thus, those interactions are not made in the transition state, which means that the region of the mutation is unfolded.

A few caveats of φ-value analysis should be noted. It is assumed that the mutation does not change the structure of the native and the denatured state. If φ = 0, this really means that the structure of the transition state was affected as much as that of the denatured state. Since the latter has little structure, it should be affected little, but this is not guaranteed. It is also assumed that the mutation affects the strength of the interactions but not the *structure* of the folded protein and the transition state. If that assumption is wrong, so is the φ-value interpretation. The structure of the mutant can be determined, but the assumption that the mutation does not affect the structure of the transition state is more difficult to assess.

If the nature of the transition state changes, or alternative folding pathways emerge because of the mutation, the situation is complicated. Ideally, the mutation is a probe, which causes very minor changes in the native, denatured, and transition states, and in the pathways between them. Ideally, φ should have only one of two values: 0 or 1. In practice, fractional φ-values are most often found. This could mean that the region of the mutation is partially folded in the transition state, or that at least one of our assumptions is wrong. For example, fractional φ-values may result from averaging the contributions of alternative folding pathways.

In conclusion, φ-value maps should be interpreted as conveying a qualitative picture of the transition state. More rigorously, they should be used as constraints with which proposed folding mechanisms must be consistent, and which folding simulations must explain. In fact, the meaning of φ-value maps can be put in a much more rigorous basis by combining them with all-atom protein folding simulations, which provide the maps of the probability of folding.

The φ-Value is Determined by Measuring the Rates of Folding and Unfolding as a Function of Denaturant Concentration

How do we determine φ-values? In Chapter 4, we studied equilibrium unfolding by denaturants, such as urea or guanidinium chloride (GuHCl). In water, the native state is favored; thus, the equilibrium Gibbs energy of folding $\Delta G_f^o < 0$. But ΔG_f^o increases linearly with denaturant concentration. Eventually, a point is reached where $\Delta G_f^o = 0$, which happens at [urea] ≈ 5 M in the example of **Figure 5.33**. Beyond that point, $\Delta G_f^o > 0$ and the denatured state is favored.

We can express this experimental observation by Equation 5.78,

$$\Delta G_f^o = \Delta G_{f,water}^o + m[\text{urea}] \tag{5.78}$$

where $\Delta G_{f,water}^o$ is the equilibrium Gibbs energy of folding in water, and the slope is the m-value. (In this definition, the sign of the m-value is positive for urea, whereas it was negative in Chapter 4, because now we are writing the reaction as folding, instead of unfolding.) Equation 5.78 is valid for urea, GuHCl, or any other denaturant.

Now let us measure the apparent rate constant of folding as a function of denaturant concentration. In this case, we will use GuHCl as the denaturant. We obtain a plot like that shown in **Figure 5.34**. This is called a *chevron* plot, because of its V-shape. The natural logarithm of the observed or apparent rate constant (ln k) first decreases and then increases with denaturant concentration, [GuHCl].

Why does the chevron plot have this V-shape? We have seen that both the folded and unfolded states change in time according to an exponential function, e^{-kt} (shown in Equations 5.56 and 5.57), where the apparent rate constant k is the sum of the rate constants of folding and unfolding, $k = k_f + k_u$. At low denaturant concentration (close to pure

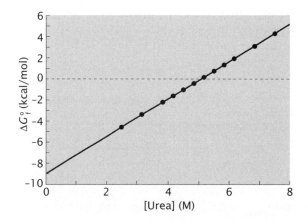

Figure 5.33 Linear dependence of the equilibrium Gibbs energy of folding (ΔG_f^o) of a protein on denaturant (urea) concentration.

Figure 5.34 Chevron plot of k ($k = k_f + k_u$) as a function of denaturant concentration (GuHCl). The solid line is the equation for two-state folding, $\ln k = \ln\left(k_f^{\text{water}} e^{-m_f[\text{GuHCl}]/RT} + k_u^{\text{water}} e^{m_u[\text{GuHCl}]/RT}\right)$, with m_f and $m_u > 0$. (Data from Jackson SE & Fersht AR [1991] *Biochemistry* 30:10428–10435.)

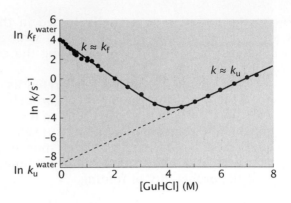

water), the native state is favored: the equilibrium constant for folding $K_F = k_f/k_u \gg 1$, or $k_f \gg k_u$. Therefore, $k \approx k_f$; the rate is dominated by the folding rate constant. Conversely, at high denaturant concentration, $[\text{GuHCl}] \geq 5$ M, the denatured state is favored: $K_F = k_f/k_u \ll 1$, or $k_f \ll k_u$. Here the apparent rate constant $k \approx k_u$; the rate is dominated by the unfolding rate constant. Thus, the linear dependence of $\ln k$ on $[\text{GuHCl}]$ at low denaturant concentration arises because of the linear decrease of $\ln k_f$ with $[\text{GuHCl}]$,

$$\ln k \approx \ln k_f = \ln k_f^{\text{water}} - \frac{m_f}{RT}[\text{GuHCl}], \tag{5.79}$$

where m_f is a positive constant. Similarly, the linear dependence of $\ln k$ on $[\text{GuHCl}]$ at high concentrations of denaturant arises because of the linear increase of $\ln k_u$ with $[\text{GuHCl}]$,

$$\ln k \approx \ln k_u = \ln k_u^{\text{water}} + \frac{m_u}{RT}[\text{GuHCl}], \tag{5.80}$$

where m_u is another positive constant. Thus, the slope of $\ln k$ is negative at low denaturant concentrations and positive at high denaturant concentrations.

How are these linear dependences related to the dependence of the equilibrium ΔG_f^o of folding on denaturant concentration, given by Equation 5.78? Very simply: take the logarithm of Equations 5.42 and 5.43,

$$\ln k_f = -\frac{\Delta G_f^{\ddagger}}{RT} + \text{constant} \tag{5.81}$$

$$\ln k_u = -\frac{\Delta G_u^{\ddagger}}{RT} + \text{constant}. \tag{5.82}$$

The Gibbs activation energies in the folding and unfolding directions, $\Delta G_f^{\ddagger}$ and $\Delta G_u^{\ddagger}$ (see Figure 5.21), respectively, also vary linearly with denaturant concentration, except with the opposite slopes of $\ln k_f$ and $\ln k_u$ because of the negative sign in Equations 5.81 and 5.82.

$$\Delta G_f^{\ddagger} = \Delta G_{f,\text{water}}^{\ddagger} + m_f[\text{denaturant}] \tag{5.83}$$

$$\Delta G_u^{\ddagger} = \Delta G_{u,\text{water}}^{\ddagger} - m_u[\text{denaturant}]. \tag{5.84}$$

Now, if we substitute Equations 5.83 and 5.84 in Equation 5.44,

$$\Delta G_f^o = \Delta G_f^{\ddagger} - \Delta G_u^{\ddagger},$$

we obtain

$$\Delta G_f^o = (\Delta G_{f,\text{water}}^{\ddagger} - \Delta G_{u,\text{water}}^{\ddagger}) + (m_f + m_u)[\text{denaturant}]. \tag{5.85}$$

In pure water, [denaturant] = 0, so

$$\Delta G^{o}_{f,water} = \Delta G^{\ddagger}_{f,water} - \Delta G^{\ddagger}_{u,water}.$$ (5.86)

Comparison of Equations 5.78 and 5.85 shows that $(m_f + m_u) = m$ (both m_f and $m_u > 0$).

The gist of the ϕ-value measurement is straightforward. You measure the apparent rate constant k as a function of denaturant concentration (see Figure 5.34). From the left side of the plot in Figure 5.34 (low-denaturant limit), you obtain m_f from the slope of the line and $\ln k_f^{water}$ from the intercept of the extrapolation of the line to [denaturant] $= 0$. From the right side of the plot in Figure 5.34 (high-denaturant limit), you obtain m_u from the slope of the line and $\ln k_u^{water}$ from the intercept of the (large!) extrapolation of the line to [denaturant] $= 0$ (dashed line). From $\ln k_f$ and $\ln k_u$ you can calculate $\Delta G^{\ddagger}_f$ and $\Delta G^{\ddagger}_u$ to within an additive constant (shown in Equations 5.81 and 5.82), but the constant cancels out when you calculate the difference between the quantities for the mutant and the wild-type protein,

$$\Delta\Delta G^{\ddagger}_f = \Delta G^{\ddagger}_{f,mutant} - \Delta G^{\ddagger}_{f,wild type}$$ (5.87)

$$\Delta\Delta G^{\ddagger}_u = \Delta G^{\ddagger}_{u,mutant} - \Delta G^{\ddagger}_{u,wild type}.$$ (5.88)

We can write a similar equation for the difference between the equilibrium Gibbs energies of folding of the mutant and the wild-type proteins,

$$\Delta\Delta G^{o}_f = \Delta G^{o}_{f,mutant} - \Delta G^{o}_{f,wild type}.$$ (5.89)

Finally, we calculate the ϕ-value from Equation 5.77:

$$\phi = \frac{\Delta\Delta G^{\ddagger}_f}{\Delta\Delta G^{o}_f}.$$

5.8 COMBINING EXPERIMENT AND COMPUTER SIMULATIONS IS THE MOST POWERFUL WAY TO UNDERSTAND PROTEIN FOLDING

Whereas ϕ-value analysis is the most used experimental method to characterize the transition state, it does not tell us the whole story about the mechanism of protein folding. However valuable, it is subject to several caveats, as we have seen. Experimental ϕ-values do not arise from single microscopic conformations, but necessarily represent averages of the conformations belonging to the transition state ensemble.

We would like to find out how the protein reaches the transition state. To clearly define the transition state and the pathway toward it, the experimental ϕ-value analysis needs to be combined with computer simulations of protein folding. The simulations allow us to rationalize the ϕ-values in terms of the structure and the folding mechanism, and to overcome many of the difficulties involved in the interpretation of ϕ-values.

Using Monte Carlo simulations with the Gō model, the structures that belong to the transition state ensemble can be identified. The transition state corresponds to the top of the Gibbs energy barrier (see Figure 5.20). As such, it has the distinctive characteristic that a molecule in the transition state has equal chances of going forward to the folded state or back to the unfolded state. That is, the probability of folding at the transition state is $p_{fold} \approx 0.5$ (p_{fold} is also called the *splitting probability*). This provides a criterion to determine whether a conformation truly belongs to the transition state ensemble.

In the next two sections we will study two small proteins for which φ-value analysis in conjunction with computer simulations has made it possible to map the crucial residues that form the *folding nucleus*. The folding nucleus is the part of the native structure that is best defined in the transition state. All conformations that belong to the transition state ensemble contain this nucleus.

The Folding Nucleus of Chymotrypsin Inhibitor 2 is Formed in the Transition State

Chymotrypsin inhibitor 2 (CI2) is a 64-residue protein, a serine protease inhibitor from barley seeds. (The wild-type protein has 83 residues, but the version used in φ-value analysis and simulations lacked the stretch of the first 19 N-terminal residues.) The structure of CI2 is shown in **Figure 5.35**. It consists of one α-helix (residues 12–24) packed onto an open β-sheet. The contact region of these two secondary-structure elements forms the hydrophobic core of CI2, consisting of 12 residues (W5, L8, A16, V19, I20, A27, I29, V47, L49, V51, I57, and P61). Strands β_3 and β_4 are the longest and best structured.

CI2 is an excellent model to study protein folding because it is simple. First, it does not have disulfide bonds, so it unfolds completely, breaking all the native contacts that will need to be formed again in the transition state. Second, it does not have *cis*-proline bonds in the native state; therefore, slow *trans–cis* isomerization is not a problem when the protein refolds. Third, CI2 folds with two-state kinetics. There are no stable intermediates that accumulate along the folding pathway, and there is only one kinetically significant transition state.

The φ-values yield a picture of the transition state of CI2. The α-helix has the largest average φ-values. For example, Ala-16 has φ = 1.0 (when mutated to Gly), indicating that it is in its native environment in the transition state. However, most other residues of the helix have φ = 0.25 to 0.75, suggesting that the helix is only partially formed or fluctuates in the transition state. β-Strands 3 and 4 (see Figure 5.35)

Figure 5.35 Structure of CI2. The folding nucleus, Ala-16, Leu-49, and Ile-57, is shown as stick models (PDB 1YPC). Strands β_3 (residues 45–52) and β_4 (residues 45–52) are indicated. (Based on Otzen DE, Itzhaki LS, elMasry NF et al. [1994] *Proc Natl Acad Sci USA* 91:10422–10425 and Itzhaki LS, Otzen DE & Fersht AR [1995] *J Mol Biol* 254:260–288.)

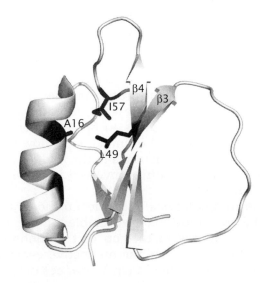

have the next highest φ-values, between ~0.2 to 0.3 for strand 3, and ~0.2 to 0.5 for strand 4. The residues in the hydrophobic core also have φ-values ~0.2 to 0.5, with a high point at φ = 0.53 for Leu-49 (mutated to Val). Most other regions of the protein have φ = 0 to 0.2, indicating that they are unstructured in the transition state. The transition state thus contains the α-helix and the core of β-strands 3 and 4, including several residues of the hydrophobic core of the protein. The folding nucleus proper consists of Ala-16, Leu-49, and Ile-57.

MD simulations of CI2 performed at the T_m show many unfolding and refolding events. In the conformations assigned to the transition state, the native contacts established agree well with the φ-values obtained experimentally for the same amino acid residues. Monte Carlo simulations of CI2 folding using the Gō model have been used to more clearly identify the transition state. The φ-values obtained experimentally were used as constraints to generate a set of possible transition state conformations, by requiring that simulated φ-values be equal to the experimental φ-values. (In the simulations, ϕ_{sim} are defined as the fraction of native contacts made in the transition state, $\phi_{sim} = n^{\ddagger}_{contacts}/n^{N}_{contacts}$.) Then, this selected set of conformations was allowed to fold or unfold under the Gō potential. To select the conformations that truly belong to the transition state ensemble, the criterion that $p_{fold} \approx 0.5$ was used.

The transition state is like an expanded native state. The simulations show that the α-helix is most structured, followed by β-strands 3 and 4. Other regions are almost completely disordered. Ala-16, which from experiment is completely structured in the transition state (φ = 1), makes most contacts to other residues of the α-helix, but also to Leu-8, located in a β-turn, and to Leu-49 and Ile-57. This agrees with experiment in establishing that the folding nucleus is formed by Ala-16, Leu-49, and Ile-57.

The Transition State may be Reached by Multiple Pathways in the Folding of Protein G$_B$

As our second example, we return to protein G$_B$, the B1 domain of protein G (**Figure 5.36**). This is a 56-residue protein whose structure consists of an α-helix packed against a four-strand, antiparallel β-sheet. Hairpin 1 is formed by strands β$_1$ (residues 1–9) and β$_2$ (13–20), and hairpin 2, by strands β$_3$ (residues 42–46) and β$_4$ (51–55). The helix comprises residues 23–37. The three residues that are different between the all-α fold in G$_A$95 and the α/β fold in G$_B$95 are shown in Figure 5.36: Ala-20, located in strand β$_2$, Phe-30, in the α-helix, and Tyr-45, in strand β$_3$, very close to the turn of hairpin 2. Mutation of those three residues in G$_B$95 to Leu-20, Ile-30, and Leu-45, changes the structure to the all-α fold of G$_A$95 (see Figure 5.12).

Protein G$_B$ displays two-state folding behavior. This is indicated by V-shaped (not curved) chevron plots, single-exponential folding and unfolding kinetics, two-state transitions observed by single-molecule fluorescence, and exponential distribution of unfolded and folded waiting times. φ-Value analysis of protein G$_B$ indicates that the most structured part of the protein in the transition state is the turn region of β-hairpin 2. In particular, residues 45–53 have φ ≈ 0.2 to 1, including Tyr-45 (mutated to Leu) with φ = 0.3. In retrospect, the mutation of Tyr-45 to Leu was an unfortunate choice for φ-value analysis because Leu-45 favors the G$_A$ fold (all-α); thus, this φ-value is probably not very meaningful. The second stretch with high φ-values comprises residues

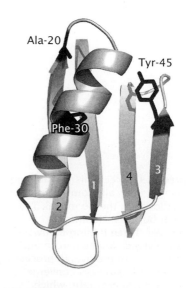

Figure 5.36 Structure of protein G (PDB 1PGA). The residues Ala-20, Phe-30, and Tyr-45 are shown as sticks. The β-strands 1–4 are labeled by their numbers.

26–34 in the α-helix, for which $\phi \approx 0.2$ to 0.55, with the notable exception of Phe-30, which has $\phi \approx 0$ (mutated to Leu). This, too, needs to be interpreted with caution, because Phe-30 is one of the most important residues in determining the fold of $G_B 95$. The same is true of Ala-20, with $\phi \approx 0$ (mutated to Gly).

These ϕ-values were determined before the fold-switching experiments between G_A and G_B were performed. We know now that the mutations A20G and F30L are not as "harmless" as intended for ϕ-value analysis of the α / β structure of protein G_B, which is probably why they result in such low ϕ-values. In addition, the beginning of strand β_1 (residues 3–7) has large ϕ-values, ≈ 0.2 to 0.38, suggesting that this region is also structured in the transition state. The variants $G_B 30$ and $G_B 77$ were also subjected to ϕ-value analysis, and the results agree with those on the wild-type protein, suggesting that the transition state does not change much in these mutants.

Complete folding of protein G_B from the denatured state was studied by all-atom Monte Carlo simulations using the Gō model (**Figure 5.37**). The simulation mapped the transition state ensemble, for which the probability of folding is $p_{\text{fold}} \approx 0.5$. The folding nucleus includes residues Y3 and L5 (hairpin 1), F30 (α-helix), and W43, Y45, and F52 (hairpin 2). The helix forms already transiently in the denatured state, as part of the initial nucleus, and is then stabilized by interactions with either hairpin 1 or 2, giving rise to two types of intermediates. Both pathways, however, lead to the same folding nucleus formed by Y3, L5, F30, W43, and Y45, in agreement with the experimental ϕ-value analysis.

MD simulations of protein unfolding performed with the variant $G_B 88$ showed that the protein folds in a two-state manner, without significantly populated intermediates. In protein $G_B 88$, the topology of the native state and some transient secondary structure, which includes the marginally stable helix and the two incipient hairpins, is already noticeable in the denatured state. In the transition state, hairpin 2 was more stable than hairpin 1. The folding nucleus appears to include the same residues as observed in Monte Carlo simulations with the Gō

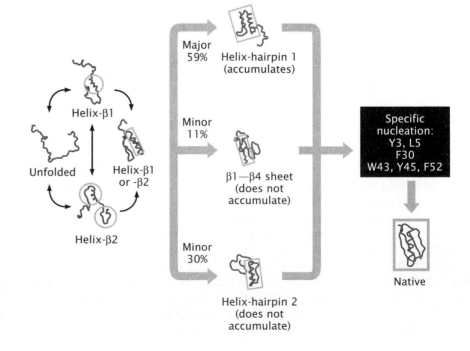

Figure 5.37 Mechanism of folding of protein G_B in an all-atom Monte Carlo simulation with the Gō model. The circle or boxed structures are native-like. Three folding pathways were observed, whose percentage contributions to the overall process are indicated. All pathways converge to a common transition state, which contains the folding nucleus. (Adapted from Shimada J & Shakhnovich EI [2002] *Proc Natl Acad Sci USA* 99:11175–11180. With permission from National Academy of Sciences.)

model, most clearly Phe-30. The mutation Y45L in G_B98 switches the fold from the α/β structure of G_B98 to the all-α structure of G_A98. Interactions with Tyr-45 may stabilize hairpin 2 and the helix.

MD simulations of protein G_B folding starting from the fully unfolded state were later performed in the Anton supercomputer. A variant of protein G_B revealed two folding pathways, which differed in the order of formation of the two β-hairpins. This is consistent with the experimental observation that in the wild-type protein G_B hairpin 2 forms first, whereas in another variant (named NuG2), hairpin 1 forms first. The sequence of the variant of protein G_B studied by MD simulations is intermediate between those of the wild-type protein G_B and NuG2. Accordingly, this variant appears to have one folding pathway more similar to that of the wild-type protein G_B and another more similar to that of NuG2. In the Gō model simulations, hairpin 1 appeared to be more stable (or form first) even for the wild-type protein G_B. However, both folding routes then converged to a transition state ensemble with a common folding nucleus. Thus, there is good agreement between the various types of simulations of protein G_B.

Folding is a Search Guided by Structure

We studied the mechanism of protein folding in two cases that are especially well understood. Is there a general folding mechanism applicable to all proteins? Two very different general mechanisms have been proposed. In the framework/diffusion-collision model (**Figure 5.38**A), the idea is that different stretches of the sequence have defined propensities to adopt certain secondary structures (α-helices, β-strands, or

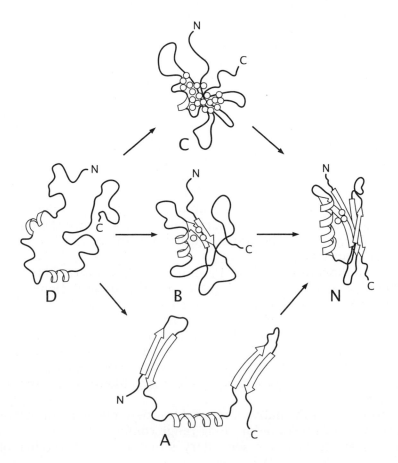

Figure 5.38 Protein folding mechanisms. (D) Denatured state, which often has residual secondary structure. (N) Native state, showing the residues that form the folding nucleus in the transition state. (A) Framework model (or diffusion–collision model); (B) nucleation-condensation model, showing the folding nucleus; (C) hydrophobic collapse model, showing the hydrophobic residues. (Based in part on Otzen DE, Itzhaki LS, elMasry NF, et al. [1994] *Proc Natl Acad Sci* USA 91:10422–10425.)

loops). Those structures form independently, early in the folding process. Although not stable by themselves, those fluctuating elements of local structure establish long-range tertiary interactions (between residues distant in the sequence), and thus come together, docking onto each other to assemble the protein tertiary structure.

The other extreme is the hydrophobic collapse model (see Figure 5.38C). The concept here is that the denatured protein rapidly collapses to a more compact structure in which the hydrophobic residues are hidden from water to a large extent. In this reduced space, native interactions are established more easily. In particular, favorable stereochemical contacts between hydrophobic residues are established, and elements of secondary structure form to satisfy hydrogen bonds inside the water-free globule. These compact but not fully structured intermediates have been called molten globules.

The difficulty with these models is that they fail to explain two-state protein folding. φ-Value analysis has revealed transition states with features from both the framework and the hydrophobic collapse models. Usually, the transition state seems like an expanded version of the native structure, with significant native-like secondary structure. The framework model fails in general because local secondary-structure preferences are usually weak in unfolded proteins. Long-range tertiary contacts appear to be essential from the outset. Those contacts usually involve hydrophobic residues, but are more specific than envisioned in the hydrophobic collapse model, and typical kinetic intermediates are not always like molten globules.

The nucleation-condensation mechanism (see Figure 5.38B) was proposed by Fersht, based on φ-value analysis, particularly of CI2. This mechanism includes concepts from the framework and the hydrophobic collapse models, but it accounts in a natural way for the cooperativity of protein folding. The idea is that a nucleus of residues forms initially, which consists primarily of *local* interactions—that is, between residues *close to each other in the sequence*. In CI2, the initial nucleus (which is not the folding nucleus) consists primarily of the fluctuating α-helix and nearby stretches of the chain. However, these initial interactions are not sufficiently stable. To reach the transition state, two additional elements are required: *long-range* interactions—that is, between residues *distant in the sequence but close in space*, which are usually hydrophobic; and a *chain topology*—that is, an overall organization of the polypeptide—that is similar to the topology of the native protein, which allows the establishment of native contacts. In CI2 the helix remains transient until stabilized by long-range interactions, which include contacts among the folding nucleus (Ala-16, Leu-49, and Ile-57). Other tertiary-stabilizing native interactions are then established and the protein structure rapidly *condenses* around the folding nucleus.

Although the folding nucleus is common to all conformations belonging to the transition state ensemble, other structural elements must also be present, which may vary slightly among those conformations. An essential feature of the transition state in the nucleation-condensation mechanism is the *concomitant* establishment of *multiple* long-range interactions. This was observed in MD simulations of the complete folding of 12 small proteins by Lindorff-Larsen et al. (2011), which produced conformations essentially identical to the native structures.

The topology of the native state and transient elements of native secondary structure form before long-range contacts, but those contacts are necessary to stabilize secondary structure. The stretches of the chain with higher propensity to adopt certain structures are those that form first in the folding pathway, and that is where the initial nucleus is located. Essential long-range contacts are then established, leading to the formation of the folding nucleus, which are sufficient to pin down the chain with much of the native topology and to stabilize transient secondary structure. In most cases, there is a well-defined folding pathway (with minor variations) and a common folding nucleus, leading to the transition state.

If nucleation involved only interactions between residues close in the sequence, folding would not be nearly as cooperative. This is what happens in a coil–helix transition. There are multiple initiation points, which arise from local interactions (hydrogen bonds) between residues close in the sequence (at positions i and $i + 4$). Thus, the coil–helix transition of a polypeptide is *not* two-state. In the folding of small globular proteins, however, distant residues must interact and the chain must adopt the correct overall topology. Therefore, all these elements must *cooperate* to reach the transition state. Nucleation is coupled with condensation of the rest of the structure to form a compact transition state. Hence, for many globular proteins, folding is an all-or-none process, or two-state, which corresponds to maximal cooperativity.

5.9 SUMMARY

The protein folding problem comprises three questions: What are the interactions that determine the structure?; how is the structure specified by the amino acid sequence?; and what is the mechanism of folding? Unlike organic polymers or random amino acid sequences, native proteins fold to unique structures. That structure is lost when the protein unfolds, but folding is an equilibrium process, determined by thermodynamics. The protein structure corresponds to a well-defined Gibbs energy minimum in conformational space, and the protein always folds to that same structure. Although it is still not possible in general to predict the structure from the sequence, computational approaches, including complete folding simulations, have become increasingly able to fold the sequence of small proteins into their correct native structure.

Protein folding is an extremely cooperative event. In fact, most small proteins approach two-state or all-or-none folding. Small polypeptides may acquire a defined secondary structure, such as an α-helix or a β-sheet, but in isolation those structures are not stable. Furthermore, the coil–helix transition is not sufficiently cooperative to be two-state. Multiple interactions between a variety of amino acid residues that are not close in the sequence are necessary to ensure two-state folding.

The Levinthal paradox arose because proteins cannot find the native structure by a random search of all possible conformations. They find it because the number of possible conformations to be searched decreases as folding proceeds, thus solving the Levinthal paradox. Formation of the initial nucleus reduces the conformational freedom of the polypeptide chain. Each native contact established restricts the possible conformations, rendering it more likely that other native contacts will be established. Transient secondary structure elements form and the chain adopts the correct topology. This brings in close proximity the residues that constitute the folding nucleus. Once the nucleus forms, the conformations of the rest of the molecule are further constrained, and the native structure rapidly condenses around this nucleus, until complete folding is achieved.

The success of simple models of protein folding (Gō, Ising-like models) indicates that the sequence specifies both the structure, which corresponds to the Gibbs energy minimum, and the pathway to reach it, downhill toward that minimum, in which the number of conformations searched decreases steadily. Essentially, protein folding combines random conformational searches with the making of crucial interactions.

5.10 PROBLEMS

5.1 Consider the following 2D-HP model of a protein in a square lattice. Black residues are hydrophobic (H) and white residues are polar (P). Can you find the conformation with the lowest possible energy? Here are the rules: a favorable contact occurs when two hydrophobic residues are on adjacent lattice sites; it is assigned an energy $E = -1$. Diagonal contacts and contacts between bonded residues don't count. For example, residue 18 makes favorable contacts with residues 1 and 15 but not with 2; and the contact between residues 1 and 2 contributes nothing because they are bonded. Hydrophobic residues exposed to the outside contribute an energy of +1. In this example, residues 3, 7, and 10, but not 12, are considered exposed. The conformation drawn has three favorable contacts and three exposed hydrophobic residues, so $E = +3 - 3 = 0$. You can place the residues in any lattice sites (square centers) but they cannot overlap.

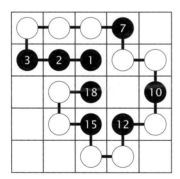

(a) Find the lowest energy conformation (Hint: $E = -7$).
(b) Find the highest energy conformation.
(c) Find another *very different conformation* whose energy differs as little as possible from the lowest one.
(d) Assuming E is given in kcal/mol, what are the relative probabilities (populations) of the two low-energy conformations at 25°C?
(e) What does this tell you about the (two-state?) nature of the transition from the denatured to the native state?

5.2 Figure 5.2 shows the fractional helicity (θ_h) as a function of temperature for the helix–coil transition of the peptide Ac–Y(AEAAKA)$_8$F–NH$_2$. The original data were measured by CD (ellipticity at 222 nm) as a function of temperature (see Figure 2 from Scholtz JM, Marqusee S, Baldwin RL et al. [1991] *Proc Natl Acad Sci USA* 88:2854–2858). Obtain the article by Scholtz et al. and read the data from the plot. Show how you convert the

original data (ellipticity) to fractional helicity, and reproduce the plot of Figure 5.2.

5.3 Consider a coil–helix transition of a 30-residue polypeptide. Assume that $\Delta h = -1.5$ kcal/mol and $T\Delta s = -1.2$ kcal/mol, per residue, for the addition of a residue to a growing helix at 0°C.
(a) Plot the fractional helicity (θ_h) as a function of temperature, for the following models. (Note: choose a temperature range that allows differences between the various models to be apparent. Use the same scale in all cases so that you can easily compare the results.)
(i) Independent residues
(ii) Two-state (all-or-none)
(iii) Zimm–Bragg model, with $\sigma = 0.1$
(b) Repeat (iii) with $\sigma = 10^{-3}$.
(c) Briefly discuss your results. Point out their physical significance to protein folding by discussing the consequences of the coil–helix transition for the formation of elements of secondary structure when a protein folds. Take into account what you have learned from the correctness of the models as applied to real proteins.

5.4 Why is protein aggregation a sign of incomplete folding or misfolding?

5.5 To design crotein G, Yuan and Clarke (1998) based their choice of mutations on several important concepts. Identify two of those concepts.

5.6 Alexander et al. (2007, 2009) and He et al. (2008) produced mutants of two proteins, G_A and G_B, with increasing levels of sequence identity, yet with with two distinct, well-defined folds. The mutant proteins were also less stable than the original ones.
(a) Does this work contradict the statement by Lattman and Rose (1993) that protein stability is independent of its fold? Briefly explain.
(b) Comparing G_A88 and G_B88, and especially mutants with higher similarity obtained later, what do you think is the most striking difference between the residue types that differ from one protein to the other (between the corresponding G_A and G_B variants)?

5.7 Why is it easier to convert a β-sheet into an α-helix than the other way around?

5.8 Consider folding of a protein from the denatured (D) to the native (N) state in the case a misfolded state (M) is formed irreversibly. To make things simpler, assume that the folding reaction is much faster than unfolding, so that the unfolding rate constant can be set to zero ($k_u \approx 0$). The reaction scheme would look like this:

$$
\begin{array}{ccc}
 & D & \xrightarrow{k_f} & N \\
k_m & \downarrow & & \\
 & M & &
\end{array}
$$

(a) Use the method of Appendix C to derive the kinetics of the three species involved in this folding reaction (D, N, and M). Suppose that initially ($t = 0$) all protein is denatured.

(b) Does folding follow one- or two-exponential kinetics? Compare the kinetics of decay of D with the kinetics of formation of N and M.

(c) What are the apparent rate constants for the kinetics of each species?

(d) Do you *understand* the results? Explain them intuitively.

5.9 Most small single-domain proteins fold thermodynamically as two-state. This means that at T_m intermediates account for at most $\sim$ 5% of the molecules.

Draw a Gibbs energy diagram for a small, single-domain protein at T_m, similar to that of Figure 5.21, appropriately modified, and including an intermediate as in Figure 5.24. Label the *y*-axis to indicate the Gibbs energy scale explicitly. Place the denatured state, the native state, and the intermediate at the correct Gibbs energy levels, quantitatively, so the intermediate accounts exactly for 5% of the proteins.

5.10 Why does GuHCl slow down folding? Hint: sketch a Gibbs energy diagram like that of Figure 5.21, with and without GuHCl.

5.11 Why is *trans*-proline $\rightarrow$ *cis*-proline isomerization slower than *cis* $\rightarrow$ *trans*? Hint: to answer, sketch again a Gibbs energy diagram (see Figure 5.21) for the peptidyl-proline isomerization reaction.

5.12 Why does folding enhance *trans*-proline $\rightarrow$ *cis*-proline conversion if the peptidyl-proline bond is *cis* in the native state?

5.11 FURTHER READING

General

Anfinsen CB (1973) Principles that govern the folding of protein chains. *Science* 181:223–230.

Cantor CR & Schimmel PR (1980) Biophysical Chemistry. Freeman.

Dill KA, Ozkan B, Shell MS & Weikl TR (2008) The protein folding problem. *Annu Rev Biophys* 37:289–316.

Dill KA & McCallum JL (2012) The protein-folding problem, 50 years on. *Science* 338:1042–1046.

Fersht A (1999) Structure and Mechanism in Protein Science. Freeman.

Gutfreund H (1995) Kinetics of the Life Sciences. Cambridge University Press.

Rose GD, Fleming PJ, Banavar JR & Maritan A (2006) A backbone-based theory of protein folding. *Proc Natl Acad Sci USA* 103:16623–16633.

Shaknnovich E (2006) Protein folding thermodynamics and dynamics: Where physics, chemistry, and biology meet. *Chem Rev* 106:1559–1588.

Sosnick TR & Barrick D (2011) The folding of single domain proteins—have we reached a consensus? *Curr Opin Struct Biol* 21:12–24.

Protein Folding Models and Simplified Alphabets

Abkevitch VI, Gutin M & Shakhnovich EI (1994) Specific nucleus as the transition state of protein folding: Evidence from the lattice model. *Biochemistry* 33:10026–10036.

Guarnera E, Pellarin R & Caflisch A (2009) How does a simplified-sequence protein fold? *Biophys J* 97:1737–1746.

Henry ER, Best RB & Eaton WA (2013) Comparing a simple theoretical model for protein folding with all-atom molecular dynamics simulations. *Proc Natl Acad Sci USA* 110:17880–17885.

Karanicolas J & Brooks III CL (2003) The importance of explicit chain representation in protein folding models: An examination of Ising-like models. *Proteins* 53:740–747.

Kaya H & Chan HS (2000) Energetic components of cooperative protein folding. *Phys Rev Lett* 85:4823–4826.

Kubelka J, Henry ER, Cellmer T, Hofrichter J & Eaton WE (2008) Chemical, physical, and theoretical kinetics of an ultrafast folding protein. *Proc Natl Acad Sci USA* 105:18655–18662.

Lau KF & Dill KE (1989) A lattice statistical mechanics model of the conformational and sequence spaces of proteins. *Macromolecules* 22:3986–3997.

Yue K, Fiebig KM, Thomas PD et al (1995) A test of lattice protein folding algorithms. *Proc Natl Acad Sci USA* 92:325–329.

Coil–Helix Transition

Baeza-Delgado C, Marti-Renom MA & Mingarro I (2013) Structure-based statistical analysis of transmembrane helices. *Eur Biophys J* 42:199–207.

Goch G, Maciejczyk M, Oleszczuk M et al (2003) Experimental investigation of initial steps of helix propagation in model peptides. *Biochemistry* 42:6840–6847.

Lopez MM, Chin D-H, Baldwin RL & Makhatadze GI (2002) The enthalpy of the alanine peptide helix measured by isothermal titration calorimetry using metal-binding to induce helix formation. *Proc Natl Acad Sci USA* 99:1298–1302.

Richardson JM, Lopez MM & Makhatadze GI (2005) Enthalpy of helix–coil transition: Missing link in rationalizing the thermodynamics of helix-forming propensities of the amino acid residues. *Proc Natl Acad Sci USA* 102:1413–1418.

Scholtz JM, Marqusee S, Baldwin RL et al (1991) Calorimetric determination of the enthalpy change for the α-helix to coil

transition of an alanine peptide in water. *Proc Natl Acad Sci USA* 88:2854–2858.

Scholtz JM, Quian H, York EJ et al (1991a) Parameters of the helix–coil transition theory for alanine-based peptides of varying chain length in water. *Biopolymers* 31:1463–1470.

Suzuki E & Robson B (1976) Relationship between helix–coil transition parameters for synthetic polypeptides and helix conformation parameters for globular proteins. A simple model. *Mol Biol* 107:357–367.

Switching Folds and the Paracelsus Challenge

Alexander PA, He Y, Chen Y et al (2007) The design and characterization of two proteins with 88% sequence identity but different structure and function. *Proc Natl Acad Sci USA* 104:11963–11968.

Alexander PA, He Y, Chen Y et al (2009) A minimal sequence code for switching protein structure and function. *Proc Natl Acad Sci USA* 106:21149–21154.

Allison JR, Bergeler M, Hansen N & van Gunsteren WF (2011) Current computer modeling cannot explain why two highly similar sequences fold into different structures. *Biochemistry* 50:10965–10973.

Dalal S, Balasubramanian S & Regan L (1997) Protein alchemy: Changing β-sheet into α-helix. *Nat Struct Biol* 4:548–552.

Hansen N, Allison JR, Hodel FH & van Gunsteren WF (2013) Relative free enthalpies for point mutations in two proteins with highly similar sequences but different folds. *Biochemistry* 52:4962–4970.

Harms MJ & Thornton JW (2013) Evolutionary biochemistry: revealing the historical and physical causes of protein properties. *Nat Rev Genet* 14:559–571.

He Y, Chen Y, Alexander PA et al (2008) NMR structures of two designed proteins with high sequence identity but different fold and function. *Proc Natl Acad Sci USA* 105:14412–14417.

He Y, Chen Y, Alexander PA et al (2012) Mutational tipping points for switching protein folds and functions. *Structure* 20:283–291.

Lattman EE & Rose GD (1993) Protein Folding—What's the Question? *Proc Natl Acad Sci USA* 90:439–441.

Rose GD (1997) Protein folding and the Paracelsus challenge. *Nat Struct Biol* 4:512–514.

Rose GD & Creamer TP (1994) Protein folding: Predicting prediction. *Proteins* 19:1–3.

Shortle D (2009) One sequence plus one mutation equals two folds. *Proc Natl Acad Sci USA* 106:21011–21012.

Yuan S-M & Clarke ND (1998) A hybrid sequence approach to the Paracelsus challenge. *Proteins* 30:136–143.

Protein Folding Simulation and Structure Prediction

Bradley P, Misura KMS & Baker D (2005) Toward high-resolution de novo structure prediction for small proteins. *Science* 309:1868–1871.

Chou PY & Fasman GD (1978) Empirical predictions of protein conformation. *Annu Rev Biochem* 47:251–276.

Das R & Baker D (2008) Macromolecular modeling with Rosetta. *Annu Rev Biochem* 77:363–382.

Kaufmann KW, Lemmon GH, DeLuca SL et al (2010) Practically useful: What the ROSETTA protein modeling suite can do for you. *Biochemistry* 49:2987–2998.

Lee J, Wu S & Zhang Y (2009) Ab Initio Protein Structure Prediction. In Protein Structure to Function with Bioinformatics (DJ Rigden ed), pp 3–25, Springer.

Lindorff-Larsen K, Piana S, Dror RO & Shaw DE (2011). How fast-folding proteins fold. *Science* 334:517–520.

Moult J, Fidelis K, Kryshtafovych A & Tramontano A (2011) Critical assessment of methods of protein structure prediction (CASP)—Round IX. *Proteins* 79:1–5.

Shimada J, Kussell EL & Shakhnovich EI (2001) The folding thermodynamics and kinetics of crambin using all-atom Monte Carlo simulation. *J Mol Biol* 308:79–95.

Protein Folding Rates

Chung HS, Louis JM & Eaton WA (2009) Experimental determination of upper bound for transition path times in protein folding from single-molecule photon-by-photon trajectories. *Proc Natl Acad Sci USA* 106:11837-11844.

Chung HS, McHale K, Louis JM & Eaton WA (2012) Single-molecule fluorescence experiments determine protein folding transition path times. *Science* 335:981–984.

Chung HS & Eaton WA (2013) Single-molecule fluorescence probes dynamics of barrier crossing. *Nature* 502:685–688.

Davis CM, Xiao S, Raleigh DP & Dyer RB (2012) Raising the speed limit for β-hairpin formation. *J Am Chem Soc* 134:14476–14482.

Davis CM & Dyer RB (2014) Dynamics of an ultrafast folding subdomain in the context of a larger protein fold. *J Am Chem Soc* 135:19260–19267.

David De Sancho D & Best RB (2011) What is the time scale for α-helix nucleation? *J Am Chem Soc* 133:6809–6816.

Ghosh K, Ozkan SB & Dill KA (2007) The ultimate speed limit to protein folding is conformational searching. *J Am Chem Soc* 129:11920–11927.

Gruebele, M. 1999. The fast protein folding problem. *Annu Rev Phys Chem* 50:485–516.

Hagen SJ, Hofrichter J, Szabo A & Eaton W (1996) Diffusion-limited contact formation in unfolded cytochrome c: Estimation of the maximum rate of protein folding. *Proc Natl Acad Sci USA* 93:11615–11617.

Huang GS & Oas TG (1995) Submillisecond folding of monomeric λ repressor. *Proc Natl Acad Sci USA* 92:6878–6882.

Prigozhin MB & Gruebele M (2013) Microsecond folding experiments and simulations: A match is made. *Phys Chem Chem Phys* 15:3372–3388.

Yang WY & Gruebele M (2003) Folding at the speed limit. *Nature* 423:193–197.

Folding Intermediates and Peptidyl-proline Isomerization

Bai Y, Sosnick TR, Mayne L & Englander SW (1995) Protein folding intermediates: Native-state hydrogen exchange. *Science* 269:192–197.

Baker D, Sohl JL & Agard DA (1992) A protein-folding reaction under kinetic control. *Nature* 356:263–265.

Brandts JF, Brennan M & Lin L-N (1977). Unfolding and refolding occur much faster for a proline-free protein than for most protein-containing proteins. *Proc Natl Acad Sci USA* 74:4178–4181.

Cecconi C, Shank EA, Bustamante C & Marqusee S (2005) Direct observation of the three-state folding of a single protein molecule. *Science* 309:2057–2060.

Creighton TE (1990) Protein Folding. *Biochem J* 270:1–16.

Englander SW, Mayne L & Krishna MG (2007) Protein folding and misfolding: Mechanism and principles. *Quart Rev Biophys* 40:287–236.

Jackson SE & Fersht AR (1991) Folding of chymotrypsin inhibitor 2: 2. Influence of proline isomerization on the folding kinetics and thermodynamic characterization of the transition state of folding. *Biochemistry* 30:10436–10443.

Kim PS & Baldwin RL (1990) Intermediates in the folding reactions of small proteins. *Annu Rev Biochem* 59:631–660.

Maki K, Ikura T, Hayano T et al (1999) Effects of proline mutations on the folding of Staphylococcal nuclease. *Biochemistry* 38:2213–2223.

Roder H, Elöve GA & Englander SW (1988) Structural characterization of folding intermediates in cytochrome c by H-exchange labelling and proton NMR. *Nature* 335:700–704.

Woodward C (1994) Hydrogen exchange rates and protein folding. *Curr Opin Struct Biol* 4:112–116.

φ-Value Analysis

Fersht AR, Matouschek A & Serrano L (1992) The folding of an enzyme. I. Theory of protein engineering analysis of stability and pathway of protein folding. *J Mol Biol* 224:771–782.

Fersht AR & Sato S (2004) φ-Value analysis and the nature of protein-folding transition states. *Proc Natl Acad Sci USA* 101:7976–7981.

Matouschek A, Kellis Jr JT, Serrano L & Fersht AR (1989) Mapping the transition state and pathway of protein folding by protein engineering. *Nature* 340:122–126.

Folding of CI2

Day R, Bennion BJ, Ham S & Daggett V (2002) Increasing temperature accelerates protein unfolding without changing the pathway of unfolding. *J Mol Biol* 322:189–203.

Day R & Daggett V (2005) Sensitivity of the folding/unfolding transition state ensemble of chymotrypsin inhibitor 2 to changes in temperature and solvent. *Protein Sci* 14:1242–1252.

Day R & Daggett V (2007) Direct observation of microscopic reversibility in single-molecule protein folding. *J Mol Biol* 366:677–686.

Itzhaki LS, Otzen DE & Fersht AR (1995) The structure of the transition state for folding chymotrypsin inhibitor 2 analysed by protein engineering methods: Evidence for a nucleation-condensation mechanism for protein folding. *J Mol Biol* 254:260–288.

Jackson SE & Fersht AR (1991) Folding of chymotrypsin inhibitor 2: 1. Evidence for a two-state transition. *Biochemistry* 30:10428–10435.

Li L & Shakhnovich EI (2001) Contructing, verifying, and dissecting the folding transition state of chymotrypsin inhibitor 2 with all-atom simulations. *Proc Natl Acad Sci USA* 98:13014–13018.

Otzen DE, Itzhaki LS, elMasry NF et al (1994) Structure of the transition state for the folding/unfolding of the barley chymotrypsin inhibitor 2 and its implications for mechanisms of protein folding. *Proc Natl Acad Sci USA* 91:10422–10425.

Folding of Protein G

Hubner IA, Shimada J & Shakhnovich EI (2004) Commitment and nucleation in the protein G transition state. *J Mol Biol* 336:745–761.

Giri R, Morrone A, Travaglini-Allocatelli C et al (2012) Folding pathways of proteins with increasing degree of sequence identities but different structure and function. *Proc Natl Acad Sci USA* 109:17772–17776.

Krantz BA, Mayne L, Rumbley J et al (2002) Fast and slow intermediate accumulation and the initial barrier mechanism in protein folding. *J Mol Biol* 324:359–371.

Kuhlman B & Baker D (2004) Exploring folding free energy landscapes using computational protein design. *Curr Opin Struct Biol* 14:89–95.

McCallister EL, Alm E & Baker D (2000) Critical role of β-hairpin formation in protein G folding. *Nat Struct Biol* 7:669–673.

Morrone A, McCully ME, Bryan PN et al (2011) The denatured state dictates the topology of two proteins with almost identical sequence but different native structure and function. *J Biol Chem* 286:3863–3872.

Shimada J & Shakhnovich EI (2002) The ensemble folding kinetics of protein G from an all-atom Monte Carlo simulation. *Proc Natl Acad Sci USA* 99:11175–11180.

Folding Mechanisms

Dill KA & Chan HS (1997) From Levinthal to pathways to funnels. *Nat Struct Biol* 4:10–19.

Jonsson A, Fersht AR & Daggett V (2012) Combining simulations and experiment to map protein folding. *Comprehensive Biophysics* 3:1–18.

Karplus M & Weaver DL (1976) Protein-folding dynamics. *Nature* 260:404–406.

Kim PS & Baldwin RL (1982) Specific intermediates in the folding reactions of small proteins and the mechanism of protein folding. *Annu Rev Biochem* 51:459–489.

Levinthal C (1968) Are there pathways for protein folding? *J Chim Phys* 65:44–45.

Levinthal C (1969) How to fold graciously. In Mössbauer Spectroscopy in Biological Systems: Proceedings of a meeting held at Allerton House, Monticello, Illinois (JTP DeBrunner & E Munck eds), pp 22–24. University of Illinois Press. (http://wwwmiller.ch.cam.ac.uk/levinthal/levinthal.html)

Binding, Allostery, and Cooperativity

6.1 BINDING OF A LIGAND TO A PROTEIN IS DESCRIBED BY A BINDING ISOTHERM

Communication in the cell or between cells is ultimately achieved through binding reactions, in which a *ligand* binds to a macromolecule. Typically, the ligand is a small organic molecule or a metal ion that binds to a protein. But the definition becomes fuzzy quite easily. In protein–protein association reactions, which protein is considered the ligand is mainly a matter of convenience or point of view. With the exception of the absorption of photons by visual pigments, interactions in biochemical systems always begin with binding: binding of a ligand to a protein, such as the binding of oxygen to hemoglobin; binding of a substrate, an inhibitor, or an activator to an enzyme; binding of a hormone or a neurotransmitter to a receptor; binding of a protein to DNA; or binding of a peptide to a lipid membrane. In pharmacology, too, binding of a drug to a target is one of the most common and important questions envisioned. The effectiveness of a drug depends critically on the affinity to its target, usually a protein. In this chapter, we will develop a solid understanding of the binding of ligands to proteins, from concepts to examples.

It is useless to answer that reality is also orderly. Perhaps it is, but in accordance with divine laws—I translate: inhuman laws—which we never quite grasp.

Jorge Luis Borges

Binding Constants in Biochemical Reactions Vary by Orders of Magnitude

Before we begin, let us briefly consider the magnitudes of binding constants typically found in biochemical reactions. We will usually formulate binding problems and determine their solutions using *binding constants*, designated by K. However, biochemists typically express ligand affinities in terms of *dissociation constants*, designated by K_d, which are the equilibrium constants for the reaction in the opposite direction. The two constants are simply related by $K_d = 1/K$. We will often write both. In Chapter 1, you learned that, strictly speaking, binding constants and activities are unitless. However, in dilute solutions we approximate activities by molar concentrations (1 M = 1 mol L^{-1}) and express binding constants in units of M^{-1}, as a reminder of the standard state used (1 M). Dissociation constants are expressed in units of concentration, M or its submultiples.

Enzymes are specific for their substrates, which is why they act only on certain compounds and not on others. Therefore, it may be surprising that enzyme–substrate binding constants are among the smallest in biological macromolecules. Typical values vary from $K = 10^3$ to $10^7\,M^{-1}$, which corresponds to $K_d = 0.1\,\mu M$ to 1 mM. The millimolar range is weak binding, whereas the micromolar range is average. The reason for the relatively weak binding of substrates to enzymes is that the protein is really optimized to bind the structure of the transition state, not the substrate. Binding the substrate too tightly would be counterproductive; it would lower its Gibbs energy, thus raising the barrier to reach the transition state, and consequently slowing down the reaction.

Interactions between receptors and their ligands are stronger, with typical $K = 10^5$ to $10^{10}\,M^{-1}$, which correspond to $K_d = 0.1\,nM$ to 10 μM. Binding in the nanomolar range is considered tight. DNA-binding proteins that recognize specific DNA sequences typically–bind with $K = 10^7$ to $10^{11}\,M^{-1}$, which corresponds to $K_d = 10\,pM$ to 0.1 μM. Antibody–antigen interactions have binding constants between $K_a = 10^8$ to $10^{12}\,M^{-1}$, corresponding to $K_d = 1\,pM$ to 10 nM. Binding in the picomolar range is considered very tight, but there are examples of yet stronger binding. Bovine pancreatic trypsin inhibitor (BPTI) binds to trypsin with $K = 10^{13}\,M^{-1}$ ($K_d = 0.1\,pM$). This is one of the largest binding constants in biochemistry. Vegetables such as potatoes and beans contain serine protease inhibitors (trypsin is a serine protease), which is why humans cannot digest raw potatoes or beans. Still larger is the binding constant of biotin (a vitamin) to avidin, which is $K = 10^{15}\,M^{-1}$. Avidin is a major protein component of egg white and its function is probably to bind biotin so tightly that no microorganism can grow in the egg, because it will be starved of biotin, as avidin does not allow any free biotin to exist. For this reason, eating raw egg whites is associated with biotin deficiency in humans. In cooked eggs, this is not a problem because avidin is denatured by heat.

The Binding Isotherm is the Fractional Saturation as a Function of Concentration

Let us begin with a very practical problem. One of the most common experiments performed by a biochemist is the determination of a binding affinity. How is this done? How do we extract the binding affinity from an experiment? Suppose you studied the binding of a ligand to a protein by measuring a physical property of the protein that changes when the ligand binds. You would obtain a data set looking somewhat like the one in **Figure 6.1**. The figure shows the change in the fluorescence emission intensity of a tryptophan residue of the protein upon binding of the ligand, plotted as a function of the *free* ligand concentration, which excludes the ligand bound to the protein.

Initially there is a rapid increase in fluorescence, which eventually reaches a plateau. If the protein concentration is much smaller than the substrate concentration, the total ligand concentration is approximately equal to the free ligand. It is the *free ligand* that is the important *thermodynamic variable*, because it sets the *chemical potential* (μ) of the ligand in the system. Now we want to extract from these data the chemical affinity of the ligand for the protein—that is, the binding constant K.

Suppose we have a protein with one binding site for a certain ligand, which is usually a small compound. A binding isotherm is a function

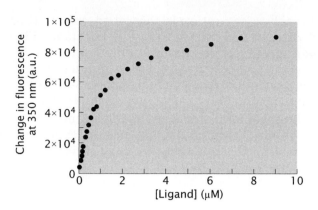

Figure 6.1 Change in fluorescence emission intensity (arbitrary units, a.u.) of an intrinsic Trp residue of a protein upon binding of a ligand as a function of free ligand concentration (simulated data).

that describes the number of ligands bound per site (or per protein molecule) as a function of the ligand concentration in solution. Strictly speaking, we should use the chemical *activity* of the ligand, but we will approximate it by the concentration for the sake of simplicity.

The number of ligands bound per site depends on two factors: the ligand concentration in solution and the binding constant K of the ligand to the site on the macromolecule. As the concentration of ligand increases, the binding sites on the proteins get progressively occupied, until saturation is reached. This phenomenon can be derived from the equilibrium constant for the binding reaction,

$$P + X \overset{K}{\rightleftharpoons} PX, \tag{6.1}$$

where P is a protein (macromolecule) and X is a ligand. We will use lower case x for the concentration of X, $x = [X]$. For a single site, the equilibrium binding constant is

$$K = \frac{[PX]}{[P]\,x}. \tag{6.2}$$

The number of ligands bound per site, also called the *fractional saturation*, is

$$\theta = \frac{[PX]}{[P] + [PX]}, \tag{6.3}$$

which, using Equation 6.2, can be written as

$$\theta = \frac{Kx}{1 + Kx}. \tag{6.4}$$

In this case, θ is also equal to the fraction of protein with one ligand bound relative to the total protein because there is only one binding site per protein. Equation 6.4 is called a *binding isotherm* (because the fractional saturation is measured at one given temperature). Note the following:

1. The binding isotherm exhibits *saturation*: when the ligand concentration is very large ($x \to \infty$), θ does not keep increasing, but tends to 1 ($\theta \to 1$).
2. When the ligand concentration goes to zero ($x \to 0$), the protein releases all ligand: $\theta \to 0$.

3. The approach to saturation is controlled *only* by the binding constant K, also called the association constant. The *dissociation constant* is $K_d = 1/K$. The dissociation constant has a simple physical meaning: K_d represents the concentration of ligand at which the protein is *half saturated*. To obtain this result, substitute $\theta = 1/2$ in Equation 6.4 and solve for x. You obtain $x = 1/K$, which is K_d.

4. This binding isotherm is a mathematical function called a rectangular hyperbola. Therefore, sometimes the expressions *hyperbolic isotherm* or *hyperbolic binding* are used. It is a fancy word with a simple meaning.

The value of the binding constant, which represents the intrinsic affinity of the ligand for its binding site on the protein, can be obtained experimentally by any method that detects a variation of some property of the protein that changes upon ligand binding. Examples of such properties are fluorescence, circular dichroism, or an NMR line. To obtain the value of the binding constant from these data, we perform a nonlinear least squares fit of the binding isotherm (shown in Equation 6.4) to the experimental data. The result is shown in the top panel of **Figure 6.2**, where the line represents Equation 6.4.

The best fit is produced with a binding constant $K = 1.0 \times 10^6 \, M^{-1}$. The bottom panel in Figure 6.2 shows a plot of the *residuals*, which are the differences between the data points and the function (in this case, Equation 6.4) fitted to the data. In a good fit, the residuals should be randomly distributed above and below the zero horizontal axis. The binding constant obtained corresponds to a dissociation constant $K_d = 1.0 \, \mu M$. We could have estimated K_d by simple inspection of the graph because K_d is the ligand concentration that results in half saturation of the protein. As shown by the dashed lines, K_d is the value on the x-axis when the signal reaches one-half of the maximum change on the y-axis. However, we must be careful because binding is only sensitive to K_d about the region of half saturation. If the [ligand] $\ll K_d$ or [ligand] $\gg K_d$, the value of K_d matters little. Note how the data in Figure 6.2 look more organized than in Figure 6.1. This is because we are now imposing on the data *a model* that shows how the data *are supposed* to behave.

Figure 6.2 The top panel shows same data as in Figure 6.1 with the fit of the binding isotherm (shown in Equation 6.4 with an amplitude factor of 1×10^5 appended). The best fit yields $K = 1.0 \times 10^6 \, M^{-1}$. The corresponding $K_d = 1.0 \, \mu M$ is shown on the concentration axis (x-axis). The dashed line drawn from the curve at half saturation intersects the concentration axis at K_d. The bottom panel shows the residuals, which are the differences between the data points and the fit line.

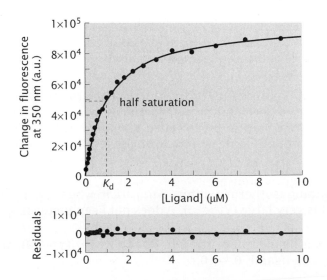

In general, if there are more than one binding sites in a protein, the fractional saturation is defined as

$$\theta = \frac{\text{Number of ligands bound}}{\text{Number of sites}} \tag{6.5}$$

$$= \frac{\text{Number of ligands bound}}{(\text{Number of binding sites per protein}) \times (\text{Number of proteins})}. \tag{6.6}$$

Suppose there are N binding sites on a protein. The total number of binding sites is the sum of the empty and occupied sites. All these numbers are proportional to the concentrations of the respective chemical species. The concentration of *proteins* with no ligand bound is [P]; with one ligand bound is [PX]; with two ligands bound is [PX$_2$]; with N ligands bound is [PX$_N$]. The total concentration of binding sites (there are N sites per protein, whether empty or occupied) is

$$[\text{Binding sites}] = N \times ([\text{P}] + [\text{PX}] + [\text{PX}_2] + [\text{PX}_3] \cdots + [\text{PX}_N]). \tag{6.7}$$

The total number of ligands *bound* is the sum of the number of ligands bound (m) in each state of the protein times the number of proteins that have m ligands bound, whose concentration is [PX$_m$]. Thus, the total concentration of ligands bound is

$$[\text{Ligands bound}] = 1 \times [\text{PX}] + 2 \times [\text{PX}_2] + 3 \times [\text{PX}_3] \cdots + N \times [\text{PX}_N]. \tag{6.8}$$

There is no term in [P] in this sum, because it has *zero* ligands. The binding isotherm is the fractional saturation (shown in Equation 6.5). To write it in terms of concentrations, we only have to divide Equation 6.8 by Equation 6.7,

$$\theta = \frac{1}{N} \frac{[\text{PX}] + 2[\text{PX}_2] + 3[\text{PX}_3] \cdots + N[\text{PX}_N]}{[\text{P}] + [\text{PX}] + [\text{PX}_2] + [\text{PX}_3] \cdots + [\text{PX}_N]}. \tag{6.9}$$

The Binding Isotherm is Easily Derived from the Partition Function

We have derived the binding isotherm quite easily for the simple case of a protein with one binding site. If we use the same approach in more complicated cases, the algebra becomes cumbersome. Instead, we will develop more systematically the partition function approach, introduced in Chapter 1. The beauty of this approach is that, while the actual experimental cases may become more complex, writing binding isotherms from the partition function does not. The power of the partition function method is only apparent when you have more than one binding site. However, even for one binding site, it allows for a different and more fundamental way of looking at the binding process.

Our work is greatly simplified by using a diagram as an aid to write the partition function. In this kind of diagram, we represent a protein state (empty or with a bound ligand) by a circle, which we call a *node*. Binding and dissociation reactions are represented by arrows, called *branches*, which link the nodes. Thus, for one binding site on a protein, we have the diagram shown in **Figure 6.3**.

We can now write the partition function Q. The first step is to choose a reference state. This choice is a matter of convenience, to make the problem clear and the calculations simple. In this case, the best is to choose the free protein P as the reference state. Next, we want to calculate the probabilities of all available states *relative* to the reference

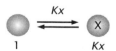

Figure 6.3 Diagram for binding of a ligand to a single site on a protein. The statistical weights are indicated below each state.

state. These relative probabilities are obtained by dividing the concentration of each state by the concentration of the reference. Thus, for the free protein, P, the relative probability is just

$$\frac{[P]}{[P]} = 1. \tag{6.10}$$

For the bound state, PX, we obtain the relative probability from the equilibrium constant (shown in Equation 6.2),

$$\frac{[PX]}{[P]} = Kx. \tag{6.11}$$

What this means is that the concentration of PX is related to the concentration of P by the factor Kx. If you were to pick a molecule at random from the solution and compare the probability that it has a ligand bound to the the probability that it has no ligand bound, the ratio of those probabilities would be Kx. In other words, the *probability* of the state PX *relative* to the state P is Kx. This *relative probability* (nonnormalized) is called a *statistical weight*.

The statistical weight is composed of two factors: K, which is the intrinsic affinity of the protein for the ligand, and x, the ligand concentration. K is a property of the *interaction* between a ligand molecule and its binding site on the protein. It depends only on how strongly they interact: by hydrogen bonds, hydrophobic interactions, or electrostatic interactions, for example. The concentration (x) is a purely *statistical factor*: the more ligand molecules you have in solution, the more likely it is that one of them will find the binding site on the protein.

The statistical weight of state P is 1 and the statistical weight of state PX is Kx. Therefore, we write the following correspondences, between each species and their statistical weights:

$$P \longrightarrow 1 \tag{6.12}$$

$$PX \longrightarrow Kx. \tag{6.13}$$

The concentrations of each species are proportional to their statistical weights, with the *same* proportionality factor. To obtain the partition function, we simply add the statistical weights,

$$Q = 1 + Kx. \tag{6.14}$$

The partition function Q is the sum of the probabilities of all states of the protein *relative* to the state chosen as reference, which is the free protein (P). Q is also the *normalization factor* needed to obtain the *absolute probabilities* of each state from the statistical weights. Thus, the probabilities of each state of the protein, with or without ligand bound, are

$$f_0 = \frac{1}{1 + Kx} \tag{6.15}$$

$$f_1 = \frac{Kx}{1 + Kx}. \tag{6.16}$$

The normalized probabilities are the *fractions* f_0 and f_1 of each form of the protein.

Now we summarize the complete procedure to obtain the binding isotherm. First, using the equations of the equilibrium constants, we write the probabilities (concentrations) of all states *relative* to the empty state. For one site, this is just $[P]/[P] = 1$ for the empty state and $[PX]/[P] = Kx$ for the bound state:

$$P \longrightarrow 1$$
$$PX \longrightarrow Kx.$$

Second, we add these terms to obtain the partition function (shown in Equation 6.14),

$$Q = 1 + Kx.$$

Third, we divide the statistical weight of each state (each term in the partition function) by the entire partition function, to obtain the actual *probabilities* (normalized) of each species, or their fractions (shown in Equations 6.15 and 6.16),

$$p_0 = \frac{1}{1 + Kx}$$
$$p_1 = \frac{Kx}{1 + Kx}.$$

Fourth, to obtain the average number of ligands bound per site, we multiply the number of ligands bound in each state (0 or 1, in this case) by its probability and add them,

$$0 \times p_0 + 1 \times p_1 = 0 \times \frac{1}{1 + Kx} + 1 \times \frac{Kx}{1 + Kx} \qquad (6.17)$$

$$= \frac{Kx}{1 + Kx}. \qquad (6.18)$$

Fifth, to obtain the binding isotherm, we just have to divide by the number of binding sites N per protein, but $N = 1$ in this case, so

$$\theta = \frac{1}{N} \frac{Kx}{1 + Kx} \qquad (6.19)$$

$$= \frac{Kx}{1 + Kx}. \qquad (6.20)$$

The last two steps are somewhat tedious. They can be replaced by taking the derivative of the natural logarithm of the partition function with respect to the natural logarithm of the ligand concentration (see Appendix B),

$$\theta = \frac{1}{N} \frac{d \ln Q}{d \ln x}. \qquad (6.21)$$

The method is especially useful when there is more than one site. To calculate the derivatives in Equation 6.21, recall that for any variable or function Y,

$$\frac{d \ln Y}{dY} = \frac{1}{Y}, \qquad (6.22)$$

which means that

$$d \ln Y = \frac{dY}{Y}. \qquad (6.23)$$

Applying Equation 6.23 to x and Q in Equation 6.21, we obtain,

$$
\begin{aligned}
\theta &= \frac{1}{N}\frac{d\ln Q}{d\ln x} \\
&= \frac{1}{N}\frac{x}{Q}\frac{dQ}{dx} \\
&= \frac{Kx}{1+Kx}.
\end{aligned}
\tag{6.24}
$$

Finally, let us take a moment to compare the result for single-site binding with that obtained in the case of protein unfolding. In both cases, we have two-state systems. In protein unfolding, we designated the folded (native) state by N and the unfolded (denatured) state by D. The equilibrium

$$
N \underset{}{\overset{K}{\rightleftharpoons}} D
\tag{6.25}
$$

is controlled by an equilibrium constant

$$
K = \frac{[D]}{[N]},
\tag{6.26}
$$

which tells us that the probability of state D relative to N is K. Thus, with N as the reference state, the partition function is

$$
Q = 1 + K
\tag{6.27}
$$

and the probabilities or fractions of each state are

$$
f_N = \frac{1}{1+K}
\tag{6.28}
$$

$$
f_D = \frac{K}{1+K}.
\tag{6.29}
$$

We see that everything is formally identical in binding and denaturation, with one exception. The ratio of the probabilities of the unfolded to folded states is given by Equation 6.26,

$$
\frac{[D]}{[N]} = K,
$$

whereas the ratio of the probabilities of the bound to the free form of the protein is given by Equation 6.11,

$$
\frac{[PX]}{[P]} = Kx.
$$

The concentration of X ($[X] = x$) now appears because the conversion of P to PX depends on the presence of the ligand molecule X. This was not the case in protein unfolding.

Oxygen Binds to a Single Site in Myoglobin

Let us study some examples. Myoglobin occupies a unique position in the history of proteins: it was the first protein whose three-dimensional structure was determined by X-ray crystallography, by John Kendrew. We talked about that history in Chapter 2, and then again discussed myoglobin in the context of the evolution of globins in Chapter 3.

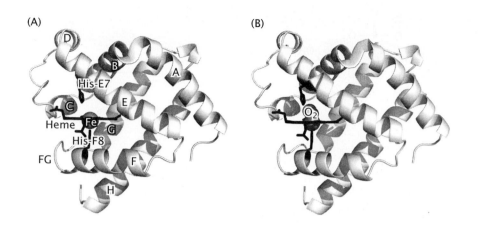

Figure 6.4 Structure of myoglobin. (A) The unliganded, deoxy state (PDB 1A6N). (B) With its ligand (oxygen) bound at the single binding site in the oxy form (PDB 1A6M). The letters A–H label the different helices according to the standard nomenclature for the globins. Turns between helices are designated with both labels (for example, FG turn). The iron metal ion (Fe^{2+}, central sphere) is coordinated by the heme group, shown in a side view, and by His-F8 (proximal histidine, below the iron) in the deoxy form. In the oxy form, the sixth coordination site is occupied by oxygen, which interacts with His E7 (distal histidine, above the iron).

The tertiary structure of myoglobin consists in an antiparallel helix arrangement called a globin fold (**Figure 6.4**).

The ligand of myoglobin is molecular oxygen (O$_2$). The binding site is at the heme group associated with the protein. In deoxymyoglobin (see Figure 6.4A), the iron metal ion (Fe^{2+}) is pentacoordinated: by four groups of the heme and by His-F8, or proximal histidine, shown below the iron in the figure. Note that the iron is somewhat below the plane of the heme, pulled down by His-F8. In oxymyoglobin (see Figure 6.4B), the iron metal ion acquires its sixth ligand, oxygen, which interacts with His-E7, or distal histidine, shown above the iron. When oxygen binds, the iron becomes hexacoordinated and moves into the plane of the heme group, because it is "pulled up" by the oxygen, which now balances the pull from His-F8.

Myoglobin follows a simple hyperbolic binding isotherm, shown in **Figure 6.5**. The oxygen binding constant $K \approx 2 \times 10^5 \, M^{-1}$; its reciprocal, the dissociation constant $K_d \approx 5 \, \mu M$ ($P_{50}(O_2) \sim 3 \, mm \, Hg$).

Two Ligands (Substrate and Inhibitor) Compete for the Same Enzyme Binding Site

In the next example we consider binding of a competitive inhibitor to an enzyme. Typically, this kind of inhibitor binds to the active site, the same site to which the substrate binds. Therefore, they *compete* for the binding site. The substrate is the normal ligand. Note that a ligand is any molecule that binds to a specific site on a protein. Both

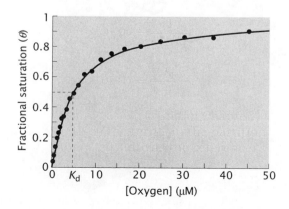

Figure 6.5 Binding isotherm of oxygen to myoglobin. The line is a fit of Equation 6.4 to the data points. The best fit yields $K_d = 5 \, \mu M$, which corresponds to the oxygen concentration at half saturation, indicated by the dashed lines. (Data simulated according to the binding affinity.)

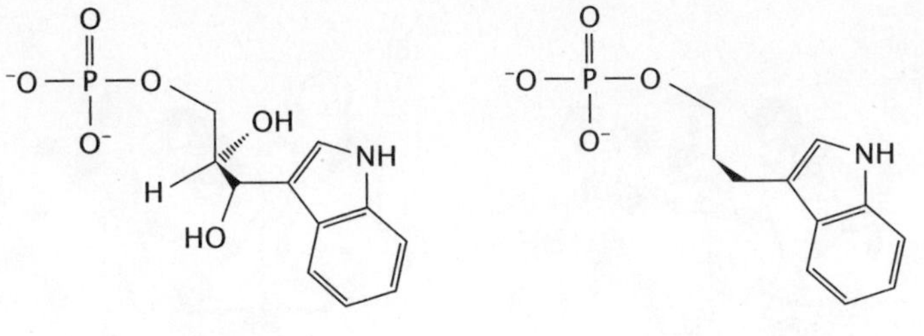

Figure 6.6 Structures of indole glycerol phosphate (IGP) and indole propanol phosphate (IPP).

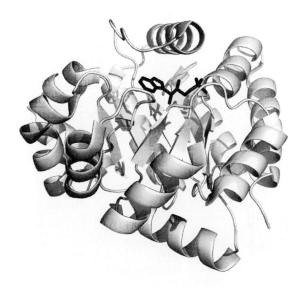

IGP

IPP

the substrate and the inhibitor are ligands. However, not all ligands are substrates. Substrates are transformed by the enzyme into different molecules, the products of the reaction.

In Chapter 3, we discussed the role of the enzyme α-tryptophan synthase (α-TrpS) in the context of the evolution of metabolic pathways. The biological substrate of α-TrpS is indole glycerol phosphate (IGP), which is cleaved by the enzyme to produce the indole component of the side chain of tryptophan and glyceraldehyde 3-phosphate. Because of strong structural similarity, indole propanol phosphate (IPP) competes with IGP for the active site (**Figure 6.6**).

The structure of the enzyme belongs to the TIM barrel fold (**Figure 6.7**). Similarly to tryptophan (to which it imparts its fluorescence properties), the indole group has an environment-dependent fluorescence emission maximum $\lambda \approx 340$–$380\,\mathrm{nm}$, which can be used to monitor binding. The binding isotherm for IPP with α-TrpS is shown in **Figure 6.8**. Also shown is the fit of Equation 6.4 to the data points, which yields $K_d \approx 50\,\mu\mathrm{M}$.

Now we want to write the partition function. Let us designate the substrate (IGP) by A and the inhibitor (IPP) by B. We begin by drawing the binding diagram for this problem (**Figure 6.9**). We are still dealing with binding to a single site for each ligand. Therefore, the partition function contains a term for the empty protein, our reference state, with statistical weight 1; a term for protein with A bound, with statistical weight $K_a[A]$; and another term for protein with B bound, with statistical weight $K_b[B]$,

$$Q = 1 + K_a[A] + K_b[B]. \tag{6.30}$$

Figure 6.7 Structure α-TrpS with its bound substrate, IGP, shown in black (PDB 1QOQ).

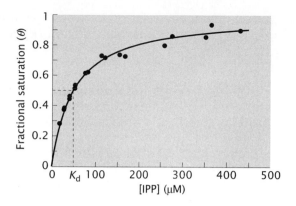

Figure 6.8 Binding isotherm of IPP to α-TrpS measured by the change in fluorescence of IPP at $\lambda = 335$ nm. (Data from Heyn MP & Weischet WO [1975] *Biochemistry* 14:2962–2968.) The line is a fit of Equation 6.4 to the data, which yields $K_d = 49$ μM.

Since each term in the partition function is proportional to the probability of occurrence of that state, the ratio of occupancies of the protein sites by the two ligands is given by the ratio of the two terms in the partition function that correspond to the probabilities that each ligand is bound,

$$\frac{p_A}{p_B} = \frac{K_a[A]}{K_b[B]}. \tag{6.31}$$

The binding constant of IGP to α-TrpS is not known exactly, but a reasonable estimate is $K_a \approx 3 \times 10^3 \, \text{M}^{-1}$ ($K_d \approx 350$ μM). The binding constant of the inhibitor (IPP) is $K_b = 2 \times 10^4 \, \text{M}^{-1}$. (It is not uncommon for an inhibitor to bind to an enzyme with greater affinity than its natural substrate.) If the two ligands are present at the same concentration, such as [A] = [B] = 100 μM, then B (IPP) is bound 7× more than A (IGP), reflecting directly the ratio of their binding constants. But if the concentration of A were increased to [A] = 100 × [B], then A would displace B from the protein, even if B had a higher intrinsic affinity for it (larger binding constant).

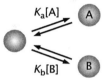

Figure 6.9 Diagram for binding of two ligands A and B that compete for a single site on a protein. The factors that convert the probability of a state to the next are indicated over the branches.

The Gibbs Energy of Binding can be Calculated from the Partition Function

In the binding reaction of a ligand X to a protein P,

$$P + X \overset{K}{\rightleftharpoons} PX, \tag{6.32}$$

the ratio of the equilibrium concentrations of PX to P, which is also the ratio of the equilibrium probabilities p_X to p_0, is given by

$$K[X] = \frac{[PX]}{[P]} \tag{6.33}$$

$$= \frac{p_X}{p_0}. \tag{6.34}$$

The Gibbs energy change in this binding reaction is

$$\Delta G = \Delta G^o_{bind} + RT \ln \frac{[PX]}{[P][X]}. \tag{6.35}$$

At equilibrium $\Delta G = 0$. Thus, the relation between the *standard Gibbs energy of binding* (ΔG^o_{bind}) and the binding constant (K) is

$$\Delta G^o_{bind} = -RT \ln K. \tag{6.36}$$

ΔG^o_{bind} *represents* the Gibbs energy of binding of 1 mole of ligand to 1 mole of protein to produce 1 mole of protein–ligand complex, when all species (free protein, free ligand, and protein–ligand complex) are in their standard states. As explained in Chapter 1, the usual standard state for a solute is a hypothetical state in which it interacts only with solvent (water), as if this were an *extremely dilute* solution, but the number of solute molecules per unit volume is such that its concentration is 1 M. You can verify, by substituting 1 M for all concentrations on the right-hand side of Equation 6.35, that you obtain $\Delta G = \Delta G^o_{bind}$, as you must, because those are the standard concentrations. Note well that 1 M concentrations are *not* the equilibrium concentrations. ΔG^o_{bind} is *not* the Gibbs energy change at equilibrium: at equilibrium that change is zero ($\Delta G = 0$).

ΔG^o_{bind} is *measured* by determining K, for example, as we just did by fitting a binding isotherm to titration data. In practice the measurements must be performed in very dilute solutions. ΔG^o_{bind} is valid only if you keep in mind the standard state that was used in obtaining K. Gibbs energies can only be compared if the same standard states are used. If nothing else is said, we will assume that the standard concentration is 1 M.

However, in Equation 6.35, the ratio [PX]/[P] matters but the result of any calculation is the same if [P] = 1 M *and* [PX] = 1 M (standard concentrations) or if they are *both equal to any other concentration*. Therefore, in practice, it is useful to think of this binding reaction as the conversion of 1 mole of free protein (P) to 1 mole of protein–ligand complex (PX) in a solution in which P *and* PX are *very dilute and in equal concentrations*, but the ligand concentration is 1 M (in standard conditions). The value of ΔG obtained in these conditions is still equal to ΔG^o_{bind}, and we do not have to worry about the effects of 1 M protein concentrations. We can represent this reaction by our usual diagram (**Figure 6.10**).

Now we would like to know the Gibbs energy change ΔG^X_{bind} for converting P to PX, when [P] = [PX] (both dilute), but the ligand concentration [X] is *not* necessarily 1 M. If we set [P] = [PX] in Equation 6.35, use Equation 6.36 for ΔG^o_{bind}, and combine the two terms on the right-hand side, we obtain

$$\Delta G^X_{bind} = -RT \ln K + RT \ln \frac{1}{[X]} \tag{6.37}$$

$$= -RT \ln K[X].$$

Now let us do the calculation using the partition function. By inspection of the diagram of Figure 6.10, we can immediately write

$$Q = 1 + K[X]. \tag{6.38}$$

Equation 6.38 gives the relative probabilities of the protein without ligand (first term, 1) and with ligand (second term, $K[X]$) at equilibrium. The Gibbs energy change in converting P to PX at a fixed concentration of ligand [X] is simply related to the equilibrium probabilities of the two states by

$$\Delta G^X_{bind} = -RT \ln \frac{p_X}{p_0}. \tag{6.39}$$

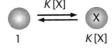

Figure 6.10 Diagram for binding of a ligand X to a single site on a protein. The factor that converts the probability of a state P to state PX of the protein is $K[X]$, indicated above the branch that connects the two states.

The relation between the probabilities p_X to p_0 is the ratio of their corresponding terms in the partition function—namely, $K[X]$ and 1,

$$\frac{p_X}{p_0} = K[X]. \tag{6.40}$$

Thus, the Gibbs energy of binding at *any concentration* of ligand, not necessarily 1 M, can be calculated from the probability of the bound state relative to the free state of the protein. ΔG_{bind}^X gives the difference between the Gibbs energy of PX and P when the ligand concentration is [X]. Substituting $K[X]$ for p_X/p_0 in Equation 6.39, we obtain again

$$\Delta G_{bind}^X = -RT \ln K[X]. \tag{6.41}$$

This expression is valid for a solution of ligand with any concentration [X]. Note that it simplifies to the standard Gibbs energy change if [X] = 1 M (shown in Equation 6.36). The main advantage of the partition function approach, however, is conceptual. You can always obtain the Gibbs energy difference between any two states by an expression of the type of Equation 6.39, and you can always write this equation easily by inspection of the partition function.

The Gibbs Energy Change is Zero if the Ligand Concentration is Equal to the Dissociation Constant

The Gibbs energy change that occurs when a ligand binds to a protein is illustrated in **Figure 6.11**. If the ligand concentration is sufficiently high—more precisely, if it is greater than the dissociation constant ([X] > K_d)—then the bound state of the protein (PX) is more favorable than the free protein state (P) (see Figure 6.11A). If you recall that $K_d = 1/K$, you can see from Equation 6.33 that [PX] > [P] when [X] > K_d. This means that the probability p_X is greater than p_0, or that PX has a lower Gibbs energy than P. Thus, the Gibbs energy of the protein decreases upon ligand binding.

On the other hand, if the ligand concentration is identical to the dissociation constant ([X] = K_d), then [PX] = [P], which means that the probabilities p_0 and p_X are identical. At this ligand concentration, the two states of the protein are *equally likely*, and their Gibbs energies are identical (see Figure 6.11B). There is no gain or loss of Gibbs energy in going from one state to the other, and $\Delta G_{bind} = 0$. You obtain this result directly from Equation 6.41 if you set [X] = $K_d = 1/K$.

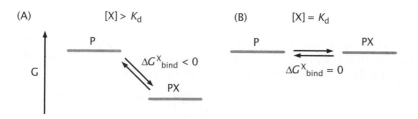

(A)   [X] > K_d

P

$\Delta G^X_{bind} < 0$

PX

(B)   [X] = K_d

P   PX

$\Delta G^X_{bind} = 0$

G

Figure 6.11 Gibbs energy changes when a ligand X binds to a protein P. (A) If [X] > K_d, the Gibbs energy decreases upon binding. (B) If [X] = K_d, the Gibbs energy does not change with binding.

If the Free Ligand is not in Large Excess the Binding Constant can be Determined by Solving a Quadratic Equation or Subtracting the Bound Ligand from the Total

We have considered binding to a single site on a protein (shown in Equation 6.1),

$$P + X \xrightleftharpoons{K} PX,$$

where K is the binding constant, and we have defined θ as the number of ligands bound per site, or fractional binding (fractional saturation, Equation 6.3),

$$\theta = \frac{[PX]}{[P] + [PX]}.$$

Here, it will be convenient to use the dissociation constant K_d instead of K ($K_d = 1/K$),

$$K_d = \frac{[P][X]}{[PX]}, \tag{6.42}$$

to obtain the equation for the binding isotherm as

$$\theta = \frac{[X]}{K_d + [X]}. \tag{6.43}$$

In the examples discussed, the total protein concentration $[P]_T$ was assumed to be much smaller than the total ligand concentration $[X]_T$ and the dissociation constant K_d. Hence, when the protein–ligand complex PX is formed, the amount of X consumed in the binding reaction is much smaller than the total ligand present, $[PX] \ll [X]_T$. Under these conditions, most ligand is free and the approximation $[X] \approx [X]_T$ is valid. Thus, we can plot the fractional binding θ (or a signal proportional to it) as a function of $[X]_T$ (assuming $[X] \approx [X]_T$), and obtain K_d from a fit of Equation 6.43 to the data, with only a very small error.

In many cases, however, when you design an experiment to measure binding, the conditions $[PX] \ll [X]_T$ or $[PX] \ll K_d$ cannot be satisfied. This is often because the protein binds the ligand with very high affinity or because you need to have a large protein concentration in solution to be able to measure an experimental signal. Thus, the protein and the ligand concentrations are of the same order of magnitude ($\sim$) as K_d, $[P] \sim [X] \sim K_d$. We can still obtain K_d from the experimental measurement of θ as a function of the total ligand concentration $[X]_T$, but we need to solve a *quadratic equation* to determine $[PX]$. To do that, we begin with the two conservation equations,

$$[P] = [P]_T - [PX] \tag{6.44}$$

and

$$[X] = [X]_T - [PX], \tag{6.45}$$

and substitute them in Equation 6.42 for the equilibrium constant. Again, we use $1/K_d$ instead of K because K_d has units of concentration and it can therefore be added to $[P]_T$ and $[X]_T$. We obtain

$$K_d = \frac{([P]_T - [PX])([X]_T - [PX])}{[PX]}, \tag{6.46}$$

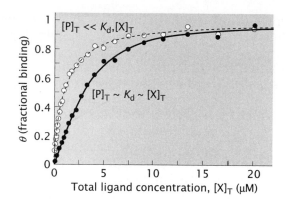

Figure 6.12 Binding isotherms of a ligand X from a protein P with a dissociation constant $K_d = 1$ μM. Open circles, data with $[P]_T \ll K_d$ and $\ll [X]_T$. The dashed line is a fit of Equation 6.43 to the data. Solid circles, data with a protein concentration $[P]_T = 4$ μM (comparable to K_d and to the ligand concentration). The solid line is a fit of Equation 6.49 to the data. (Data simulated for the experimental conditions.)

which can be rearranged to the canonical form of a quadratic equation,

$$[PX]^2 - ([P]_T + [X]_T + K_d)[PX] + [P]_T[X]_T = 0. \qquad (6.47)$$

The physically meaningful solution of Equation 6.47 for [PX] is

$$[PX] = (1/2)\left(([P]_T + [X]_T + K_d) - \sqrt{([P]_T + [X]_T + K_d)^2 - 4[P]_T[X]_T}\right). \qquad (6.48)$$

If we divide Equation 6.48 by $[P]_T$, we obtain the fractional binding, or binding isotherm θ,

$$\theta = \frac{[PX]}{[P]_T} = \frac{1}{2}\left[\left(1 + \frac{[X]_T}{[P]_T} + \frac{K_d}{[P]_T}\right) - \sqrt{\left(1 + \frac{[X]_T}{[P]_T} + \frac{K_d}{[P]_T}\right)^2 - 4\frac{[X]_T}{[P]_T}}\right]. \qquad (6.49)$$

You know what $[P]_T$ is because in this kind of experiment you add increasing ligand concentrations to a measured amount of protein. Therefore, Equation 6.49 can be used to fit the binding data and determine K_d.

Suppose that you measure binding of a ligand X to a protein P, and that the dissociation constant is $K_d = 1$ μM. You perform the experiment with a protein concentration $[P]_T \ll K_d$ (for example, $[P]_T \approx 0.1$ μM). In this case, the free $[X] \approx [X]_T$. The data you obtain are shown by the open circles in **Figure 6.12**. The dashed line is a fit of the binding isotherm to the data (shown in Equation 6.43).

However, if you perform the same experiment with $[P]_T = 4$ μM, which is of the same order of magnitude of K_d and of the ligand concentrations used, the plot of θ as a function of *total* ligand concentration looks like the data shown by the solid circles in Figure 6.12. The solid line is a fit of Equation 6.49 to the data; the fit yields $K_d = 1$ μM.

Alternatively, instead of fitting the data to the binding isotherm given by Equation 6.49, we can use the data to calculate the free ligand concentration at each point. Note that the y-axis gives us the fraction of protein with ligand bound. Because we know the total protein concentration $[P]_T$, we can calculate the concentration of bound protein by

$$[PX] = \theta[P]_T. \qquad (6.50)$$

Since the stoichiometry of the reaction is 1 : 1 this is also the concentration of bound ligand. To obtain the free ligand, we simply subtract the bound ligand from the total ligand,

$$[X] = [X]_T - \theta[P]_T. \qquad (6.51)$$

Finally, we re-plot the fractional binding θ as a function of the calculated free ligand, and fit the data to the binding isotherm of Equation 6.43.

You may think that the direct fit of Equation 6.49 to the data is better because it is "exact." However, the subtraction method has one big advantage: if you know the number of binding sites on the protein, you can use this method for a protein with any number of binding sites, whereas the quadratic equation that leads to the binding isotherm of Equation 6.49 can only be solved for the case of one binding site. This is a serious limitation of the "exact" approach. To use the subtraction method, however, we must obtain data until the plateau region (where the curve levels off), so that the protein is completely saturated ($\theta \approx 1$); otherwise we cannot accurately calculate the fractional binding.

Binding Measurements at High Protein Concentration can Provide Information on Stoichiometry

Now suppose the protein concentration in your experiment is *very large* compared to K_d. A small K_d means that binding is strong. Since its concentration is large, the protein binds *all the ligand* at each successive ligand addition. The protein is consumed in direct proportion to the amount of ligand added until all protein is bound. Therefore, the data points fall on a straight line with a positive slope (**Figure 6.13**).

However, when there is *no more free protein* available, no more ligand can bind, and the data fall on a horizontal line. The point where the two straight lines intersect corresponds to the concentration at which the ratio $[X]_T/[P]_T$ equals the stoichiometric ratio of the reaction, which is 1:1 in this case. This experiment cannot be used to determine the binding constant, but it can be used to determine the stoichiometry of the complex formed in the binding reaction.

Isothermal Titration Calorimetry is an Excellent Method to Measure Binding

To measure binding, all we need is a signal that changes when a ligand binds to a protein. This is often a change in protein fluorescence, in the CD spectrum, or in an NMR signal. However, isothermal titration calorimetry (ITC) is especially suitable to measure binding because, in a single experiment, it yields the binding constant, the enthalpy of binding, and the stoichiometry of the complex, or binding ratio. In a typical ITC experiment, a protein solution (≈ 1 mL) is placed in a cell inside the

Figure 6.13 Ligand binding to a protein with $K_d = 0.5$ μM at high protein concentration ($[P]_T = 20$ μM). The data points fall on two straight lines that intersect when the ratio $[X]_T/[P]_T$ equals the stoichiometric ratio of the reaction, which is 1:1 in this case. (Data simulated for the experimental conditions.)

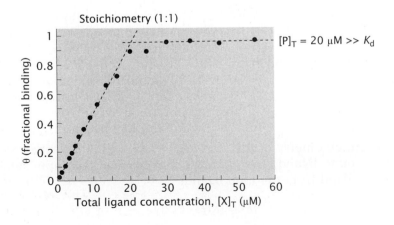

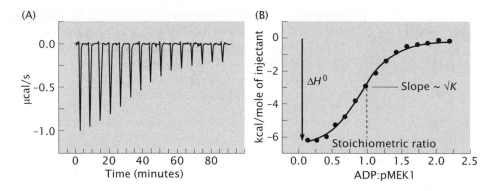

Figure 6.14 Binding of a ligand (ADP) to a protein (MEK1 kinase) measured by isothermal titration calorimetry. (A) Individual heats released ($\Delta H < 0$) upon each ligand injection as a function of time. (B) Integrated heat released per mole of ligand injected, as a function of the total ligand-to-protein ratio ($[X]_T/[P]_T$). The plot in (B) is obtained from panel (A) by calculating the areas within each peak in (A), correcting for dilution effects, and normalizing the heat per mole of ligand. (Adapted from Smith CK & Windsor WT [2007] *Biochemistry* 46:1358–1367. With permission from American Chemical Society.)

calorimeter, and the ligand (the titrant) solution is injected in small ($\approx 10\ \mu L$), successive aliquots, in short time intervals (≈ 10 minutes). The calorimeter measures the heat (enthalpy) released or absorbed at each injection of ligand (**Figure 6.14**A).

Initially, the heat is large because most protein is free and can bind the ligand. After a few injections, however, the protein binding sites become progressively occupied, and the heat measured decreases in absolute value (the peaks become smaller). Eventually, no protein is available to bind additional ligand: the protein binding sites are saturated. Figure 6.14A shows the raw data obtained by ITC. To extract the binding parameters (K_d, ΔH^o, and the stoichiometry) from these data, first we integrate the individual peaks. That is, we calculate the area within each peak, which corresponds to the heat released (or absorbed) upon each ligand binding event. Second, we calculate the total ligand-to-protein ratio ($[X]_T/[P]_T$) from the rate of injection and the concentrations (and volumes) of the protein and ligand solutions. Third, we normalize the heats by dividing them by the number of moles of ligand injected each time. (In practice, it is also necessary to subtract heats of dilution, which give rise to the peaks visible on the right side of Figure 6.14A.) The result is the plot of the heat released (or absorbed) per mole of ligand as a function of the total ligand-to-protein ratio, shown by the data points in Figure 6.14B.

In the beginning all the protein is free. If the binding constant K is large, in the first injection all ligand is bound by the protein. Thus, the heat per mole of ligand measured in the first addition is the enthalpy of binding ΔH^o, which is indicated in Figure 6.14B (for practical reasons, the second injection is the one usually used). To understand how K_d and the stoichiometry of the protein–ligand complex are obtained from these data, we need to take a moment to learn how the heats measured by the calorimeter are related to a binding isotherm.

The increment dq_i of heat measured at each injection is proportional to the increment $d[PX]$ in complex formed,

$$dq = \Delta H^o V_0 d[PX], \qquad (6.52)$$

where ΔH^o is the binding enthalpy change per mole of ligand and V_0 is the volume of the calorimeter cell. The ordinate (y-axis) in the plot of Figure 6.14B is a *differential* or *incremental* heat obtained by dividing

Equation 6.52 by the increment in the total ligand concentration $[X]_T$,

$$\frac{1}{V_0}\frac{dq_i}{d[X]_T} = \Delta H^{\circ}\frac{d[PX]}{d[X]_T}. \tag{6.53}$$

The derivative $d[PX]/d[X]_T$ on the right-hand side of Equation 6.53 is calculated by differentiating Equation 6.49 with respect to $[X]_T$. The result can be expressed as a function of the total ligand-to-protein ratio $[X]_T/[P]_T$,

$$\frac{1}{V_0}\frac{dq_i}{d[X]_T} = \Delta H^{\circ}\left(\frac{1}{2} + \frac{1 - (1 + [X]_T/[P]_T + K_d/[P]_T)/2}{\sqrt{(1 + [X]_T/[P]_T + K_d/[P]_T)^2 - 4[X]_T/[P]_T}}\right). \tag{6.54}$$

Equation 6.54 is called the *Wiseman isotherm*. It is the line plotted in Figure 6.14B along with the data points. Its shape is different from a regular (hyperbolic) binding isotherm (shown in Equation 6.43) for two reasons. First, it represents a differential binding; it is a derivative of $[PX]$, rather than proportional to $[PX]$. Second, it is plotted as a function of the ratio $[X]_T/[P]_T$, rather than as a function of free ligand $[X]$. Other than that, it represents the usual increase in fractional binding, eventually leading to saturation, just like a regular binding isotherm. The usefulness of the Wiseman isotherm is that the thermodynamic parameters can be easily obtained from the plot of Figure 6.14B. The initial (maximum) amplitude of the change gives ΔH°. For sufficiently strong binding, the *slope* of the curve at the inflection point (midpoint of the "S") is proportional to $\sqrt{K}$ (binding constant $K = 1/K_d$): the steeper the slope, the stronger the binding (larger K). Finally, the ratio $[X]_T/[P]_T$ (*x*-axis) at the inflection point gives the *stoichiometry* of the PX complex (in this case, 1 : 1).

6.2 MOST PROTEINS BIND MORE THAN ONE LIGAND

Two Binding Sites for the Same Ligand may not be Equivalent

Consider the binding data shown in **Figure 6.15**, which you could have obtained for another protein. If we analyze these data using a fit to a simple binding isotherm, the result is the line shown in Figure 6.15A. The fit looks reasonably good and yields $K = 3.8 \times 10^6\,\text{M}^{-1}$ (or $K_d = 1/K = 0.26\,\mu\text{M}$). However, closer inspection of the beginning of the curve shows that the isotherm does not fit the data accurately (see Figure 6.15B). The data points are *systematically* above the fitted line in the beginning of the curve, up to [ligand] $= 0.2\,\mu\text{M}$, and then systematically below the fitted line. This systematic deviation of the fit from the data points is especially apparent in the plot of the residuals in Figure 6.15B.

We tried the simplest binding model, for a single site, but it did not work very well. Perhaps the protein has more than one binding site with different affinities for the ligand. The next simplest possibility is two different binding sites. To test this idea, we need to derive the binding isotherm for two sites on the protein. We will do this using the diagram method and the partition function. Then we will try to fit the new model to the data.

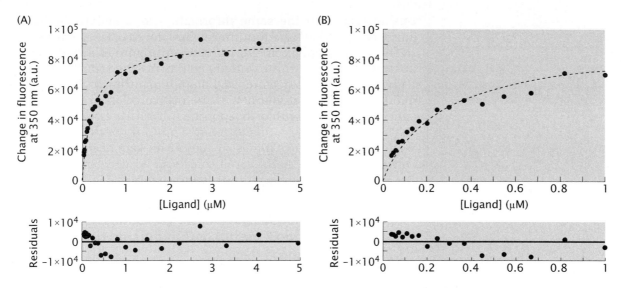

Figure 6.15 A second data set for ligand binding to a protein. (A) Full concentration range showing the data and the fit of a single-site binding isotherm (shown in Equation 6.4, with an amplitude factor of 1×10^5 appended) to the data. (B) Low-ligand concentration range from panel (A) enlarged. The bottom panels show the residuals. (Simulated data.)

Two Sites are Independent if Binding to One does not Affect Binding to the Other

Consider a protein P with *two different and independent sites*, A and B, to which a ligand X binds, with association constants K_a and K_b. We say the sites are independent because each one binds the ligand as if the other site were not present. In other words, whether a ligand is bound to site A has no influence on the binding to site B, and vice versa. The corresponding diagram is shown in **Figure 6.16**, where $x = [X]$.

Before we proceed, here is a quick note on notation: to indicate that a site is empty, we will use an empty circle, like this ◯. To indicate an occupied site, we will use ⊗. We will write site A always on the left side and site B on the right. For example, ⊗◯ Means that site A is occupied and B is free. To get from the completely empty state to the fully occupied state in Figure 6.16, you can *either* go by the upper branch *or* by the lower branch of the diagram. (You can take either path but not both; the path indicates the order in which the binding events happen.) If you choose the upper branch, the probability of going from ◯◯ to ⊗◯ is $K_a x$. If you choose the lower branch, the probability of going from ◯◯ to ◯⊗ is $K_b x$. The probability of going from ◯◯ to ⊗⊗ is $K_a x \times K_b x = K_a K_b x^2$, because this is the product of going (by the upper branch) from ◯◯ to ⊗◯ *and* from ⊗◯ to ⊗⊗. Thus, to obtain the probability of reaching the end of a branch, you just multiply the probability factors *along* that branch. You could also have chosen the lower branch: the final result would be the same. This is an illustration of the important principle of thermodynamic state functions (equilibrium

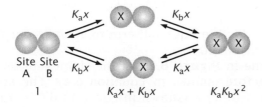

Figure 6.16 Diagram for binding of a ligand to two independent sites on a protein.

constants contain the same information as standard Gibbs energies) that *only the initial and final states, but not the path, matter.*

Now let us apply our general procedure to generate the partition function and, from it, the binding isotherm. The statistical weights (relative probabilities) of the empty state, ○○, the singly occupied state, ⊗○ + ○⊗, and the doubly occupied state, ⊗⊗, are

$$\frac{[P]}{[P]} = 1 \tag{6.55}$$

$$\frac{[PX]}{[P]} = K_a x + K_b x = (K_a + K_b)x \tag{6.56}$$

$$\frac{[PX_2]}{[P]} = K_a x \times K_b x = K_a K_b x^2. \tag{6.57}$$

The partition function is their sum,

$$Q = 1 + (K_a + K_b)x + K_a K_b x^2 \tag{6.58}$$

$$= (1 + K_a x)(1 + K_b x). \tag{6.59}$$

Therefore, the (absolute) probabilities of each state are

$$p_0 = \frac{1}{Q} \tag{6.60}$$

$$p_1 = \frac{(K_a + K_b)x}{Q} \tag{6.61}$$

$$p_2 = \frac{K_a K_b x^2}{Q}. \tag{6.62}$$

The average number of ligands bound per protein is

$$0 \times p_0 + 1 \times p_1 + 2 \times p_2. \tag{6.63}$$

Finally, the binding isotherm is

$$\theta = \left(\frac{1}{2}\right)\frac{(K_a + K_b)x + 2K_a K_b x^2}{1 + (K_a + K_b)x + K_a K_b x^2}. \tag{6.64}$$

The factor of $(1/2)$ in front arises from dividing by the number of sites per protein ($N = 2$ sites in this case). It is included to obtain the isotherm *per site* instead of the isotherm per protein (see Equation 6.5). Again, the last two steps can be simplified by using Equation 6.21,

$$\theta = \frac{1}{N}\frac{d \ln Q}{d \ln x}$$

$$= \frac{1}{N}\frac{x}{Q}\frac{dQ}{dx}$$

$$= \left(\frac{1}{2}\right)\frac{(K_a + K_b)x + 2K_a K_b x^2}{1 + (K_a + K_b)x + K_a K_b x^2}. \tag{6.65}$$

We can now use this binding isotherm to analyze our data and see if there is an improvement.

The dashed line in **Figure 6.17** is the single-site binding isotherm that we used before (shown in Equation 6.4). The solid line is the two-site binding isotherm (shown in Equation 6.64). You can see that

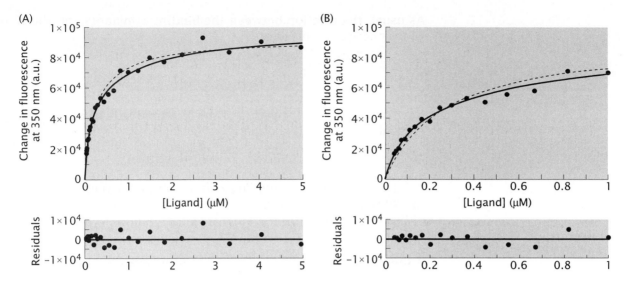

Figure 6.17 Ligand binding to two different sites on a protein (same data as in Figure 6.15). (A) Full concentration range showing the data and the fit of a single-site binding isotherm (dashed line, Equation 6.4) and a two-site binding isotherm (solid line, Equation 6.64) to the data (an amplitude factor of 1×10^5 is appended to the fit equations). (B) Low-ligand concentration range from panel (A) enlarged. The bottom panels show the residuals for the two-site isotherm fit.

this fit with the two-site isotherm is better in the full range if you compare the residual plot in the bottom panel of Figure 6.17A with that in Figure 6.15A, although it might not have been apparent how poorly the single-site isotherm did without comparison with the two-site isotherm. The improvement in the fit is especially apparent in the initial part of the curve if you compare the residual plots in Figures 6.17B and 6.15B. Now the residuals are not all above the zero line in the beginning, but essentially fall on it. The values of the two binding constants obtained from the fit of Equation 6.64 to the data are $K_a = 1.0 \times 10^6$ and $K_b = 1.0 \times 10^7 \, \text{M}^{-1}$ (K_d's of 1.0 and 0.10 μM, respectively). The single-site model had replaced these two binding constants by one, with an "average" value of $K = 3.8 \times 10^6 \, \text{M}^{-1}$ ($K_d = 0.26 \, \mu$M).

The Coupling Gibbs Energy Measures the Mutual Dependence of Binding Sites

In the diagram of Figure 6.16, we used the same symbol, K_a, to indicate binding to site A, no matter whether site B was empty or occupied because the sites were independent. In this section, however, we want to be specific about the order of binding. Thus, we use K_a to indicate binding to site A if B is *free* (the first step in the upper branch of Figure 6.16), and $K_{a(b)}$ to indicate binding to site A if B is *occupied* (second step in the lower branch). Similarly, we use K_b to indicate binding to site B if A is *free*, and $K_{b(a)}$ to indicate binding to site B if A is *occupied*. If the sites are independent, $K_a = K_{a(b)}$ and $K_b = K_{b(a)}$. Then the binding constant to site A on the empty protein is the same as the binding constant to site A on a protein that already has site B occupied. Thus, the criterion for independent binding to two sites A and B is

$$\frac{K_{a(b)}}{K_a} = 1. \tag{6.66}$$

As usual, the relation between the binding constants and the standard Gibbs energy changes are

$$\Delta G_a^o = -RT \ln K_a \tag{6.67}$$

$$\Delta G_{a(b)}^o = -RT \ln K_{a(b)}. \tag{6.68}$$

Now, if we take the logarithm of the left-hand side of Equation 6.66 and multiply it by $-RT$, we obtain

$$\Delta G_{a(b)}^o - \Delta G_a^o = \Delta G(a|b). \tag{6.69}$$

$\Delta G(a|b)$ is called the *coupling Gibbs energy* between sites A and B. If the sites are independent, $\Delta G(a|b) = 0$ ($\ln 1 = 0$ on the right-hand side of Equation 6.66).

A Ligand can Change the Apparent Binding Constant of Another

Let us consider an example in which two ligands A and B bind to a protein, each at their own site, but ligand B only binds if A binds first. This is the case of enzymes that follow a sequential ordered mechanism; they have a mandatory order in binding their substrates. A classical example is the reaction catalyzed by the enzyme lactate dehydrogenase (**Figure 6.18**). NAD$^+$ binds first, then lactate binds. In the opposite direction, NADH binds first, then pyruvate binds. Sometimes the order is not as strict; then, binding of substrate B to the free enzyme may occur, but it improves if A is already bound. This kind of situation occurs when binding of the first ligand (A) induces a conformational change in the protein whereby the binding site for B is made available or completed. The diagram for the sequential ordered case is shown in **Figure 6.19**.

First, let us write the partition function,

$$Q = 1 + K_a[A] + K_a K_b[A][B]. \tag{6.70}$$

Q can also be written as

$$Q = 1 + K_a[A](1 + K_b[B]), \tag{6.71}$$

where we recognize the last term as a partition function for B only.

The binding constant of A to the free protein is

$$K_a = \frac{[PA]}{[P][A]}, \tag{6.72}$$

which we can write as

$$K_a[A] = \frac{[PA]}{[P]}. \tag{6.73}$$

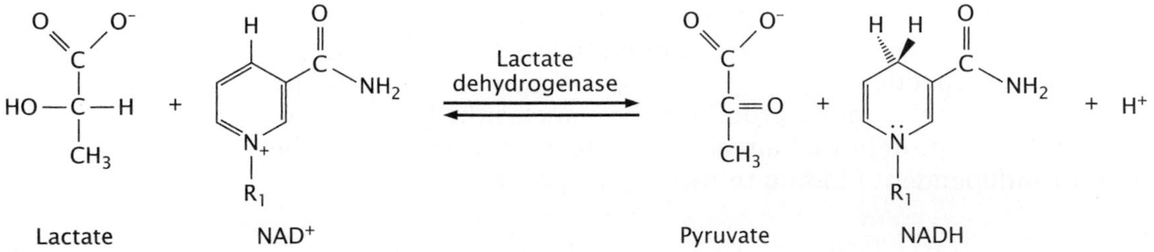

Figure 6.18 The reaction catalyzed by lactate dehydrogenase.

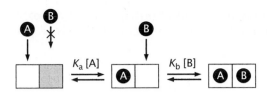

This equation tells us that the probability of protein with A bound (PA) relative to the probability of empty protein (P, the reference state) is $K_a[A]$.

Now suppose you are studying binding of A to the protein, but don't know if B is present or not. You can write an *apparent binding constant* of A to the protein as

$$K_a^{app} = \frac{[\text{Protein with A bound}]}{[\text{Protein without A bound}][A]}. \qquad (6.74)$$

This is an *apparent* constant because we don't know if the protein that has A bound has also something else bound (such as B), and we don't know if the protein that is free of A has anything else bound. If we write the equivalent of Equation 6.73,

$$K_a^{app}[A] = \frac{[\text{Protein with A bound}]}{[\text{Protein without A bound}]}, \qquad (6.75)$$

the right-hand side is the ratio of two concentrations, which is the same as the ratio of probabilities. We know how to get that from the partition function: it is the ratio of the terms of the protein with A bound (independently of whether anything *else*—B in this case—is bound) to the terms of protein without A. If you look at Equation 6.70, this is simply

$$\frac{[\text{Protein with A bound}]}{[\text{Protein without A bound}]} = \frac{K_a[A] + K_a K_b[A][B]}{1}$$
$$= K_a(1 + K_b[B])[A]. \qquad (6.76)$$

Comparing Equation 6.76 with Equation 6.75, we see that

$$K_a^{app} = K_a(1 + K_b[B]), \qquad (6.77)$$

which differs from the true K_a by the factor $1 + K_b[B]$. Thus, if the concentration of B increases, K_a^{app} for A increases because B is pulling the equilibrium to the right (see Figure 6.19).

Similarly, we can write the apparent binding constant of B in the presence of A as

$$K_b^{app}[B] = \frac{[\text{Protein with B bound}]}{[\text{Protein without B bound}]} \qquad (6.78)$$

and obtain the ratio of concentrations from the partition function (shown in Equation 6.70) by selecting the terms that have B bound and the terms that don't have B bound, but may have A or nothing else bound:

$$\frac{[\text{Protein with B bound}]}{[\text{Protein without B bound}]} = \frac{K_a K_b[A][B]}{1 + K_a[A]}$$
$$= K_b[B] \frac{K_a[A]}{1 + K_a[A]}. \qquad (6.79)$$

If we did not know anything about the presence of A (or its requirement), but thought that only protein and B were present, we would write this ratio as

$$K_b[B] = \frac{[PB]}{[P]}. \tag{6.80}$$

Comparing Equation 6.78 with Equation 6.79, we see that

$$K_b^{app} = K_b \frac{K_a[A]}{1 + K_a[A]}. \tag{6.81}$$

Thus, the apparent binding constant of B to the protein depends on the presence of A. If $[A] = 0$, then $K_b^{app} = 0$, which just says that B does not bind to the protein without A. As the concentration of A increases, K_b^{app} increases, but it shows saturation: for very large $[A]$,

$$\frac{K_a[A]}{1 + K_a[A]} \longrightarrow 1, \tag{6.82}$$

and K_b^{app} becomes the true K_b.

6.3 COOPERATIVE BINDING IS A CONSEQUENCE OF INTERACTING SITES

Consider now the data shown in **Figure 6.20** for binding of a substrate X to a protein P, measured by the change in Trp fluorescence of the protein upon ligand binding. These data appear different from those in the previous cases. They seem to follow an "S-like" or *sigmoidal* shape. By now you would probably predict that a fit with a single-site binding isotherm would not work (and it does not), as shown by the dashed line in Figure 6.20 in the upper panel and even more clearly by the residual plot in the lower panel. A model with two independent binding sites would not work either.

Figure 6.20 Binding data for a cooperative case. Top panel: the dashed line is a fit of a single-site binding isotherm to the data points (solid symbols). (An amplitude factor of 1×10^5 is appended to Equation 6.4.) The fit yields $K = 5.6 \times 10^5 \, M^{-1}$, or $K_d = 1.8 \, \mu M$. Bottom panel: residuals of the fit in the top (difference between the points and the fit).

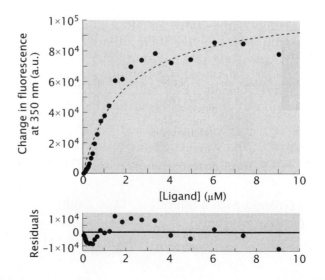

The Minimal Model for Cooperative Binding Requires Two Sites and an Interaction Between Them

To describe the binding data of Figure 6.20, we need a model for a protein with two binding sites that are *not* independent, but *communicate* with each other. We say that the sites *interact*. If binding of a ligand to one site increases its affinity to the next site, the binding is *cooperative*. **Figure 6.21** shows the diagram of the simplest cooperative binding model. The statistical weights are written below each state.

The way the model works is by *decreasing* the probability of states that have one site occupied and the other empty, so that the protein has a tendency to bind two ligands or none. If one subunit has a ligand bound and the other is empty, there is an *unfavorable interaction* between the two subunits. The protein acquires an *extra free energy* $\varepsilon > 0$ (see Figure 6.21). We define the *cooperativity parameter* σ by the Boltzmann factor,

$$\sigma = e^{-\varepsilon/RT}. \tag{6.83}$$

Since $\varepsilon > 0$, $\sigma \le 1$ and often $\sigma \ll 1$ in a cooperative system.

The statistical weight of the empty state is 1 (reference). The intrinsic binding constant to an *isolated* protein monomer is K. The binding constant to the first site in the dimer is σK (see Figure 6.21). The probability of binding to the first site depends on three factors (σKx): the intrinsic affinity (K), the ligand concentration (x), *and* the factor σ, which arises from the additional Gibbs energy ε *gained* as a consequence of the unfavorable interaction between an empty and an occupied site on the protein. Thus, σ represents the probability by which binding to a site is reduced if the *other site* is empty. When both sites are occupied, there is no unfavorable interaction, so the statistical weight of the doubly occupied state must be K^2x^2. Therefore, binding of a second ligand is associated with the probability factor Kx/σ. The second ligand binds with a *larger* binding constant $K/\sigma > K$ (because $\sigma < 1$). Hence, binding is cooperative.

Figure 6.22 shows a Gibbs energy diagram comparing the cooperative system ($\sigma \ll 1$) with the corresponding noncooperative case ($\sigma = 1$, $\varepsilon = 0$). The solid arrows indicate the path of this cooperative model; the dashed arrows indicate the path of the noncooperative model. Let us call ΔG the Gibbs energy of binding to an isolated monomer. Keep in mind that in binding reactions, ΔG depends on x (the concentration of X) according to $\Delta G = -RT \ln Kx$ (shown in Equation 6.41). If $x = 1\,\text{M}$ (standard concentration), this equation yields $\Delta G^0 = -RT \ln K$. The Gibbs energy level of the semi-occupied state $\otimes\bigcirc$ in the diagram of Figure 6.22 depends on the values of ΔG and ε. In turn, ΔG is determined by the binding constant K and the concentration x. Depending on the values of those parameters, the Gibbs energy of the state $\otimes\bigcirc$ may lie above or below the Gibbs energy of the empty protein. In any case, cooperativity arises because ε *increases* the Gibbs energy of the

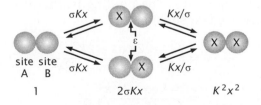

site site
A   B

Figure 6.21 Diagram for a two-site binding with cooperativity. When one subunit is occupied and the other is empty, there is an unfavorable interaction Gibbs energy $\varepsilon > 0$ between the two subunits. The statistical weights are indicated below each state.

Figure 6.22 Comparison of the Gibbs energy changes in cooperative (solid arrow path) and noncooperative binding (dashed arrow path).

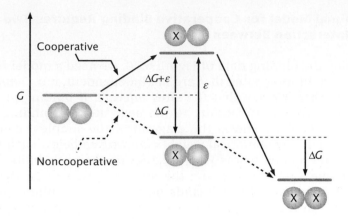

intermediate state. In the cooperative example of Figure 6.22, for a sufficiently large ε, the first binding step corresponds to an *increase* in Gibbs energy, which is compensated by a much larger *decrease* in the second step. This is why binding of the second ligand is much easier if the first one is already bound.

To conclude this section, let us calculate the coupling Gibbs energy $\Delta G(a|b)$ between two X ligands bound to the two sites (we call the sites a and b to distinguish them, but they are identical in this case). We need to compare the Gibbs energy $\Delta G^{o}_{a(b)}$ for binding to one site when the other is occupied with the Gibbs energy ΔG^{o}_{a} for binding to the free protein. From Figure 6.21, those Gibbs energies are

$$\Delta G^{o}_{a} = -RT \ln(\sigma K x)$$

$$\Delta G^{o}_{a(b)} = -RT \ln(K x / \sigma).$$

Hence,

$$\Delta G(a|b) = \Delta G^{o}_{a(b)} - \Delta G^{o}_{a} \tag{6.84}$$

$$= RT \ln \sigma^2 \tag{6.85}$$

$$= -2\varepsilon. \tag{6.86}$$

The coupling Gibbs energy is negative ($\varepsilon > 0$), which means that binding is rendered more favorable by the previously bound ligand.

A Sigmoidal Binding Isotherm is the Signature of a Cooperative System

We now derive the binding isotherm for this two-site cooperative system. The partition function can be written by looking at Figure 6.21,

$$Q = 1 + 2K\sigma x + K^2 x^2. \tag{6.87}$$

The factor of 2 in the statistical weight $2\sigma K x$ arises because there are two ways of producing a semi-occupied dimer: with the ligand on the left site or on the right site. The binding isotherm per site (factor of 1/2 because there are two sites) is

$$\theta = \left(\frac{1}{2}\right) \frac{2K\sigma x + 2K^2 x^2}{1 + 2K\sigma x + K^2 x^2} \tag{6.88}$$

$$= \frac{K\sigma x + K^2 x^2}{1 + 2K\sigma x + K^2 x^2}. \tag{6.89}$$

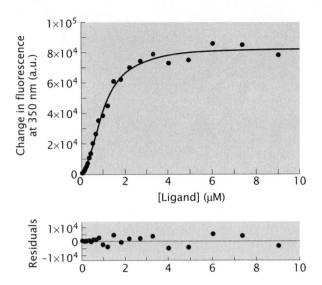

Figure 6.23 Fit of the two-site cooperative binding isotherm to the same data as in Figure 6.20. The solid line in the top panel represents Equation 6.89 (with an amplitude factor of 8.3×10^4 appended). The lower panel shows a plot of the residuals.

For the corresponding noncooperative system, the binding isotherm *per site* is

$$\theta = \frac{Kx}{1 + Kx}.$$

(6.90)

The ligand concentration x at the midpoint of the binding curves ($\theta = 1/2$) is the dissociation constant, $x = K_d = 1/K$. This is always valid for the noncooperative system (shown in Equation 6.4). It is also valid for this cooperative model with two binding sites, but not in general.

Now, let us use our model to describe the experimental data by fitting the binding isotherm to the data points. The plot of this cooperative binding isotherm is shown by the solid line in the top panel of **Figure 6.23**. The curve is *sigmoidal*. A sigmoidal isotherm plotted as a function of ligand concentration is the signature of cooperative binding. The fit yields $K = 1.0 \times 10^6 \, \mathrm{M}^{-1}$ ($K_d = 1.0 \, \mu\mathrm{M}$) and a cooperativity factor $\sigma = 0.01$. The slope of the curve initially increases, then there is an *inflection point* (at $x = 1 \, \mu\mathrm{M}$ in this case), and finally the slope decreases. In contrast, the slope of a hyperbolic (noncooperative) binding isotherm decreases continuously (see dashed line in Figure 6.20). In the residual plot in the bottom panel of Figure 6.23, the residues fall *randomly* about the zero line. Thus, the two-site cooperative model correctly describes the experimental data. This, however, does *not prove* that the model is correct, but only that it is *consistent* with the data.

Two-site Cooperative Binding can be Described by Another Model

Often there is more than one way of modeling a process. Sometimes models are equivalent, other times they are not. We will see later, in the case of hemoglobin, that some models are better than others. Because the case we are discussing is simple, this is a good occasion to make this point by writing an alternative model for cooperative two-site binding. Suppose that, instead of an unfavorable interaction between an occupied and an empty site, there is a *favorable* interaction when *both* sites are occupied. The diagram for this alternative model is shown in **Figure 6.24**.

Figure 6.24 Graph for an alternative model for two-site binding with cooperativity. Binding to the second site when the first is occupied releases an extra Gibbs energy ε ($\varepsilon > 0$, $\alpha = \exp(\varepsilon/RT) > 1$). The statistical weights are indicated below each state.

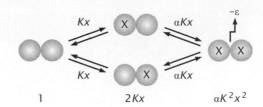

The partition function is

$$Q = 1 + 2Kx + \alpha K^2 x^2. \tag{6.91}$$

This time, binding to either subunit of the dimer when the other is empty occurs with the same binding constant K as for the isolated monomer. But if a second ligand binds, it does so with higher affinity, with a binding constant αK, where $\alpha = e^{\varepsilon/RT} > 1$. Mathematically, the two cooperative models are equivalent. We cannot distinguish them by a fit to experimental data. However, the underlying physical concepts are different. In the first model, the initial binding event changes the protein conformation or its flexibility or rigidity (order). In the second model, it does not. If we can examine the partially bound state $\otimes\bigcirc$, we can decide which model is more correct. Next, we study such an example in a protein with two Ca^{2+} binding sites.

Proteins Whose Structures Contain EF-hand Motifs Bind Calcium Ions Cooperatively

In Chapter 2 we described the EF-hand motif, which is a helix–loop–helix motif that binds a Ca^{2+} ion in the loop region (**Figure 6.25**). This motif occurs in a number of calcium-binding proteins, including parvalbumin, calpain, troponin C, and calbindin, but the best known is calmodulin.

The functional domain of EF-hand proteins consists of two EF-hand motifs. Calmodulin actually contains four EF-hand motifs (**Figure 6.26**). Calmodulin is a Ca^{2+} sensor of primary importance in

(A)   (B)

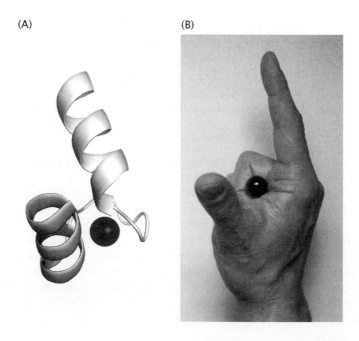

Figure 6.25 (A) The EF-hand motif of parvalbumin. The sphere represents a Ca^{2+} ion in its binding site (PDB 4CPV). (B) The author's EF-hand (B, courtesy of Antje Almeida).

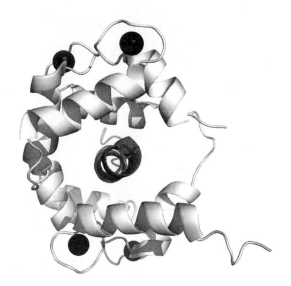

Figure 6.26 Upon cooperative binding of Ca^{2+} ions, calmodulin "wraps itself" around a helical segment of its target protein (darker gray, in the center). The spheres represent Ca^{2+} ions in the EF-hand motifs (PDB 2BBM).

signal transduction. Ca^{2+} is released into the cytoplasm from intra- or extracellular reservoirs. Calmodulin binds Ca^{2+} cooperatively, which allows it to switch from a Ca^{2+}-free to a Ca^{2+}-bound state within about 100-fold change of Ca^{2+} concentration, from $\sim 0.1\,\mu M$ to $\sim 10\,\mu M$. This constitutes the signal. Concomitant with Ca^{2+} binding, calmodulin undergoes a conformational change, exposing a hydrophobic patch of residues that bind to helical segments of its targets, which are other proteins down the signal transduction pathway. In essence, calmodulin "wraps itself" around a helical segment of its target (see Figure 6.26).

A simpler example of cooperative binding by an EF-hand motif is provided by the protein calbindin D_{9k}. Calbindin is not a sensor but a Ca^{2+} signal modulator. These proteins bind Ca^{2+} and control its concentration, modulating the intensity and duration of the Ca^{2+} concentration changes in the cytosol. Calbindin D_{9k} has two EF-hand motifs and binds Ca^{2+} cooperatively at two sites (**Figure 6.27**).

The two binding sites in calbindin D_{9k} are not identical. Site I is closer to the N-terminus of the protein, whereas site II is closer to the C-terminus. The EF-hand motif of site I is slightly different from the canonical type, whereas that of site II is canonical. The binding constants to the two sites, however, are not too different. As a consequence, it is not possible to study the bound states of each site separately in the wild-type protein. However, the mutant N56A behaves similarly to the wild type but its Ca^{2+}-binding constant to site II is ~ 100 times smaller. This allowed the determination of the structure of calbindin D_{9k}, with Ca^{2+} bound only to site I, by NMR. The structure revealed fascinating aspects. Upon binding of Ca^{2+} to site I, site II *becomes organized*, even though it has no Ca^{2+} bound, acquiring a conformation much more similar to the Ca^{2+}-bound state than to the empty state. The protein backbone in the loop region of site II *loses flexibility* when Ca^{2+} binds to site I.

This is the essence of cooperativity in calbindin D_{9k}. Binding of the first Ca^{2+} to site I is more difficult than it would be to the isolated site because this is accompanied by a loss of conformational flexibility in site II. That is, there is reduction of conformational entropy, which corresponds to $T\Delta S = -1.5\,\text{kcal/mol}$ at room temperature. Once that unfavorable negative entropy change has been paid, binding of the second Ca^{2+}, to site II, is rendered much easier. Therefore, binding of the first Ca^{2+} at site I enhances the Ca^{2+} affinity of site II.

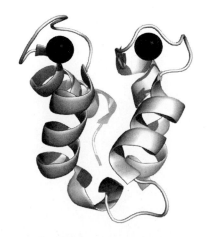

Figure 6.27 Structure of calbindin D_{9k}, with its two EF-hand motifs. The spheres represent Ca^{2+} ions (PDB 4ICB).

Figure 6.28 Diagram of Ca^{2+} binding to calbindin D_{9k}. The protein has two EF-hand motifs, each with a Ca^{2+}-binding site, called site I (N-terminal) and site II (C-terminal). The spheres represent Ca^{2+} ions. The binding constants and the Gibbs energy changes are indicated.

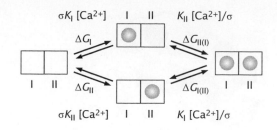

Let us translate this information into a diagram for the system and write the partition function. The diagram for Ca^{2+} binding to calbindin D_{9k} is shown in **Figure 6.28**. Now we write the partition function. Because binding is cooperative, we can choose one of the two models we developed, which are mathematically equivalent. But conceptually the first model seems more appropriate because we know from NMR that the structure of site II actually changes upon binding of Ca^{2+} to site I. Therefore, we write the partition function as

$$Q = 1 + \sigma(K_I + K_{II})[Ca^{2+}] + K_I K_{II}[Ca^{2+}]^2. \tag{6.92}$$

Measuring the fractional binding as a function of Ca^{2+} would not allow us to uniquely determine all three parameters (σ, K_I, and K_{II}). However, because $K_I \gg K_{II}$ in the mutant N56A, it was possible to estimate the product $\sigma K_I \approx 10^7 \, M^{-1}$. Further, measurements of Ca^{2+} binding to the wild-type protein and binding of other metal ions (Ca^{2+} mimics) to the mutant protein allow the estimate of the coupling free energy (shown in Equation 6.84) as

$$\Delta G(I|II) = \Delta G_{II(I)} - \Delta G_{II} = -1.5 \text{ kcal/mol}. \tag{6.93}$$

This is a small energy: cooperativity does not require very large interactions. With $\Delta G(I|II) = -1.5 \text{ kcal/mol}$, we obtain $\sigma = 0.28$. Based on experimental binding data at low salt conditions, and using $\sigma = 0.28$, the binding constants are approximately $K_I = 4.4 \times 10^6 \, M^{-1}$ and $K_{II} = 1.6 \times 10^6 \, M^{-1}$ for the wild-type protein. Thus, the two *microscopic* binding constants to sites I and II of calbindin D_{9k} are not very different.

A significant contribution to the unfavorable interaction ε is the decrease in entropy arising from the ordering of site II of the protein. Thus, the unfavorable "interaction" ε may represent *a decrease in entropy*. In fact, there are many examples in which the interactions that give rise to cooperativity stem from changes in *conformational entropy*, manifested structurally by increases or decreases in protein flexibility or backbone order.

Now we examine the physiological consequences of the binding affinity of Ca^{2+} to calbindin D_{9k}. In a signaling event, the Ca^{2+} concentration in the cytosol changes from $\sim 10^{-7}$ to $\sim 10^{-5} \, M$, which completely changes the distribution of calbindin D_{9k} (**Figure 6.29**). When $[Ca^{2+}]$ changes from 10^{-7} to $10^{-6} \, M$, the protein essentially switches from empty to fully bound over a 10-fold change in Ca^{2+} concentration.

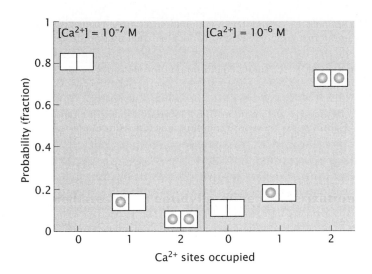

Figure 6.29 Probabilities (or fractions) of the different species of calbindin D_{9k}, with zero, one, or two Ca^{2+} ions bound, when the cytosol Ca^{2+} concentration changes from $\sim 10^{-7}$ to $\sim 10^{-6}$ M.

6.4 MACROSCOPIC BINDING CONSTANTS ARE COMBINATIONS OF MICROSCOPIC CONSTANTS AND COOPERATIVITY FACTORS

When analyzing experimental binding data for proteins with two or more binding sites, we have two choices. First, we may have some evidence that a certain microscopic binding model is appropriate, which expresses binding in terms of *microscopic* binding constants and interaction parameters defined for each step. This could be one of the two cooperative binding models derived earlier. Then we fit the corresponding binding isotherm to the data. The second choice is to fit the data using *macroscopic* binding constants, and later worry about what those constants mean. If practical, the first option is preferable, but often the second approach is used. We are then left with the problem of interpreting those macroscopic constants. In Chapter 2, we encountered this problem when writing partition functions for proton binding to amino acids. We said, then, that the first and second macroscopic dissociation constants were, to a very good approximation, identical to the *microscopic* dissociation constants (pK_a). We return to the problem of proton binding to amino acids here and will have an opportunity to examine our assumption. But this is a good point to consider macroscopic binding constants more formally. We will illustrate it with the binding of Ca^{2+} to calbindin D_{9k}.

Let us define the two macroscopic binding constants for Ca^{2+} binding to calbindin D_{9k} (CBD), for the two equilibria

$$CBD + Ca^{2+} \underset{}{\overset{K_1}{\rightleftharpoons}} CBD \cdot Ca^{2+} \tag{6.94}$$

$$CBD \cdot Ca^{2+} + Ca^{2+} \underset{}{\overset{K_2}{\rightleftharpoons}} CBD \cdot (Ca^{2+})_2, \tag{6.95}$$

by

$$K_1 = \frac{[CBD \cdot Ca^{2+}]}{[CBD][Ca^{2+}]} \tag{6.96}$$

$$K_2 = \frac{[CBD \cdot (Ca^{2+})_2]}{[CBD \cdot Ca^{2+}][Ca^{2+}]}. \tag{6.97}$$

Using these equations, we can write something that looks like a partition function as

$$P = 1 + K_1[Ca^{2+}] + K_1K_2[Ca^{2+}]^2. \qquad (6.98)$$

We call this a *binding polynomial*. It has the same form of the binding partition function, but uses macroscopic binding constants. Indeed, you can derive it in exactly the same way as a partition function, by building the diagram of this system from Equations 6.94 and 6.95. The binding polynomial contains the same information as the partition function, but there is no interpretation (no model) attached to it; it is just empirical. Yet you can use it to write a binding isotherm and fit data just as we did with partition functions. In this case we would write

$$\theta = \left(\frac{1}{2}\right) \frac{K_1[Ca^{2+}] + 2K_1K_2[Ca^{2+}]^2}{1 + K_1[Ca^{2+}] + K_1K_2[Ca^{2+}]^2}. \qquad (6.99)$$

The first macroscopic binding constant is $K_1 = 1.8 \times 10^6 \, M^{-1}$ and the second is $K_2 = 3.9 \times 10^6 \, M^{-1}$. Macroscopic binding constants are obtained experimentally as the apparent constants for binding the first and the second Ca^{2+} ions, but we do not know which site each ion binds to. Next, we use the partition function to interpret the macroscopic binding constants in terms of microscopic binding constants and interactions between sites. If we compare the partition function (shown in Equation 6.92),

$$Q = 1 + \sigma(K_I + K_{II})[Ca^{2+}] + K_I K_{II}[Ca^{2+}]^2,$$

with the binding polynomial (shown in Equation 6.98), we see what the macroscopic binding constants are in terms of the microscopic ones,

$$K_1 = \sigma(K_I + K_{II}) \qquad (6.100)$$

$$K_1 K_2 = K_I K_{II}. \qquad (6.101)$$

However, from this information alone we cannot determine the three parameters σ, K_I, and K_{II} (we have three unknowns and only two equations).

Now suppose the microscopic binding constants are identical ($K_I = K_{II} = K$). Using the first cooperative model, the partition function is

$$Q = 1 + 2\sigma K[Ca^{2+}] + K^2[Ca^{2+}]^2. \qquad (6.102)$$

Comparing the binding polynomial (shown in Equation 6.98) with the partition function (shown in Equation 6.102), we obtain

$$K_1 = 2\sigma K \qquad (6.103)$$

$$K_1 K_2 = K^2. \qquad (6.104)$$

Now σ and K can be determined uniquely.

Finally, suppose $\sigma = 1$, as if binding were not cooperative. Then,

$$K_1 = 2K \qquad (6.105)$$

$$K_1 K_2 = K^2, \qquad (6.106)$$

which yields

$$K_1 = 2K \qquad (6.107)$$

$$K_2 = K/2. \qquad (6.108)$$

The microscopic constant K refers to binding to an individual site. The first macroscopic constant is twice the microscopic one because there are two ways of binding (to site I or to site II) and only one of dissociating. The second macroscopic constant is one-half of the microscopic constant because there is only one way of binding and two ways of dissociating (from site I or II). Note, however, that in this case a titration experiment would yield only one constant, K.

6.5 NEGATIVE COOPERATIVITY OCCURS WHEN BINDING OF THE FIRST LIGAND IMPAIRS BINDING OF THE SECOND

The kind of cooperativity we have discussed so far is called *positive cooperativity*: binding of a first ligand makes binding of a second *more favorable*. There is also *negative cooperativity*, in which binding of the first ligand makes binding of a second *less favorable*. This kind of cooperativity is not as common in proteins, but it is common in the familiar case of proton binding to amino acids.

Consider proton (H^+) binding to glycine in the diagram of **Figure 6.30**. There are two proton binding sites: the amino group (NH_2) and the carboxylate group (COO^-). We call the amino group site A and the carboxylate group site B. The corresponding *binding constants* for the proton are K_A and K_B. When we discussed acid–base reactions of amino acids in Chapter 2, we used *dissociation constants* instead of binding (association) constants because the pK_a is defined as $-\log K_a$, where K_a is the *acidity* constant, or the proton dissociation constant. Note that K_a is the *reciprocal* of the binding constant (K_A or K_B). It makes no difference whether we use binding or dissociation constants. You just need to be sure of which kind are used. Look at the units if you are not sure: M^{-1} for binding constants versus M (or more often its submultiples, mM, μM, nM) for dissociation constants. This is one reason it is useful to indicate units on equilibrium constants: it immediately tells you what kind of constants they are and what the standard state is (1 M in this case).

The *intrinsic* affinity of the NH_2 group for the proton ($K_A = 5 \times 10^9\,M^{-1}$) is much larger than that of the COO^- group ($K_B = 2 \times 10^4\,M^{-1}$). The probability is thus overwhelming that the first proton will bind to the amino group. This is why we said in Chapter 2 that the first macroscopic binding constant ($K_1 = K_A + K_B$) is essentially the microscopic constant of the amino group: $K_1 \approx K_A$.

The pK_a of the NH_3^+ group in glycine is ≈ 9.7. This is equal to $\log K_A = \log(5 \times 10^9)$, as expected. However, the pK_a of the COOH group in glycine is about 2.3, which is equal to $\log(2 \times 10^2)$, not $\log(2 \times 10^4)$,

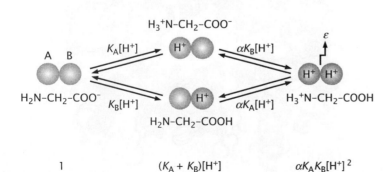

Figure 6.30 Diagram for proton binding to the amino acid glycine.

as would be expected for the microscopic pK_a of COOH. They differ by a factor of 100. Why? Binding of H^+ to the amino group makes it 100 times more difficult for a second H^+ to bind, to the carboxylate. This is because binding of the first H^+ converts NH_2 to NH_3^+, which now exerts an electrostatic repulsion on the second H^+. Thus, the apparent binding constant to the COO^- group is 100 times smaller than its intrinsic microscopic binding constant. We can write the partition function for proton binding to glycine as

$$Q = 1 + (K_A + K_B)[H^+] + \alpha K_A K_B [H^+]^2, \qquad (6.109)$$

where $\alpha = 10^{-2}$ in this case. This parameter is related to the unfavorable electrostatic free energy by a Boltzmann factor, as in Equation 6.83,

$$\alpha = e^{-\varepsilon/RT}. \qquad (6.110)$$

The binding constant to the COO^- group when a proton is already bound on the NH_3^+ group is $\alpha K_B = 2 \times 10^2 \, M^{-1}$ (upper branch in Figure 6.30).

To conclude, let us calculate the coupling Gibbs energy of a proton bound at site A and another at site B (shown in Equation 6.84):

$$\Delta G(A|B) = -RT(\ln(\alpha K_B[H^+]) - \ln(K_B[H^+])) \qquad (6.111)$$

$$= -RT \ln \alpha \qquad (6.112)$$

$$\approx 2.7 \, \text{kcal/mol} \qquad (6.113)$$

at room temperature (using $\alpha = 10^{-2}$). The coupling energy is positive, because binding of the first proton makes binding of the second *less* favorable.

6.6 HEMOGLOBIN BINDS OXYGEN COOPERATIVELY

Hemoglobin is the most studied and best understood cooperative system in biochemistry. Hemoglobin is a tetramer, $\alpha_2 \beta_2$, of two types of subunits, α and β (**Figure 6.31**).

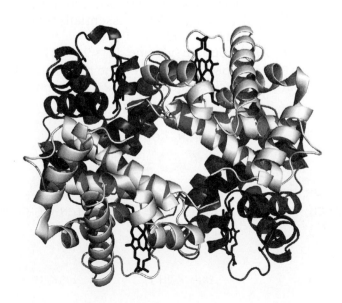

Figure 6.31 Hemoglobin tetramer in the deoxy T state, showing the two α subunits (light) and the two β subunits (dark). The heme groups are shown as stick models, in black (PDB 1A3N).

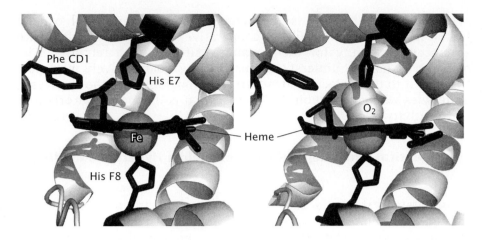

Figure 6.32 Center of the α chain of human hemoglobin, showing the heme group with its iron metal ion. (A) Deoxyhemoglobin, T state; (B) oxyhemoglobin, R state. His F8 directly coordinates the iron, whereas His E7 binds oxygen from the opposite side of heme in oxyhemoglobin. Phe CD1 interacts with the heme group (PDB 1A3N and 1HHO).

The α and β subunits are not identical but very similar. Each subunit binds one oxygen molecule (O_2) at the heme group. As we have seen when we studied the evolution of globins, the tertiary structure of each hemoglobin protomer (monomer) is extremely similar to that of myoglobin (see Figure 6.4). In deoxyhemoglobin, the iron ion (Fe^{2+}) is pentacoordinated by four groups of the heme and by the proximal histidine (His F8) from helix F (**Figure 6.32**A). The iron is out of the plane of the heme, "pulled down" by His F8. In oxyhemoglobin (see Figure 6.32B), O_2 occupies the sixth coordination position of Fe^{2+} and binds to His E7 from helix E (distal histidine) on the other side.

The Binding Isotherm of Oxygen to Hemoglobin is Sigmoidal

Christian Bohr (father of the physicist Niels Bohr) first measured oxygen binding to hemoglobin. He discovered that the binding isotherm was sigmoidal (**Figure 6.33**). At that time, however, it was not known how many oxygen binding sites existed in hemoglobin. Bohr thought there were two and on this basis proposed a model to explain the sigmoidal curve. It was Gilbert Adair who later found that each hemoglobin contains four binding sites, which he inferred from a determination of the stoichiometry of the number of iron metal ions per hemoglobin molecule.

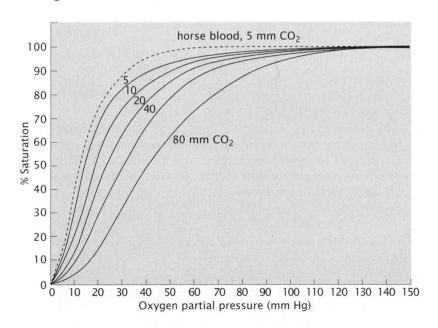

Figure 6.33 Binding of oxygen to hemoglobin as a function of the partial pressure of oxygen, in the presence of various fixed partial pressures of CO_2. The percent saturation was determined by measuring the mass of gas bound by the blood. (Adapted from Bohr C, Hasselbalch K & Krogh A [1904]. *Skand Arch Physiol* 16:402–412. With permission from John Wiley & Sons.)

The sigmoidal binding isotherm indicates that binding of oxygen to the hemoglobin molecule is cooperative: binding to the first site enhances its affinity for the remaining sites in each tetramer. The question is how to understand this binding quantitatively, and how to relate it to the molecular structure of hemoglobin and the changes it undergoes when oxygen binds. That is, what is the molecular origin of cooperativity?

Four Identical and Independent Sites Yield a Hyperbolic Binding Isotherm

To understand the molecular origin of cooperativity in the binding of oxygen to hemoglobin, we will need a plausible binding model. We will also need to relate the cooperativity embodied in the model to its physical origin in the molecular structure of the protein. However, before we study the cooperative models, we will take one step back and examine the corresponding case without cooperativity. What would happen if hemoglobin had four binding sites that were identical and independent? (In truth, the sites are almost identical, because the α and β subunits are very similar, but not independent.) **Figure 6.34** shows the diagram for this noncooperative case.

Using the empty state as reference, and writing x for the oxygen concentration and K for the binding constant, we obtain the partition function by inspection of the diagram (see Figure 6.34):

$$Q = 1 + 4Kx + 6K^2x^2 + 4K^3x^3 + K^4x^4. \tag{6.114}$$

The numbers in front of each term in the partition function (1, 4, 6, 4, 1) are the *multiplicities*, or *degeneracies*, of each state. They are the numbers of ways of drawing each state (not all shown in Figure 6.34): There is only one way of drawing the empty state; there are four ways of drawing the state with one ligand bound; six ways of drawing the state with two ligands; four ways for the state with three; and only one way for the state with four ligands.

These degeneracies are the coefficients of the *binomial expansion* of $[1 + (Kx)]^4$. Indeed, this is just another way of writing the partition function,

$$Q = (1 + Kx)^4 \tag{6.115}$$

$$= (1 + Kx)(1 + Kx)(1 + Kx)(1 + Kx). \tag{6.116}$$

The last equality shows explicitly that the partition function is a product of four individual partition functions—one for each site. This applies whenever the binding sites are independent.

The multiplicities are obtained from the *combinatorial formula*, which we encountered in Chapter 1 in connection with the number of ways of arranging black and clear marbles in a given number of slots. Here it represents the number of ways of arranging p ligands in n binding sites. That number is

$$\Omega_{n,p} = \frac{n!}{(n-p)!p!}, \tag{6.117}$$

Figure 6.34 Model for binding of oxygen (X) to hemoglobin if the sites were identical and independent.

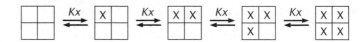

where n is the total number of sites ($n = 4$ in this case) and p is the number of occupied sites. $\Omega_{n,p}$ gives the number in front of each term in Equation 6.114 as p varies ($p = 0$, 1, 2, 3, or 4, in this case). You can also obtain the multiplicities from *Pascal's triangle*:

$$
\begin{array}{ccccccccc}
 & & & & 1 & & & & \\
 & & & 1 & & 1 & & & \\
 & & 1 & & 2 & & 1 & & \\
 & 1 & & 3 & & 3 & & 1 & \\
1 & & 4 & & 6 & & 4 & & 1 \\
 & & & & \cdots & & & &
\end{array}
$$

Each term in a line of the triangle is obtained by adding the terms above it, to its left and to its right, in the preceding line. For example, in the fifth line, $4 = 1 + 3$.

The binding isotherm for this case is (shown in Equation 6.21),

$$
\begin{aligned}
\theta &= \left(\frac{1}{4}\right) \frac{d \ln Q}{d \ln x} \qquad\qquad (6.118) \\
&= \left(\frac{1}{4}\right) \frac{x}{Q} \frac{d Q}{d x} \\
&= \left(\frac{1}{4}\right) \frac{x}{[1 + (Kx)]^4} \frac{d\,[1 + (Kx)]^4}{d x} \\
&= \left(\frac{1}{4}\right) \frac{x}{[1 + (Kx)]^4} \left(4K\,[1 + (Kx)]^3\right) \\
&= \frac{Kx}{1 + Kx}.
\end{aligned}
$$

Thus, it simplifies to exactly the same as the binding isotherm for one site, which is hyperbolic. This is because the sites are independent. Since they *do not interact*, it does not matter if they are isolated or part of a tetramer.

The Model of Pauling or Koshland–Nemethy–Filmer Explains Cooperativity by Unfavorable Interactions between Occupied and Empty Sites

Linus Pauling (1935) proposed a molecular mechanism that explained the cooperativity observed in oxygen binding by an unfavorable inter-action between adjacent subunits that differ in having an O_2 molecule bound or not. This is the same concept we used in the first model of cooperativity for a protein with two binding sites, now applied to a protein with four sites. Thirty years later, this model was further developed by Koshland, Nemethy, and Filmer, and is now usually des-ignated by KNF or *sequential model*. The diagram of the Pauling/KNF model is shown in **Figure 6.35**.

Each *subunit* can exist in two states, T (for taught or tense) and R (for relaxed). When a subunit binds oxygen, *that* subunit changes

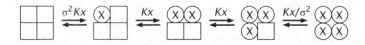

Figure 6.35 The Pauling/Koshland–Nemethy–Filmer (KNF) model for binding of oxygen (X) to hemoglobin. The T state is represented by squares and the R state by circles.

conformation from the T state to the R state. The interaction between adjacent subunits in *different* conformations is unfavorable, because the *interface* is associated with a free energy $\varepsilon > 0$, which increases the Gibbs energy of this mixed-conformation state. The unfavorable interaction at the interface brings in the Boltzmann factor (shown in Equation 6.83),

$$\sigma = e^{-\varepsilon/RT},$$

which lowers the probability of mixed states (T/R).

In the particular model used here, we consider that each ligand binds with an intrinsic affinity given by the binding constant K; when a T/R interface appears, an additional factor σ is included to represent the higher energy of contact between unlike conformations. Because $\sigma \ll 1$, states with σ factors have a much lower probability. This applies to all bound states except the fully occupied state. Going from zero to one ligand brings in a factor of σ^2, whereas no *new* σ factors arise when a second and a third ligand are bound (the number of interfaces remains constant). At high concentration of oxygen, the factors Kx favor states with two and three ligands relative to that with only one ligand, but the full benefit only appears when the protein is fully occupied. As the unfavorable interfaces disappear, so do the σ factors in the final state.

The partition function can easily be written by inspection of the diagram of Figure 6.35 using the empty T state as reference, in a form analogous to Equation 6.114,

$$Q = 1 + 4\sigma^2 Kx + 6\sigma^2 K^2 x^2 + 4\sigma^2 K^3 x^3 + K^4 x^4. \tag{6.119}$$

Strictly speaking, the middle term, $6\sigma^2 K^2 x^2$, is not correct because if the two X (oxygen molecules) are placed across the diagonal, then we have four interfaces, not two. This term should then be replaced by $(4\sigma^2 K^2 x^2 + 2\sigma^4 K^2 x^2)$. However, the expression depends on the exact geometrical arrangement of the subunits. For example, square (shown in Figure 6.35) and tetrahedral arrangements yield different mathematical expressions. But this is not important for the concept and we will ignore these complications. Furthermore, because this state is so unfavorable, this term is small and the difference between the correct and the approximate terms we used is very small compared to the larger terms in the partition function. The approximation makes a negligible difference in the fit of the binding isotherm to the experimental data.

The partition function is obtained by multiplying the factors written above the arrows in the diagram of Figure 6.35 as you move from one state to the next. There is no *new* σ factor added in going from the state with one ligand to the state with two ligands, but there is a σ^2 factor in the third term in the partition function, which carries over from the state with two ligands. This is because, to get to the third state, you multiply $1 \times \sigma^2 Kx \times Kx = \sigma^2 K^2 x^2$, and so on. In the end, after you figure out the energetic part of the statistical weights, you need to append the multiplicities (or degeneracies) of the states that have the same Gibbs energy, as in Equation 6.114. Comparing Equations 6.114 and 6.119, you can see that they differ only in the factors of σ^2, present in the middle terms of Equation 6.119, which decrease the probability of the intermediate states.

The binding isotherm is obtained by choosing the terms that have ligand bound and multiplying each one by its number of ligands (shown in Equation 6.8). Alternatively you can use the differentiation method (shown in Equation 6.21), to obtain

$$\theta = \left(\frac{1}{4}\right) \frac{4\sigma^2 Kx + 12\sigma^2 K^2 x^2 + 12\sigma^2 K^3 x^3 + 4K^4 x^4}{1 + 4\sigma^2 Kx + 6\sigma^2 K^2 x^2 + 4\sigma^2 K^3 x^3 + K^4 x^4} \qquad (6.120)$$

$$= \frac{\sigma^2 Kx + 3\sigma^2 K^2 x^2 + 3\sigma^2 K^3 x^3 + K^4 x^4}{1 + 4\sigma^2 Kx + 6\sigma^2 K^2 x^2 + 4\sigma^2 K^3 x^3 + K^4 x^4}. \qquad (6.121)$$

In summary, cooperativity arises in the Pauling/KNF model because the σ factors reduce the probability of the intermediate binding states, so the most likely are the totally empty and the fully bound states (which have no σ factors).

The Model of Monod–Wyman–Changeux Explains Cooperativity by an Oxygen-Induced Quaternary Change that Converts Low-affinity Sites to High-affinity Sites

Perutz (1964) showed that the quaternary structures of deoxyhemoglobin (Hb) and oxyhemoglobin (Hb(O$_2$)$_4$) differ by the mutual orientation of the two αβ dimers in such a way that the β subunits move closer to each other (**Figure 6.36**). Thus, there is a conformational change, but unlike in the Pauling model, it involves the *entire tetramer*. The individual monomers may *also* change their conformation, but this was not apparent in the initial low-resolution structures obtained by Perutz.

Monod, Wyman, and Changeux recognized this *symmetry* in the structures of oxy- and deoxyhemoglobin as a switch between two quaternary conformations, from a tetramer with low affinity to a tetramer with high affinity for oxygen. The model they proposed is designated by MWC, symmetry model, or concerted model. It produces cooperativity by a mechanism that is completely different from the Pauling/KNF model. It considers that the protein has two *quaternary conformational states*, T and R. Now *all subunits* are either in one or the other state. The quaternary conformational change between T and R is called the *allosteric transition*.

The term *allosteric* refers to the effect that the binding of a ligand at a given site has on *another site* in the protein. In hemoglobin, binding of an oxygen molecule to the heme in one of the subunits changes (increases) the affinity of the hemes at the other subunits for other oxygen molecules. Since this *communication* at a distance must happen through a conformational change of the protein, the T-to-R conformational transition in hemoglobin is called an allosteric transition. Because the ligand that induces the affinity change (oxygen) is chemically identical to the ligand whose binding to another site is affected, we call this a *homotropic* interaction. If the two ligands are different, it is called *heterotropic*.

Binding of oxygen to one of the subunits makes it more likely that they all switch from T to R. If some of the subunits in that same hemoglobin molecule are empty, their affinity for oxygen increases. Therefore, other oxygen molecules are much more likely to bind to the sites that are still free. The essence of the model is that when the allosteric transition occurs, *high affinity sites are created* from low-affinity ones. This is how cooperativity enters the model. Conversely, release of oxygen by one subunit makes it more likely that they all switch from R to T, reversing the process in an equally cooperative manner. The MWC model is represented in **Figure 6.37**, where X represents O$_2$.

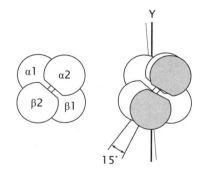

Figure 6.36 Hemoglobin is a tetramer of two types of protomers, two α and two β. When oxygen binds, the protein undergoes a quaternary conformational change, with a rotation of one αβ unit relative to the other by 15° around an axis perpendicular to the paper. Y is the twofold symmetry axis. (Adapted from Baldwin J & Chothia C [1979] *J Mol Biol* 129:175–220. With permission from Elsevier.)

Figure 6.37 The Monod–Wyman–Changeux (MWC) model for binding of oxygen (X) to hemoglobin. The T state is represented by squares and the R state by circles.

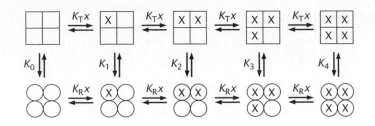

The model looks complicated but the partition function is easily written. (Note that the multiplicities of the microstates are *not* shown in Figure 6.37.) Let us choose the empty T state (T_0) as the reference and assign it a statistical weight of 1. Binding within the T state is simple binding to identical and independent sites, characterized by a binding constant K_T. We have studied this case before. The partition function *within* the T state is just a product of single-site, identical partition functions, $(1 + K_T x)^4$, where x is the oxygen concentration. Similarly, *within* the R state, binding occurs also to identical and independent sites, but with a *larger* binding constant K_R. The contribution of binding within the R state to the partition function is $(1 + K_R x)^4$.

The transition between empty T and R states (T_0 and R_0) is described by the allosteric transition constant K_0,

$$K_0 = \frac{[R_0]}{[T_0]}. \tag{6.122}$$

To go from the T- to the R-branch, you first go down, then horizontally (see Figure 6.37). The complete partition function is

$$Q = (1 + K_T x)^4 + K_0 (1 + K_R x)^4. \tag{6.123}$$

(You could choose other paths to write the partition function, but it would look more complicated.) If you expand the terms in parentheses in Equation 6.123, you can identify each resulting term as representing a different binding state. The binding isotherm is most easily obtained by differentiating the partition function according to Equation 6.21,

$$\theta = \left(\frac{1}{4}\right) \frac{d \ln Q}{d \ln x}$$

$$= \left(\frac{1}{4}\right) \frac{x}{Q} \frac{dQ}{dx}$$

$$= \left(\frac{1}{4}\right) \frac{4 K_T x (1 + K_T x)^3 + 4 K_0 K_R x (1 + K_R x)^3}{(1 + K_T x)^4 + K_0 (1 + K_R x)^4}$$

$$= \frac{K_T x (1 + K_T x)^3 + K_0 K_R x (1 + K_R x)^3}{(1 + K_T x)^4 + K_0 (1 + K_R x)^4}. \tag{6.124}$$

At low concentration of oxygen (x), the protein is mainly in the empty T state because K_0 is small. Initially, oxygen binding is difficult because the T state has low affinity for it (K_T is small). However, if one or two sites become occupied, there is a higher chance that the protein will undergo the allosteric transition, changing to the R conformation. The *apparent* allosteric transition constant between all forms of the T state and all forms of the R state in the presence of oxygen is

$$K_0^{\text{app}} = \frac{[\text{All R-state forms}]}{[\text{All T-state forms}]}. \tag{6.125}$$

The terms corresponding to the T and R forms can be obtained from the partition function of the MWC model (shown in Equation 6.123):

$$\text{All T-state forms} \rightarrow (1 + K_T x)^4$$

$$\text{All R-state forms} \rightarrow K_0(1 + K_R x)^4.$$

Therefore,

$$K_0^{\text{app}} = K_0 \frac{(1 + K_R x)^4}{(1 + K_T x)^4}. \tag{6.126}$$

Since $K_R \gg K_T$, K_0^{app} increases with the oxygen concentration (x), thus making the allosteric transition to the R state more likely. Once this happens, the newly transformed empty sites have a high affinity for oxygen (K_R is large) and they get promptly occupied. This is why the system exhibits cooperativity.

Comparison with Experiment is the Test of Competing Models

Figure 6.38 shows an isotherm for oxygen binding to hemoglobin. The data points are experimental, the solid line represents the MWC model, and the dashed line represents the Pauling/KNF model (the difference is only noticeable in the beginning of the curves).

In each case the parameters have been chosen to obtain the best fit to the data points. In the MWC model, $K_0 \approx 10^{-5}$, which is very small because in the absence of oxygen the T state is much more favorable than the R state. Oxygen binds much better to the R state than to the T state; hence, $K_T \approx 10^4 \, \text{M}^{-1}$ and $K_R \approx 10^6 \, \text{M}^{-1}$. In the Pauling/KNF model, binding to the tertiary R state conformation is characterized by $K \approx 10^5 \, \text{M}^{-1}$ and the unfavorable interaction between an empty (T) and an occupied (R) subunit is represented by the cooperativity factor $\sigma \approx 0.2$, corresponding to an interaction Gibbs energy of $\varepsilon = -RT \ln \sigma \approx 1 \, \text{kcal/mol}$ at room temperature. This is a small value, of the order of magnitude of the thermal energy (RT). However, it is sufficient to ensure cooperativity. We have seen this pattern repeatedly: interaction free energies are small in biochemical systems.

We have obtained the partition function and evaluated its parameters (binding constants and interaction energies). Now, what is the distribution of the various binding states of hemoglobin? **Figure 6.39** shows the probabilities (fractions) of each state for the MWC and Pauling/KNF models in the middle of the transition between T_0 and R_4, at

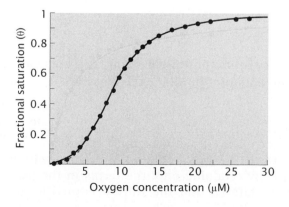

Figure 6.38 Binding isotherm of oxygen to hemoglobin. The fractional saturation was obtained by measuring the change in the absorption of the solution as a function of oxygen pressure. The solid line is a fit of the MWC model (shown in Equation 6.124) to the data ($K_0 = 9.75 \times 10^{-6}$, $K_T = 1.12 \times 10^4 \, \text{M}^{-1}$, $K_R = 2.07 \times 10^6 \, \text{M}^{-1}$). The dashed line is a fit of the Pauling/KNF model (shown in Equation 6.120) to the data ($K = 1.14 \times 10^5 \, \text{M}^{-1}$, $\sigma = 0.21$). (Data from Mills FC, Johnson ML & Ackers GK [1976] *Biochemistry* 15:5350–5362 [not all data points are shown for clarity].)

Figure 6.39 Distributions of the binding states of hemoglobin at the midpoint of the transition in the MWC and Pauling/KNF models. The probabilities of each state are calculated with the parameters given in the legend of Figure 6.38 at $\theta = 1/2$.

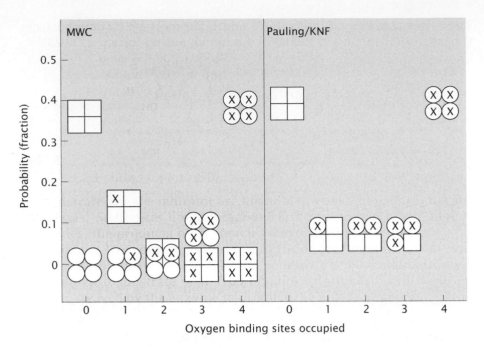

the midpoint of the curve in Figure 6.38. In either model the intermediate binding states are almost absent. The population consists almost exclusively of proteins in the T state with no ligand bound and proteins in the R state with four ligands bound. That is, the transition is *almost all-or-none*.

The All-or-none Approximation

The all-or-none model represents the extreme case of the MWC and KNF models (**Figure 6.40**). Here only the empty T state and the fully bound R state are preserved. All the other states are considered so unfavorable that they are ignored. The system then goes from the empty T to the saturated R state, $T_0 \rightleftharpoons R_4$.

The partition function of the all-or-none model is especially simple. To derive it from the MWC model, we consider that the protein undergoes a transition from T to R and binds all four ligands. Using the T state as reference, K_0 represents the constant for the equilibrium $T \rightleftharpoons R$. We use K_R to represent the binding constant of oxygen to each site on hemoglobin, in the R state, as in the MWC model. The partition function can be written as

$$Q = 1 + K_0 K_R^4 x^4. \tag{6.127}$$

Since K_0 always appears multiplied by K_R^4 and these are all constants, we can define an apparent binding constant by $K = K_R \sqrt[4]{K_0}$ and write

$$Q = 1 + K^4 x^4. \tag{6.128}$$

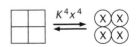

Figure 6.40 The all-or-none model for binding of oxygen (X) to hemoglobin. There are no intermediate states. The T state is represented by squares and the R state by circles.

This is also the result we would obtain if we started with the KNF model, and simplified it by retaining only the terms in the partition function for the empty T state (1) and the fully occupied R state ($K^4 x^4$). The binding isotherm is also very simple:

$$\theta = \frac{(Kx)^4}{1 + (Kx)^4}. \tag{6.129}$$

A.V. Hill proposed a variation on this model that has been extensively used, because of its simplicity. Hill's approximation consists essentially in reinterpreting the exponent in Equation 6.129 (4 in this case, but in general equal to the number of binding sites n), which becomes a phenomenological parameter n_H, called the *Hill coefficient*. The partition function becomes

$$Q = 1 + (Kx)^{n_H} \tag{6.130}$$

and the binding isotherm is

$$\theta = \frac{(Kx)^{n_H}}{1 + (Kx)^{n_H}}. \tag{6.131}$$

The Hill coefficient is used as a parameter in fitting this binding isotherm to experimental data; n_H can take any value, including non-integer values, up to the number of binding sites on the protein. This model yields binding isotherms similar to those generated by the MWC and Pauling/KNF models. The closer the value of n_H to the number of binding sites, the more cooperative the system is. In the case of hemoglobin, $n_H = 4$ would correspond to maximal cooperativity, whereas $n_H = 1$ would corresponds to no cooperativity. Because of its simplicity, the model is very popular. However, as a phenomenological model, it lacks the ability to provide a detailed interpretation of data in terms of interactions at the molecular level. The Hill coefficient n_H does not have a rigorous physical meaning, unlike σ (shown in Equation 6.83); it is not simply related to an energy such as ε.

If we rearrange Equation 6.131 to

$$\frac{\theta}{1 - \theta} = x^{n_H} K^{n_H}, \tag{6.132}$$

and take the logarithm of both sides,

$$\log \frac{\theta}{1 - \theta} = n_H \log x + n_H \log K, \tag{6.133}$$

we obtain the so-called Hill equation (shown in Equation 6.133). A plot of the Hill equation as a function of $\log x$ yields a straight line with slope n_H. In practice, the slope is not constant over the entire range of $\log x$. Its value is 1 at the beginning and at the end of the binding isotherm (why?), and acquires its maximum value at the midpoint of the curve. The value at the midpoint indicates the degree of cooperativity of the binding process.

We can also replace the ligand concentration (x) by the pressure of oxygen (pO_2), and replace K by $1/P_{50}$. P_{50} is the pressure of oxygen that results in half saturation of the hemoglobin, and is equivalent to a dissociation constant K_d, but is expressed in terms of oxygen partial pressure, instead of concentration. The Hill equation becomes

$$\log \frac{\theta}{1 - \theta} = n_H \log pO_2 - n_H \log P_{50}. \tag{6.134}$$

The Structural Changes and Oxygen Affinities Observed Experimentally Indicate that the MWC Model is Essentially Correct

Now, which model, MWC or Pauling/KNF, is correct for hemoglobin? The fit of the MWC model to the data is slightly better (see Figure 6.38), but this is hardly convincing. Besides, the MWC model has one more parameter than the Pauling/KNF model, and adding parameters always improves the ability of a model to fit the experiment. To answer this question we must turn to the protein structure.

Perutz (1970) proposed a molecular interpretation of cooperativity in hemoglobin, which is still largely valid. In the absence of oxygen, the quaternary T state is the most stable. The movements that occur in each subunit when oxygen binds relative to the T state are indicated by the arrows in **Figure 6.41**A. The Fe^{2+} Moves into the plane of the heme, pulling His F8 with it. This moves helix F and the FG corner. The resulting conformation is shown in Figure 6.41B. Those movements change the interactions between dimers $\alpha_1\beta_1$ and $\alpha_2\beta_2$ at the α_1/β_2 and α_2/β_1 interfaces. The packing at these protein–protein interfaces becomes unfavorable; they became high-energy regions. This unfavorable interaction is relieved when the entire tetramer undergoes the allosteric transition, from the T quaternary state to the R quaternary state. Then the oxygen binding sites change from low to high affinity.

The *communication* between one subunit that just bound oxygen and another that is still empty, such as the α_1 and β_2 subunits, occurs at their contact interface and has two physical origins. The first is steric strain due to poor packing at the interface when one subunit has oxygen bound and the other does not. Ligand binding at the heme induces strain at the subunit interface through the movement of the F helix and the FG corner, thus transmitting the information from one subunit to another. The second means of communication is through a set of *salt bridges* (**Figure 6.42**) that exist in the T quaternary state but not in the R state. Those salt bridges stabilize the T state, but are broken in the R state, when the allosteric transition occurs.

A tertiary conformational change occurs in each subunit when oxygen binds to it. However, the oxygen affinity of the subunit to which oxygen binds does not increase at that moment: it is still low as long as the hemoglobin molecule remains in the T quaternary state. This is contrary to the prediction of the Pauling/KNF model. The oxygen affinity of the subunit that first binds oxygen and the affinity of the other subunits only increases when the quaternary T → R transition occurs. This is in agreement with the essence of the MWC model: the

Figure 6.41 Tertiary changes in the α chain of human hemoglobin upon oxygen binding to the heme group. (A) Deoxyhemoglobin, T state. (B) Oxyhemoglobin, R state. His F8 directly coordinates the iron, whereas His E7 binds oxygen from the opposite side of the heme in oxyhemoglobin (PDB 1A3N and 1HHO).

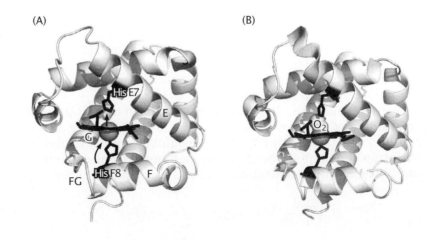

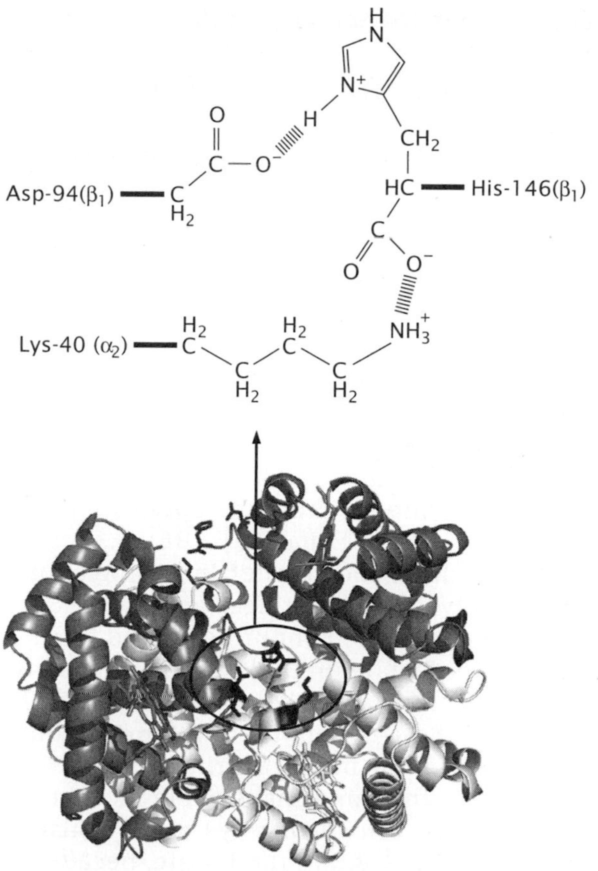

Figure 6.42 Salt bridges in deoxyhemoglobin stabilize the T state (they are absent in the R state). The residues involved in salt bridges at the α_2/β_1 interface (α subunits, light; β subunits, dark) are shown as stick models (black). The situation is identical at the α_1/β_2 interface. His-146 of each β chain makes a salt bridge, through its protonated imidazole ring, to the carboxylate of the side chain of Asp-94 of the same β chain. In addition, because His-146 is the C-terminal residue of the β_1 (β_2) chain, its (terminal) carboxylate makes another salt bridge to Lys-40 of the α_2 (α_1) chain (PDB 1A3N).

oxygen affinity of hemoglobin is determined by the quaternary state. Conversely, if the protein is in the R state, with oxygen bound at its four sites, and one oxygen molecule dissociates, the affinity of that subunit for oxygen remains high; that subunit does not concomitantly switch to the T state. Again, this is in agreement with the MWC model. Finally, oxygen binding to hemoglobin crystals in the T state, where the quaternary conformation is "frozen," is not cooperative, but follows a simple hyperbolic binding isotherm (shown in Equation 6.118). This is predicted by the MWC model, corresponding to the upper branch of the diagram in Figure 6.37, when the transition to the R state is not allowed (in this case, by the physical constraints of the crystal). Thus, slightly modified to include tertiary conformational changes in the subunits upon oxygen binding and pH-induced changes in the T state, the MWC model remains correct in its essence.

The Bohr Effect: When Hemoglobin Binds Oxygen it Releases Protons and Vice Versa

Christian Bohr, in addition to determining that the oxygen binding isotherm of hemoglobin is sigmoidal, made yet another important observation: in the presence of CO_2, the affinity of hemoglobin for oxygen decreases. This is important physiologically because oxygen is necessary for molecular respiration and CO_2 is produced in the tissues as a result of respiration. Thus, where CO_2 is produced, hemoglobin releases oxygen, which is the desired physiological outcome.

Most of this effect is mediated by protons. Protons (H^+) and bicarbonate (HCO_3^-) are generated by the reaction of CO_2 with water, catalyzed by carbonic anhydrase in the red blood cell:

$$CO_2 + H_2O \rightleftharpoons H^+ + HCO_3^-. \tag{6.135}$$

(It is in the form of HCO_3^- that most carbon dioxide is transported in the blood, from the tissues to the lungs.) Protons generated by this reaction bind to hemoglobin in the T state but not in the R state. This is called the *Bohr effect*: at low pH (high H^+ concentration), hemoglobin binds H^+ and releases O_2; at high pH (low H^+ concentration), it releases H^+ and binds O_2. In a simplified way, we may write

$$\text{(T state)} \qquad \text{(R state)}$$

$$Hb \cdot (H^+)_n + 4O_2 \rightleftharpoons Hb \cdot (O_2)_4 + nH^+. \tag{6.136}$$

The Bohr effect follows directly from the application of Le Châtelier's principle to the equilibrium of Equation 6.136.

Where do protons bind in the T state? In hemoglobin several amino acid residues are involved in salt bridges across the α_1/β_2 and α_2/β_1 interfaces (see Figure 6.42). The Bohr protons bind to residues involved in the salt bridges in the T state. In the R state, however, the salt bridges break and those protons are released by hemoglobin. His β-146, the C-terminal residue of the β subunits, is the main contributor to the Bohr effect. In the T state, the imidazole ring of His β-146 makes a salt bridge with Asp β-94 of the same chain, while its terminal carboxylate group makes another salt bridge with Lys α-40. Consequently, His β-146 has an anomalous $pK_a' = 8.0$ in the T state, because the salt bridge with Asp β-94 stabilizes the proton on the imidazole side chain (see Figure 6.42). But in the R state, in which this salt bridge is absent, the pK_a of His β-146 reverts to the normal value for histidine ($pK_a = 6.5$). Since there are two β subunits in each hemoglobin tetramer, each one containing His-146, there are two Bohr protons released when the transition to the R state occurs: $n = 2$ in Equation 6.136.

In reality there are other residues involved in the Bohr effect, but they are of lesser importance and their effects almost cancel out because some have an anomalously low pK_a, whereas others have an anomalously high pK_a. (The effect we are discussing is more accurately named *alkaline Bohr effect* because it is manifested at pH > 7. There is also the acid Bohr effect, manifested at pH $\approx$ 4.5, which involves mainly His β-143.)

The proton is an *allosteric effector* of oxygen binding to hemoglobin. H^+ binds at the imidazole ring of His β-146, which is different from the binding site of oxygen. Proton binding changes (decreases) the affinity of hemoglobin for oxygen, because it shifts the T $\rightleftharpoons$ R equilibrium to the T state (shown in Equation 6.136). The interaction of the two ligands is called *heterotropic* because the proton and oxygen are chemically different.

The Physiological Consequences of the Bohr Effect can be Understood with the All-or-none Model

When the blood travels from the lungs to the peripheral tissues, its pH changes from 7.6 to 7.2. This appears to be a small change, but has a large effect on oxygen binding by hemoglobin. To understand the effect of pH, we will use the all-or-none model for oxygen binding to hemoglobin. The MWC model would be more accurate, but the concept is fully embodied in this simpler model. We designate the proton dissociation constant of His β-146 by K_a' in the T state and by K_a in the R state.

The situation is summarized in the diagram of **Figure 6.43**, where we added the proton binding equilibria to the all-or-none diagram of Figure 6.40. Using the empty T state (T_0) as reference, the partition function for the allosteric transition (vertical transition in Figure 6.43), is

$$Q = 1 + (K[O_2])^4. \tag{6.137}$$

The partition function for proton binding to the T state is

$$\begin{aligned} Q &= 1 + 2[H^+]/K_a' + [H^+]^2/K_a'^2 \\ &= (1 + [H^+]/K_a')^2. \end{aligned} \tag{6.138}$$

Similarly, the partition function for proton binding to the R state, using the fully oxygen-bound R state as the reference, is

$$\begin{aligned} Q &= 1 + 2[H^+]/K_a + [H^+]^2/K_a^2 \\ &= (1 + [H^+]/K_a)^2. \end{aligned} \tag{6.139}$$

Note that (a) the acidity constants (K_a or K_a') are proton *dissociation* constants, so we divide $[H^+]$ by K_a or K_a' instead of multiplying; (b) the factor of 2 in the second term appears because there are two ways for the proton to bind (to either one of the His β-146 residues); and (c) oxygen binds cooperatively but protons bind independently of other protons *within* each state (T or R). The proton binding partition functions are just products of single-site partition functions (shown in Equations 6.138 and 6.139). Recall that the partition function for independent sites, just like the probabilities of independent events, are products of the individual ones.

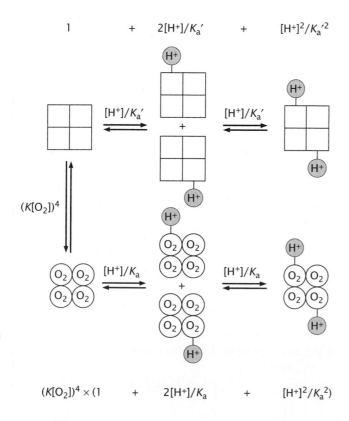

$$
\begin{array}{ccccc}
1 & + & 2[H^+]/K_a' & + & [H^+]^2/K_a'^2
\end{array}
$$

$$(K[O_2])^4$$

$$[H^+]/K_a' \qquad [H^+]/K_a'$$

$$[H^+]/K_a \qquad [H^+]/K_a$$

$$
\begin{array}{ccccc}
(K[O_2])^4 \times (1 & + & 2[H^+]/K_a & + & [H^+]^2/K_a^2)
\end{array}
$$

Figure 6.43 Diagram of the Bohr effect in the all-or-none model for oxygen binding. The T state is represented by squares and the R state by circles. Proton binding to the β subunits is indicated. The terms of the partition function are written above and below each branch.

Finally, we combine proton binding (shown in Equations 6.138 and 6.139) and the allosteric transition (shown in Equation 6.137), with *reference to one common state*, the empty T state, to obtain the overall partition function. To do so, we multiply the terms in the partition function for the allosteric transition (1 for T the state and $(K[O_2])^4$ for the R state) by their respective proton-binding partition functions:

$$Q = (1 + [H^+]/K_a')^2 + (K[O_2])^4 (1 + [H^+]/K_a)^2. \qquad (6.140)$$

The terms $(1 + [H^+]/K_a')^2$ and $(K[O_2])^4 (1 + [H^+]/K_a)^2$ correspond to the T and R states, respectively, with 0 to 2 protons bound, one per β subunit.

The *apparent* allosteric transition constant between all forms of the T state and all forms of R state is given by Equation 6.125,

$$K_0^{app} = \frac{[\text{All R-state forms}]}{[\text{All T-state forms}]}.$$

K_0^{app} is the ratio of the terms corresponding to the R and T states in the partition function (shown in Equation 6.140). Those terms are proportional to (α) the equilibrium concentrations of each state.

$$\text{Oxygen-free (T state)} \propto (1 + [H^+]/K_a')^2$$

$$\text{Oxygen-bound (R state)} \propto K^4[O_2]^4 (1 + [H^+]/K_a)^2.$$

In the all-or-none model, the T state is free (of oxygen) and the R state is fully bound, as $R(O_2)_4$. The oxygen binding constant K in Equation 6.128 is now replaced by an apparent constant, which we call K'. It is an *apparent* oxygen binding constant because it depends on $[H^+]$ (or pH). K' is determined by the ratio

$$K'^4[O_2]^4 = \frac{[\text{All R-state forms}]}{[\text{All T-state forms}]}. \qquad (6.141)$$

If there were no proton binding involved, we would simply have the equilibrium (corresponding to Figure 6.40)

$$Hb + 4O_2 \rightleftharpoons Hb(O_2)_4, \qquad (6.142)$$

and Equation 6.141 would reduce to

$$K^4[O_2]^4 = \frac{[Hb(O_2)_4]}{[Hb]}. \qquad (6.143)$$

If we include proton binding, using the partition function for the probabilities (concentrations) of the T and R forms, we obtain the explicit version of Equation 6.141,

$$K'^4[O_2]^4 = \frac{K^4[O_2]^4 (1 + [H^+]/K_a)^2}{(1 + [H^+]/K_a')^2} \qquad (6.144)$$

$$K'^4 = K^4 \left(\frac{1 + [H^+]/K_a}{1 + [H^+]/K_a'}\right)^2. \qquad (6.145)$$

The effect of pH is determined by the ratio

$$\gamma = \left(\frac{1 + [H^+]/K_a}{1 + [H^+]/K_a'}\right)^2, \qquad (6.146)$$

which is boosted by the power of 2 (because we assumed two proton binding sites, one to each β chain). With the substitutions $K_a = 10^{-pK_a}$ and $[H^+] = 10^{-pH}$, we can write

$$\gamma = \left(\frac{1 + 10^{(pK_a - pH)}}{1 + 10^{(pK_a' - pH)}}\right)^2 . \tag{6.147}$$

Using Equation 6.147 we can see what happens when the blood pH changes from 7.6, in the lungs, to 7.2, in the tissues. His β-146 has $pK_a = 6.5$ (R) and $pK_a' = 8$ (T). Thus, $\gamma = 0.027$ at pH 7.2 and $\gamma = 0.095$ at pH 7.6. When the pH changes from 7.6 to 7.2, a change of only 0.4 pH units, the concentration of the T conformation increases 3.5 times relative to R (this is only due to pH and does not include the effect of oxygen concentration). In the tissues, where the pH is lower, the T state is greatly favored: hemoglobin binds protons and releases oxygen concomitantly. The Bohr effect is an excellent example of the functional importance of pH.

Bisphosphoglycerate (BPG) Binds to the T State and Hampers Oxygen Binding by Hemoglobin

Protons are not the only heterotropic allosteric effectors of hemoglobin. 2,3-Bisphosphoglycerate (BPG) binds to the T state only, at the center of a hole that is present in the T state (see Figure 6.31) but closes when the quaternary conformational change to the R state occurs. BPG is a small polyanionic molecule, which associates with several cationic residues in hemoglobin—namely, with His-2, His-43, and Lys-82 of the β chains (**Figure 6.44**). The effect of BPG is similar to that of protons, but there is only one binding site to the T state and none to the R state.

Red blood cells contain about 5 mM BPG. Like protons, BPG binding reduces the affinity of hemoglobin for oxygen because it shifts the T ⇌ R equilibrium to the T state. This effect is extremely important physiologically: without BPG, hemoglobin would bind oxygen *too well*; so well, in fact, that it would not release sufficient of its bound oxygen in the tissues. BPG is also a major player in altitude adaptation. Higher concentrations of BPG (~8 mM) are accumulated by red blood cells when humans move from low to high altitudes, ensuring an even more complete release of oxygen in the tissues.

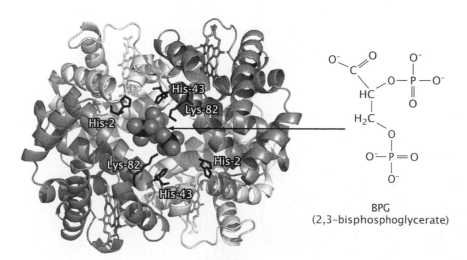

BPG
(2,3–bisphosphoglycerate)

Figure 6.44 Binding site of BPG in a hole at the center of the hemoglobin tetramer. The residues that interact with BPG, positively charged histidines and lysines, are indicated (PDB 1A3N).

6.7 LIGAND BINDING CHANGES PROTEIN STABILITY

We now turn our attention to the effects of ligand binding on protein stability. We will consider two types of ligands—specific and nonspecific. *Specific ligands*, such as substrates, activators, or inhibitors, that bind to a well-defined stereochemical binding site on the protein always stabilize the *native state*. Nonspecific ligands, such as protons, may favor either state, depending on their binding constants to the native and the denatured states.

Specific Ligands Stabilize the Folded State

Suppose that a protein binds a ligand X. This could be the substrate of an enzyme binding at the active site. The structure of the active site only exists in the native state, so binding requires that the protein be folded. The problem of the effect of a ligand on protein stability is entirely analogous to the effect of a ligand on the binding of another, such as in the allosteric heterotropic interactions of protons or BPG with oxygen in hemoglobin. In all cases, the ligand affects *another* equilibrium.

To study the effect of a ligand on protein stability, we begin by drawing the diagram (**Figure 6.45**). The folded protein binds the ligand (X), but the unfolded protein does not. Let us write the partition function with the free folded state (F) as the reference. The relative probabilities are $p_f = 1$ for free folded, $p_u = K_u$ for unfolded, and $p_x = K_X[X]$ for bound (folded):

$$Q = 1 + K_u + K_X[X]. \tag{6.148}$$

To calculate the equilibrium constant between any two states from the partition function, you need to look for the terms of Q that correspond to those states. In the absence of ligand, we have

$$K_u = \frac{[U]}{[F]} = \frac{p_u}{p_f} = \frac{K_u}{1}. \tag{6.149}$$

Now, we want to calculate the apparent equilibrium constant K_u^{app} between the folded and unfolded states in the *presence of ligand*. K_u^{app} is an *apparent* constant because it depends on the ligand concentration. Both terms 1 and $K_X[X]$ in the partition function represent folded protein; the term K_u represents the unfolded protein. The apparent equilibrium constant for unfolding in the presence of the ligand X is the ratio of the relative probabilities,

$$K_u^{app} = \frac{p_u}{p_f + p_x}, \tag{6.150}$$

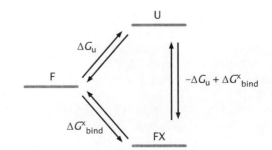

Figure 6.45 Binding of a ligand X to the folded state (F) of a protein changes the equilibrium between the folded and unfolded (U) states.

but one must include all forms of folded and unfolded states in those probabilities. Therefore,

$$K_u^{app} = \frac{K_u}{1 + K_X[X]}. \qquad (6.151)$$

The equilibrium constant for unfolding is smaller by a factor of $(1 + K_X[X])$ in the presence of ligand. The protein is stabilized against unfolding by the binding of its ligand. We still have an equilibrium between the folded and unfolded states, but the folded state now comprises two microstates, one free and one with ligand bound (**Figure 6.46**).

The standard Gibbs energy change for unfolding is given by $-RT\ln(p_u/(p_f + p_x)) = -RT\ln K_u^{app}$. In the absence of ligand, we have the usual result:

$$\Delta G_u^o = -RT\ln K_u. \qquad (6.152)$$

In the presence of ligand X, we have

$$\Delta G_u^X = -RT\ln K_u^{app}. \qquad (6.153)$$

The *change* in the Gibbs energy change for unfolding caused by the ligand is

$$\Delta\Delta G_u^X = \Delta G_u^X - \Delta G_u^o = -RT\ln\frac{1}{1 + K_X[X]} = RT\ln(1 + K_X[X]). \qquad (6.154)$$

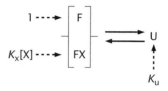

Figure 6.46 The unfolded (U) state has only one microstate in this level of approximation (in reality, an unfolded protein has a huge number of conformations, but we are lumping them all together here), whereas the folded state has two microstates, free (F) and with ligand bound (FX).

Changes of pH may Stabilize Either the Folded or the Unfolded State

When we discussed protein stability in Chapter 4, we postponed consideration of the effect of pH. The reason is that only now do we have the necessary tools to understand it. You will see that understanding the effect of pH requires only that the proton be treated as a ligand that binds to certain sites on the protein. Those sites are the terminal amino and carboxylic groups, and the ionizable side chains of the amino acid residues.

There are only two important things you need to know to understand the effect of pH on protein stability. First, the effect of pH usually arises because some functional groups in the protein bind the proton in the folded state but not in the unfolded state, or vice versa, at the pH of the experiment. For example, the carboxylate group of glutamic acid may have a proton bound in the folded state (COOH), but may lose it in the unfolded state (to form COO^-).

Second, in the unfolded state of the protein, the pK_a of the ionizable groups are normal, but in the folded state they are sometimes *anomalous*. Their values may deviate considerably from the normal pK_a, because ionizable groups may reside in special environments in the folded protein. We will designate the anomalous acidity constants in the folded state by K_a' and the corresponding negative logarithms by pK_a'. A glutamate residue in the protein surface or in the denatured state will typically be deprotonated (negatively charged) at pH 7; but if it is buried inside the protein, it is often protonated, because a charge in a nonpolar environment is very unfavorable. For example, Glu-35 in the active site of the enzyme lysozyme has $pK_a' = 6.5$ in the native state, whereas the normal pK_a of the side chain of glutamic acid is 4.3. Similarly, Lys-115 in the active site of acetoacetate decarboxylase has $pK_a' = 6.0$, whereas its normal $pK_a = 10.5$ (see Table 2.3).

In the previous section we discussed protein stabilization by ligand binding. The ligand bound to the folded state but not to the unfolded state. That was the case because the three-dimensional structure of the binding site exists only in the folded protein. In the case of proton binding, the ligand (H^+) binds both to the folded and to the unfolded states, but with different affinities (pK_a). Let us write the partition function, again with the free folded state as reference. Since K_a is a dissociation constant, we divide $[H^+]$ by K_a when writing Q. K_u is the equilibrium constant for unfolding in the deprotonated state.

$$Q = 1 + [H^+]/K_a' + K_u + K_u[H^+]/K_a. \qquad (6.155)$$

The terms that correspond to folded protein are 1 and $[H^+]/K_a'$; those that correspond to unfolded protein are K_u and $K_u[H^+]/K_a$. Let us group them like this:

$$Q = (1 + [H^+]/K_a') + K_u (1 + [H^+]/K_a). \qquad (6.156)$$

The apparent equilibrium constant for unfolding is given by the ratio of the statistical weights of unfolded to folded states, from Equation 6.156,

$$K_u^{app} = \frac{K_u (1 + [H^+]/K_a)}{1 + [H^+]/K_a'}.$$

Hence, the unfolding constant K_u is modified by the ratio

$$\gamma = \frac{1 + [H^+]/K_a}{1 + [H^+]/K_a'} = \frac{1 + 10^{pK_a - pH}}{1 + 10^{pK_a' - pH}}. \qquad (6.157)$$

The effect of pH is contained in γ:

$$K_u^{app} = \gamma K_u. \qquad (6.158)$$

Again, we can write the change in the Gibbs energy of unfolding caused by pH as

$$\Delta\Delta G_u^H = \Delta G_u^H - \Delta G_u^o \qquad (6.159)$$

$$= -RT \ln K_u^{app} - (-RT \ln K_u) \qquad (6.160)$$

$$= -RT \ln \gamma. \qquad (6.161)$$

Let us calculate the change in the Gibbs energy of unfolding. In acetoacetate decarboxylase, Lys-115 has $pK_a' = 6.0$ in the folded state and $pK_a = 10.5$ in the unfolded state. First, consider the problem qualitatively. Since the pK_a' of this lysine is lower in the folded state, more protons are bound to Lys-115 when the protein unfolds,

$$\text{Folded} + H^+ \rightleftharpoons \text{Unfolded–}H^+. \qquad (6.162)$$

According to Le Châtelier's principle, decreasing the pH (increasing $[H^+]$) shifts the equilibrium to the unfolded state. To determine what happens quantitatively, we calculate $\Delta\Delta G_u$ at pH 3, 7, and 10 using Equations 6.157 and 6.161. We find that $\Delta\Delta G_u = -6.2$ kcal/mol at pH 3, -4.2 kcal/mol at pH 7, and ≈ 0 at pH 10. Thus, as the pH decreases, $\Delta\Delta G_u$ becomes more negative, favoring unfolding: the protein is destabilized by low pH. The effect is maximal in acidic solution, but almost vanishes in alkaline solution, because the proton concentration is so small that the protonation levels of lysine in the folded and unfolded states are both small.

Glu-35 in lysozyme has an anomalous $pK_a' = 6.5$ in the folded state, but the normal $pK_a = 4.3$ in the unfolded state. In this case, the folded protein binds more protons than the unfolded one,

$$\text{Folded–H}^+ \rightleftharpoons \text{Unfolded} + \text{H}^+. \qquad (6.163)$$

According to Le Châtelier's principle, decreasing the pH (increasing $[H^+]$) shifts the equilibrium to the folded state. Again, let us calculate $\Delta\Delta G_u$ at pH 3, 7, and 10. Using Equations 6.157 and 6.161, we find that $\Delta\Delta G_u = +3.0$ kcal/mol at pH 3, $+0.2$ kcal/mol at pH 7, and ≈ 0 at pH 10. Thus, $\Delta\Delta G_u$ becomes more positive (favoring folding) as the pH decreases. Low pH stabilizes the native protein.

6.8 SUMMARY

Ligand binding to a protein is described by a binding isotherm, which is the fractional saturation of the binding sites as a function of ligand concentration. The binding isotherm is most conveniently derived from the partition function. Most proteins bind more than one ligand. If binding of a ligand does not affect binding of another, binding is independent. This case is described by a hyperbolic binding isotherm,

$$\theta = \frac{Kx}{1 + Kx},$$

where x is the free ligand concentration and K is the binding constant. If the ligand is in large excess compared to the protein, we can approximate the free ligand by the total ligand when plotting the binding isotherm to determine the binding constant. If not, the binding constant can be determined by solving a quadratic equation or subtracting the bound ligand from the total using the experimental fractional binding.

Essentially any physical technique, such as fluorescence, the CD spectrum, or an NMR line, that exhibits a difference in signal for the protein with and without ligand bound can be used to measure binding. However, ITC is unique in this respect, because it yields not only the binding constant but also the stoichiometry and the enthalpy of binding, all in a single experiment.

If the binding of a ligand A to a protein affects the binding of another ligand B, we say the binding sites interact. The coupling Gibbs energy $\Delta G(A|B)$ measures the mutual dependence of the binding sites. If $\Delta G(A|B) < 0$, binding of the first ligand enhances binding of the second. If $\Delta G(A|A) < 0$ for two or more identical ligands, we say that binding is cooperative. Cooperative binding yields a sigmoidal binding isotherm when plotted against the concentration of free ligand. The essence of cooperativity is that intermediate binding states are disfavored compared to the free and the fully bound protein. Thus, the protein has a tendency to switch between the empty and the completely occupied states.

Binding of oxygen to hemoglobin is cooperative. It is best described by the MWC model. The protein exists in two quaternary states, T with low affinity for oxygen and R with high affinity. Binding of the first oxygen molecules to the T state increases the probability that hemoglobin will change conformation to the R state. This is called the allosteric transition. The essence of cooperativity in hemoglobin is that when the allosteric transition occurs, the oxygen binding sites change from low- to high-affinity, and oxygen binding occurs more readily. Protons and BPG are allosteric effectors of hemoglobin: they bind to different sites than oxygen on the protein and reduce its affinity for oxygen. Both H^+ and BPG have important physiological effects.

Ligand binding changes protein stability. Specific ligands, such as a substrate or an inhibitor of an enzyme, bind only to the folded protein and therefore always stabilize the native state. Changes of pH may stabilize either the folded or the unfolded state. Proton binding affects protein stability primarily if a side chain or the N- or C-terminus have different pK_a's in the folded and the unfolded states. In this case, the protein conformation (folded or unfolded) with higher affinity for H^+ (higher pK_a) is stabilized relative to the other conformation at low pH (high $[H^+]$).

6.9 PROBLEMS

6.1 Suppose you collected the data shown in Table 6.1 for ligand binding to a protein and you know there is only one binding site. Plot the data. Determine the binding (K) and dissociation (K_d) constants by fitting the equation of a binding isotherm to the data. You will need an initial guess of the value of K_d. Then plot the binding isotherm you calculate with the best-fit K_d on the same diagram.

Present K in units of M^{-1}, with necessary powers of 10 as required, and present K_d in the subunits of M that are most appropriate for the enzyme. Include the plots in your answers.

Table 6.1 Binding isotherm for a substrate to an enzyme: the fraction of sites occupied (θ) as a function of the free substrate concentration ([X])

[X] (μM)	θ
0	0.000
30	0.251
60	0.397
90	0.495
120	0.580
150	0.632
180	0.651
210	0.717
240	0.765
270	0.751
300	0.794
330	0.804
360	0.822
390	0.842

6.2 A protein has two conformations, A and B, that exist in equilibrium. It binds a ligand X, but only in conformation B.
(a) Draw the diagram for this case.
(b) Write the partition function.
(c) Write the isotherm for binding of X.

6.3 The following curves represent three types of two-site ligand binding to a protein: identical and independent,

different but independent, and cooperative. Identify which curve corresponds to which type.

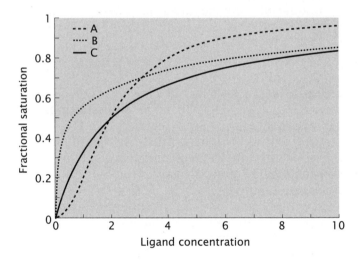

6.4 Table 6.2 shows ligand binding data for three different proteins. Determine which binding isotherm is best to analyze each set of data, and calculate the dissociation constants. You have to figure out how many sites are needed in each case and which model to use.

6.5 You collected the data in Table 6.3 for the binding of a ligand to a protein. There is only one binding site on each protein. The data represent the fluorescence intensity (in arbitrary units) of the protein at 340 nm as a function of total ligand concentration $[X]_T$ (in μM). The total protein concentration is $[P]_T = 4$ μM. Use these data and the subtraction method we learned in Section 6.1 (shown in Equation 6.51) to determine K_d as outlined in the text. Note that before you do any calculations, you need to plot the data and convert the fluorescence intensity data to fractional binding (θ). Hint: The total amplitude of the change in the y-axis must correspond to the range of θ, from ≈ 0 to ≈ 1.

6.6 In Problem 6 of Chapter 4, we considered thermal unfolding of lysozyme at pH 7. We calculated ΔG_u^o for unfolding at 25°C, as well as the corresponding equilibrium constant. That is an *apparent* equilibrium constant for unfolding (K_u^{app}). The apparent constant is what you actually measure if you do an experiment at 25°C and pH 7. It is called apparent because it depends on pH. Here we revisit this problem to consider the effect of pH.

(a) One of the active site residues of lysozyme is Glu-35. In the denatured (unfolded) state, $pK_a = 4.3$, which is normal for the Glu side chain. But in the native state, its value is anomalously high, $pK_a \approx 6.5$. What does this suggest about the local environment of Glu-35 in the native state?

(b) If the pK_a of Glu-35 were the only anomalous one in the protein, it would indicate that lysozyme releases protons upon unfolding at pH 7,
Lysozyme–Glu-35-H (folded) $\rightleftharpoons$
Lysozyme–Glu-35$^-$ (unfolded) + H$^+$.
Assuming that there are no other anomalous pK_a's in lysozyme, what does Le Châtelier's principle tell you about the effect of lowering the pH on the unfolding equilibrium of this enzyme?

(c) Under these assumptions, make a quantitative prediction of the effect of pH on lysozyme stability, by calculating the apparent equilibrium constant and ΔG_u^o at pH 5 and 25°C. You will need to write the partition function for the protein, including the ionization effects of

Glu-35. Begin by drawing a diagram that includes unfolding and protonation of the folded and unfolded states explicitly. You know K_u^{app} (25°C) at pH 7. Calculate it at pH 5. Think about your result. (Note that at pH 5 there is not much of an effect on stability arising from the ionization of groups with normal pK_a values, so the effect should arise only from Glu-35.)

Table 6.3 Fluorescence intensity data for the binding of a ligand X to a protein P, with $[P]_T = 4$ μM

$[X]_T$ (μM)	Fluorescence
0.0	100
0.11	117
0.30	140
0.55	169
0.82	192
1.00	209
1.22	233
1.49	267
1.82	296
2.23	318
2.72	373
3.32	416
4.06	457
4.95	510
6.05	549
7.39	558
9.03	583
11.0	599
13.5	617
16.5	635
20.9	633
24.5	652
30.0	653

Table 6.2 Binding isotherms for three proteins (L1, L2, and L3)

Concentration	Fractional occupancy (θ)		
(μM)	L1	L2	L3
0.041	0.023	0.133	0.001
0.050	0.028	0.153	0.001
0.061	0.034	0.172	0.002
0.074	0.040	0.207	0.002
0.091	0.050	0.225	0.003
0.111	0.061	0.265	0.005
0.135	0.070	0.291	0.007
0.165	0.087	0.329	0.010
0.202	0.106	0.341	0.016
0.247	0.125	0.397	0.023
0.301	0.151	0.423	0.032
0.368	0.178	0.460	0.046
0.449	0.219	0.468	0.067
0.549	0.234	0.507	0.095
0.670	0.271	0.536	0.130
0.819	0.331	0.612	0.191
1.000	0.365	0.626	0.261
1.221	0.415	0.650	0.338
1.492	0.483	0.707	0.417
1.822	0.513	0.712	0.547
2.226	0.589	0.753	0.652
2.718	0.638	0.822	0.685
3.320	0.678	0.794	0.775
4.055	0.709	0.843	0.846
4.953	0.749	0.840	0.861
6.050	0.789	0.867	0.905
7.389	0.798	0.885	0.983
9.025	0.849	0.914	1.000

6.7 In the absence of oxygen, hemoglobin behaves as a two-state system. Write its partition function. If the protein were 99.0% in the T state at room temperature, what would be the Gibbs energy difference between the T and R states?

6.8 Derive the binding isotherm for the alternative cooperative model whose partition function is given by Equation 6.91. Plot the binding isotherm. Use $K = 5 \times 10^4 \, M^{-1}$ and $\alpha = 100$. Choose a ligand concentration range that allows you to see the full binding isotherm (until $\theta \approx 1$), but in such a way that the inflection point corresponds to a ligand concentration at $\approx 1/5$ of the range.

6.9 The red blood cell contains millimolar concentrations of 2,3-bisphosphoglycerate (BPG). Regulation of the concentration of BPG in the erythrocyte is the most important component of the high-altitude adaptation mechanism. The normal BPG concentration is 5 mM at sea level. But when a person moves to high altitude, this concentration increases to ≈ 8 mM. In the complete absence of BPG, the affinity of hemoglobin for oxygen is actually too high, and the protein would not release sufficient oxygen in the tissues. BPG binding to

hemoglobin reduces the affinity of the protein for oxygen, which allows easier oxygen release in the tissues. BPG binds to the T state but not to the R state. One molecule of BPG binds to the tetramer of the hemoglobin molecule. Use the all-or-none model for oxygen binding to hemoglobin to answer the following questions.
(a) Sketch the diagram for this problem, including the binding of oxygen and BPG (but not the effect of pH).
(b) Based on your diagram, write the binding partition function for hemoglobin.
(c) Write the oxygen binding isotherm, including the effect of BPG.

(d) Treating oxygen binding with the all-or-none model, write the expression for the apparent oxygen binding constant (K_{app}) in the presence of BPG. Hint: recall that you can get the probabilities of the two states directly from the partition function.
(e) In the absence of BPG, the P_{50} for oxygen binding is 0.5 kPa. In the presence of 5 mM BPG, the apparent P_{50} for oxygen binding is 3.5 kPa. Using this information and your expression for K_{app}, calculate the dissociation constant K_d^{BPG} of BPG in the T state.

6.10 FURTHER READING

General

Ben-Naim A (2001) Cooperativity and Regulation in Biochemical Processes. Kluwer Academic/Plenum Publishers.

Creighton TE (1993) Proteins. Structure and Molecular Properties, 2nd ed. Freeman.

Hill TL (1985) Cooperativity Theory in Biochemistry. Springer.

Wyman J & Gill SJ (1990) Binding and Linkage. University Science Books.

Isothermal Titration Calorimetry

Wiseman T, Williston S, Brandts JF & Lin L-N (1989) Rapid measurements of binding constants and heats of binding using a new titration calorimeter. *Anal Biochem* 179:131–137.

Leavitt S & Freire E (2001) Direct measurement of protein binding energetics by isothermal titration calorimetry. *Curr Opin Struct Biol* 11:560–566.

Freire E, Mayorga OL & Straume M (1990) Isothermal titration. *Anal Chem* 62:950–959.

Calcium Binding to EF-hands

Chazin WJ (2011) Relating form and function of EF-hand calcium binding proteins. *Acc Chem Res* 44:171–179.

Linse S, Johansson C, Brodin P et al (1991) Electrostatic Contributions to the Binding of Ca^{2+} in Calbindin D_{9kt}. *Biochemistry* 30:154–162.

Linse S & Chazin WJ (1995) Quantitative measurements of the cooperativity in an EF-hand protein with sequential calcium binding. *Protein Sci* 4:1038–1044.

Mäler L, Blankenship J, Rance M & Chazin WJ (2000) Sitesite communication in the EF-hand Ca^{2+}-binding protein calbindin D_{9k}. *Nature Struct Biol* 7:245–250.

Conformational Entropy and Allostery

Hilser VJ & Thompson EB (2007) Intrinsic disorder as a mechanism to optimize allosteric coupling in proteins. *Proc Natl Acad Sci USA* 104:8311–8315.

Reinhart GD, Hartleip SB & Symox MM (1989) Role of coupling entropy in establishing the nature and magnitude of allosteric response. *Proc Natl Acad Sci USA* 86:4032–4036.

Stone MJ (2001) NMR relaxation studies of the role of conformational entropy in protein stability and ligand binding. *Acc Chem Res* 34:379–388.

Tzeng S-R & Kalodimos CG (2012) Protein activity regulation by conformational entropy. *Nature* 488:236–240.

Wand AJ (2013) The dark energy of proteins comes to light: Conformational entropy and its role in protein function revealed by NMR relaxation. *Curr Opin Struct Biol* 23:75–81.

Wrabl JO, Gu J, Liu T et al (2011) The role of protein conformational fluctuations in allostery, function, and evolution. *Biophys Chem* 159:129–141.

Hemoglobin, Allostery, and Cooperativity

Ackers GK & Holt JM (2006) Asymmetric cooperativity in a symmetric tetramer: Human hemoglobin. *J Biol Chem* 281:11441–11443.

Adair GS (1925) The hemoglobin system. VI. The oxygen dissociation curve of hemoglobin. *J Biol Chem* 63:529–545.

Baldwin J & Chothia C (1979) Haemoglobin: The structural changes related to ligand binding and its allosteric mechanism. *J Mol Biol* 129:175–220.

Bellelli A & Brunori M (2011) Hemoglobin allostery: Variations on a theme. *Biochim Biophys Acta* 1807:1262–1272.

Bohr C, Hasselbalch K & Krogh A (1904) Über einen in biologischer Beziehung wichtigen Einfluss, den die Kohlensäurespannung des Blutes auf dessen Sauerstoffbindung übt. *Skand Arch Physiol* 16:402–412. [Wiley Online Library: http://onlinelibrary.wiley.com/doi/10.1111/apha.1904.16.issue-2/issuetoc. Available in translation at: http://www.udel.edu/chem/white/C342/Bohr%281904%29.html]

Bohr C (1904) Theoretische Behandlung der quantitativen Verhältnisse bei der Sauerstoffaufnahme des Hämoglobins. *Zentralblatt Physiol* 17:682–688.

Cui Q & Karplus M (2008) Allostery and cooperativity revisited. *Protein Sci* 17:1295–1307.

Eaton WA, Henry ER, Hofrichter J & Mozzarelli A (1999) Is cooperative oxygen binding by hemoglobin really understood? *Nature Struct Biol* 6:351–358.

Fang T-Y, Zou M, Simplaceanu V, Ho NT & Ho C (1999) Assessment of roles of surface histidyl residues in the molecular basis of the Bohr effect and of β143 histidine in the binding of 2,3-bisphosphoglycerate in human normal adult hemoglobin. *Biochemistry* 38:13423–13432.

Henry ER, Jones CM, Hofrichter J & Eaton WA (1997) Can a two-state MWC allosteric model explain hemoglobin kinetics? *Biochemistry* 36:6511–6528.

Imai K & Yonetani T (1975) pH dependence of the Adair constants of human hemoglobin. *J Biol Chem* 250: 2227–2231.

Johnson M (2000) Mathematical modeling of cooperative interactions in hemoglobin. *Methods Enzymol* 323:124–155.

Laskowski RA, Gerick F & Thornton JM (2009) The structural basis of allosteric regulation in proteins. *FEBS Lett* 583:1692–1698.

Koshland DE, Nemethy G & Filmer D (1966) Comparison of experimental binding data and theoretical models in proteins containing subunits. *Biochemistry* 5:365–385.

Mills FC, Johnson MJ & Ackers GK (1976) Oxygenation-linked subunit interactions in human hemoglobin: Experimental studies on the concentration dependence of oxygenation curves. *Biochemistry* 15:5350–5362.

Monod J, Wyman J & Changeux JP (1965) On the nature of allosteric transitions: A plausible model. *J Mol Biol* 12: 88–118.

Perutz MF (1964) The hemoglobin molecule. *Sci Am* 211:64–76.

Perutz MF (1970) Stereochemistry of cooperative effects in hemoglobin. *Nature* 228:726–734.

Perutz MF (1998) The stereochemical mechanism of the cooperative effects in hemoglobin revisited. *Annu Rev Biomol Struct* 27:1–34.

Zheng G, Schaefer M & Karplus M (2013) Hemoglobin Bohr effects: Atomic origin of the histidine residue contributions. *Biochemistry* 52:8539–8555.

Enzyme Kinetics

7.1 ENZYMES ENHANCE REACTION RATES BY STABILIZING THE TRANSITION STATE

In this last chapter we will examine the fundamental concepts of the kinetics of enzyme-catalyzed reactions. We studied kinetics of protein folding (Chapter 5) and ligand binding to proteins (Chapter 6). The concepts learned then are especially applicable to enzyme kinetics. We will use the partition function to understand several aspects of enzyme kinetics. The Michaelis–Menten equation or the kinetics of enzyme inhibition can easily be derived using the partition function method. We close the chapter with a detailed study of two examples of the kinetics and catalytic mechanism of enzyme reactions that are particularly well understood and provide excellent illustrations of the various concepts learned.

Chemical Reactions Proceed through a Transition State, or Activated Complex

Arrhenius (1889) studied experimentally the effect of temperature on the rate of inversion of cane sugar catalyzed by acid (the same reaction that Michaelis and Menten studied a few years later, catalyzed by invertase). He found that the rate constant of a reaction (k_r) depends exponentially on temperature. We express this temperature dependence by writing

$$k_r = A_0 e^{-Ea/RT}, \tag{7.1}$$

where E_a is the *activation energy* that the reactants must acquire for the reaction to take place. E_a determines the temperature dependence of k_r. A_0 is a pre-exponential factor, which may vary with temperature but much less than the exponential factor. In a first-order reaction, A_0 has units of frequency (s^{-1}).

The modern formulation of *transition state* theory is due to Eyring (1935). The transition state (‡), or *activated complex*, is the chemical species with the highest Gibbs energy ($\Delta G^{\ddagger}$) in the pathway from reactants to products (**Figure 7.1**).

Figure 7.1 The course of a chemical reaction. The transition state corresponds to the top of the Gibbs energy barrier (‡).

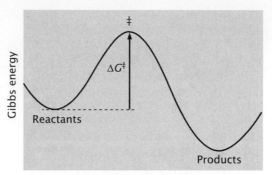

A simple example is the transition state of an S_N2 nucleophilic substitution (see Figure 5.18):

In Eyring's theory, the transition state ($AB^{\ddagger}$) is assumed to be in equilibrium with the reactants,

$$A + B \rightleftharpoons AB^{\ddagger}, \tag{7.2}$$

and is controlled by an equilibrium constant

$$K^{\ddagger} = \frac{[AB^{\ddagger}]}{[A][B]}. \tag{7.3}$$

The rate constant is written as

$$k_{\mathrm{r}} = \kappa \frac{kT}{h} e^{-\Delta G^{\ddagger}/RT} \tag{7.4}$$

$$= \left(\kappa \frac{kT}{h} e^{-\Delta S^{\ddagger}/R} \right) e^{-\Delta H^{\ddagger}/RT}.$$

The activation enthalpy $\Delta H^{\ddagger} \approx E_a$, and thus Equations 7.1 and 7.4 are equivalent. The factor kT in the pre-exponential is the product of the Boltzmann constant (k) and the temperature. The factor κ is called the *transmission coefficient*, which accounts for the probability of the transition state converting to products instead of going back to reactants (κ varies between 0 and 1).

Kramers (1940) formulated transition state theory in a slightly different but equivalent manner. The main difference is that the pre-exponential factor contains a *friction* $\eta^{\ddagger}$, which slows down the motion in the neighborhood of the transition state:

$$k_{\mathrm{r}} = \frac{kT}{\eta^{\ddagger}} e^{-\Delta G^{\ddagger}/RT}. \tag{7.5}$$

For motion in a liquid, $\eta^{\ddagger}$ corresponds to the viscosity encountered by the reactants approaching the energy barrier of the transition state. For a reaction in a protein, as in enzyme catalysis but also in protein folding, this friction reflects the dynamics of the protein, as it transitions from one conformation to the next, about the transition state. If the chemical reaction were activationless, this intrinsic protein internal viscosity would determine the rate. However, in most cases, the transition state barrier completely dominates the rate, because the effect of the exponential factor is much more important than that of the pre-exponential.

Enzymes are Optimized to Bind the Transition State

Enzymes, like chemical catalysts in general, enhance the rates of chemical reactions but do not alter the equilibrium between reactants, which we call *substrates*, and products. One of the first ideas to explain how enzymes function was the *lock-and-key* mechanism proposed by Fischer (1894). According to this idea, enzymes would be optimized to bind their substrates very well, hence explaining specificity. This idea was later modified by Koshland (1960) in the *induced-fit* mechanism. According to this concept, the protein undergoes a conformational change in which it adapts to improve binding of the substrate. Indeed, many enzymes, such as hexokinase and triose phosphate isomerase, undergo large conformational changes upon binding of a substrate. The same is true in many cases upon binding of other ligands that enhance or suppress the enzyme activity, which we call *activators* or *inhibitors*, respectively. However, this does not mean that substrate binding is tight. The binding constants of *substrates* are actually *not* very large. Their reciprocal, the dissociation constants (K_d) are typically in the range of μM to mM. Too tight binding would in fact be counterproductive: lowering the Gibbs energy of the substrate too much would *increase* the activation barrier to reach the transition state (**Figure 7.2**).

Kinetically, an enzyme works by lowering the Gibbs energy of the transition state of a chemical reaction. It does so because the *active site*, where the substrate binds and the chemistry takes place on the enzyme, is *complementary* to the transition state in the position and organization of the chemical groups. Binding stabilizes the transition state. The concept of transition-state stabilization by complementarity was clearly formulated by Pauling in 1948: "I think that enzymes are molecules that are complementary in structure to the activated complexes of the reactions that they catalyze.... The attraction of the enzyme for the activated complex would thus lead to a decrease in its energy, and hence... to an increase in the rate of the reaction."

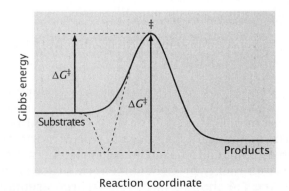

Figure 7.2 The effect of improving substrate binding (dashed line) without a concomitant decrease in the Gibbs energy of the transition state would be to increase the activation energy and reduce the rate of the reaction compared to reaction in the absence of enzyme (solid line).

Figure 7.3 The course of a reaction in the absence (solid line) and the presence (dashed line) of an enzyme.

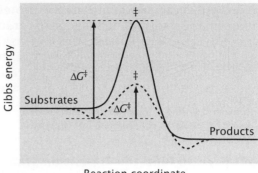

Therefore, we must redraw the Gibbs energy profile of the reaction to include the effect of the enzyme on the activation energy (**Figure 7.3**).

The enzyme also binds the product about as well as it binds the substrate, thus lowering the Gibbs energy of both, as indicated by the dashed line in Figure 7.3. However, the enzyme binds the transition state *better* than the substrate or the product, lowering its Gibbs energy much more. Finally, note that the enzyme lowers the activation energy in *both directions*, which means that it catalyzes the reaction product → substrate, as well as substrate → product.

Stabilization of the transition state is achieved through favorable interactions with the protein, including hydrogen bonds, electrostatic interactions, and optimal packing of nonpolar groups (London dispersion forces). These interactions help to "pay" for the *reduction in entropy* that is incurred when "forcing" the substrates to a small location in the enzyme (the active site), instead of being allowed to occupy any position in solution. The *position* and *orientation* of the substrates in the active site are optimized in the transition state for the chemical reaction to take place. Thus, the compensation of translational and rotational entropic costs of the reaction by interactions with the enzyme (called the binding energy) are facets of transition state stabilization. Binding of the substrate then appears as a necessary consequence of binding of the transition state, because of the structural similarity shared by the transition state and the substrate.

With a smaller activation barrier, the *rate constant k_r* of the reaction increases dramatically because it varies exponentially with $\Delta G^{\ddagger}$ (shown in Equation 7.4). This has been well understood. Chemical catalysts work in this way, too. But enzymes are much better catalysts than chemical ones. Enzymes typically enhance the rates of reactions by factors of 10^8 to 10^{12} compared to the same reaction in solution (enhancements are not always easy to measure; some probably reach $\sim 10^{15}$). These amazing rate enhancements have fascinated biochemists for decades. An increase in rate by 10^{10} corresponds to a decrease in $\Delta G^{\ddagger}$ by 14 kcal/mol.

How can enzymes possibly achieve this? Attempts to design catalysts as efficient as enzymes on the basis of optimal interactions with the transition state have generally failed. This is because the optimization of interactions with a particular transition state structure ignores the role of protein *conformational dynamics* in enzymatic catalysis. Indeed, the energy diagram of Figure 7.3 is a very simplified picture of what enzymes do. The dynamical aspects of enzyme conformations lie at the root of why proteins need to be such large molecules. Hence, we need to redraw the Gibbs energy profile of an enzyme-catalyzed reaction yet once more. **Figure 7.4** shows a more realistic representation.

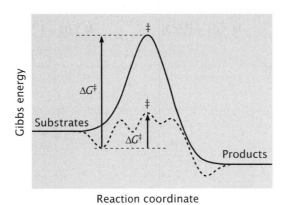

Figure 7.4 A more realistic picture of the effect of an enzyme (dashed line) is the stabilization of several transition states and intermediates in the reaction mechanism, by adopting various conformations along the reaction pathway.

Enzymes Provide a Reaction Pathway through a Series of Stabilized Intermediates

Enzymes must stabilize the transition states in all steps of a multi-step mechanism. They must also provide a *pathway* for the reaction to take place in a smoother way. Hence the need for intermediates. Those intermediates may be covalent or noncovalent. The path between them often involves acid–base catalysis or electrostatic interactions as a means to stabilize intervening transition states and intermediates. For example, in the case of serine proteases (see Figure 7.32), a good nucleophile is obtained on Ser-195 in the active site when His-57 accepts its proton; and a good leaving group is produced when the same histidine residue donates a proton to the amide nitrogen of the substrate.

There is a well-defined, covalent acyl-enzyme *intermediate* in the reaction catalyzed by serine proteases. In fact, there are many more intermediates in this reaction. Those intermediates have different structures. Thus, the enzyme must change conformation, if subtly, to complement each one. Hammond's (1955) postulate tells us that the energies and structures of a high-energy intermediate and an adjacent transition state are *linked*: "If two states, as for example, a transition state and an unstable intermediate, occur consecutively during a reaction process and have nearly the same energy content, their interconversion will involve only a small reorganization of the molecular structures." As the energy of the transition state decreases, so does the energy of an intermediate preceding or succeeding that transition state in the mechanism (see Figure 7.4). Acid–base catalysis or electrostatic interactions are particular (and common) examples of ways by which enzymes create lower energy pathways, but they are not special principles of enzyme catalysis. The principle is the stabilization of the transition state and intervening intermediates.

The *same* enzyme must be able to bind all those intermediates and transition states, which means it must be flexible enough to adopt different conformations along the reaction pathway. We will learn, with the example of dihydrofolate reductase, that the dynamics of the conformational changes are themselves important. Enzymes sample the available conformations quickly and cooperatively. That means the entire protein structure, not just the binding site, is involved in the conformational changes. This *plasticity* of the structure, to fit various intermediates and transition states that are separated by small Gibbs energies, requires that enzymes be large molecules.

7.2 LIGANDS WITH MULTIPLE CONTACTS REDUCE THE UNFAVORABLE ENTROPY LOSS UPON BINDING

One of the most important and least obvious ways by which enzymes achieve the stabilization of transition states is by minimizing the entropy loss that occurs in bimolecular reactions. By binding and pre-organizing the substrates, in the appropriate position and orientation for reactions to take place, enzymes achieve rate enhancements of many orders of magnitude.

Many Ligands Interact with Proteins at Multiple Sites

Many ligands in biochemical reactions bind to proteins through multiple interaction sites. For example, in the serine proteases, the substrate is an *extended* polypeptide chain. The polypeptide binds in a cleft on the surface of the enzyme and interacts with several residues beyond the active site, where the most important residues for catalysis are His-57, Asp-102, and Ser-195, called the "catalytic triad" (**Figure 7.5**).

The polysaccharides hydrolyzed by lysozyme also bind to the enzyme at a cleft. The same is true of the binding of RNA to ribonuclease A. However, the active site, where the chemistry takes place (hydrolysis in these cases), is only one component of the binding site. Indeed, binding sites in proteins are often extended. Extended binding sites exhibit high affinity without strong interactions at each contact between the substrate and the enzyme.

Binding to multiple sites is more favorable for the same reason that the binding of metal ions to multivalent organic molecules is stronger. This enhancement in binding obtained in a complex of a multidentate ligand with a metal ion is called the *chelate effect*. Examples are the binding of Ca^{2+} by ethylenediaminetetraacetic acid (EDTA) or Mg^{2+} by ATP (**Figure 7.6**). Chelating agents interact with the ions through multiple binding groups. The fundamental concept in the chelate effect is that the entropy lost in the binding reaction is paid mainly when the first interaction is established between the metal ion and the organic molecule. Subsequent interactions result in a much smaller entropy loss while contributing a similar favorable enthalpy as the first interaction.

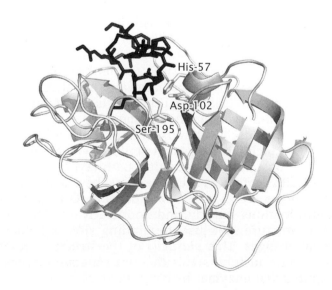

Figure 7.5 Trypsin with a proteolytically resistant polypeptide substrate analog (black) bound to the surface cleft. The residues of the catalytic triad are indicated: His-57, Asp-102, and Ser-195 (PDB 4HGC).

(A)   (B)

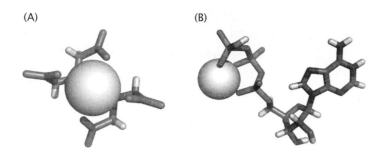

Figure 7.6 Structures of chelates. The multidentate (multivalent) complexes of (A) Ca^{2+}-EDTA complex (CSD MUJFEC) and (B) Mg^{2+}-ATP complex (CSD DECDIY01). The metal ions are shown as spheres. (CSD, Cambridge Structural Database.)

This problem is akin to that of the difference between bimolecular and intramolecular (unimolecular) reactions, which lies at the root of the extremely large acceleration of the rates of reactions by enzymes. Consider the two reactions shown in **Figure 7.7**.

On top we have a bimolecular reaction between two molecules A and B. This reaction is disfavored by the loss of entropy that occurs when the two molecules are brought together. They lose the freedom of being both anywhere in the solution. To make a bond, one molecule is forced to stay next to the other. On the bottom of Figure 7.7, we have an intramolecular reaction between the same molecular groups A and B. Now, however, A and B are already connected by other bonds, and are already forced to be next to each other. The entropic penalty of bringing B to the side of A has already been paid to a large extent when the intervening covalent bonds were made. Therefore, establishing the direct bond between A and B is now much more favorable than in the bimolecular reaction. How much more favorable?

Let us compare a reaction in which A and B are both in standard concentrations (1 M) in water to a reaction in which A is at 1 M but is entirely surrounded by B. The concentration of water is 55 M; if all of it could be replaced by B, the concentration of B would be 55 M. This is 55 times larger than the standard concentration of 1 M. If the intramolecular reaction brought B to the proximity of A as much as replacing all the water (solvent) by B, then we may say that the "effective concentration" of B is 55 M, because the reaction rate would be enhanced 55 times, whereas the concentration is actually only 1 M (standard conditions). Since the solvent concentration is about the maximum a solute of similar size could possibly achieve, we might reasonably expect a factor of 55 to be the maximum possible enhancement of the rate of an intramolecular over the corresponding bimolecular reaction at standard concentrations.

This idea, however, is incorrect. The enhancements in the rate and equilibrium and constants in intramolecular reactions are much greater than 55 times. Typical enhancements of reaction rates by enzymes are of the order of 10^8, and may be as large as 10^{15}. Enzymes position the reactants in orientations close to optimal, paying for the entropy loss of the substrate with the Gibbs energy of binding—that is, the Gibbs energy released when interactions, such as hydrogen bonds, hydrophobic interactions, ion pairing, and van der Waals contacts, are established with the substrates. This same kind of enhancement in the rate and the equilibrium constant is obtained when comparing bimolecular and intramolecular reactions.

Figure 7.8 shows the same chemical reaction, the formation of an anhydride from a carboxylate and its phenyl ester, in three different cases. In Figure 7.8A, acetic anhydride is formed from two molecules in a bimolecular reaction. In Figure 7.8B, the formation of succinic

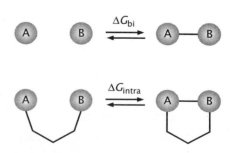

Figure 7.7 Bimolecular (top) and intramolecular (bottom) reactions have very different Gibbs energy changes.

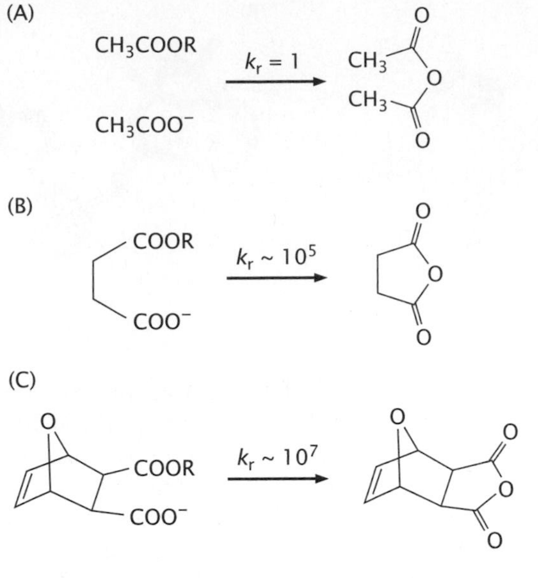

Figure 7.8 Relative rate constant (k_r) for formation of anhydrides in (A) a bimolecular reaction, (B) an intramolecular reaction, and (C) an intramolecular reaction in a rigid substrate (R = phenyl; k_r is set to 1 for the acetate reaction). (Data from Bruice TC & Pandit UK [1960] *J Am Chem Soc* 82:5858–5865, renormalized according to Kyte J [2006] Structure in Protein Chemistry, 2nd ed. Garland.)

(A)

$$CH_3COOR$$
$$CH_3COO^-$$
$k_r = 1$

(B)
$k_r \sim 10^5$

(C)
$k_r \sim 10^7$

anhydride occurs with a relative rate constant $k_r \sim 10^5$ times greater than the formation of acetic anhydride. The chemical reaction is the same, but A is bimolecular whereas B is intramolecular. The effective concentration of the carboxylate group in the succinic anhydride reaction is 10^5 M, because 10^5 times is the enhancement observed in the intramolecular reaction when the concentration is actually only 1 M (standard conditions). In the intramolecular reaction, we have effectively already paid for the loss of translational entropy—by pre-making the bonds that connect the two carboxylic acid groups in the reactant. Therefore, the reaction happens much faster than in A, where this penalty must still be paid. However, the loss of rotational entropy still has to be paid in the reaction of Figure 7.8B, which diminishes the maximum possible rate constant. This is because only a particular set of rotational angles is conducive to reaction.

As shown in **Figure 7.9**, to obtain the appropriate conformation of the molecule for the reaction, several rotational degrees of freedom must be "frozen." This freezing results in loss of entropy, reducing the rate of reaction. But in the reaction of Figure 7.8C, the rigid structure of the bicyclic reactant effectively freezes those rotations already. Consequently, the rate constant increases by another factor of $\sim 10^2$ relative to reaction B, or a total of $k_r \sim 10^7$ times relative to the bimolecular reaction A.

The reduction in translational entropy when two reactants are brought together to be combined is much larger than that which would result from a simple concentration effect, because the restrictions on the position and orientation of the reactants are much more severe than that simple estimate suggests. If it were just a matter of effective concentration, the entropy loss of bringing B close to A under standard conditions would simply be the difference between the reaction occurring at [B] = 1 M or 55 M. That is, the translational entropy loss would be $\Delta S_{trans} = -R \ln(55/1) = -8$ cal/K/mol, which corresponds to an increase in Gibbs energy of $\Delta G_{trans} = -T\Delta S_{trans} = 2.4$ kcal/mol at room temperature. However, as explained by Page and Jencks in their classic 1971 paper, restricting the position of a molecule in space by freezing its three translational degrees of freedom actually results in $\Delta S_{trans} \approx -30$ cal/K/mol, which corresponds to $\Delta G_{trans} = -T\Delta S_{trans} = 9$ kcal/mol at room temperature. In addition, fixing the orientation, by

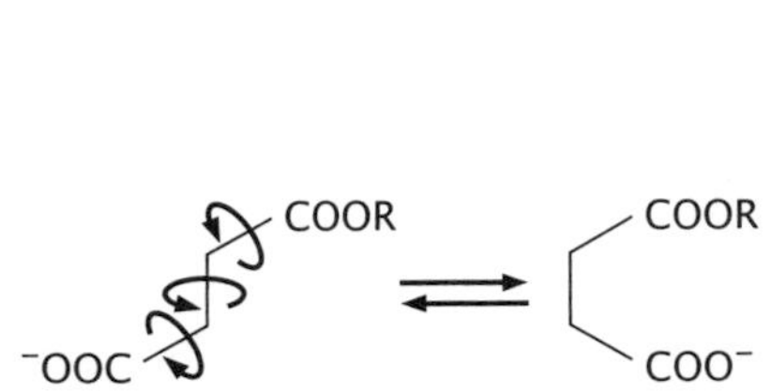

Figure 7.9 The freedom of rotation around three bonds is lost when the substrate assumes the conformation required for the reaction to occur.

freezing the three rotational degrees of freedom, contributes $\Delta S_{rot} \approx -20$ cal/K/mol, which corresponds to $\Delta G_{rot} = -T\Delta S_{rot} = 6$ kcal/mol at room temperature. These terms are slightly offset by the gain in internal degrees of freedom of rotation and vibrations in the product, but this amounts only to $\Delta S_{rot} \approx 10$ cal/K/mol. Thus, the Gibbs energy change corresponding to the combined loss of three translational and three rotational degrees of freedom is ~10–12 kcal/mol. This corresponds to a change in the rate or in the equilibrium constant for an intramolecular reaction compared to a bimolecular reaction by a factor of ~10^8, not 55.

The Coupling Gibbs Energy Depends on the Number of Intervening Bonds

To gain a deeper understanding of the effect of extended ligands in enzymatic reactions, let us consider binding of a ligand that contains two groups A and B, which bind to two different sites on a protein. We will examine the effect of an increasing degree of linkage: first, A and B are separate molecules; second, A and B are connected by a long flexible linker; and third, A and B are connected by a short linker.

When the two ligands are separate, we have the same case as when a ligand X bound to two different sites, A and B, on the protein. We studied this case in Section 6.2. We measured how binding of X to site A depends on whether site B is occupied by the *coupling* Gibbs energy between the two sites (shown in Equation 6.69). Now the ligands themselves are different (A and B), but the problem is essentially the same. The situation is shown in **Figure 7.10**. A and B bind to two different sites on the protein (each ligand binds only to its site). Binding of one ligand does not affect the binding of the other, so the ligands bind independently. The partition function is

$$Q = 1 + K_A[A] + K_B[B] + K_A K_B[A][B]. \tag{7.6}$$

Because binding is independent, Q can also be written as a product of two independent partition functions, one for A and one for B:

$$Q = (1 + K_A[A])(1 + K_B[B]) \tag{7.7}$$

$$= Q_A Q_B. \tag{7.8}$$

The coupling Gibbs energy is expressed in terms of the Gibbs energies of binding defined in Figure 7.10. ΔG_A is the Gibbs energy of binding of group A to its site when site B is free, and $\Delta G_{A(B)}$ is the Gibbs energy of binding of A to its site when site B is occupied. Similarly, ΔG_B

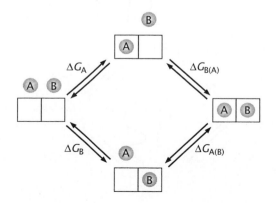

Figure 7.10 Binding of two different ligands to two different sites on a protein.

is the Gibbs energy of binding of group B to its site when site A is free, and $\Delta G_{B(A)}$ is the Gibbs energy of binding of B to its site when site A is occupied. The corresponding equilibrium binding constants are related to the standard Gibbs energy changes by $\Delta G_A^o = -RT \ln K_A$, $\Delta G_{A(B)}^o = -RT \ln K_{A(B)}$, $\Delta G_B^o = -RT \ln K_B$, and $\Delta G_{B(A)}^o = -RT \ln K_{B(A)}$. In this case, $K_{A(B)} = K_A$ because binding of B does not affect binding of A (similarly, $K_{B(A)} = K_B$). The coupling Gibbs energy is given by

$$\Delta G(A|B) = \Delta G_{A(B)} - \Delta G_A \qquad (7.9)$$

$$= \Delta G_{B(A)} - \Delta G_B. \qquad (7.10)$$

Here $\Delta G(A|B) = 0$, because the sites are independent.

Consider now an extended ligand that binds to two sites on a protein (**Figure 7.11**). The ligand A–B contains the same two binding groups A and B, both of which interact with the protein, but these two groups are now connected by a *long flexible linker*. If group A is already bound to its site, the effective concentration of B is likely to be larger close to its own binding site, because of the linker. Binding is no longer independent. The Gibbs energy of binding of B in the absence of A is again designated by ΔG_B, and if A is already bound, by $\Delta G_{B(A)}$. Going through the upper branch of the diagram of Figure 7.11, we see that the Gibbs energy ΔG_{AB} of binding of the entire ligand A–B to the enzyme is

$$\Delta G_{AB} = \Delta G_A + \Delta G_{B(A)}, \qquad (7.11)$$

or, going through the lower branch,

$$\Delta G_{AB} = \Delta G_B + \Delta G_{A(B)}. \qquad (7.12)$$

The Gibbs energies $\Delta G_{B(A)}$ and ΔG_B (or the equilibrium constants $K_{B(A)}$ and K_B) are no longer necessarily identical (the same applies to $\Delta G_{A(B)}$ and ΔG_A, or $K_{A(B)}$ and K_A). Therefore, ΔG_{AB} is *not* necessarily the sum of the contributions ΔG_A and ΔG_B from the binding of A and B individually. Thus, in general

$$\Delta G_{AB} \neq \Delta G_A + \Delta G_B, \qquad (7.13)$$

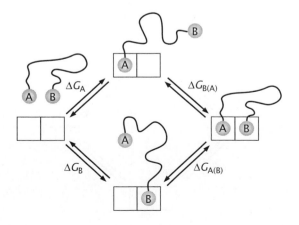

Figure 7.11 Binding of an extended flexible ligand to two sites on a protein.

or

$$K_{AB} \neq K_A K_B, \tag{7.14}$$

where K_{AB} is the binding constant for the entire A–B ligand ($\Delta G_{AB}^o = -RT \ln K_{AB}$). Rather, if we combine Equations 7.9, 7.10, and 7.11, we find that in general,

$$\Delta G_{AB} = \Delta G_A + \Delta G_B + \Delta G(A|B), \tag{7.15}$$

or equivalently,

$$\Delta G(A|B) = \Delta G_{AB} - \Delta G_A - \Delta G_B. \tag{7.16}$$

Thus, $\Delta G_{AB} = \Delta G_A + \Delta G_B$ only if $\Delta G(A|B) = 0$.

Now compare this case with the previous one. If A and B are present at small concentrations, typically <10 mM, then most likely $\Delta G(A|B) = \Delta G_{B(A)} - \Delta G_B < 0$, because the entropic loss in the bimolecular case (ΔG_B) is large. However, if we compare *standard* Gibbs energy changes, which refer to 1 M concentrations, it is *not* necessarily true that $\Delta G^o(A|B) < 0$. This is because 1 M concentrations are fairly high, which means that, on average, B groups are already fairly close to their binding site even without the linker. The value of $\Delta G^o(A|B)$ depends on the details of the linker.

This is a good point to note the following. First, to meaningfully interpret Gibbs energy changes, it is essential to clearly specify the standard state. If we had adopted a different standard state, such as a mole fraction of 1 instead of a concentration of 1 M, $\Delta G^o(A|B)$ would have a completely different value. The reason is physical: our choice of standard state affects the value of $\Delta G^o(A|B)$ mainly because it affects the value of the translational entropy loss at the standard concentration.

Second, it has been often stated that the equality $K_{AB} = K_A K_B$ could *not* be true *because* the units of K_A and K_B are M^{-1}, and therefore K_{AB} would have units of M^{-2}, whereas in fact it has units of M^{-1}, as any other binding constant. This argument is incorrect. Indeed, the equality $K_{AB} = K_A K_B$ *may* be valid in special cases. It is *not* valid in general, but not because of units. We saw in Sections 1.5 and 6.1 that biochemists traditionally write units in binding constants as a reminder of the standard state, and as a matter of convenience, because then dissociation constants have units of concentration. However, strictly speaking, binding constants are ratios of activities, so they are *unitless*. The reason $K_{AB} \neq K_A K_B$ in general is that the presence of the linker between the groups A and B alters the translational and rotational entropy losses upon binding of one of the groups if the other is already bound; it also alters low-frequency motions of the complexes formed. The final manifestations of these effects in aqueous solutions can be quite complex.

Finally consider the case where the link between the parts A and B of the ligand is short but still allows the two groups to bind optimally, as illustrated in **Figure 7.12**. Now, when A binds to its site, B is very close to its own site and it is much more likely to bind as well. In the top branch of Figure 7.12, the translational entropy loss must be paid for the binding of A, but not much for the binding of B, because, after A binds, B is already very close to its site. Therefore, the Gibbs energy of binding of B if A is already bound ($\Delta G_{B(A)}$) is more negative (more favorable) than if B were the first to bind (ΔG_B). That is, $\Delta G_{B(A)} < \Delta G_B$, and the coupling Gibbs energy is favorable (negative), $\Delta G(A|B) = \Delta G_{B(A)} - \Delta G_B < 0$.

Figure 7.12 Binding of a rigid ligand containing the groups A and B to two sites on a protein.

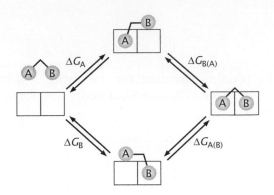

There are many examples of improved binding due to multiple interactions of ligands with enzymes. The classical case is the binding of the peptide Ac-Pro-Ala-Pro-Ala-CHO to the serine protease elastase, an enzyme of the trypsin family. This peptide is modified by acetylation (Ac, CH_3CO) of the N-terminus (see Section 2.1) and reduction of C-terminal carboxylic acid (COOH) to an aldehyde (CHO). The amino acid residues of the peptide interact with several sites close to the active site of the enzyme; they correspond to our A group. The C-terminal aldehyde makes a covalent (but reversible) bond to the Ser-195 residue on the active site of elastase; it corresponds to our B group. Binding of the A group alone has $\Delta G_A^o = -4.5$ kcal/mol; binding of the aldehyde (B group) alone has $\Delta G_B^o = +2.5$ kcal/mol; but binding of the entire peptide aldehyde (A–B ligand) has $\Delta G_{AB}^o = -8.9$ kcal/mol. Thus, from Equation 7.16, the coupling Gibbs energy is $\Delta G(A|B) = -8.9 + 4.5 - 2.5 = -6.9$ kcal/mol. This is a large favorable coupling Gibbs energy caused by the linker. It corresponds to a Boltzmann factor of $e^{-\Delta G(A|B)/RT} = 10^5$, which means that the binding constant for the full A–B ligand is enhanced by a factor of 10^5 relative to the independent binding of the A and B groups: $K_{AB} = 10^5 K_A K_B$.

As a second example, we examine a case where the linker causes an unfavorable $\Delta G(A|B)$. This is the binding of bivalent synthetic inhibitors to the enzyme peptidyl-prolyl *cis/trans* isomerase (PPIase), which catalyzes the isomerization of a peptide bond preceding a proline residue. The human enzyme PPIase has two domains, the WW domain (residues 5–39) and the PPIase domain (residues 52–163), connected by a flexible loop. PPIase inhibitors were synthesized, which consist of two peptides connected by the *inflexible* linker $(Pro)_5$: Ac-[Phe-DGlu-Pip-Nal-Gln]-$(Pro)_5$-[Ala-Bth-Thr-(PO_3^{2-})-Pro-Cha-Gln]-NH_2. (Note: the peptides contain a phosphorylated (PO_3^{2-}) threonine and several nonstandard amino acids: Bth (benzothienylalanine), Cha (cyclohexylalanine), DGlu (D-glutamic acid), Nal (naphtylalanine), and Pip (piperidine-2-carboxylic acid). The N-terminus is acetylated and the C-terminus is amidated.) The sequence before the proline linker is group A, which binds to the PPIase domain with $\Delta G_A = -5.6$ kcal/mol. The sequence after the linker is group B, which binds to the WW domain with $\Delta G_B = -6.8$ kcal/mol (the linker by itself binds negligibly to the enzyme). Binding of the entire A–B peptide occurs with $\Delta G_{AB}^o = -8.9$ kcal/mol. Thus, the coupling Gibbs energy is $\Delta G(A|B) = -8.9 + 6.8 + 5.6 = +3.5$ kcal/mol. This is an *unfavorable* coupling Gibbs energy, probably caused by the rigidity of the linker, which allows parts A and B to bind to their respective sites on the enzyme, but not optimally.

7.3 THE DISTINCTIVE CONCEPT OF ENZYME KINETICS IS THE FORMATION OF AN ENZYME–SUBSTRATE COMPLEX

Formation of a Michaelis–Menten Complex Leads to Saturation in the Reaction Kinetics

In 1913 Michaelis and Menten published their classic paper on the kinetics of enzyme catalysis. They studied the reaction catalyzed by the enzyme invertase, the hydrolysis of the disaccharide sucrose to the monosaccharides glucose and fructose, shown in **Figure 7.13**.

The enzyme was called *invertase* because the *optical rotation* of the carbohydrate solution was inverted, from positive to negative, when sucrose is hydrolyzed to glucose and fructose. We can write this reaction in a simplified form, using S for sucrose, G for glucose, and F for fructose, as

$$S \quad \underset{\text{Invertase}}{\overset{H_2O}{\rightleftharpoons}} \quad G + F. \tag{7.17}$$

The reaction is reversible. The enzyme does not change the equilibrium concentrations of the three saccharides, but drastically reduces the time needed to reach equilibrium.

Michaelis and Menten envisioned the invertase-catalyzed reaction as

$$E + S \;\rightleftharpoons\; ES \;\rightleftharpoons\; E + G + F, \tag{7.18}$$

where E is the *free* enzyme, S is the *substrate* (sucrose), and ES is the *enzyme–substrate complex*. The products are glucose and fructose. This reaction scheme means that before the chemical reaction takes place, a noncovalent association between enzyme and substrate *must* occur. Further, it means that the enzyme–substrate complex has a defined stoichiometry. This simple concept has far-reaching consequences for the kinetics. It is the main feature that distinguishes enzyme kinetics from regular chemical kinetics. Indeed, the postulation of the existence of an enzyme–substrate complex is the most important contribution of Michaelis and Menten to our understanding of enzyme kinetics. We honor them by calling it the *Michaelis–Menten complex*.

Measuring the kinetics of the reaction of Equation 7.18 is complicated by the binding of products to the enzyme. However, this complication is eliminated if we start from a solution of sucrose and enzyme, and measure the *initial rate*, *v*, of the reaction (or initial velocity). In the beginning there are essentially no products (glucose and fructose) present that can bind to the enzyme, and the second step is

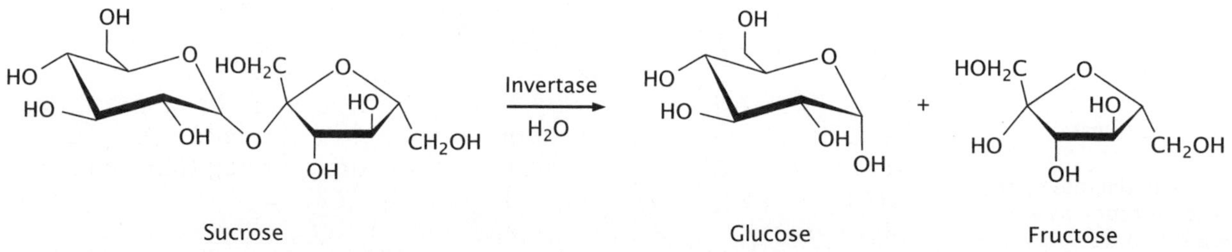

Figure 7.13 The reaction catalyzed by invertase.

effectively irreversible. We can therefore simplify Equation 7.18 to the *Michaelis–Menten mechanism*,

$$E + S \underset{k_{-1}}{\overset{k_1}{\rightleftharpoons}} ES \overset{k_2}{\longrightarrow} E + P, \tag{7.19}$$

where S stands for substrate and P for product. The rate constant for the binding of substrate to the enzyme is k_1, that for dissociation of the enzyme–substrate complex is k_{-1}, and the rate constant of the chemical step, the transformation of the substrate to product *on the enzyme*, is k_2. Michaelis and Menten postulated that the rate of the reaction, measured by the rate of product formation, is proportional to the concentration of the enzyme–substrate complex,

$$v = \frac{d[P]}{dt} = k_2[ES]. \tag{7.20}$$

The problem now is to calculate [ES].

Suppose you measured the initial rate of the reaction catalyzed by invertase, with concentrations of substrate and enzyme similar to those used by Michaelis and Menten. Then if you plotted the observed initial rate as a function of the initial substrate concentration, you would obtain a curve similar to that shown in **Figure 7.14**.

The shape of this curve should be very familiar to you; it is the shape of a simple binding isotherm. Like the binding isotherm, the initial rate of an enzymatic reaction exhibits *saturation*: beyond a certain substrate concentration, the rate approaches its maximum value (V_{max}). This is because at that point all enzyme molecules are bound with substrate, in form of the ES complex, so no increase in rate is possible by increasing the substrate concentration further. The total enzyme concentration [E0] is the sum of the free enzyme concentration [E] and bound enzyme concentration [ES],

$$[E_0] = [E] + [ES]. \tag{7.21}$$

When all the enzyme has substrate bound, $[E] \approx 0$ and $[ES] = [E_0]$. This means that no enzyme molecule is idle: the reaction proceeds at the maximum possible rate, V_{max}. Thus, from Equation 7.20,

$$V_{max} = k_2[E_0]. \tag{7.22}$$

The equation that describes the curve in Figure 7.14 is the *Michaelis–Menten equation*,

$$v = \frac{V_{max}[S]}{K_M + [S]}, \tag{7.23}$$

where K_M is called the Michaelis constant.

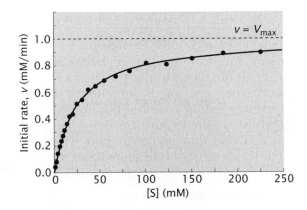

Figure 7.14 The initial rate of an enzyme-catalyzed reaction as a function of free substrate concentration. V_{max} (dashed line) is the maximum rate. (Simulated data.)

The Michaelis–Menten Equation Describes Steady-State Enzyme Kinetics

The modern version of the Michaelis–Menten equation was derived some 10 years later by Briggs and Haldane (1925). They calculated [ES], to use in Equation 7.20, by assuming that a *steady state* is reached shortly after the reaction begins. Let us try to understand what this means. We begin with the kinetic scheme of Equation 7.19,

$$E + S \underset{k_{-1}}{\overset{k_1}{\rightleftharpoons}} ES \overset{k_2}{\longrightarrow} E + P,$$

and write the differential equations that express the *rate of change* of each species with time,

$$\frac{d[E]}{dt} = -k_1[E][S] + (k_{-1} + k_2)[ES] \tag{7.24}$$

$$\frac{d[S]}{dt} = -k_1[E][S] + k_{-1}[ES] \tag{7.25}$$

$$\frac{d[ES]}{dt} = k_1[E][S] - (k_{-1} + k_2)[ES] \tag{7.26}$$

$$\frac{d[P]}{dt} = k_2[ES]. \tag{7.27}$$

There is a brief initial period of rapid change in [P], [S], and [ES]. The changes in these concentrations in this period are called *transient kinetics*. The ES complex forms rapidly in the beginning of the reaction, but its concentration remains small because the amount of total enzyme is much smaller than that of substrate. After that, [ES] remains essentially constant for a while. This second period is the *steady state.*

The steady-state concentration of ES is *not* its equilibrium concentration; but like the equilibrium concentration, it does not change with time during this period. Thus, the *steady-state condition* applied to the ES complex is

$$\frac{d[ES]}{dt} = 0. \tag{7.28}$$

After the transient kinetics, during the steady state of [ES], [P] increases as a linear function of time. This means that the *rate* of product formation is constant. This rate is what we measure as the initial velocity of the reaction in the steady state (shown in Equation 7.20). It is constant as long as [ES] is constant.

To determine [ES] we use the steady-state condition (shown in Equation 7.28). The change of [ES] with time is given by Equation 7.26, which is zero in the steady state,

$$k_1[E][S] - (k_{-1} + k_2)[ES] = 0. \tag{7.29}$$

We do not know the concentration of free enzyme [E], but only the total enzyme concentration, $[E_0] = [E] + [ES]$ (shown in Equation 7.21). Substituting this expression for $[E_0]$ in Equation 7.29, we can solve for [ES] to obtain

$$[ES] = \frac{[E_0][S]}{\frac{k_2+k_{-1}}{k_1} + [S]}. \tag{7.30}$$

If we substitute [ES] from Equation 7.30 in Equation 7.20, the rate of the reaction becomes

$$v = \frac{k_2[E_0][S]}{\frac{k_2+k_{-1}}{k_1} + [S]}.$$ (7.31)

Thus, the *Michaelis constant* K_M is given by

$$K_M = \frac{k_2 + k_{-1}}{k_1}.$$ (7.32)

Finally, we substitute V_{max} for $k_2[E_0]$ in Equation 7.31 and use the definition of K_M to obtain the Michaelis–Menten equation,

$$v = \frac{V_{max}[S]}{K_M + [S]}.$$ (7.33)

K_M is the Concentration of Substrate at which the Rate is Half-maximum

Intuitively, we know that the reaction rate is maximal when *all enzyme* has substrate bound, because no enzyme molecule is idle. This conclusion also follows from the Michaelis–Menten Equation 7.33 under the condition that the substrate concentration is so large that the enzyme binding sites approach *saturation* by the substrate: they are all occupied. This happens when $[S] \gg K_M$. Then, $K_M + [S] \approx [S]$ in the denominator of Equation 7.33, and $v = V_{max}$. The horizontal line in **Figure 7.15** indicates V_{max}; its value is approached closer and closer by v as the initial substrate concentration increases.

We defined K_M by Equation 7.32

$$K_M = \frac{k_2 + k_{-1}}{k_1}.$$

Two important extremes of K_M arise depending on the relation between the magnitudes of k_2 and k_{-1} in the numerator. If $k_2 \ll k_{-1}$, then

$$K_M = \frac{k_{-1}}{k_1}$$ (7.34)

$$= K_S,$$ (7.35)

where K_S is the equilibrium dissociation constant of the ES complex,

$$K_S = \left(\frac{[E][S]}{[ES]} \right)_{eq}.$$ (7.36)

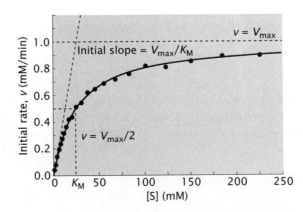

Figure 7.15 The initial rate of an enzyme-catalyzed reaction as a function of substrate concentration. The horizontal dashed line shows V_{max}. The inclined dashed line shows the initial slope of the curve. K_M (25 mM) is indicated on the x-axis; it corresponds to the point where $v = V_{max}/2$ on the curve.

On the other hand, if $k_2 \gg k_{-1}$, then

$$K_M = \frac{k_2}{k_1}. \tag{7.37}$$

There is an additional, important meaning of K_M, which we will use when applying the partition function method to the calculation of enzyme reaction rates. We can treat K_M as an *apparent dissociation constant* in the *steady state*. To see why this is valid, consider again the application of the steady-state condition $d[ES]/dt = 0$ (shown in Equation 7.29),

$$k_1[E][S] - (k_{-1} + k_2)[ES] = 0.$$

Rearranging this equation, we obtain

$$\left(\frac{[E][S]}{[ES]}\right)_{ss} = \frac{k_{-1} + k_2}{k_1}, \tag{7.38}$$

where we have used the subscript ss to emphasize that this is valid in the *steady state*. The right-hand side of Equation 7.38 is K_M. Thus, K_M has exactly the *form* of an equilibrium dissociation constant. However, the concentrations on the left-hand side of Equation 7.38 are *not* equilibrium concentrations: they are *concentrations in the steady state*. In more complicated reaction mechanisms, with more than one enzyme-bound intermediates, Equation 7.38 is still valid as long as we *interpret* [ES] in the denominator as the sum of the concentrations of *all* enzyme-bound species.

The Catalytic Constant k_{cat} Measures the Rate at which the Enzyme Transforms the Bound Substrate

In the simple Michaelis–Menten reaction mechanism, k_2 is the rate constant for the conversion of the ES complex to products (and free enzyme).

$$ES \overset{k_2}{\longrightarrow} E + P.$$

In other words, k_2 is the rate constant for the transformation of the substrate *on the enzyme*. It is the rate constant for the *catalytic step* in the mechanism, the step in which the chemical transformation to product takes place. Therefore, *in this simple reaction mechanism, k_2 is the catalytic constant*, abbreviated as k_{cat}. We emphasize that $k_2 = k_{cat}$ only applies to *this* scheme, because in more complex enzyme reaction mechanisms, k_2 is *not* identical to k_{cat}.

The rate of the reaction is $v = k_2[ES]$. Under saturating concentrations of S, all enzyme is bound ($[ES] = [E_0]$), and the reaction proceeds at its maximum velocity. If we substitute k_{cat} for k_2 and $[E_0]$ for [ES] in Equation 7.20, we obtain

$$V_{max} = k_{cat}[E_0]. \tag{7.39}$$

This is a first-order rate equation (it depends linearly on concentration). Thus, k_{cat} is the first-order rate constant of the enzyme-catalyzed reaction *observed at saturation*. Think of it this way: if you give the enzyme all the substrate it can possible handle, the rate is now determined only by the catalysis itself—by how good the enzyme is—that is, k_{cat}. The units of k_{cat} are s^{-1} (or min^{-1}). Rearranging, this equation, we have

$$k_{cat} = \frac{V_{max}}{[E_0]}.$$ (7.40)

Equation 7.40 *defines* k_{cat}. That is, k_{cat} is the maximum rate divided by the total enzyme concentration. This definition applies to *any* enzyme-catalyzed reaction that obeys Michaelis–Menten kinetics, not just to those that obey this simple scheme. Therefore, in general, we can write the Michaelis–Menten equation as

$$v = \frac{k_{cat}[E_0][S]}{K_M + [S]}.$$ (7.41)

The Specificity Constant k_{cat}/K_M Measures the Catalytic Efficiency of the Enzyme

At sufficiently low substrate concentration, we reach the condition $[S] \ll K_M$. In this case, we can disregard $[S]$ in the denominator of the Michaelis–Menten Equation 7.33 to obtain

$$v = \frac{V_{max}}{K_M}[S].$$ (7.42)

Thus, when the enzyme is far from saturation, the rate depends linearly on substrate concentration. The initial slope of the graph of Figure 7.15 is V_{max}/K_M. In the regime of low $[S]$, $[P]$ increases linearly with the concentration of substrate.

Furthermore, if $[S] \ll K_M$, almost all the enzyme is free, $[E_0] \approx [E]$. Then, if we substitute $V_{max} = k_{cat}[E_0]$ and $[E_0] = [E]$ in Equation 7.42, we obtain

$$v = \frac{k_{cat}}{K_M}[E][S].$$ (7.43)

This is a second-order rate equation—it depends on two concentrations. So, you see, k_{cat}/K_M is an *apparent second-order* rate constant for the reaction between free enzyme E and substrate S to produce P (and free enzyme). The units of k_{cat}/K_M are $M^{-1}s^{-1}$ (or $M^{-1}min^{-1}$). The constant k_{cat}/K_M incorporates binding specificity, included in K_M, and catalytic ability, in k_{cat}. It increases as K_M decreases (better binding) and as k_{cat} increases (better catalysis). Because k_{cat}/K_M weighs in the contributions of how good an enzyme is at catalyzing the transformation of a particular substrate and how well it binds that substrate, it embodies the specificity of the enzyme for that substrate. Therefore, k_{cat}/K_M is called the *specificity constant*. It is the most useful in comparing the same enzyme-catalyzed reaction on different substrates. Rather than the pseudo-equilibrium constant K_M, enzymologists now consider the kinetic constants k_{cat} and k_{cat}/K_M as the most fundamental steady-state parameters.

We derived Equation 7.43 under the condition that $[S] \ll K_M$. However, it is actually valid under *any* steady-state conditions. The rate of the reaction is given by

$$v = k_{cat}[ES],$$ (7.44)

which is the general form of our initial Equation 7.20, with k_{cat} replacing k_2. But in the steady state, we can always write $[ES] = [E][S]/K_M$; substitution of [ES] in Equation 7.44 immediately yields Equation 7.43.

The Constant k_{cat}/K_M Can Never Exceed the Diffusion Limit

How large can k_{cat}/K_M be? There is a limit to the rate of any second-order reaction: it can never happen in a time shorter than the time it takes for the two reactants to find each other by random motion in solution. This largest possible rate is called the *diffusion limit*. Reactions that occur at this rate are said to be *diffusion-controlled*. An estimate of the diffusion-limited rate constant k_s can be obtained from the rate of random collisions of two spheres with radii r_A and r_B and diffusion coefficients D_A and D_B, which is given by the Smoluchowski equation,

$$k_s = 4\pi RD, \tag{7.45}$$

where $R = r_A + r_B$ and $D = D_A + D_B$. The physical reason for this equation is simple. There is collision (contact) when the centers of two spheres are at a distance R of each other (**Figure 7.16**). This means that the center of B is located on the surface of a sphere of radius $R = r_A + r_B$ centered on A (or vice versa).

The surface of this sphere is $4\pi R^2$. This is the origin of the factor $4\pi R$ in Equation 7.45. (The factor $4\pi R$ contains R instead of R^2 because the rate constant is determined by the *gradient* of the concentration of B, $(d[B]/dr)_{r=R}$ at the surface ($r = R$), which brings in a factor of $1/R$.) The factor D is the relative diffusion coefficient of spheres A and B, which tells us how fast they approach each other by random motion in solution. (The sum of D_A and D_B is dominated by the larger of the two diffusion coefficients. Small molecules diffuse faster than large ones, so if the substrate is a small molecule its diffusion coefficient largely determines D. Diffusion coefficients of small molecules and small proteins in water are $\approx 5 \times 10^6$ and $0.5 \times 10^6 \, cm^2 s^{-1}$, respectively.) With a typical $D \approx 10^6 \, cm^2 s^{-1}$ in water, we obtain $k_s \approx 10^{10} \, M^{-1} s^{-1}$. Now, not all the surface of the enzyme is reactive—only the active site. If we make the rough assumption that 1% of the collisions occur close enough to the active site, we obtain a diffusion limit of $k_s \approx 10^8$ $M^{-1} s^{-1}$ for enzyme-catalyzed reactions. The value of k_{cat}/K_M for many enzymes approaches this limit, which means that they are as efficient as possible.

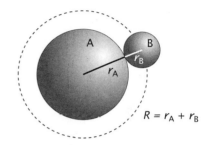

Figure 7.16 A collision of two solutes (spheres) occurs when the center of B is on a sphere of radius $r_A + r_B$ centered on A.

The Double-Reciprocal Lineweaver–Burk Plot is Useful in Qualitative Diagnostic of Enzyme Mechanisms and Inhibition

In the days before computers, which include the development of enzyme kinetics more than 50 years after the Michaelis–Menten paper, only linear models could easily be fitted to experimental data. The Michaelis–Menten equation is not linear. Lineweaver and Burk (1934) showed that it can easily be converted into a mathematically equivalent equation by taking the reciprocal of both sides. The Lineweaver–Burk double-reciprocal equation is linear and allows easy fitting to data by linear regression to obtain the kinetic parameters. Today, with the widespread use of computers, this aspect is of little importance. The Michaelis–Menten equation should be fitted to experimental data directly using a nonlinear fitting program with a computer. In fact, the errors incurred in using the Lineweaver–Burk equation are large.

However, the double-reciprocal plot is still very useful in qualitative or semi-quantitative analysis, particularly in enzyme inhibition and multisubstrate enzyme reactions because of the easy diagnostic it provides of the type of inhibitor or mechanism.

To derive the Lineweaver–Burk equation we begin with the Michaelis–Menten Equation 7.33,

$$v = \frac{V_{max}[S]}{K_M + [S]}$$

and take the reciprocal of each side,

$$\frac{1}{v} = \frac{K_M + [S]}{V_{max}[S]}.$$

Rearranging yields the standard form of the *Lineweaver–Burk equation*,

$$\frac{1}{v} = \frac{K_M}{V_{max}}\frac{1}{[S]} + \frac{1}{V_{max}}. \tag{7.46}$$

This is the equation of a straight line with slope K_M/V_{max}, y-intercept $1/V_{max}$, and x-intercept $-1/K_M$. The Lineweaver–Burk plot of the data of Figure 7.15 is shown in **Figure 7.17**.

This seems all fine, but you must be aware of some important caveats with the Lineweaver–Burk plot. Points located at low [S] and low v in the Michaelis–Menten plot appear at high $1/[S]$ and high $1/[v]$ in the Lineweaver–Burk plot. Conversely, points originally at high [S] and high v appear at low $1/[S]$ and low $1/[v]$ in the Lineweaver–Burk plot. Normally, the data should be collected to cover the Michaelis–Menten *line* uniformly, which means collecting more points in the interval of greatest change, about K_M. However, when the Lineweaver–Burk transformation is performed, the data get squeezed on the left side of the plot (see Figure 7.17). The accumulation of data close to the y-axis helps to determine the y-intercept accurately ($1/V_{max}$). Conversely, though, the data are sparse on the right side of the plot. This is a serious problem because those data have the greatest influence on the slope of the line, which determines K_M. Furthermore, the data at high $1/[S]$ correspond to low concentrations, and thus carry the largest experimental error, both in the concentration and the rate. Yet they carry the largest weight in the determination of the slope (and the x-intercept).

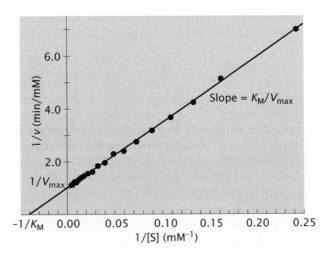

Figure 7.17 The Lineweaver–Burk plot of the initial rate of an enzyme-catalyzed reaction. The data are replotted from Figure 7.15.

7.4 THE MICHAELIS–MENTEN EQUATION APPLIES TO MOST ENZYME REACTION MECHANISMS

The K_M and k_{cat} are in General Combinations of Rate Constants and Concentrations

You might think that the Michaelis–Menten equation is of limited use if it only applies to such simple mechanisms. It turns out that it applies to the most complex mechanisms, even those involving many intermediates or multiple substrates, as long as binding of each substrate follows a hyperbolic binding isotherm. In other words, it works if binding is *not cooperative*. There is, however, an important change. In more complex cases, K_M is *not* given by $(k_2 + k_{-1})/k_1$. Similarly, k_{cat} is no longer equal to k_2. Instead, both K_M and k_{cat} become combinations of multiple rate constants. Those expressions can be complex, depending on the reaction mechanism, but the general form of the Michaelis–Menten equation is still valid.

The Rates of Complex Mechanisms can be Determined by the King–Altman Method

Before we examine more complicated cases, let us summarize the steps in solving the rate equations for the simple Michaelis–Menten scheme. First, we wrote the rate of the reaction as $v = d[P]/dt = k_2[ES]$ (shown in Equation 7.20). Second, we solved the differential rate equations by applying the steady-state condition to the enzyme-bound complex, $d[ES]/dt = 0$ (shown in Equation 7.28). Third, once we obtained an expression for [ES], we substituted it in Equation 7.20. Solving the differential equations in complex cases, however, is more complicated, prone to mistakes, and tedious.

To circumvent these problems, King and Altman (1956) devised a graphical method to solve enzyme rate equations in the *steady state*. The King–Altman method resembles the partition function method in several ways. In both cases, we seek to determine the fraction f_{ES} of the ES complex, or another enzyme-bound complex. However, in the partition function method, we obtain the fractions of complexes at *equilibrium*, whereas in the King–Altman method we obtain the fractions in the *steady state*. Now we will derive the Michaelis–Menten equation for a few enzyme mechanisms of increasing complexity. In each case, we want to determine K_M and k_{cat}. We will illustrate the King–Altman method with our first example.

Consider a scheme with one more enzyme-bound complex, EP, formed irreversibly from ES before the release of product,

$$\text{E} + \text{S} \underset{k_{-1}}{\overset{k_1}{\rightleftharpoons}} \text{ES} \overset{k_2}{\longrightarrow} \text{EP} \overset{k_3}{\longrightarrow} \text{E} + \text{P}. \qquad (7.47)$$

In this example, $f_{EP} = [EP]/[E_0]$, where $[E_0]$ is the total concentration of the enzyme, which is the *sum* of the concentrations of all the different enzyme species. The rate is given by

$$v = k_3[EP] \qquad (7.48)$$

$$= k_3 f_{EP}[E_0]. \qquad (7.49)$$

$[E_0]$ is known; thus, if we calculate f_{EP}, we obtain the rate. Let us then calculate the fraction f_{EP} in the steady state.

The important concept in the King–Altman method is that the concentration of each enzyme species in the *steady state* is proportional to its *rate of formation* by all possible pathways leading to that enzyme species. A graphical method is used to calculate the rates of formation and therefore the fractions of the enzyme species. We now describe the method, with reference to **Figure 7.18**.

First we write the mechanism of Equation 7.47 as a cycle, as shown in Figure 7.18, top. This is always possible because the enzyme is always regenerated, like any catalyst. Then, we write the effective rate constants for each step next to each arrow. These effective constants are first-order (k_{-1}, k_2, k_3) or pseudo first-order rate constants ($k_1[S]$); the latter are products of true rate constants and concentrations. This procedure is similar to that used in the partition function method, but now we are using rate constants instead of equilibrium constants. Next, we draw the geometrical figure that corresponds to the reaction cycle—in this case, a triangle. Each vertex corresponds to an enzyme species.

Second, we draw all possible linear diagrams obtained from the triangle by removing one side at a time, so that there are *no closed loops*. There are three such diagrams in this case. Then, we draw these open diagrams for each enzyme species, E, ES, and EP, identified by the vertex to which the arrows converge (see Figure 7.18, lines 2–4). Now, we need to indicate all possible *directional paths* that lead to each enzyme species on each diagram. The possible directions are given by the reaction mechanism, on top. Thus, referring to the cycle, we draw the arrows on each diagram and write the corresponding rate constants next to the arrows. For example, for the free enzyme E, there are two directional diagrams that end up in E. The third diagram for E in Figure 7.18 is not viable because *no arrow combination* from the mechanism can be used to reach E in that diagram. Similarly, for ES and EP, only one of the diagrams is viable. In summary, there are two directional diagrams for E, but only one for ES and EP.

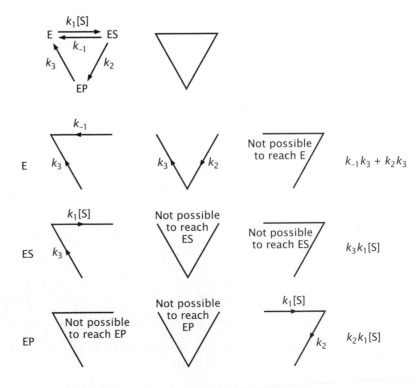

Figure 7.18 The King–Altman method to obtain steady-state concentrations. The mechanism, written as a reaction cycle, and the corresponding geometrical figure are shown on top. The directional diagrams for each enzyme species, E, ES, and EP, provide the recipes to write the rates of formation, which are written in the last column.

Third, for each directional diagram, we write the rates of formation of each species, which are proportional to the *products of the rate constants in each diagram*. The total rate of formation for each species is obtained by adding the rates along all viable diagrams. The rates of formation are written in the last column of Figure 7.18. There are two terms for E, and one for ES and EP. Now, the total concentration of enzyme $[E_0]$ in the steady-state is proportional (symbol $\propto$) to the *sum* of all rates of formation of all species,

$$[E_0] \propto k_{-1}k_3 + k_2k_3 + k_3k_1[S] + k_2k_1[S], \qquad (7.50)$$

whereas [EP] is proportional to $k_2k_1[S]$ (with the same proportionality constant),

$$[EP] \propto k_2k_1[S]. \qquad (7.51)$$

Therefore, the fraction of EP in the steady state is

$$f_{EP} = \frac{k_2k_1[S]}{k_{-1}k_3 + k_2k_3 + k_3k_1[S] + k_2k_1[S]} \qquad (7.52)$$

$$= \frac{k_2/(k_2 + k_3)[S]}{[k_3/(k_2 + k_3)][(k_2 + k_{-1})/k_1] + [S]}. \qquad (7.53)$$

Finally, we substitute f_{EP} from Equation 7.53 in the rate Equation 7.49 to obtain

$$v = \frac{k_2k_3/(k_2 + k_3)[E_0][S]}{[k_3/(k_2 + k_3)][(k_2 + k_{-1})/k_1] + [S]}. \qquad (7.54)$$

Now compare this result with the Michaelis–Menten Equation 7.41,

$$v = \frac{k_{cat}[E_0][S]}{K_M + [S]}.$$

The combination $[k_3/(k_2 + k_3)][(k_2 + k_{-1})/k_1]$ in the denominator of Equation 7.54 occupies the place of K_M. Thus, the Michaelis constant is

$$K_M = \left(\frac{k_2 + k_{-1}}{k_1}\right)\left(\frac{k_3}{k_2 + k_3}\right). \qquad (7.55)$$

The expression of K_M appears modified by the factor $k_3/(k_2 + k_3)$. Similarly, by comparison of Equations 7.54 and 7.41, we see that

$$k_{cat} = k_2k_3/(k_2 + k_3). \qquad (7.56)$$

As our second complex example, let us make the conversion between ES and EP reversible:

$$\text{E} + \text{S} \underset{k_{-1}}{\overset{k_1}{\rightleftharpoons}} \text{ES} \underset{k_{-2}}{\overset{k_2}{\rightleftharpoons}} \text{EP} \overset{k_3}{\longrightarrow} \text{E} + \text{P}. \qquad (7.57)$$

Using the King–Altman method, you can find that the steady-state rate is given by

$$v = \frac{k_2k_3/(k_2 + k_3 + k_{-2})[E_0][S]}{[k_3(k_2 + k_{-1}) + k_{-1}k_{-2}]/[k_1(k_2 + k_3 + k_{-2})] + [S]}, \qquad (7.58)$$

and by comparison with Equation 7.41, we have

$$K_M = \frac{k_3(k_2 + k_{-1}) + k_{-1}k_{-2}}{k_1(k_2 + k_3 + k_{-2})} \qquad (7.59)$$

$$k_{cat} = \frac{k_2k_3}{k_2 + k_3 + k_{-2}}. \qquad (7.60)$$

If the last step in Equation 7.57 is also reversible, then the rate of this reverse reaction needs to be included in the denominator of Equation 7.58 and subtracted from the rate of the forward reaction in the numerator to obtain the *net* rate of product formation. We have omitted this complication in the examples examined, but it presents no difficulty. For more details and examples, see the original paper by King and Altman (1956) or the book by Cornish-Bowden (2012).

7.5 MOST ENZYME REACTIONS INVOLVE TWO SUBSTRATES

The reactions we have been considering involve only one substrate. However, enzyme reactions with only one substrate are actually not very common; those with two substrates are the most common. We will concentrate on the essential concept: despite the apparent complexity of multisubstrate reactions, the Michaelis–Menten equation still applies *to each substrate*, keeping the concentration of the other substrates constant.

In Sequential Reactions all Substrates Bind Before Products are Formed

Consider an ordered sequential, two-substrate reaction. It is called *sequential* because all substrates bind before *any* product is formed; it is *ordered* because the substrates bind in a mandatory order. If there is no mandatory binding order, the reaction is called random sequential.

We call the substrates S_1 and S_2, and the products P_1 and P_2. To make matters simple, assume that the product concentrations in the beginning of the steady-state period are so small that we can set them to zero; thus, we do not have to worry about reactions of products binding back to the enzyme. A somewhat simplified ordered sequential mechanism can be written as

$$E \underset{k_{-1}}{\overset{k_1[S_1]}{\rightleftharpoons}} ES_1 \underset{k_{-2}}{\overset{k_2[S_2]}{\rightleftharpoons}} ES_1S_2 \overset{k_3}{\underset{+P_1}{\longrightarrow}} EP_2 \overset{k_4}{\longrightarrow} E + P_2. \qquad (7.61)$$

You may have noticed that we included the concentrations of S_1 and S_2 above the arrows in the binding reactions of Equation 7.61. This is because $k_1[S_1]$ is the effective rate constant for the binding of S_1 to E (and similarly for $k_2[S_2]$). Note that k_1 is a second-order rate constant (units of $M^{-1}s^{-1}$), which when multiplied by $[S_1]$ (units of M) yields a *pseudo* first-order rate constant ($k_1[S_1]$, with units of s^{-1}). Since $[S_1] \approx$ constant in the beginning of the reaction, so is $k_1[S_1]$.

As before, we begin with the rate equation

$$v = k_4[EP_2] = k_4 f_{EP_2}[E_0], \qquad (7.62)$$

and apply the steady-state condition to $[EP_2]$ to obtain

$$v = \frac{k_1 k_2 k_3 k_4 [E_0][S_1][S_2]}{k_{-1}k_4(k_3 + k_{-2}) + k_1 k_4(k_3 + k_{-2})[S_1] + k_2 k_3 k_4[S_2] + k_1 k_2(k_3 + k_4)[S_1][S_2]}. \qquad (7.63)$$

The denominator contains a constant term, a term in $[S_1]$, a term in $[S_2]$, and a term in $[S_1][S_2]$. The numerator depends on both substrates because both need to bind before product can be made. If we keep $[S_2]$ fixed and vary $[S_1]$, we can write Equation 7.63 in the Michaelis–Menten form, as

$$v = \frac{k_{\mathrm{cat}}^{S_1}[E_0][S_1]}{K_{\mathrm{M}}^{S_1} + [S_1]}, \tag{7.64}$$

where $K_{\mathrm{M}}^{S_1}$ and $k_{\mathrm{cat}}^{S_1}$ relate to substrate S_1 but are now combinations of the rate constants *and* of the fixed concentration $[S_2]$. ($K_{\mathrm{M}}^{S_1} = (a_0 + a_1[S_2])/(a_2 + a_3[S_2])$ and $k_{\mathrm{cat}}^{S_1} = (a_4[S_2])(a_2 + a_3[S_2])$, where the constants a_i are functions of the rate constants.) Similarly, if we keep $[S_1]$ fixed and vary $[S_2]$, we can write Equation 7.63 as

$$v = \frac{k_{\mathrm{cat}}^{S_2}[E_0][S_2]}{K_{\mathrm{M}}^{S_2} + [S_2]}, \tag{7.65}$$

where $k_{\mathrm{cat}}^{S_2}$ and $K_{\mathrm{M}}^{S_2}$ are functions of the rate constants *and* of the fixed concentration $[S_1]$.

The Lineweaver–Burk plot of the rate of the reaction as a function of $[S_1]$ for several, fixed concentrations of $[S_2]$ is shown in **Figure 7.19**. The slopes of the lines decrease as $[S_2]$ increases, but the change becomes progressively smaller because the slope contains a factor proportional to $1/[S_2]$.

In Ping-Pong Reactions Substrate Binding Alternates with Product Release

As our second example of two-substrate reactions, we consider the ping-pong mechanism. In this mechanism one substrate binds and a product is released, leaving the enzyme in a chemically modified state (E'). Then, the second substrate binds, and finally the second product is released. The name *ping-pong* arises because the substrates and the products alternate in binding and dissociating from the

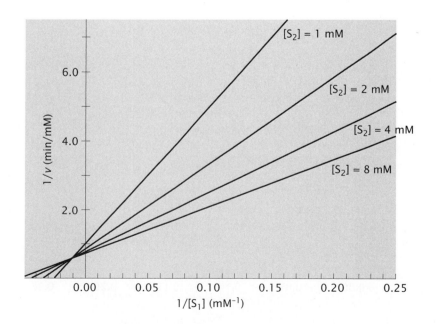

Figure 7.19 The Lineweaver–Burk plot for a two-substrate, ordered sequential reaction, with the concentration of substrate S_1 as the variable ($1/[S_1]$), at four different concentrations of the second substrate, $[S_2]$.

enzyme. We can write a ping-pong mechanism as

$$E \underset{k_{-1}}{\overset{k_1[S_1]}{\rightleftharpoons}} ES_1 \overset{k_2}{\longrightarrow} E' \underset{k_{-3}}{\overset{k_3[S_2]}{\rightleftharpoons}} E'S_2 \overset{k_4}{\longrightarrow} E + P_2. \qquad (7.66)$$

$$+ P_1$$

The rate of the reaction is given by $v = k_4[E'S_2] = k_4 f_{E'S_2}[E_0]$, which, in the steady state, becomes

$$v = \frac{k_1 k_2 k_3 k_4 [E_0][S_1][S_2]}{k_1 k_2 (k_4 + k_{-3})[S_1] + k_3 k_4 (k_2 + k_{-1})[S_2] + k_1 k_3 (k_4 + k_{-3})[S_1][S_2]}. \qquad (7.67)$$

What is different in this case is that there is *no constant term* in the denominator. We can write this equation in the Michaelis–Menten form as

$$v = \frac{k_{cat}^{S_1}[E_0][S_1]}{K_M^{S_1} + [S_1]}, \qquad (7.68)$$

where the concentration $[S_1]$ is variable, and $[S_2]$ is fixed at a constant value, or as

$$v = \frac{k_{cat}^{S_2}[E_0][S_2]}{K_M^{S_2} + [S_2]}, \qquad (7.69)$$

where $[S_2]$ varies and $[S_1]$ is fixed. In either case, K_M and k_{cat} are functions of the rate constants and of the fixed substrate concentration.

The Lineweaver–Burk plot as a function of $[S_1]$ for several fixed concentrations of $[S_2]$ shows a distinctive feature in the ping-pong mechanism: a series of *parallel lines* (**Figure 7.20**). The slope is constant and the lines are spaced along the y-axis with intervals that decrease as $[S_2]$ increases. The plot looks similar if you vary $[S_2]$ and keep $[S_1]$ constant, because Equation 7.67 is essentially symmetrical in $[S_1]$ and $[S_2]$.

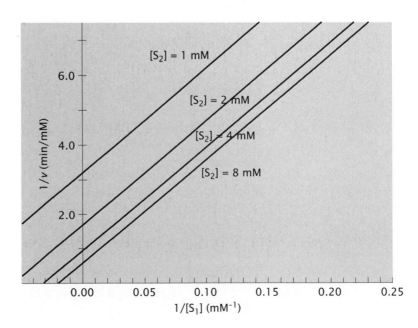

Figure 7.20 The Lineweaver–Burk plot for a two-substrate ping-pong reaction, with the concentration of substrate S_1 as the variable ($1/[S_1]$), at four different concentrations of the second substrate, $[S_2]$.

7.6 THE PARTITION FUNCTION CAN BE USED TO OBTAIN STEADY-STATE KINETICS IN SPECIAL CASES

The Michaelis Constant Behaves as an Apparent Dissociation Constant

Throughout this book, we have used the partition function to obtain the probabilities or fractions of the various chemical species present in a sample. The partition function method, however, applies to equilibrium, and does *not* work *in general* for steady-state situations. But it *works* if we replace the equilibrium dissociation constant of the substrate, K_S, by K_M, provided ES is the last enzyme complex before products are formed and all previous binding reactions are equilibria (except the one controlled by K_M). This may seem too restrictive to be useful. However, is applies to the simple Michaelis–Menten scheme and to many other important cases, such as *reversible* enzyme inhibition. Essentially, what we are doing is to replace the actual situation, depicted in **Figure 7.21**A, with that in **Figure 7.21**B. The approach works because K_M is an *apparent dissociation constant*, valid to calculate the concentrations of E, S, and ES *in the steady state* (shown in Equation 7.38). It includes the ES complex formation (through k_1) and two ways of its breakdown (through k_{-1} and k_2).

Let us apply the partition function method to the general case of the simple Michaelis–Menten mechanism. We will designate this partition function by Q' as a reminder that it includes one constant, K_M, that is not a proper equilibrium constant,

$$Q' = 1 + [S]/K_M. \qquad (7.70)$$

Here, 1 is the relative probability of the free enzyme (E), and $[S]/K_M$ is the relative probability of the ES complex. The fraction of E is simply the term for E in the partition function divided by the entire Q',

$$f_E = \frac{1}{1 + [S]/K_M}, \qquad (7.71)$$

and the fraction of ES is simply the term for ES divided by Q',

$$f_{ES} = \frac{[S]/K_M}{1 + [S]/K_M}. \qquad (7.72)$$

Finally, f_{ES} can be substituted in the rate equation ($v = k_2 f_{ES}[E_0]$) to yield

$$v = \frac{k_2[E_0][S]}{K_M + [S]}, \qquad (7.73)$$

which is the Michaelis–Menten equation. In summary, we use the partition function to calculate f_{ES} in the *steady state*. Then, the fractional

(A)   (B)

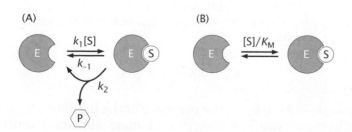

Figure 7.21 Diagram of the Michaelis–Menten mechanism. (A) The actual situation. (B) The model used to write the partition function where K_M is the apparent dissociation constant in the steady state.

velocity (v relative to V_{max}) is identical to the fraction of ES relative to E_0:

$$\frac{v}{V_{max}} = \frac{[ES]}{[E_0]} = f_{ES}. \tag{7.74}$$

Enzyme Inhibition Kinetics can be Derived from the Partition Function

The simplicity of the partition function method is well illustrated by its use in the steady-state kinetics of enzyme inhibition. In this section, we will reserve the symbol K_M for its meaning in the standard Michaelis–Menten mechanism, $K_M = (k_2 + k_{-1})/k_1$, and use K_M^{app} for the cases in the presence of inhibitors. We will now examine the three classical cases of reversible enzyme inhibition.

Competitive inhibition is represented in the diagram of **Figure 7.22**. The substrate (S) and the inhibitor (I) *compete* for the free enzyme. Usually they have similar structures and bind to the same site on the enzyme (the active site). But the only condition for competitive inhibition is that S and I bind to the free enzyme in a *mutually exclusive* fashion.

Here we define K_I as the equilibrium dissociation constant for the inhibitor,

$$K_I = \frac{[E][I]}{[EI]}. \tag{7.75}$$

By inspection of the diagram of Figure 7.22 we can immediately write the partition function,

$$Q' = 1 + [S]/K_M + [I]/K_I. \tag{7.76}$$

The fraction of ES complex is the term corresponding to ES divided by the entire partition function,

$$f_{ES} = \frac{[S]/K_M}{1 + [S]/K_M + [I]/K_I} \tag{7.77}$$

$$= \frac{[S]}{K_M(1 + [I]/K_I) + [S]}. \tag{7.78}$$

The initial rate in the steady state is proportional to f_{ES} (shown in Equation 7.74),

$$v = V_{max} f_{ES}. \tag{7.79}$$

Therefore, using f_{ES} from Equation 7.78 we obtain

$$v = \frac{V_{max}[S]}{K_M(1 + [I]/K_I) + [S]}. \tag{7.80}$$

By comparing Equation 7.80 with the Michaelis–Menten equation (shown in Equation 7.33), we see that V_{max} remains the same as without inhibitor but K_M changes:

$$K_M^{app} = K_M \left(1 + \frac{[I]}{K_I}\right), \tag{7.81}$$

where $K_M = (k_2 + k_{-1})/k_1$. In the presence of the inhibitor, V_{max} remains constant because you can always add more substrate until it wins

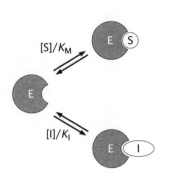

[S]/K_M

[I]/K_I

Figure 7.22 Competitive inhibition. The enzyme E binds a substrate S (apparent dissociation constant K_M) or an inhibitor I (dissociation constant K_I), but never both simultaneously.

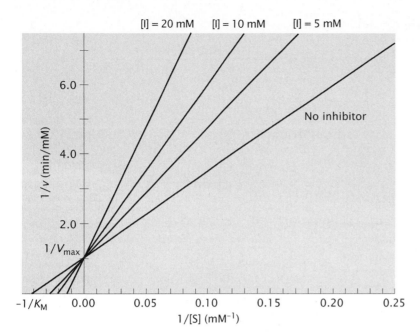

No inhibitor

$1/v$ (min/mM)

$1/V_{max}$

$-1/K_M$ 0.00   0.05   0.10   0.15   0.20   0.25

$1/[S]$ (mM^{-1})

Figure 7.23 Lineweaver–Burk plot of the rate of an enzyme-catalyzed reaction, in the absence and in the presence of a competitive inhibitor ($K_M = 25$ mM, $K_I = 10$ mM).

the competition with the inhibitor for the active site of the enzyme. Eventually, you reach a point where the enzyme is saturated with the substrate, and you recover V_{max}. But the apparent K_M increases in the presence of the inhibitor. It appears that the substrate binds worse now. This is because of the competition with inhibitor: more substrate is needed to reach the same degree of binding that existed when inhibitor was absent.

The Lineweaver–Burk plot is shown in **Figure 7.23** in the absence and in the presence of increasing concentrations of a competitive inhibitor. Note the characteristic signature of competitive inhibition: the lines have a *common y-intercept* at $1/V_{max}$ (because V_{max} does not change).

Uncompetitive inhibition is represented in **Figure 7.24**. Here, the inhibitor binds only to the ES complex and not to the free enzyme. Inhibition occurs because the ESI ternary complex is not productive.

The dissociation constant of the inhibitor from the ES complex is K_I',

$$K_I' = \frac{[ES][I]}{[ESI]}. \tag{7.82}$$

Again, we first write the partition function using the diagram of Figure 7.24 to determine the relative probability of each state,

$$Q' = 1 + [S]/K_M + ([S]/K_M)\,([I]/K_I') \tag{7.83}$$

$$= 1 + ([S]/K_M)(1 + [I]/K_I'). \tag{7.84}$$

Now the fraction of ES can be immediately obtained by dividing the term that represents ES in the partition function ($[S]/K_M$) by the

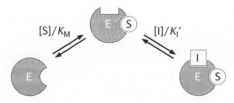

$[S]/K_M$ $[I]/K_I'$

Figure 7.24 Uncompetitive inhibition. The enzyme E binds the inhibitor I, but only if S is already bound. The ES complex is productive but the ESI complex is not.

entire Q':

$$f_{ES} = \frac{[S]/K_M}{1 + [S]/K_M + ([S]/K_M)\,([I]/K_I')} \tag{7.85}$$

$$= \frac{[S]}{K_M + [S](1 + [I]/K_I')}. \tag{7.86}$$

The velocity is proportional to f_{ES} ($v = V_{max}f_{ES}$, Equation 7.79) and we obtain

$$v = \frac{(V_{max}/(1 + [I]/K_I'))\,[S]}{K_M/(1 + [I]/K_I') + [S]}. \tag{7.87}$$

Comparing this equation with the Michaelis–Menten Equation 7.33 we see that

$$K_M^{app} = \frac{K_M}{1 + [I]/K_I'} \tag{7.88}$$

$$V_{max}^{app} = \frac{V_{max}}{1 + [I]/K_I'}. \tag{7.89}$$

Now the apparent K_M *decreases* in the presence of the inhibitor, as if the enzyme would bind the substrate *better*. This is because, by Le Châtelier's principle, binding of the inhibitor to the ES complex shifts the equilibrium $E + S \rightleftharpoons ES$ to the right. Indeed the apparent binding is better, but a non-productive ESI complex is formed and V_{max} decreases.

The Lineweaver–Burk plot in the absence and in the presence of increasing concentrations of an uncompetitive inhibitor is shown in **Figure 7.25**. The characteristic signature of uncompetitive inhibition is a *series of parallel lines*. The more uncompetitive inhibitor you add, the more the line shifts up, because $1/V_{max}$ increases (V_{max} decreases). The spacing between successive lines is proportional to the inhibitor concentration.

This plot is reminiscent of that for a two-substrate ping-pong reaction, but the cases are *not* mechanically related. Furthermore, in the

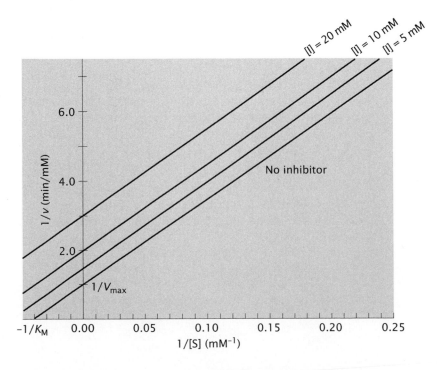

Figure 7.25 Lineweaver–Burk plot of the rate of an enzyme-catalyzed reaction, in the absence and in the presence of an uncompetitive inhibitor ($K_M = 25$ mM, $K_I' = 10$ mM).

ping-pong reaction, the parallel lines *shift down* as the addition of the second substrate (S_2) increases V_{max}, and the spacing between the lines does *not* change in proportion to [S_2] but to $1/[S_2]$.

Noncompetitive inhibition occurs when the inhibitor binds to both the free enzyme E and the ES complex, to yield EI and ESI complexes, respectively, neither of which is productive, as shown in the diagram of **Figure 7.26**. The class of noncompetitive inhibitors is particularly important because most *allosteric inhibitors* are noncompetitive. The dissociation constants K_I and K_I' are different in general. Many textbooks call this type of inhibition *mixed* because it looks like a mixture of competitive and uncompetitive inhibition. However, as Cleland (1970) pointed out, "there is no good theoretical reason for doing this."

By inspection of Figure 7.26 we write the partition function. Since we are free to choose which branch of the diagram we take to get to the final product, we take the upper branch,

$$Q' = 1 + [S]/K_M + ([S]/K_M)\,([I]/K_I') + [I]/K_I \qquad (7.90)$$

$$= 1 + [S]/K_M(1 + [I]/K_I') + [I]/K_I. \qquad (7.91)$$

The fraction of ES (f_{ES}) is obtained by choosing the term that corresponds only to ES (without inhibitor) and dividing this term by the entire partition function,

$$f_{ES} = \frac{[S]/K_M}{1 + [S]/K_M + [I]/K_I + ([S]/K_M)\,([I]/K_I')} \qquad (7.92)$$

$$= \frac{[S]}{K_M(1 + [I]/K_I) + [S](1 + [I]/K_I')}. \qquad (7.93)$$

The rate of the reaction is $v = V_{max}f_{ES}$, in which we substitute the expression for f_{ES} to obtain

$$v = \frac{V_{max}/(1 + [I]/K_I')[S]}{K_M(1 + [I]/K_I)/(1 + [I]/K_I') + [S]}. \qquad (7.94)$$

Finally, comparing this result with the general Michaelis–Menten Equation 7.33, we find

$$K_M^{app} = \frac{K_M(1 + [I]/K_I)}{(1 + [I]/K_I')} \qquad (7.95)$$

$$V_{max}^{app} = \frac{V_{max}}{(1 + [I]/K_I')} \qquad (7.96)$$

where $K_M = (k_2 + k_{-1})/k_1$.

The Lineweaver–Burk plot in the absence and in the presence of increasing concentrations of a noncompetitive inhibitor is shown in **Figure 7.27**. In this case, both the *y*-intercept and the slope of the line

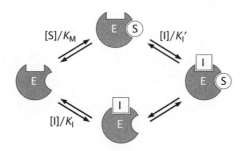

Figure 7.26 Noncompetitive inhibition. The inhibitor I binds both to the free enzyme E and to the enzyme–substrate complex ES. Only the ES complex is productive.

Figure 7.27 Lineweaver–Burk plot of the rate of an enzyme-catalyzed reaction in the absence and in the presence of a noncompetitive inhibitor ($K_M = 25$ mM, $K_I = 10$ mM and $K_I' = 20$ mM).

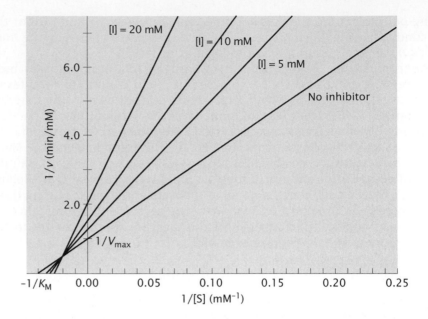

are affected by the inhibitor. The lines are *not* parallel (as in uncompetitive inhibition), but they do *not* have a common *y*-intercept either (as in competitive inhibition). Rather, they intersect at a point to the left of the *y*-axis. In the special (rare) case when $K_I = K_I'$, the lines cross on the horizontal axis and thus have a common *x*-intercept, which is $-1/K_M$. But otherwise they can cross anywhere to the left of the *y*-axis and may not even cross all at the same point.

7.7 SERINE PROTEASES USE ACID–BASE CATALYSIS TO HYDROLYZE PEPTIDE BONDS

We learned about the evolution of serine proteases in Chapter 3. The serine protease family includes trypsin (**Figure 7.28**A), chymotrypsin, and many other proteins, such as *S. griseus* protease A. These proteins are *homologous*—they descend from a common ancestor—and therefore their sequences are related by evolution. Their three-dimensional structures are very similar. These enzymes catalyze the *same reaction*, with the *same mechanism*, but on different substrates. The substrate is always a polypeptide, but each enzyme hydrolyzes a particular peptide bond in the substrate, a bond that is flanked by a specific residue. The serine proteases of the trypsin family also share a conserved active

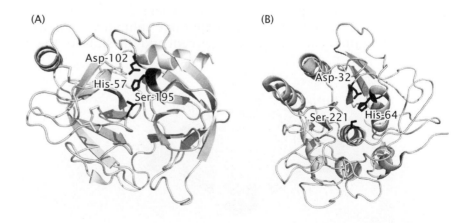

Figure 7.28 Structures of two analogous serine proteases. (A) Bovine trypsin (PDB 3OTJ) and (B) subtilisin (PDB 1SBT). The catalytic triads, His-57, Asp-102, and Ser-195, in trypsin, and His-64, Asp-32, and Ser-221, in subtilisin, are shown as stick models.

site: the *catalytic triad*, which consists of His-57, Asp-102, and Ser-195. It is to this serine residue that these enzymes owe their name.

We also learned in Chapter 3 that there are other serine proteases that do not belong to the trypsin family. For example, subtilisin from *Bacillus subtilis* is *not* homologous to trypsin and chymotrypsin, but *analogous*. It descends from a different ancestor, its amino acid sequence is not related to that of trypsin, and its structure is also completely different (see Figure 7.28B). Yet the catalytic mechanism of subtilisin is the same as that of trypsin, and the active site of subtilisin contains the same catalytic triad: His-64, Asp-32, and Ser-221. The numbers of the triad residues are different, but their spatial organization in the active site and their roles in catalysis are the same as in the trypsin family.

Now we want to understand how these enzymes function. How does the enzyme catalyze the reaction? What happens at the molecular level? What are the roles of the catalytic residues? How do the molecular steps of catalysis relate to the enzyme kinetics?

Substrate Specificity in Trypsin and Chymotrypsin is Conferred by the Residue Preceding the Bond Hydrolyzed

The mechanism of the enzyme-catalyzed hydrolysis of the peptide bond is the same in all serine proteases. However, the substrate specificity is different: binding determines specificity. In chymotrypsin, trypsin, and to a large extent, subtilisin, the substrate specificity is determined by the type of residue preceding the peptide bond hydrolyzed (**Figure 7.29**).

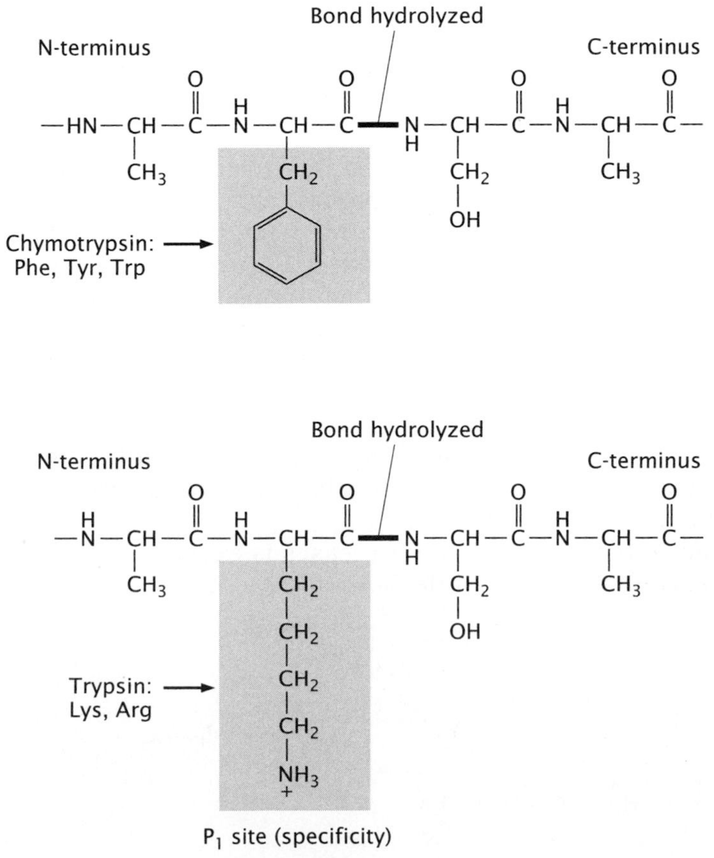

Figure 7.29 Substrate specificity of chymotrypsin and trypsin. Chymotrypsin cuts the polypeptide substrate after a Phe, Tyr, or Trp residue (P_1 site). Trypsin cuts after Lys or Arg.

Figure 7.30 The specificity pockets of chymotrypsin and trypsin. (A) In chymotrypsin, the pocket is lined by hydrophobic residues that interact favorably with the large aromatic side chains of Phe, Tyr, and Trp (PDB 7GCH). (B) In trypsin, the pocket is nonpolar as well, but Asp-189, in the bottom, interacts favorably with the NH_3^+ group of Lys or the guanidinium group of Arg (PDB 3OTJ).

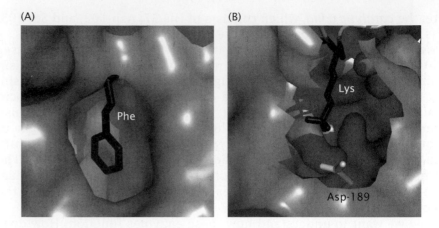

(A)   (B)

Phe   Lys

                           Asp-189

This residue of the substrate is called the P_1 site (P, for polypeptide; the P-sites are numbered from the cleaved bond: to the N-terminus, as P_1, P_2, P_3, etc.; to the C-terminus, as P'_1, P'_2, P'_3, etc.). Trypsin is specific for the hydrolysis of a peptide bond preceded by Lys or Arg, whereas chymotrypsin is specific for a bond preceded by an aromatic residue, but their mechanisms are identical. The side chain of the P_1 residue of the substrate, which precedes the target bond, fits in a pocket on the enzyme surface. It is the nature of this pocket that confers the specificity. In chymotrypsin, the pocket is large and is lined by backbone groups and nonpolar side chains (**Figure 7.30**A). In trypsin, a negatively charged residue, Asp-189, lies at the bottom of the binding pocket and interacts with the lysine or arginine residue of the substrate in the binding pocket. This interaction is mediated by a water molecule, which bridges the Asp and Lys (or Arg) side chains by making hydrogen bonds with both (see Figure 7.30B).

The Residues of the Catalytic Triad are the Most Important in the Mechanism of Serine Proteases

We turn now to the mechanism of catalysis itself at the molecular level by examining the action of the catalytic triad—His-57, Asp-102, and Ser-195—in the hydrolysis of a peptide bond by trypsin. The catalytic triad of the serine proteases is shown in **Figure 7.31**.

Ser-195 is the primary nucleophile: it makes the first nucleophilic attack on the carbonyl group of the target peptide bond. The role of His-57 is to accept the proton from the –OH group of the serine (normally $pK_a \approx 14$), making it equivalent to the much more powerful alkoxide (R–O⁻) nucleophile. The role of Asp-102 is the most subtle: by maintaining a hydrogen bond with the $\delta 1$ N–H of the imidazole ring of His-57, Asp-102 keeps the $\varepsilon 2$ nitrogen with a lone pair of electrons available to accept the proton from Ser-195.

The mechanism of the serine proteases is exemplified by trypsin. In **Figure 7.32**A, the nucleophile Ser-195 attacks the carbonyl group of

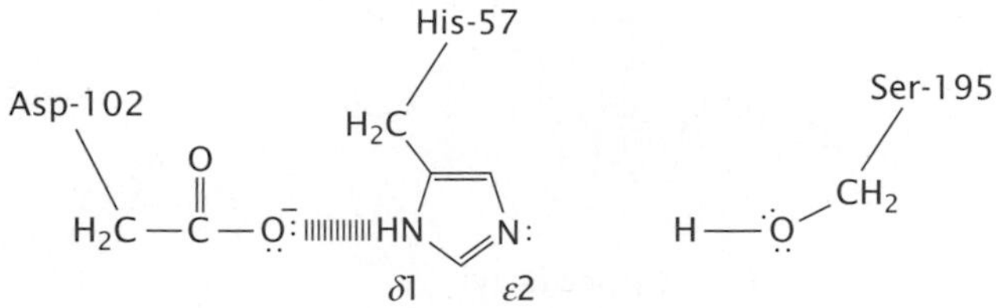

Figure 7.31 The catalytic triad of the serine proteases (trypsin family): Asp-102, His-57, and Ser-195. The Nδ1 of His makes a hydrogen bond to Asp-102.

His-57

Asp-102   Ser-195

H_2C

$H_2C-C-O^{\cdot-}$ ‖‖‖‖‖‖‖ HN N: $H-\overset{\cdot\cdot}{O}$ CH_2

$\delta 1$ $\varepsilon 2$

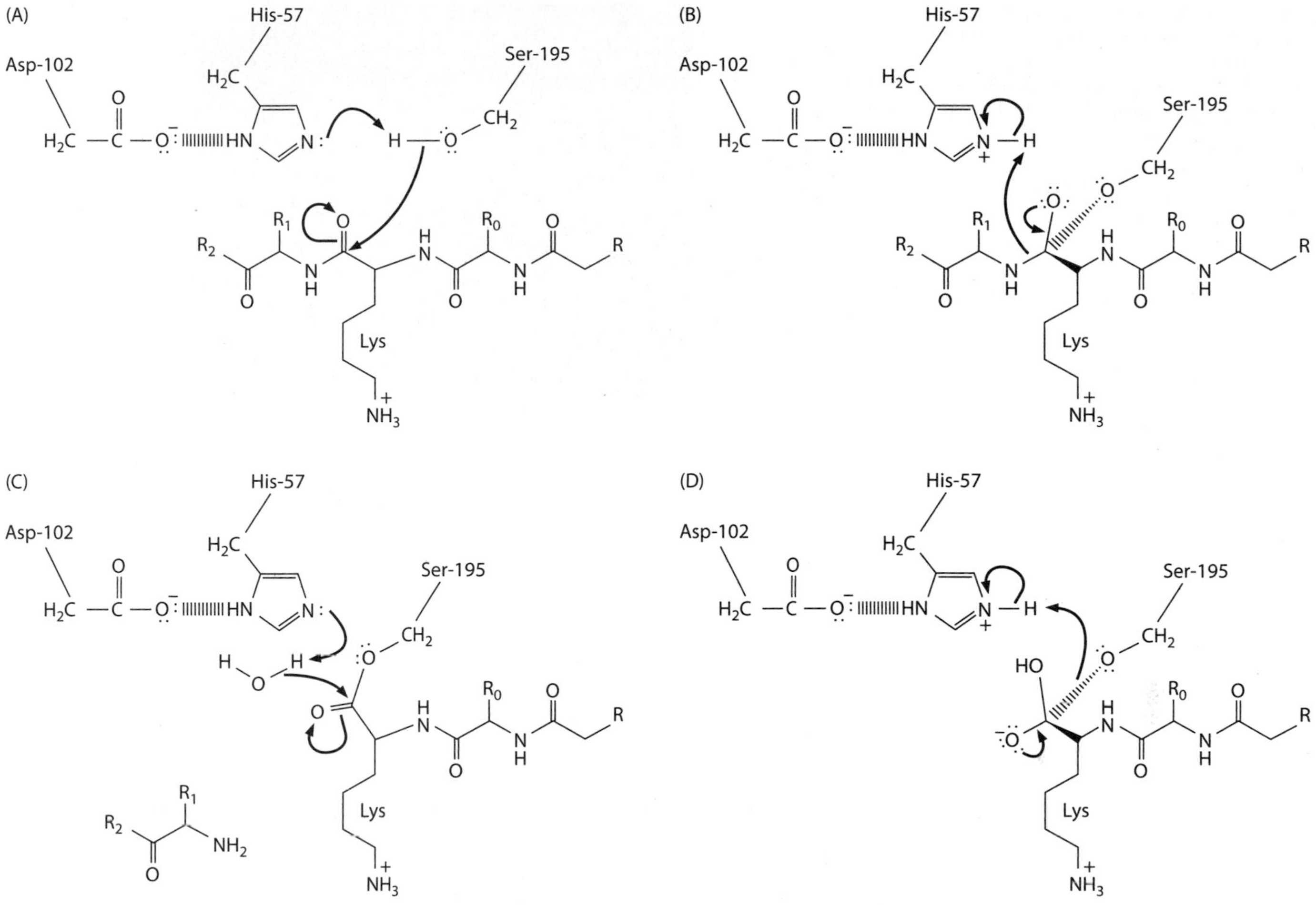

Figure 7.32 Mechanism of the serine proteases. (A) Nucleophilic attack by Ser-195 on the carbonyl, with His-57 accepting the proton from Ser-195. (B) The tetrahedral intermediate is formed. The oxyanion is stabilized by hydrogen bonds to the amides of Gly-193 and Ser-195, which lie in the oxyanion hole. The C-terminal part of the polypeptide chain leaves, capturing the proton from His-57. The target peptide bond is broken and an acyl-enzyme intermediate is formed. (C) The acyl-enzyme intermediate undergoes a nucleophilic attack by water and His-57 accepts the proton. (D) The original state of the enzyme is regenerated. The second product, the N-terminal part of the polypeptide chain, leaves the enzyme.

the peptide bond. His-57 acts as a base, accepting the proton and thus increasing the nucleophilicity of Ser-195. As a consequence, a *tetrahedral intermediate* is formed (see Figure 7.32B). The oxygen atom of the original carbonyl group now has a negative charge: it becomes an oxyanion. The negatively charged oxygen lies in a pocket in the enzyme called the *oxyanion hole* (**Figure 7.33**), where it is stabilized by hydrogen bonding to the backbone amide (N–H) of residues Gly-193 and Ser-195. The leaving group (the C-terminal stretch of the polypeptide chain) is improved by accepting a proton from His-57, which now works as an acid. The target peptide bond is broken and an *acyl-enzyme intermediate* is formed.

In Figure 7.32C, the acyl-enzyme intermediate undergoes another nucleophilic attack, this time from a water molecule. His-57 now accepts a proton from water, which concomitantly attacks the carbonyl group of the peptide bond (the water becomes equivalent to the powerful nucleophile hydroxide ion, OH^-). This step is similar to that in Figure 7.32A, except that water (the second reactant in the serine proteases) is the nucleophile instead of Ser-195. In Figure 7.32D, the original state of the enzyme is regenerated. The carbonyl group is

Figure 7.33 The oxyanion hole of chymotrypsin. The negatively charged oxygen in the tetrahedral intermediate is stabilized by hydrogen bonds to the backbone N–H groups of Gly-193 and Ser-195. The phenylalanine side chain of the substrate analog (Ac–Leu–Phe–CF$_3$) is shown in its binding pocket (PDB 7GCH).

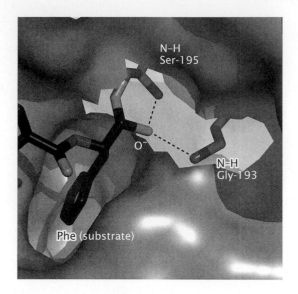

restored, as the extra lone pair in the anionic oxygen moves to the carbon atom, and the bond to Ser-195 is broken. Ser-195 is now the leaving group, which is again improved by capturing the proton from the acid form of His-57. Finally, the second product, the N-terminal stretch of the polypeptide chain, leaves the enzyme.

The Catalytic Triad Accelerates Peptide Hydrolysis a Million Times

We have learned the roles of each residue of the catalytic triad—His-57, Asp-102, and Ser-195—in the mechanism of the serine proteases. But how much does each residue of the triad contribute to enhance the rate of peptide bond hydrolysis? This question was addressed in subtilisin. Although evolutionarily unrelated to chymotrypsin (not homologous), subtilisin has the same catalytic triad (His-64, Asp-32, and Ser-221) and functions in the same manner as chymotrypsin, with similar specificity, cleaving the polypeptide chain after an aromatic residue.

To determine the importance of each residue of the triad for catalysis, His-64, Asp-32, and Ser-221 were mutated to alanine one at time, then two at a time, and finally all three. Replacement of only Ser-221 by Ala resulted in a decrease of k_{cat} by a factor of 10^6. Similarly, replacement of only His-64 by Ala decreased k_{cat} by 10^6. The double mutation H64A/S221A, however, did not result in additional decrease in catalytic efficiency compared to the two single mutants. The replacement of Asp-32 by Ala had a comparatively smaller effect, with a decrease in k_{cat} by $\sim 10^4$. This indicates that the functions of His and Ser are the most important, whereas Asp has a somewhat lesser role, consistent with the mechanism of serine proteases (see Figure 7.32). In all mutants the K_M increases slightly, from 180 μM to ≈200–400 μM. This increase, however, is small, which indicates that substrate binding is not significantly impaired by the mutations to Ala. The effect of the mutations is on the catalysis.

What happens if all three residues of triad are mutated to alanine? The activity drops considerably, but not more than if only His or Ser were mutated to Ala. What's more, the triple alanine mutant still catalyzes hydrolysis, increasing the rate by a factor of 10^3 compared to the same reaction in solution. Thus, elimination of the three residues of the triad does not completely abolish catalysis. Why? Because subtilisin retains the ability to bind and therefore stabilize the transition state to some extent.

7.8 TRANSIENT KINETICS ALLOW DETERMINATION OF THE MOLECULAR RATE CONSTANTS

We have learned much about enzyme catalysis from steady-state kinetics. However, to really understand the detailed mechanism of the reaction, and determine the individual molecular rate constants, it is essential to study the *transient kinetics*, which are observed before the steady state is established. Therefore, we need to return to the simple Michaelis–Menten mechanism (shown in Equation 7.19),

$$E \underset{k_{-1}}{\overset{k_1\,[S]}{\rightleftharpoons}} ES \xrightarrow{k_2} E + P,$$

and obtain a *time-dependent* solution of the corresponding rate equations (shown in Equations 7.24–7.27):

$$\frac{d[E]}{dt} = -k_1[E][S] + (k_{-1} + k_2)[ES]$$

$$\frac{d[S]}{dt} = -k_1[E][S] + k_{-1}[ES]$$

$$\frac{d[ES]}{dt} = k_1[E][S] - (k_{-1} + k_2)[ES]$$

$$\frac{d[P]}{dt} = k_2[ES].$$

If the mechanism is more complicated, this task is not simple in general, but we will break down the problem into smaller steps and solve it under conditions that render our task easier.

Binding Kinetics can be Determined Exactly in Favorable Circumstances

The first step in any enzyme-catalyzed reaction is *substrate binding* to the enzyme,

$$E \underset{k_{-1}}{\overset{k_1[S]}{\rightleftharpoons}} ES, \tag{7.97}$$

where k_1 is also called the on-rate constant (k_{on}). This is a second-order rate constant, with units of $M^{-1}s^{-1}$. If the substrate concentration is much larger than that of the enzyme, $[S] \gg [E]$, then $[S]$ does not change appreciably when a few molecules bind to the enzyme. Therefore, in the beginning we can consider $[S] \approx$ constant, and the product $k_1[S]$ is a pseudo first-order rate constant (with units of s^{-1}). The substrate dissociates from the enzyme with a true first-order rate constant k_{-1}, also called the off-rate constant (k_{off}). Now suppose a binding equilibrium is rapidly established. If the catalytic step controlled by k_2 in the Michaelis–Menten mechanism (shown in Equation 7.19) is much slower than dissociation ($k_2 \ll k_{-1}$), the binding equilibrium can be treated separately from the overall reaction.

Since $k_1[S] \approx$ constant in the beginning, Equation 7.97 is formally identical to the chemical equation for protein folding, which we already solved in Chapter 5 (see also Appendix C). After the reaction is started by mixing the enzyme and the substrate, the fraction of free enzyme decreases and the fraction of the ES complex increases as exponential functions of time,

$$f_E \sim e^{-k_{app}t} \tag{7.98}$$

$$f_{ES} \sim 1 - e^{-k_{app}t}. \tag{7.99}$$

The rate of binding is controlled by an *apparent rate constant*, k_{app}, given by

$$k_{app} = k_1[S] + k_{-1}. \tag{7.100}$$

The apparent rate constant depends on the substrate concentration [S] used in the experiment. This dependence allows us to separate the two components of k_{app}. If you perform the binding experiment with different substrate concentrations, k_{app} increases with [S] because the binding step, controlled by $k_1[S]$, becomes faster the larger the concentration [S].

To measure binding kinetics (**Figure 7.34**A), we need to use a rapid mixing technique, such as stopped-flow, or a relaxation method, as discussed in Chapter 5 in the context of protein folding kinetics. Measuring binding by the change in an experimental signal of the enzyme (for example, fluorescence, absorbance, or CD), you would obtain kinetic traces similar to those shown in Figure 7.34A. The curves are exponential functions of the form $f_{ES} \sim 1 - e^{-k_{app}t}$ (shown in Equation 7.99). A nonlinear least-squares fit of this equation to those

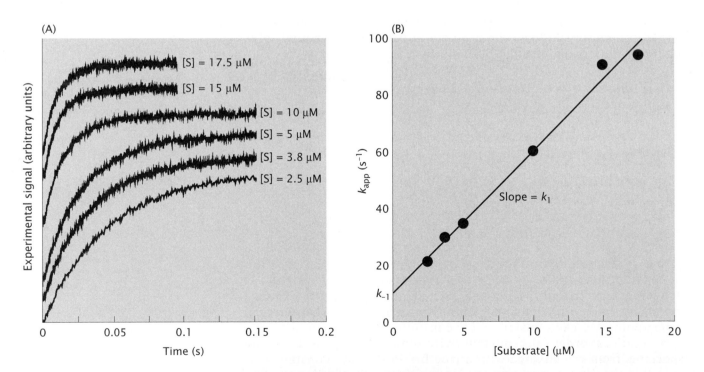

Figure 7.34 Binding kinetics of a substrate to an enzyme (hypothetical experiment). (A) Kinetic traces of a fluorescence signal measured for different substrate concentrations. Each curve yields one value of k_{app}. (B) Plot of k_{app} as a function of substrate concentration. The equation of the line is $k_{app} = k_1[S] + k_{-1}$. The y-intercept is $k_{-1} = 10$ s^{-1} and the slope is $k_1 = 5 \times 10^6$ M^{-1}s^{-1}. (Data from the author's laboratory, adapted for this illustration from a different binding experiment, in Gregory SM, Cavenaugh A, Journigan V, Pokorny A & Almeida PFF [2008] *Biophys J* 94:1667–1680.)

experimental traces allows you to determine k_{app} for each substrate concentration used. Then, if you plot those values of k_{app} as a function of [S] in each assay, you obtain a plot similar to that in Figure 7.34B. Note that k_{app} depends linearly on [S], according to Equation 7.100. The slope of the line is k_1 (k_{on}) and the y-intercept is k_{-1} (k_{off}).

If Binding is not Rapid Compared to the Chemical Step, Transient Kinetics Still Allow Determination of Rate Constants

What happens if binding is *not* fast compared to the subsequent steps—for example, if k_2 is large compared to k_{-1} in the Michaelis–Menten mechanism? Then we cannot isolate the first step of the Michaelis–Menten reaction (shown in Equation 7.19),

$$E \underset{k_{-1}}{\overset{k_1\,[S]}{\rightleftharpoons}} ES \overset{k_2}{\longrightarrow} E + P,$$

and the solution of the differential rate equations is more complicated.

Nevertheless, if the substrate concentration is sufficiently large ([S] $\gg$ [E]), then [S] still changes very little when a small fraction of S reacts with the enzyme. Thus, in the *very beginning* of the reaction, we can still consider $k_1[S] \approx$ constant. During this period we can still solve the rate equations by the general method described for protein folding (see Chapter 5 and Appendix C). However, since there are now *two steps* and we cannot ignore the second, the solution of these equations contains *two* exponential functions. For a reaction that begins with substrate only ($[P] = 0$ at $t = 0$), we obtain

$$f_E = A_1 e^{-k_1^{app}t} + A_2 e^{-k_2^{app}t} \tag{7.101}$$

$$f_{ES} = -B_1 e^{-k_1^{app}t} + B_1 e^{-k_2^{app}t} \tag{7.102}$$

$$f_P \approx 1 + C_1 e^{-k_1^{app}t} - (1 + C_1)e^{-k_2^{app}t}, \tag{7.103}$$

where the k^{app} are apparent rate constants, and A_1, A_2, B_1, and C_1 are the *amplitudes* associated with the exponential functions. The signs of these amplitudes are explicitly indicated in Equations 7.101–7.103 (that is, A_1, A_2, B_1, and $C_1 > 0$). The amplitudes are functions of the molecular rate constants and the substrate concentration. The fractions of E and ES (f_E and f_{ES}) in these equations are expressed relative to the total amount of enzyme. The fraction of P (f_P) is relative to the maximum possible amount of product formed, which is identical to the initial amount of substrate (f_P only approaches 1 if the reaction goes nearly to completion, $K_{eq} = [P]/[S] \gg 1$).

Let us see what the plots of the concentrations of E, ES, S, and P look like as a function of time. **Figure 7.35** shows the numerical (exact) solution of the differential rate equations 7.24–7.27. In this example, the true *molecular rate constants* are $k_1 = 1.0 \times 10^4$ M^{-1}s^{-1}, $k_{-1} = 1.0$ s^{-1}, and $k_2 = 10.0$ s^{-1}; the initial concentrations are [E] = 1.0 μM and [S] = 1.0 mM.

Figure 7.35A shows the very beginning of the reaction (first 0.5 seconds). In the initial *transient* period (first ~0.2 seconds), the concentration of E decreases exponentially and the concentration of ES increases exponentially. Then, the *steady state* is reached and the curves for [E] and [ES] level out. Similarly, the formation of product P accelerates in the very beginning, but once the steady state is established, [P] increases at a constant rate, as shown by the constant slope of its

(A)

(B)

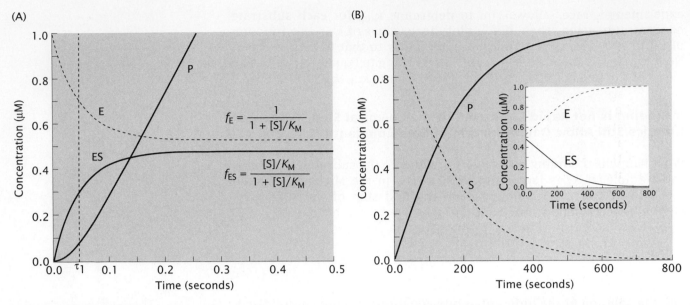

Figure 7.35 Kinetics of the reaction $E + S \rightleftharpoons ES \rightarrow E + P$, showing the concentration of E, ES, S, and P as a function of time. (A) The beginning of the reaction shows the concentrations of E, ES, and P in the initial approach to the steady state. The larger apparent rate constant $k_1^{app} = 21 \text{ s}^{-1}$ corresponds to $\tau_1 = 0.0474$ s (indicated on the x-axis). (B) The full range of the reaction shows the long-time behavior of the concentrations of S and P, and those of E and ES (inset). Note the very different concentration scales (y-axis) and the very different time scales (x-axis) in A and B. The plots are numerical solutions of the Michaelis–Menten reaction scheme (shown in Equation 7.19) with $k_1 = 1.0 \times 10^4 \text{ M}^{-1}\text{s}^{-1}$, $k_{-1} = 1.0 \text{ s}^{-1}$, $k_2 = 10.0 \text{ s}^{-1}$, and initial concentrations $[E_0] = 1.0 \ \mu\text{M}$ and $[S_0] = 1.0 \ \text{mM}$.

line in Figure 7.35A. The fractions of E and ES in the steady state are determined by K_M. They can be easily calculated from the modified partition function, $Q' = 1 + [S]/K_M$ (shown in Equation 7.70), according to Equations 7.71–7.72,

$$f_E = \frac{1}{1 + [S]/K_M}$$

$$f_{ES} = \frac{[S]/K_M}{1 + [S]/K_M}.$$

In this example,

$$K_M = \frac{k_{-1} + k_2}{k_1}$$

$$= \frac{1 \text{ s}^{-1} + 10 \text{ s}^{-1}}{1 \times 10^4 \text{ M}^{-1}\text{s}^{-1}} = 1.1 \ \text{mM}.$$

The concentrations of E and ES in the *steady state* are $[E] = f_E[E_0] = 0.524 \ \mu\text{M}$ and $[ES] = f_{ES}[E_0] = 0.476 \ \mu\text{M}$ (see Figure 7.35A). Note that these are *not* equilibrium concentrations (which are given by Equations 7.71 and 7.72 if we replace K_M by K_S, the substrate *equilibrium* dissociation constant). In this example, $K_S = k_{-1}/k_1 = 0.1$ mM, 10 times smaller than K_M. Since $k_2 \gg k_{-1}$, the approximation of treating the initial binding of S to E as a rapid equilibrium is not valid here. However, the concentrations of E and ES in the steady state can be calculated accurately from the partition function Q' (shown in Equations 7.71 and 7.72).

Equations 7.101–7.103 are only valid in the very beginning (see Figure 7.35A). Beyond that, $k_1[S]$ is *not* constant and the solution of the rate equations given by Equations 7.101–7.103 is *not* valid. As [S] varies with time, the rate Equations 7.24–7.27 must be solved numerically, and in general the results are not exponential functions. Yet, in the appropriate time limits, the *shapes* of the functions that describe the change of [E], [ES], [S], and [P] over time are qualitatively similar to those given by exponential functions and it is worthwhile to spend a moment to gain a feeling for what those equations mean.

There are two *apparent* first-order rate constants in Equations 7.101–7.103. We call the larger one k_1^{app} and the smaller one k_2^{app}. A first-order rate constant has units of s^{-1}; thus, $1/k^{app}$ has units of time (s). The reciprocal of each k^{app} is a *characteristic time* that we call τ. It is on times in the neighborhood of its corresponding τ that the effect of each rate constant is most prominent.

If $k_1^{app} \gg k_2^{app}$, we can set $k_2^{app} \approx 0$ in the beginning, and therefore $e^{-k_2^{app}t} \approx 1$. Hence, we can simplify Equations 7.101 and 7.102 to

$$f_E \approx A_1 e^{-k_1^{app}t} + A_2 \tag{7.104}$$

$$f_{ES} \approx B_1 \left(1 - e^{-k_1^{app}t}\right). \tag{7.105}$$

If k_1^{app} is large, the corresponding characteristic time $\tau_1 = 1/k_1^{app}$ is *small*, which corresponds to the beginning of the reaction. In Figure 7.35A, $k_1^{app} = 21 \ s^{-1}$, or $\tau_1 = 0.0474 \ s$. The larger apparent rate constant k_1^{app} dominates the kinetics in the short time. Thus, k_1^{app} determines the initial decay of [E], the initial rise of [ES], and the initial acceleration of the increase of [P] (see Figure 7.35A). Note that, in general, the k^{app} are functions of the *molecular rate constants* (k_1, k_{-1}, and k_2) for *all steps* and the substrate concentration. Neither k_1^{app} nor k_2^{app} coincides with any of the true molecular rate constants. In this example, $k_1^{app} = k_1[S] + k_{-1} + k_2$.

The behavior of [S] and [P] at long times is shown in Figure 7.35B. The functions look roughly like exponentials, but they are not because [S] is not constant beyond the beginning. The inset shows [E] returning to 1 μM and [ES] decaying back to zero as the reaction ends.

As discussed in Chapter 5, it is essential to determine the amplitudes of intermediates in order to assign the molecular rate constants. This is because often the *shape* of the kinetic function remains identical (or almost identical) upon interchange of molecular rate constants, but the *amplitudes* do not. In the case of enzyme kinetics, this problem is somewhat alleviated by the dependence of the pseudo first-order rate constant $k_1[S]$ on the initial substrate concentration, whose variation affords extra help in establishing the mechanism. A positive amplitude means that the curve is *concave up* ($\smile$) in the time period dominated by the corresponding k^{app}. In Figure 7.35A, A_1 and C_1 are positive (shown in Equations 7.101–7.103), and therefore the curves of E and P are concave up in the period dominated by k_1^{app} (beginning). Conversely, a negative amplitude means that the curve is *concave down* ($\frown$) in the time period dominated by the corresponding k^{app}. In Figure 7.35A, $-B_1$ is negative, and therefore the curve of ES is concave down in the period dominated by k_1^{app}.

In mechanisms with several steps, such as in the scheme

$$\mathrm{E} \underset{k_{-1}}{\overset{k_1 \ [S]}{\rightleftharpoons}} \mathrm{ES} \overset{k_2}{\longrightarrow} \mathrm{ES} \overset{k_3}{\longrightarrow} \mathrm{E + P},$$

the number of exponentials in the solutions increases. A solution of the rate equations for this mechanism that is valid for short times ($k_1[S] \approx$ constant) has three exponential functions (one apparent rate constant for each step in the mechanism). The amplitudes become complicated functions of the rate constants and initial concentrations, in practice too complex to determine analytically if there are more than three steps. However, the k^{app} (eigenvalues) are always simpler than the amplitudes and provide very useful information about the mechanism, as they relate to the characteristic τ's that dominate the kinetics at different times. In many cases, approximate solutions can be obtained analytically, or numerically with a computer. If the transient kinetics can be measured experimentally, then the individual rate constants can be determined, whereas this is rarely possible with steady-state kinetics.

The book by Gutfreund (1995) is an excellent source to continue to learn the methods of transient kinetics and how to design experiments to extract the most from them. We will not study more complicated cases here, however, because they add no new concepts. Instead, we will turn to an essential aspect of transient kinetics and their importance in establishing enzyme mechanisms: the detection of kinetic intermediates.

Intermediates in the Mechanism of Enzyme-Catalyzed Reactions can be Detected in Transient Kinetics

To fully establish a mechanism for an enzyme-catalyzed reaction, the *intermediate* states occurring between the formation of the ES complex and the release of the final product must be identified. The ability to detect intermediates is one of the greatest advantages of transient kinetics. The most famous intermediate of enzyme kinetics is the acyl-enzyme covalent intermediate that occurs in the mechanism of serine proteases (see Figure 7.32). The natural substrates of serine proteases are peptides. Amides are also substrates; they can be viewed as shorter versions of peptides, in which the amino acid residue following the target peptide bond is replaced by an amide on the C-terminus (**Figure 7.36**).

Serine proteases are also active on esters, obtained by replacing the last amide with an ester bond. Esters are convenient artificial substrates and were used in the early studies on serine proteases. In particular, *p*-nitrophenyl esters are especially useful because their hydrolysis releases the bright yellow ion *p*-nitrophenolate (PNP⁻, **Figure 7.37**), which is used to monitor the course of the reaction by recording its absorbance at $\lambda = 400$ nm as a function of time. The hydrolysis of peptides, amides, and esters by serine proteases proceeds through the formation of an *acyl-enzyme intermediate* (see Figures 7.32 and 7.37). In the case of peptides and amides, the acyl-enzyme intermediate is very short-lived, whereas in the case of esters, it accumulates and is easily detected in pre-steady-state kinetics.

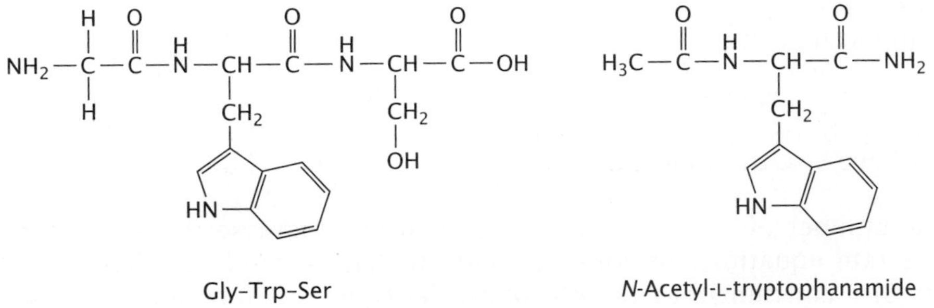

Figure 7.36 Substrates of chymotrypsin: left, the tripeptide Gly-Trp-Ser; right, *N*-acetyl-L-tryptophanamide.

Gly–Trp–Ser

N-Acetyl-L-tryptophanamide

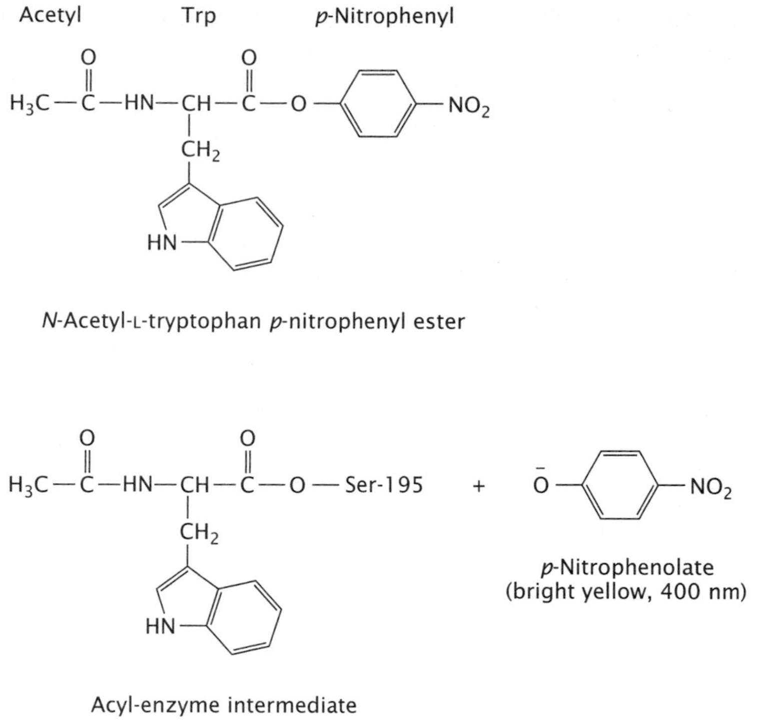

N-Acetyl-L-tryptophan p-nitrophenyl ester

Acyl-enzyme intermediate

Figure 7.37 Top, the chymotrypsin artificial substrate N-acetyl-L-tryptophan p-nitrophenyl ester (Ac-Trp-PNP). Bottom, the acyl-enzyme intermediate and the p-nitrophenolate ion, which is bright yellow (strong absorbance at $\lambda = 400$ nm), used for detection of the activity of chymotrypsin.

The Serine Proteases Follow a Ping-Pong Mechanism with an Acyl-Enzyme Covalent Intermediate

The mechanism of serine proteases (see Figure 7.32) belongs to the ping-pong type. Let us match the chemical mechanism of chymotrypsin to the kinetic mechanism, to attach a concrete physical meaning to the states in the ping-pong mechanism (shown in Equation 7.66). The mechanism of chymotrypsin can be written as

$$
\begin{array}{ccc}
& k_1[\text{Ac-Trp-PNP}] & & k_2 \\
\text{E} & \rightleftharpoons & \text{E} \cdot \text{Ac-Trp-PNP} \longrightarrow & \text{Ac-Trp–E} \\
& k_{-1} & & + \text{PNP}^-
\end{array}
$$

$$(7.106)$$

$$
\begin{array}{ccc}
k_{3a}[\text{H}_2\text{O}] & & k_{3b} \\
\rightleftharpoons & \text{Ac-Trp–E} \cdot \text{H}_2\text{O} \longrightarrow & \text{E} + \\
k_{-3a} & & \text{Ac-Trp-COOH}
\end{array}
$$

where the substrate is N-acetyl-L-tryptophan p-nitrophenyl ester (Ac-Trp-PNP, Figure 7.37). The first step, as always, is binding. If the substrate is in large excess relative to the enzyme, binding occurs with a pseudo first-order rate constant given by $k_1[\text{Ac-Trp-PNP}]$. The substrate dissociates with a first-order rate constant k_{-1}. Upon binding, the Michaelis–Menten complex is formed (E · Ac-Trp-PNP). Then, the amide bond is cleaved, the first product (PNP$^-$) leaves, and the acyl-enzyme intermediate is formed, with a covalent bond to Ser-195 (Ac-Trp–E, see Figure 7.37 bottom left). This is the *acylation* step, controlled by k_2.

After that, the second substrate (water) comes in. It binds with another pseudo first-order rate constant, $k_{3a}[\text{H}_2\text{O}]$, which is really constant because the concentration of water (the solvent) does not change. Finally, the acyl-enzyme bond is hydrolyzed, and the last product leaves (Ac-Trp-COOH). Because water is in large excess, the rate of the last two steps in Equation 7.106 is determined by k_{3b}. Therefore, these two steps can be combined as a *deacylation* step, controlled by the apparent rate constant k_3. The mechanism can then be written in a more compact fashion:

$$E \underset{k_{-1}}{\overset{k_1[\text{Ac-Trp-PNP}]}{\rightleftharpoons}} E \cdot \text{Ac-Trp-PNP} \overset{k_2}{\longrightarrow} \overset{\text{Ac-Trp-E}}{\underset{+ \text{PNP}^-}{}} \overset{k_3}{\longrightarrow} \overset{E +}{\underset{\text{Ac-Trp-COOH}}{}}.$$

$$(7.107)$$

This is the same mechanism as in Equation 7.47. The expressions for K_M and k_{cat} (shown in Equations 7.55 and 7.56) are

$$K_M = \frac{k_2 + k_{-1}}{k_1} \frac{k_3}{k_2 + k_3}$$

$$k_{cat} = k_2 k_3 / (k_2 + k_3).$$

In esters, the acylation step (k_2) is fast compared to the deacylation step (k_3). Thus, since $k_2 \gg k_3$, we can simplify K_M and k_{cat} to $K_M = (k_3/k_2)(k_2 + k_{-1})/k_1$ and $k_{cat} = k_3$. In amides, the acylation step is slow, and $k_2 \ll k_3$. Therefore, K_M and k_{cat} reduce to $K_M = (k_2 + k_{-1})/k_1$ and $k_{cat} = k_2$. In either case, k_{cat} is the slower of k_2 or k_3 (rate-limiting). **Table 7.1** lists rate and equilibrium constants for the hydrolysis of N-acetyl-L-Trp- and L-Phe-esters and amides by chymotrypsin.

"Burst" Kinetics Indicate that an Intermediate is Formed

The use of p-nitrophenyl esters as artificial substrates was a fortunate choice. Hartley and Kilby (1954) observed that the steady-state period of the hydrolysis of p-nitrophenyl esters by chymotrypsin as a function of time did *not extrapolate to zero* on the y-axis (**Figure 7.38**). Rather, the number of moles of p-nitrophenolate product released extrapolated to a value *equal to the number of moles of chymotrypsin*. There is an initial "burst" of activity resulting in the release of 1 mole of p-nitrophenolate per mole of protein, followed by a much slower, steady-state period. The type of transient kinetics observed here are called *burst kinetics*.

The observation of burst kinetics corresponding to the hydrolysis of 1 mole p-nitrophenyl ester per mole of enzyme indicates that a number of p-nitrophenyl ester molecules *exactly equal to the number of enzyme molecules* present is rapidly hydrolyzed by the enzyme.

Table 7.1 Rate and equilibrium constants for the hydrolysis of N-acetyl-L-Trp- and L-Phe-esters and amides by chymotrypsin at pH $\approx 7^a$

Substrate	$K_M{}^b$ (mM)	$\dfrac{k_2+k_{-1}}{k_1}{}^c$ (mM)	K_S (mM)	$k_{cat}{}^d$ (s^{-1})	k_2 (s^{-1})	k_3 (s^{-1})	k_1 (M^{-1}s^{-1})	k_{-1} (s^{-1})
N-Ac-L-Trp-PNP	0.001	2	~ 1	30	4×10^4	30	6×10^7	6×10^4
N-Ac-L-Trp-ethyl ester	0.1	2	2	27	7×10^2	28		
N-Ac-L-Trp-amide	7	7	7	0.026	0.026	30		
N-Ac-L-Phe-PNP	0.02	8	7.4	77	2×10^4	77		
N-Ac-L-Phe-ethyl ester	~ 1	7	7	63	3×10^2	80		
N-Ac-L-Phe-amide	37	37	37	0.04	0.04	72		

^{a}(Combined data from Renard M & Fersht AR [1973] *Biochemistry* 12:4713–4718, and Zerner B, Bond RPM & Bender M [1964] *J Am Chem Soc* 86:3674–3679.)
^{b} $K_M = [(k_2 + k_{-1})/k_1] [k_3/(k_2 + k_3)]$.
^{c}Calculated assuming k_1 is identical to that of N-acetyl-L-Trp-p-nitrophenyl ester.
^{d} $k_{cat} = k_2 k_3 / (k_2 + k_3)$.

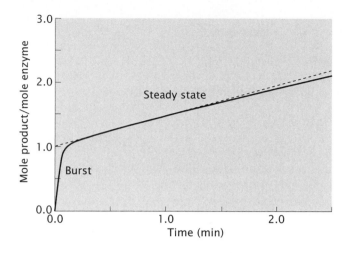

Figure 7.38 Burst kinetics. The pre-steady-state kinetics of chymotrypsin hydrolysis of Ac-L-Trp-*p*-nitrophenyl ester revealed an initial burst of product formation followed by a slower, steady rate.

After that, hydrolysis of additional molecules of *p*-nitrophenyl ester has to wait for the enzyme to be regenerated for a second round of reaction. The regeneration is the hydrolysis of the acyl-enzyme covalent intermediate. The intermediate forms rapidly but decays slowly. This experiment also demonstrates a way of *titrating the active site of the enzyme*. Indeed, this is a method to determine the enzyme concentration exactly.

Burst kinetics in chymotrypsin are observed in the hydrolysis of esters but not of amides. In both cases an acyl-enzyme intermediate forms, but it is only in the case of esters that it forms fast and breaks down slowly ($k_2 \gg k_3$), so the intermediate accumulates. In amides and peptides, $k_2 \ll k_3$, so there is no accumulation of the acyl-enzyme intermediate.

7.9 THE ACTIVITY OF ENZYMES DEPENDS ON PH

The activity of most enzymes depends significantly on pH. This is because the pH determines the state of protonation of the residues directly involved in catalysis, or residues whose state of protonation affects substrate binding or the stability of the enzyme. We will examine this problem in some detail using chymotrypsin as an example.

Enzymes are Often Synthesized as Inactive Precursors that are Post-Translationally Activated

To understand the effect of pH on the activity of chymotrypsin, we need to learn a little about its activation process. Many enzymes are synthesized as inactive precursors, which are post-translationally activated to produce the mature enzyme. Precursors are generally called proenzymes, but in the case of proteolytic enzymes they are called *zymogens*. Chymotrypsin is synthesized as the inactive precursor *chymotrypsinogen* (trypsin, as *trypsinogen*). In the serine proteases, as in many secreted enzymes, disulfide bonds are made between specific cysteine residues after translation of the polypeptide. The disulfide bonds are established in the zymogens, which are then activated by specific proteolytic cleavage.

The proteolytic activation of chymotrypsinogen is illustrated in **Figure 7.39**. Chymotrypsin consists of three chains, A, B, and C, connected by disulfide bonds between cysteine residues (numbered

Figure 7.39 The activation of chymotrypsinogen to produce mature chymotrypsin, catalyzed by trypsin and chymotrypsin. The disulfide bonds are formed in chymotrypsinogen, but for the sake of clarity, they are only shown in α-chymotrypsin (top).

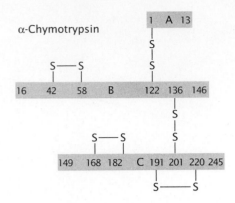

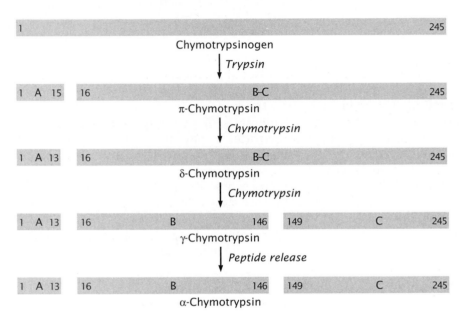

in Figure 7.39, top). Originally, the chains A, B, and C are part of a single chymotrypsinogen polypeptide. After its synthesis, the disulfide bonds are established. Then, chymotrypsinogen undergoes several proteolytic cleavages until it becomes the mature form of the enzyme, α-chymotrypsin. The first cleavage is by trypsin, between residues 15 and 16, to generate π-chymotrypsin. Then there are two cleavages catalyzed by chymotrypsin: the first removes residues 14–15, to generate δ-chymotrypsin, and the second removes residues 147–148 to generate γ-chymotrypsin. Dissociation of a peptide from γ-chymotrypsin finally yields the mature enzyme, α-chymotrypsin.

The importance of this post-translational modification for the effect of pH on chymotrypsin activity is that residue Ile-16 of chymotrypsinogen becomes the N-terminus of the B chain of chymotrypsin. Its α-amino group is then free to accept a proton (to become NH_3^+). Normally, the N-terminus of a protein has $pK_a \approx 7.8$. However, in α-chymotrypsin (as well and in π- and δ-chymotrypsin, which are also active), the protonated Ile-16 of chain B establishes a *salt bridge* with Asp-194 of chain C (**Figure 7.40**). This hydrogen bond stabilizes the proton on Ile-16, changing the pK_a of its α-amino group to 10.0. When Ile-16 is deprotonated, Asp-194 makes a salt bridge with His-40 instead, and the enzyme acquires a conformation in which the substrate binding site is not fully accessible, as in chymotrypsinogen. Further, in this conformation, the amide N–H bond of Gly-193 in the active site points in the wrong direction to establish a hydrogen bond with the

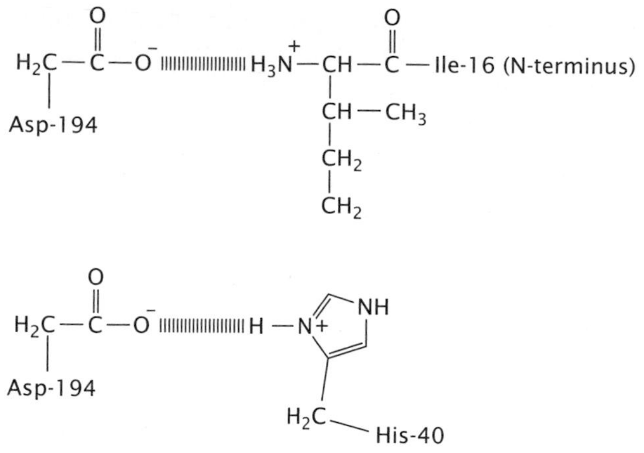

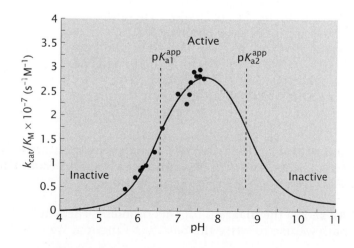

Figure 7.40 Salt bridges in chymotrypsin. Top: In the active conformation of chymotrypsin (pH ≤ 8), Asp-194 makes a salt bridge with the N-terminus of chain B, Ile-16. Bottom: At high pH, Ile-16 becomes deprotonated and Asp-194 makes a salt bridge with His-40 instead, as in chymotrypsinogen. In this conformation, access of the substrate to the active site of the enzyme is blocked and chymotrypsin is inactive.

substrate. When the Ile-16···Asp-194 salt bridge is formed, the protein undergoes a small conformational change that completes the formation of the substrate binding site.

The pH Dependence of Enzyme Activity is Often Bell-Shaped

The activity of enzymes as a function of pH is often described by a bell-shaped curve. For example, a plot of k_{cat}/K_M of chymotrypsin against pH yields the curve shown in **Figure 7.41**. There appear to exist two titratable groups whose protonation affects enzyme activity. These two groups appear to have pK_a's about the *inflection points* in the bell-shaped curve: one at $pK_{a1}^{app} \approx 6.5$ and another at $pK_{a2}^{app} \approx 8.7$ (indicated by the dashed vertical lines in Figure 7.41).

What gives rise to this pH dependence? Things are a little more complicated when we measure activity than when we measure an equilibrium property, and not all measures of activity—the rate (v), k_{cat}, K_M, or k_{cat}/K_M—provide the same information. Rather, their dependence on pH is usually different. But they are all based on the same concept. Let us understand this concept. To do so, we will again use the partition function. We will use chymotrypsin as our example, but the same ideas apply to any enzyme.

In Chapter 6 we studied the effect of pH on protein stability (see Section 6.7). We had two states in equilibrium: native (N) or folded, and denatured (D) or unfolded. We learned that pH affects the folding equilibrium if at least one group of the protein, typically an amino

Figure 7.41 The activity of chymotrypsin measured by k_{cat}/K_M for the hydrolysis of Ac-L-Trp-p-nitrophenyl ester as a function of pH. The curve is calculated from Equation 7.122. The two apparent pK_a's (≈ 6.5 and 8.7) are indicated by the dashed lines. (Data from Renard M & Fersht AR [1973] *Biochemistry* 12:4713–4718.)

acid side chain, has an *anomalous* pK_a in the folded state. The pK_a in the unfolded state is usually normal. However, residues in the active sites of enzymes often have anomalous pK_a's in the folded state. In chymotrypsin, the pK_a of the active site residue His-57 is close to normal (≈ 6.5). It is the pK_a of the N-terminus (Ile-16) of chain B that is anomalous.

Here, as in the folding equilibrium, we have two states of the enzyme: free (E) and in complex with substrate (ES). The pH will affect substrate binding if the pK_a of a side chain is different in E and ES. Further, regardless of whether the pK_a is normal or anomalous, the protonation of a residue will affect the rate of the enzyme-catalyzed reaction if it affects the *fraction of active enzyme*.

The pH Determines How Much of the Active form of the Enzyme Exists

The basic concept you need to understand the effect of pH on enzyme activity is that the pH determines how much of the *active form of the enzyme* exists. We might be interested in the fraction of active enzyme under equilibrium conditions, in which case the equilibrium pK_a's of the enzyme determine the pH dependence of the fraction of interest. Or we might be interested in the active fraction under steady-state conditions, in which case *apparent* pK_a's (also called *kinetic* pK_a's in this case) determine the active fraction. The problem is essentially reduced to the calculation of the fraction of active enzyme as a function of pH.

What is the active form of the enzyme? In chymotrypsin, there are two important pK_a's that we need to take into account. First, His-57 needs to be deprotonated (in the neutral form) for the enzyme to function. This is because its lone pair must be able to accept the proton from Ser-195 in the first step of the catalytic mechanism (see Figure 7.32A). For a typical His residue, p$K_a \approx 6.5$, which is fairly close to the value for His-57. Second, Ile-16, which is the N-terminal residue, needs to be protonated to interact through a salt bridge with Asp-194, so that chymotrypsin is in a conformation in which the active site is accessible. The N-terminus typically has p$K_a \approx 7.8$, but because of the salt bridge with Asp-194 (see Figure 7.40), Ile-16 has an anomalous p$K_a = 10.0$ in the active conformation.

We want to calculate the fraction f_a of chymotrypsin that has His-57 *deprotonated* and Ile-16 *protonated*. Let us designate those two pK_a's by pK_{a1} (His) and pK_{a2} (Ile). The acidity constant $K_{a2} \ll K_{a1}$, by a factor of $\sim 10^3$. Note well, though, that K_a is the proton *dissociation* constant (the binding constant is $1/K_a$). Thus, the first proton to bind will almost always go to Ile-16, never to His-57. Therefore, we can safely simplify the partition function by ignoring states with His protonated and Ile deprotonated. We will use the proton binding partition function to calculate f_a. The partition function method works here, greatly simplifying the calculations, because the active enzyme–substrate complex is the last enzyme species before product formation and all previous steps are rapid equilibria (none is irreversible).

The Fraction of the Active form of Chymotrypsin and the Kinetic Parameters k_{cat}, K_M, and k_{cat}/K_M can be Calculated from the Partition Function

Figure 7.42 shows the diagram for this problem. The shaded box indicates the path we use to write the partition function. We could use any

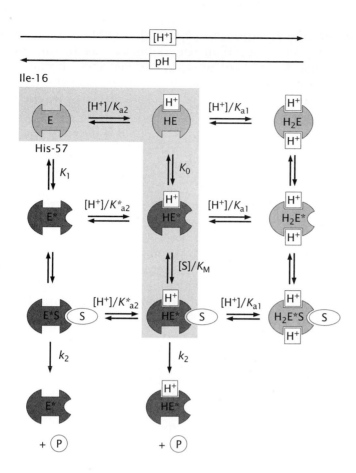

Figure 7.42 Diagram of proton and substrate binding by chymotrypsin. Enzyme forms in the active *conformation* are labeled with an asterisk ($*$) and represented by an ellipse. Of those forms, only those with His-57 deprotonated are active, which are shown in dark gray. Forms in light gray are inactive. The shaded box indicates the main path used to write the partition function. The states outside the box are obtained from those in the box by moving along the equilibria to the left or to the right. K_M includes k_2. The values of the constants are $pK_{a1} = 6.56$, $pK_{a2} = 7.9$, $pK_{a2}^* = 10.0$, and $K_0 = 5.6$.

path but this is the most convenient because we know the values of the constants along it. We begin with the free enzyme (E), which we choose as our *reference state*, and assign its relative probability as 1. We learned in Chapter 6 how to obtain the probabilities of the other states. The relative probability of any form of the enzyme is obtained by starting with the reference (E) and moving along the selected path in the diagram, multiplying the equilibrium factors in the forward direction as we go.

Before we proceed, let us be clear about some definitions. In this section, we designate by an asterisk ($*$) the forms of the enzyme that are in the active *conformation*. Furthermore, we use the symbol K_M for the Michaelis constant of the substrate defined by $K_M = (k_2 + k_{-1})/k_1$, where k_1 and k_{-1} are the rate constants for binding and dissociation of the substrate, and k_2 is the catalytic constant in the active form of the enzyme (see Figure 7.42). The apparent Michaelis constant is designated by K_M^{app}.

We are now ready to apply the method. We start at the reference, with relative probability $p_E = 1$. First, we protonate Ile-16, which brings in the factor $[H^+]/K_{a2}$, to get to HE; $p_{HE} = [H^+]/K_{a2}$. Second, the conformational change takes place, as we go down to HE*. The conformational change HE $\rightleftharpoons$ HE* brings in the equilibrium constant K_0,

$$K_0 = \frac{[HE^*]}{[HE]} = 5.6. \tag{7.108}$$

Thus, the relative probability of HE* is $p_{HE^*} = K_0[H^+]/K_{a2}$. Third, HE* binds the substrate to produce HE*S. This is the active form of the enzyme that we need for the kinetics, because it is the form that appears in the rate equation, $v = k_2[\text{HE*S}]$. To obtain the relative probability of HE*S, we proceed in the same way as for an equilibrium situation but use K_M instead of K_S in the last step, because we need to include here not only substrate binding but also the reaction controlled by k_2, HE*S $\to$ HE* + P. This step brings in the factor $[S]/K_M$. Thus, we obtain the relative probability of state HE*S (relative to E), as

$$p_{\text{HE*S}} = K_0 \frac{[\text{H}^+]}{K_{a2}} \frac{[\text{S}]}{K_M}. \tag{7.109}$$

Now here is a slight complication: the form E*S is *also active*, to the same degree as HE*S. There is just much less of it present because the conformational change is very unlikely if the enzyme is not protonated on the N-terminus of Ile-16. Still, we need to include its contribution to the rate of the reaction, which becomes dominant at very high pH, when the protonation of Ile-16 essentially does not occur. The relative probability of E*S is obtained by moving *backward* from HE*S (to the left), on the diagram of Figure 7.42. This is because, going from E*S to HE*S (from left to right), we know that

$$p_{\text{HE*S}} = \frac{[\text{H}^+]}{K_{a2}^*} p_{\text{E*S}}. \tag{7.110}$$

Hence, from right to left, we must multiply $p_{\text{HE*S}}$ by the reciprocal of the factor above the arrow, to obtain

$$p_{\text{E*S}} = K_0 \frac{K_{a2}^*}{[\text{H}^+]} \frac{[\text{H}^+]}{K_{a2}} \frac{[\text{S}]}{K_M}. \tag{7.111}$$

The relative probability of all the forms of active enzyme (p_a^*) is the sum of the two terms, which we can group together as

$$p_a^* = K_0 \frac{[\text{H}^+]}{K_{a2}} \frac{[\text{S}]}{K_M} \left(1 + \frac{K_{a2}^*}{[\text{H}^+]}\right). \tag{7.112}$$

In the last factor, the term 1 corresponds to state HE*S, and the term $K_{a2}^*/[\text{H}^+]$ corresponds to state E*S. K_{a2}^* is the acidity constant of Ile-16 in the active conformation of the enzyme. To calculate the fraction of active enzyme, we just need to divide Equation 7.112 by the partition function.

To write the partition function of chymotrypsin, including protonation of Ile-16 and His-57, and substrate binding, we again use E as the reference, with $p_E = 1$, and move along any convenient path in the diagram of Figure 7.42 to reach all possible states. Using the shaded path as our main path, which branches to the sides to reach the other states, we obtain

$$Q = 1 + \frac{[\text{H}^+]}{K_{a2}} \left[1 + \frac{[\text{H}^+]}{K_{a1}} + q^* K_0 \left(1 + \frac{[\text{S}]}{K_M}\right)\right], \tag{7.113}$$

where q^* corresponds to the states in the second and third rows in Figure 7.42: q^* is the part of the partition function that takes into account proton binding to the conformation labeled with an asterisk (*). It is given by

$$q^* = 1 + \frac{K_{a2}^*}{[H^+]} + \frac{[H^+]}{K_{a1}}, \qquad (7.114)$$

where the term $K_{a2}^*/[H^+]$ again corresponds to going from HE* to E* by losing a proton (from right to left), and the term $[H^+]/K_{a1}$ corresponds to gaining a proton to yield H$_2$E*. The velocity *relative* to V_{max}, or *fractional velocity* v/V_{max}, is equal to the probability of the active form of the enzyme complexed with the substrate relative to Q—that is, the *fraction of bound active enzyme*,

$$f_a = \frac{v}{V_{max}}$$
$$= \frac{([H^+]/K_{a2})K_0([S]/K_M)(1 + K_{a2}^*/[H^+])}{Q}. \qquad (7.115)$$

To write the rate in the form of the Michaelis–Menten equation, we isolate the terms containing [S] in the partition function:

$$Q = 1 + \frac{[H^+]}{K_{a2}}\left(1 + \frac{[H^+]}{K_{a1}} + K_0 q^*\right) + \frac{[H^+]}{K_{a2}}K_0 q^* \frac{[S]}{K_M}. \qquad (7.116)$$

The rate is $v = k_2 f_a[E_0]$, where k_2 is the catalytic constant in the active conformation. Thus, we obtain

$$v = \frac{k_2[E_0]([H^+]/K_{a2})K_0([S]/K_M)(1 + K_{a2}^*/[H^+])}{Q}. \qquad (7.117)$$

We want to compare this equation with the general Michaelis–Menten form,

$$v = \frac{k_{cat}[E_0][S]}{K_M^{app} + [S]}, \qquad (7.118)$$

where we wrote K_M^{app} for the apparent Michaelis constant. To this end, we multiply the numerator and the denominator of Equation 7.117 by $K_M K_{a2}/(q^* K_0[H^+])$ to obtain

$$v = \frac{(k_2(1 + K_{a2}^*/[H^+])/q^*)[E_0][S]}{K_M[1 + [H^+]/K_{a2}(1 + [H^+]/K_{a1} + K_0 q^*)]/(q^* K_0[H^+]/K_{a2}) + [S]}. \qquad (7.119)$$

Now you can see that

$$k_{cat} = k_2(1 + K_{a2}^*/[H^+])/q^*$$
$$= \frac{k_2(1 + K_{a2}^*/[H^+])}{1 + K_{a2}^*/[H^+] + [H^+]/K_{a1}}, \qquad (7.120)$$

the apparent K_M is

$$K_M^{app} = K_M \frac{1 + [H^+]/K_{a2}(1 + [H^+]/K_{a1} + q^* K_0)}{q^* K_0[H^+]/K_{a2}}, \qquad (7.121)$$

Figure 7.43 Variation of k_{cat} and K_M^{app} for the hydrolysis of Ac-L-Trp-p-nitrophenyl ester by chymotrypsin as a function of pH. These are the two separate components of the bell-shaped curve of k_{cat}/K_M^{app} as a function of pH shown in Figure 7.41. K_M^{app} is given by Equation 7.121 and k_{cat} is given by Equation 7.120.

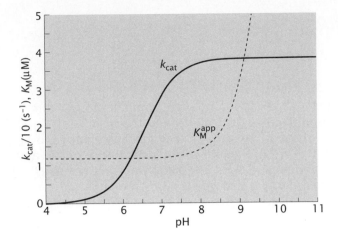

and the ratio

$$\frac{k_{cat}}{K_M^{app}} = \left(\frac{k_2}{K_M}\right) \frac{K_0\,([H^+]/K_{a2})\,(1 + K_{a2}^*/[H^+])}{1 + ([H^+]/K_{a2})\,(1 + [H^+]/K_{a1} + q^*K_0)}. \tag{7.122}$$

This is the equation of the curve plotted in Figure 7.41.

The two separate constants, K_M^{app} (shown in Equation 7.121) and k_{cat} (shown in Equation 7.120), are plotted as a function of pH in **Figure 7.43**. Their ratio, k_{cat}/K_M^{app}, yields the bell-shaped curve of Figure 7.41. We can now understand the dependence of the activity of chymotrypsin on pH. The left, rising part of the bell curve is caused by an increase in k_{cat} as His-57 becomes deprotonated and therefore able to take part in catalysis. The right, falling part of the bell curve is caused by an increase in K_M^{app} as Ile-16 becomes deprotonated. Then the enzyme conformation changes and the substrate no longer binds to the active site.

The apparent $pK_{a1}^{app} \approx 6.8$, which controls the rising part of the bell curve in Figure 7.41, is about that of His-57 ($pK_{a1} = 6.56$), in both the free enzyme and the ES complex. So, the rising part of the bell curve is easy to understand. However, the *apparent* $pK_{a2}^{app} \approx 8.7$, which controls the decay of the bell curve, is neither the pK_{a2} of Ile-16 in the free enzyme (E) in the inactive conformation ($pK_{a2} = 7.9$) nor the pK_{a2} of Ile-16 in the ES complex in the active conformation ($pK_{a2}^* = 10.0$). Why? Look carefully at Equation 7.121 for K_M^{app}. At moderate to high pH (pH $\approx$ 7–10, or $[H^+] = 10^{-10}$–10^{-7} M), K_M^{app} is simplified because $q^* \approx 1$ and becomes

$$K_M^{app} \approx K_M \frac{1 + K_0[H^+]/K_{a2}}{K_0\,[H^+]/K_{a2}}. \tag{7.123}$$

This equation shows that K_M^{app} is controlled by an apparent proton dissociation constant equal to K_{a2}/K_0. Thus, $pK_{a2}^{app} = -\log(K_{a2}/K_0) = pK_{a2} + \log K_0 = 7.9 + 0.8 = 8.7$. You can see this directly from the diagram in Figure 7.42 because the probability of state HE* is obtained by multiplying the factors $[H^+]/K_{a2}$ and K_0.

In summary, the enzyme species responsible for almost all the activity is HE*. The species E* can bind substrate and make product, but its concentration is always very small (unless [S] is very high and drives the equilibrium down the first column of the diagram of Figure 7.42). Moreover, the form E*S could only be important at high pH (pH > 10), because at moderate to low pH, [HE*S] $\gg$[E*S]. In fact, in the absence of substrate, the pH dependence of the concentration of HE* follows a bell-shaped function almost identical to that of k_{cat}/K_M^{app} (**Figure 7.44**).

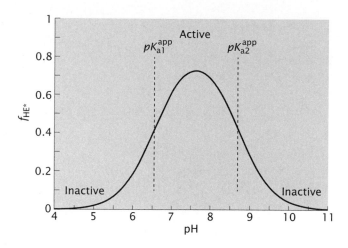

Figure 7.44 The bell-shaped curve of the fraction of chymotrypsin in the active state HE*, in the absence of substrate. The curve is almost identical to that of k_{cat}/K_M^{app} as a function of pH (see Figure 7.41).

7.10 CONFORMATIONAL DYNAMICS ARE ESSENTIAL IN THE MECHANISM OF DIHYDROFOLATE REDUCTASE

We have seen that chymotrypsin has two major conformations. One of them is inactive, which predominates when Ile-16 is deprotonated; the other is active, which predominates when Ile-16 is protonated. We also saw that there is one well-established acyl-enzyme covalent intermediate. However, there are probably many more intermediate states, with subtly different conformations of the enzyme. Most enzyme mechanisms involve a series of intermediates, which often correspond to different conformations of the enzyme. Some of those conformations are very different and are therefore well characterized in terms of mechanism and structure. Others differ more subtly from the better known structures. We chose dihydrofolate reductase (DHFR) as our second enzyme to study in detail, because it provides one of the best examples of the role of *conformational dynamics* in enzyme catalysis.

Dihydrofolate Reductase Catalyzes a Hydride Transfer Reaction

Dihydrofolate reductase (DHFR) is a fundamental enzyme in the metabolism of several amino acids and in the biosynthesis of the nucleotides adenosine, guanosine, and thymidine monophosphates (AMP, GMP, and dTMP). These metabolic processes use tetrahydrofolate (THF, **Figure 7.45**), a key substrate in transfers of *one-carbon units* in several biochemical reactions.

The methyl group that differentiates thymine from uracil (Chapter 3) is donated by N^5, N^{10}-methylene tetrahydrofolate. In this reaction, THF is oxidized to dihydrofolate (DHF). DHFR is necessary to reduce DHF back to THF. The enzyme uses a cofactor, the coenzyme nicotinamide adenine dinucleotide phosphate (NADPH) as the source of hydride (H:$^-$) to reduce DHF. DHFR catalyzes the hydride transfer from NADPH to 7,8-dihydrofolate (DHF) to produce 5,6,7,8-tetrahydrofolate (THF) and NADP$^+$, the oxidized form of the nicotinamide dinucleotide (**Figure 7.46**). The pro-*R* hydrogen atom of NADPH is specifically transferred as a hydride to carbon number 6 of DHF. For this reaction to occur, DHF and NADPH must be close to each other and with the correct orientation in the enzyme. In solution, the bimolecular reaction between DHF and NADPH would entail a very large translational

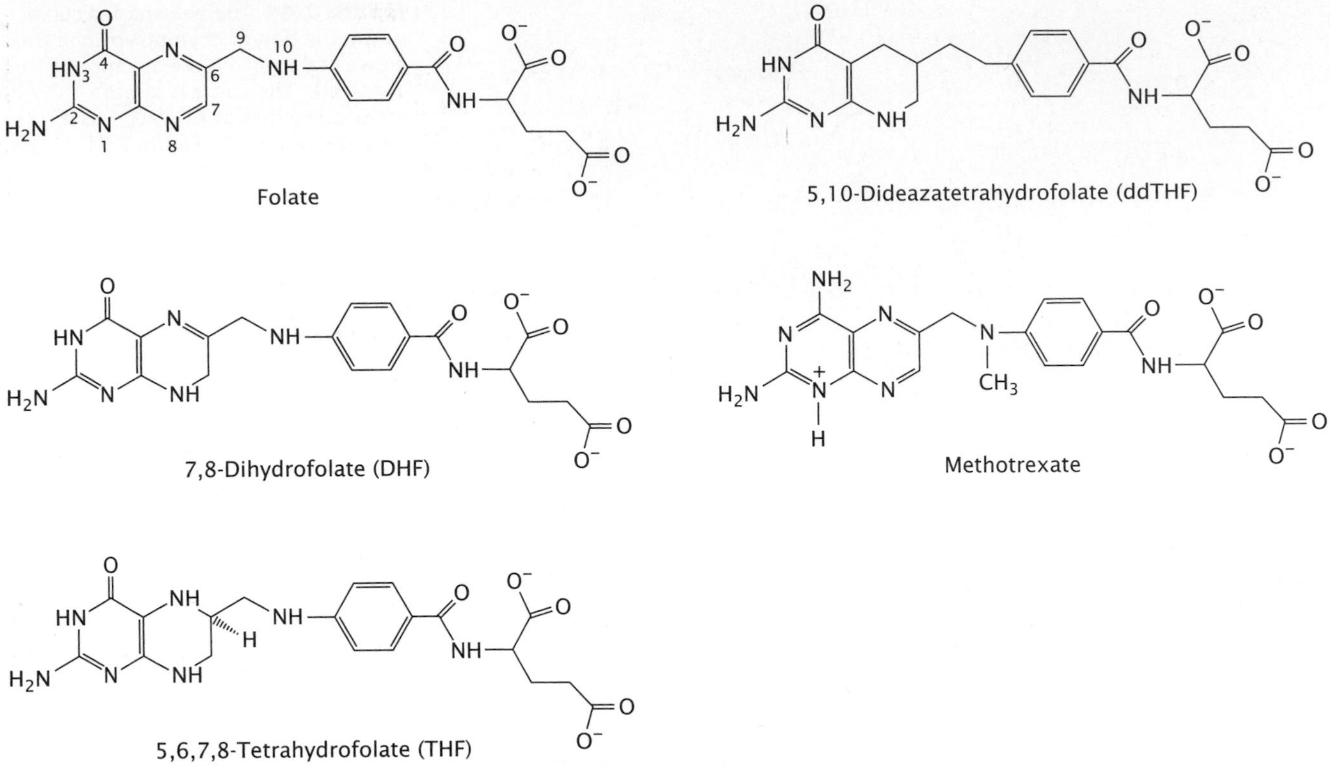

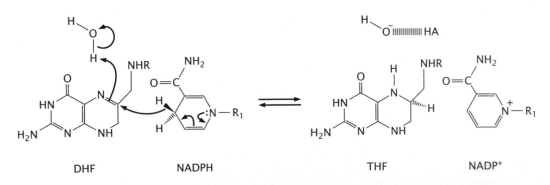

Folate

5,10-Dideazatetrahydrofolate (ddTHF)

7,8-Dihydrofolate (DHF)

Methotrexate

5,6,7,8-Tetrahydrofolate (THF)

Figure 7.45 Structures of folate, dihydrofolate (DHF), tetrahydrofolate (THF), its analog ddTHF, and the enzyme inhibitor methotrexate.

and rotational entropy loss, but in the enzyme this entropy loss is minimized and the reaction is much more favorable (see Section 7.2).

Because of its fundamental role in the synthesis of dTMP, DHFR is critical for rapidly dividing cells. As such, it is the target of several antibacterial and anticancer drugs, such as methotrexate. Methotrexate is a structural analog of folic acid (see Figure 7.45) and therefore a competitive inhibitor of the binding of DHF and THF to DHFR, but binding is so tight that it is essentially irreversible. The binding constant of methotrexate in the ternary complex with DHFR and $NADP^+$ is $K = 1.3 \times 10^{12}$ M^{-1}, and the off-rate constant is $k_{-1} = 3 \times 10^{-5}$ s^{-1} (Birdsall B, Burgen ASV & Roberts GCK [1980] *Biochemistry* 19:3723–3731).

DHF   NADPH   THF $NADP^+$

Figure 7.46 The reaction catalyzed by DHFR. The hydride (the proton and the pair of electrons that constitute the C–H bond) is transferred from NADPH to DHF. The reaction is stereospecific for the pro-*R* hydrogen of NADPH, shown in the figure as protruding out of the plane of the paper (solid wedge). The pro-*S* hydrogen on the same carbon atom of NADPH does not react (dashed wedge). The proton donor is probably a water molecule in the active site, hydrogen bonded to a group of the enzyme (HA). The complete structures of the DHF and THF are shown in Figure 7.45.

DHFR Cycles through Several Conformations During Catalysis

DHFR occurs in three major conformations: closed, occluded, and open (**Figure 7.47**). The enzyme changes its structure between these three conformations during the catalytic cycle. The *closed conformation* is adopted in the Michaelis–Menten complex, with DHF (substrate) and NADPH (cofactor) bound to the enzyme (see Figure 7.47A). The name *closed* refers to the position of the loop containing the residue Met-20 (M20 loop), which appears close to the active site in this conformation. The *occluded conformation* (see Figure 7.47B) occurs in the initial product complex between the enzyme, NADP⁺, and THF, which is established after hydride transfer from NADPH to DHF. In the occluded conformation, the M20 loop prevents binding of the nicotinamide by occluding its binding pocket.

The open conformation is common in most complexes—namely, in the folate complex (see Figure 7.47D), and also in the free enzyme (apoenzyme). Rather than a stable state in the catalytic cycle, the open conformation appears to be an intermediate state between the closed and occluded conformations. In the transition state, the enzyme switches between closed and open conformations, until it finally adopts the occluded conformation in the product complex. The closed

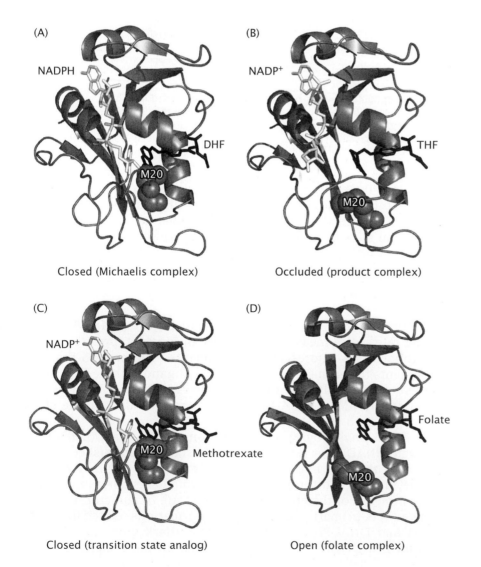

Closed (Michaelis complex)

Occluded (product complex)

Closed (transition state analog)

Open (folate complex)

Figure 7.47 Conformations of DHFR. (A) Closed conformation, Michaelis ES complex. NADP⁺ is shown in place of NADPH and folate in place of DHF (PDB 1RX2). (B) Occluded conformation, initial product complex, with ATP-ribose in place of NADP⁺ and ddTHF in place of THF (PDB 1RX4). (C) Closed conformation (transition state) in complex with NADPH and methotrexate (PDB 1RX3). (D) Open conformation, folate complex (PDB 1RD7). NADP(H), or its analogs, are shown in white; folate and its derivatives or analogs, in black; Met-20, at the center of the M20 loop is shown in space-filling representation. The names of the physiological ligands are indicated, but in most X-ray structures cofactor and substrate analogs were used instead of the true ligands. (X-ray crytallography structures by Sawaya MR & Kraut J [1997] *Biochemistry* 36:586–603.)

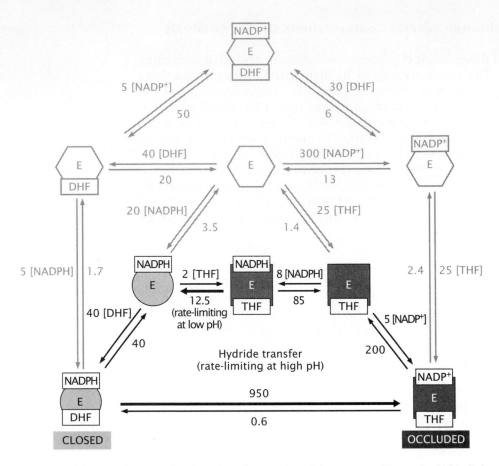

Figure 7.48 The kinetic mechanism of DHFR. The closed conformations of the enzyme (E) are shaded in light gray, whereas the occluded conformations are dark. The products of the on-rate (binding) constants (in $\mu M^{-1}s^{-1}$) by the ligand concentrations and the off-rate (dissociation) constants (in s^{-1}) for each step are indicated next to each arrow. The rate constants for the hydride transfer step (in s^{-1}) at low pH are also indicated (bottom). (Based on Fierke CA, Johnson KA & Benkovic SJ [1987] *Biochemistry* 26:4085–4092.)

conformation in the transition state is believed to be very similar to that adopted by DHFR complexed with the inhibitor methotrexate (see Figure 7.47C), even though methotrexate does not actually resemble the transition state of the DHF $\rightleftharpoons$ THF conversion.

The Rates of All DHFR Reactions were Determined by Transient Kinetics Experiments

The reactions of DHFR are represented in **Figure 7.48**. The corresponding rates were measured by transient kinetics (pre-steady state) using stopped-flow fluorescence experiments. The pseudo first-order rate constants for each step, of the form k_1[ligand], where k_1 is the on-rate constant, are indicated over the arrows in Figure 7.48 (the value of k_1 is given in $\mu M^{-1}s^{-1}$). Also indicated are the values of the off-rate constants for each step (k_{-1} in units of s^{-1}).

At pH 7 and physiological concentrations of substrate (DHF or THF) and cofactor (NADPH or NADP⁺), a steady state is established in which DHFR cycles through five different intermediates. Figure 7.48 shows this cycle, in dark (bottom). The five intermediates correspond to different complexes of the enzyme with substrate or cofactor. Each intermediate adopts primarily one of two conformations: *closed* (light gray: E·NADPH and E·NADPH·DHF) and *occluded* (dark gray: E·NADPH·THF, E·THF, and E·NADP⁺·THF). The chemical step in the cycle, the hydride

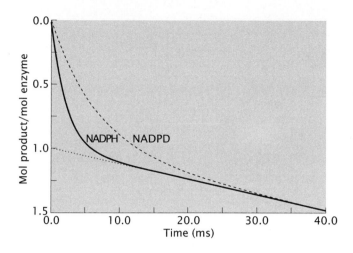

Figure 7.49 The initial rapid phase (burst) of the DHFR reaction, followed by the slower steady-state phase (rate of 12 s^{-1}) at pH 6.5. The rapid phase is an exponential function of the form $F(t) = e^{-kt}$, where the rate constant $k = 450$ s^{-1} for NADPH and $k = 150$ s^{-1} for NADPD at pH 6.5. The curves are calculated from the rate constants. (Based on Fierke CA, Johnson KA & Benkovic SJ [1987] *Biochemistry* 26:4085–4092.)

transfer, occurs concomitant with the conformational change closed $\rightleftharpoons$ occluded.

Let us concentrate on the catalytic cycle shown in dark in Figure 7.48. When the complex E·NADPH is mixed with DHF in the stopped-flow apparatus at low pH (≤ 6.5), an initial rapid reaction is observed (**Figure 7.49**) in which NAPDH is oxidized to NADP$^+$ by transfer of a hydride to DHF. This rapid phase has a maximal apparent rate constant of $k_{hyd} = 950$ s^{-1} (from hydride transfer) and ends when 1 mole of product is formed per mole of enzyme. It is followed by a much slower, steady-state phase, which has a rate constant of 12.5 s^{-1}. This slow process is the release of THF from the E·NADPH·THF complex, and it is the rate-limiting step in the steady state. That is, in order for another round of oxidation–reduction reaction to occur, the product (THF) must dissociate first from the enzyme. The dissociation is slow and limits the rate of the overall reaction cycle.

As the pH is increased, the rate of the hydride transfer step decreases progressively (**Figure 7.50**). Eventually hydride transfer becomes slower than THF dissociation and therefore rate-limiting. The pH dependence of the kinetics arises because protonation of the active site, with p$K_a = 6.5$, must take place for the hydride transfer step to occur. As the pH increases, the fraction protonated decreases and the

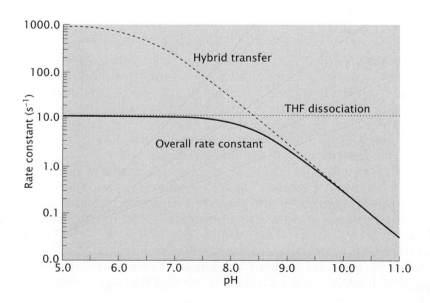

Figure 7.50 Dependence of the rate constant for hydride transfer on pH (dashed line) in comparison with the rate constant for THF dissociation (dotted line) from the complex E·NADPH·THF. The overall rate constant (solid line) is the smallest of the two: the rate is limited by THF dissociation at low pH and by hydride transfer at high pH. The curves are calculated from the pH-dependence of the rate constants. (Based on Fierke CA, Johnson KA & Benkovic SJ [1987] *Biochemistry* 26:4085–4092.)

Figure 7.51 Deuterium isotope effect. The frequency—and therefore the energy—of the vibration of the C–H bond is higher than that of the C–D bond in the ground state. This results in a smaller energy difference to the transition state when the C–H bond is broken and, therefore, in a larger rate constant.

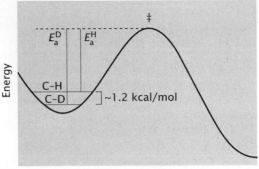

apparent rate constant of hydride transfer decreases to $k_{hyd} \approx 200$ s^{-1} at pH 7.

In summary, the effect of pH on the kinetics of the DHFR reaction arises because of a change in rate-limiting step (see Figure 7.48). The rate-limiting step changes with pH because the hydride transfer is pH-dependent. At low pH (≤ 6.5), the rate-limiting step is the release of THF from the E·NADPH·THF complex. At high pH (≥ 8.5), hydride transfer becomes rate-limiting (see Figure 7.50).

A Kinetic Isotope Effect Occurs when a C–H Bond is Broken in the Rate-Limiting Step

How do we know that the rapid initial phase in the kinetics shown in Figure 7.49 reflects hydride transfer? If you repeated this experiment, replacing the coenzyme NADP*H* by NADP*D*, which is labeled with deuterium (D or ^{2}H) at the bond broken during the hydride transfer step (C–D instead of C–H in NADPH, see Figure 7.46), you would see that with NADPD the reaction is *slower*. Compare the curves for NADPH and NADPD in Figure 7.49. This change in rate is called a primary *kinetic isotope effect*. In this case, it is a *deuterium* isotope effect because H is replaced by D. It is a *primary* effect because the isotope replacement is directly at the bond broken in the mechanism.

The deuterium kinetic isotope effect is defined as the *ratio* of rate constants, k^H/k^D, for a reaction with a substrate with a C–H bond relative to that with a C–D bond. To understand the kinetic isotope effect, you need to examine **Figure 7.51**. Because of the larger mass of deuterium, the zero-point energy (ground state) of the vibrational frequency of the C–D bond is lower than that of the C–H bond. This frequency difference corresponds to an energy difference of about 1.2 kcal/mol: the C–H bond has a higher energy already in the ground state. However, the height of the transition barrier for bond breaking is the same for both bonds. Therefore, the activation energy (E_a) is *smaller* for breaking the C–H bond than the C–D bond by $\Delta E_a = -1.2$ kcal/mol. Since the rate constant $k \sim e^{-E_a/RT}$, then $k^H/k^D = e^{-\Delta E_a/RT}$. With $\Delta E_a = -1.2$ kcal/mol, we obtain a ratio of rate constants $k^H/k^D = 7$. However, smaller values are often observed, even if breaking of the C–H bond occurs in the rate-limiting step. Larger deuterium isotope effects are also possible, particularly if tunneling plays a significant role in proton transfer. On the other hand, $k^H/k^D \approx 1$ indicates that the bond is *not* broken in the rate-limiting step.

In this example, we compare the apparent rate constant of the burst (k_{burst}) with NADPH and with NADPD. We find that k_{burst} is significantly affected by the H/D substitution, with $k^H/k^D \approx 3.5$ at pH 7. Therefore, we conclude that the burst reflects the hydride transfer and $k_{hyd} \approx k_{burst}$. Indeed, the deuterium kinetic isotope effect in k_{burst} at pH 7 is about the same as that determined independently in k_{hyd}, confirming that the burst measures the hydride transfer step, when THF release is rate-limiting (at pH 7, see Figure 7.50).

Protein Conformational Dynamics Determine the Rates of Hydride Transfer and Ligand Release

The dynamics of conformational changes in DHFR were measured using FRET (Förster resonance energy transfer; see Section 5.6), either using the energy transfer between the intrinsic Trp residues and NADPH, or by attaching a donor and acceptor pair of fluorescent probes at different residues on the protein. FRET efficiency increases as the donor–acceptor distance decreases, which makes it sensitive to conformational changes. The apparent rate constant of the conformational changes k_{conf} for the step E·NADPH·DHF $\rightleftharpoons$ E·NADP$^+$·THF and k_{hyd} were similar at pH 7; both rate constants $\approx$150–200 s^{-1}, which corresponds to a time of ~5 ms.

The similarity of the rates of the chemical step (hydride transfer) and the conformational changes might suggest that the conformational change and the chemical step are *coupled*, as if the rate of hydride transfer were coupled with the frequency of the protein motions, like in a sort of "resonance" phenomenon, which would facilitate the chemical reaction. Vibrations (and bond breaking and formation) occur in femtoseconds to picoseconds, but those motions are coupled over the enzyme structure, resulting in slower motions that occur in the micro- to millisecond time scale. The idea of "protein-promoting vibrations" or "coherent protein motions" that may transfer the conformational energy to a chemical reaction step is thought-provoking as a biochemical analog of a physical resonance phenomenon. However, when the experimental conditions are changed or mutations are introduced in the protein, k_{conf} and k_{hyd} do *not* vary in the same manner. In wild-type (wt) DHFR, k_{hyd} depends on pH between 6 and 8.5, and exhibits a kinetic isotope effect of $\approx$3–3.5. But k_{conf} is independent of pH and does not exhibit any effect at pH 7 ($k^H/k^D \approx 1$). Furthermore, when mutations are introduced in the protein, k_{conf} and k_{hyd} are affected in very different ways.

Another approach to examine the importance of conformational dynamics is to produce a "heavy" enzyme, in which the common isotopes are replaced by heavy ones: ^{1}H by ^{2}H (D), ^{12}C by ^{13}C, and ^{14}N by ^{15}N. Heavy enzymes have *slower vibrations*. If the enzyme motions were coupled to breaking and making bonds of the substrates, then the corresponding rate constants should be affected (k_{hyd} in DHFR). However, k_{hyd} and its kinetic isotope effect are identical in both heavy and light enzymes. Thus, heavy DHFR catalyzes hydride transfer as fast as light DHFR.

In the case of heavy DHFR the change in mass appears to have a greater effect on the conformational ensemble of the *ground state* than on the transition state. However, when amino acid mutations are introduced in DHFR, the conformational ensemble of the transition state seems to be affected the most. Those mutations slow down catalysis, reducing the hydride transfer rate. The mutations reduce the flexibility of the protein in the transition state, reducing its access to some of the conformations that were available to wt DHFR. Thus, the *conformational entropy* of the transition state decreases in the mutant. The transition state is destabilized by the mutation relative to the ground state, the Gibbs energy of activation increases, and the hydride transfer rate constant consequently decreases.

In conclusion, the temporal coincidence of the chemical step and the conformational transitions indicates that hydride transfer rate is limited by the time the enzyme needs to *sample the conformations* available, until it finds the right one. It does *not* mean that motions in the enzyme speed up the actual transfer of the hydride. This situation is very similar to that encountered in protein folding. If we say that a protein folds in a time $\tau = 1$ ms, which is the inverse of the folding rate constant $k = 1/\tau = 10^3$ s^{-1}, this means that a *population of proteins* takes about 1 ms to change from unfolded to folded. The actual transition, unfolded $\rightleftharpoons$ folded, in an *individual protein* takes place *much faster*: once it begins, it is completed on the order of microseconds (transition path time, τ_{tp}). Similarly, the hydride transfer itself occurs almost instantaneously in comparison to the millisecond time scale of the enzyme-catalyzed reaction. What the millisecond time represents is the time the enzyme takes to search the conformational space until it randomly finds the best in terms of distance and relative orientation between the donor (NADPH) and acceptor (DHF) for the hydride transfer to take place. The dynamics of the protein—the rate at which it samples conformations—therefore determine the rate of catalysis.

7.11 SUMMARY

The Michaelis–Menten equation describes steady-state enzyme kinetics and applies to most mechanisms, even complex, one substrate at a time. The rates of complex mechanisms can be determined by the King–Altman method. In general, the Michaelis constant K_M and the catalytic constant k_{cat} are combinations of molecular rate constants and concentrations of ligands other than the substrate (activators, inhibitors, or other substrates in multisubstrate reactions). To assign molecular rate constants, however, it is usually necessary to examine the transient kinetics that occur before the steady state is established. Transient kinetics allow detection of reaction intermediates and a much better understanding of the mechanism. They require faster kinetics methods, such as stopped-flow or a perturbation (jump) technique. The partition function method allows determination of the rates of steady-state kinetics if the reactions preceding the catalytic step are all reversible (equilibria). The method is powerful because many common cases, such as reversible enzyme inhibition and the effect of pH on enzyme kinetics, fall in this category.

The first step in a reaction catalyzed by an enzyme is always binding of the substrate. However, the enzyme is optimized to bind the transition state, not the ligand. Enzymes enhance reaction rates by as much as 10^8 to 10^{15} times. This is mainly achieved by binding and therefore lowering the Gibbs energy of the transition state. Enzymes are large molecules—they must be—for three reasons. First, the large size of an enzyme allows great binding improvement, because it allows binding of extended substrates and transition states at multiple points of interaction, while each interaction remains weak and therefore reversible at physiological temperature. Second, the enzyme needs to bind not only the substrate and the transition state, but a series of intermediates in the reaction path. Hence, the need to adapt its conformation. Third, the enzyme needs to be flexible to allow reorientation of bound substrates, until the optimal positioning is achieved for the reaction to take place. Small molecules are rigid. Only large molecules are flexible, because they contain a large number of bonds over which rotations can occur. The rate-limiting step in DHFR catalysis, and probably many other enzymes, is a search of the conformational space of the enzyme for optimal substrate positioning.

7.12 PROBLEMS

7.1 The Michaelis–Menten equation has the same *form* as a hyperbolic binding isotherm. Explain why. (Note: this question is not about the *formal* aspect of the equations, but about the *concept*.)

7.2 The data in Table 7.2 show the activity of enzyme A as a function of substrate concentration [S] in the absence and in the presence of an inhibitor I.
Generate a Lineweaver–Burk (double reciprocal) plot from the data in Table 7.2. What is the type of inhibition?

7.3 Use the King–Altman method to obtain the steady-state rate for the reaction mechanism,

$$E + S \underset{k_{-1}}{\overset{k_1}{\rightleftharpoons}} ES \underset{k_{-2}}{\overset{k_2}{\rightleftharpoons}} EP \overset{k_3}{\longrightarrow} E + P.$$

7.4 The Michaelis–Menten equation does not apply to an enzyme that binds its substrate cooperatively. However, we can derive the rate of a reaction catalyzed by a cooperative enzyme under the conditions in which the partition function method can be used to determine the rate.

Phosphofructokinase-1 (PFK-1) catalyzes a phosphoryl transfer from ATP to fructose-6-phosphate (F6P), to produce fructose-1,6-bisphosphate (F-1,6-bisP) and ADP:

$$\overset{\text{PFK-1}}{\underset{}{F6P + ATP \rightleftharpoons F\text{-}1,6\text{-bisP} + ADP.}}$$

PFK-1 exhibits cooperativity for binding of F6P, and therefore the initial rate of the reaction catalyzed by PFK-1 has a sigmoidal dependence on the initial concentration of F6P. Binding of ATP to PFK-1, however, is not cooperative. PFK-1 is the most important control point of glycolysis. As such, its activity is regulated by a number of allosteric effectors. PFK-1 is inhibited by citrate and high concentrations of ATP, and it is activated by AMP or ADP.

Table 7.2 Enzyme activity in the absence and presence of an inhibitor

[S] (μM)	V_0 (μM/min) (No inhibitor)	V_0' (μM/min) (With inhibitor)
2	275	180
5	490	350
10	650	520
20	790	685
40	880	810

Table 7.3 PFK-1 activity as function of [F6P], with fixed [ATP] = 1.0 mM

[F6P] (mM)	v/V_{max}
0.0	0.0
0.50	0.073
0.75	0.165
1.0	0.334
1.5	0.759
2.0	0.846
3.0	0.957
4.0	1.021
5.0	1.00

(Data from Brüser A, Kirchberger J, Kloos M et al. [2012] *J Biol Chem* 287:17546–17553. Courtesy of Torsten Schöneberg, University of Leipzig, Germany.)

The binding of ATP is interesting. In addition to being a substrate, ATP is also an allosteric inhibitor of PFK-1. Thus, at low concentrations, ATP increases the rate (substrate concentration increases), but at high concentrations, ATP lowers the rate because it binds as an inhibitor. In this problem, we will try to understand the activity of PFK-1 quantitatively, with respect to its two substrates by using the partition function method.

Tables 7.3 and 7.4 show experimental data for the relative rate (v/V_{max}) of the reaction catalyzed by PFK-1 as a function of [F6P] and [ATP]. In all experiments, the

Table 7.4 PFK-1 activity as function of [ATP], with fixed [F6P] = 2.0 mM[a]

[ATP] (mM)	v/V_{max}
0.0	0.0
0.02	0.308
0.05	0.491
0.10	0.691
0.25	0.859
0.50	0.891
0.75	0.894
1.0	0.873
1.5	0.696
2.0	0.477
3.0	0.038

[a]The original data were slightly renormalized for this problem. (Data from Brüser A, Kirchberger J, Kloos M et al. [2012] *J Biol Chem* 287:17546–17553. Courtesy of Torsten Schöneberg, University of Leipzig, Germany.)

Table 7.5 Equilibrium constants for PFK-1 conformational change and ligand binding

Constant	Value
K_0	1.3×10^{-5}
K_T	$1.0 \times 10 \ M^{-1}$
K_R	$4.1 \times 10^4 \ M^{-1}$
K_A	$2.2 \times 10^4 \ M^{-1}$
K_I	$1.9 \times 10^3 \ M^{-1}$

concentration of the enzyme is very small compared to the concentrations of the substrates.

PFK-1 in mammals and bacteria like *E. coli* is a homotetramer, which can exist in two quaternary conformational states, an active R state and an inactive T state, very much like hemoglobin. The quaternary allosteric transition T $\rightleftharpoons$ R is controlled by an equilibrium constant $K_0 = [R]/[T]$ (defined in the absence of ligands). PFK-1 has four catalytic binding sites for F6P and ATP, and four regulatory (inhibitory) sites for ATP, which are different from the catalytic sites. Binding of F6P is cooperative: as F6P binds, the T $\rightarrow$ R allosteric transition becomes increasingly more likely. The binding constant of F6P to the T state is K_T and to the R state is K_R. ATP, however, binds to its catalytic site with a hyperbolic binding isotherm controlled by a binding constant K_A. Binding of ATP to the inhibitory site occurs only in the T state and is controlled by a binding constant K_I. The values of the equilibrium constants listed in Table 7.5 are consistent with the experimental data.

(a) Plot the activity data (v/V_{max} as a function of each concentration) in Tables 7.3 and 7.4.
(b) Recall that the relative rate v/V_{max} is equal to the ratio of the concentration of the active enzyme–substrate complex in the steady state relative to the sum of the concentrations of all enzyme forms E_0 (shown in Equation 7.74). In this case, the enzyme–substrate complex ($E_R \cdot F6P \cdot ATP$) has both F6P and ATP bound to the active (catalytic) site in the R state of the enzyme (E_R),

$$\frac{v}{V_{max}} = \frac{[E_R \cdot F6P \cdot ATP]}{[E_0]}.$$

This ratio is just the fraction of the active enzyme–substrate complex in the steady state. We want to calculate the fraction of active complex using the partition function method. Therefore, begin by writing the partition function for PFK-1 binding of F6P and ATP (to the catalytic and to the inhibitory sites) using the MWC model (see Section 6.6).

Hints: the partition function will be similar to Equation 6.123, but will contain additional factors corresponding to the binding of ATP. Recall that in the MWC model binding *within* each conformational state (T or R) is hyperbolic (not cooperative). That is, the partition function for binding to each type of site is of the form $q = 1 + K[X]$, where K is the appropriate binding constant and $[X]$ is the appropriate ligand concentration. Cooperativity arises because of the quaternary allosteric transition T $\rightleftharpoons$ R that is more likely upon ligand binding and creates high-affinity sites from low-affinity ones. This simplifies our work because it means that *within each state* binding of a ligand is independent of any other ligand. In turn, this means that

the partition function of the T or the R state is a product of four (identical) partition functions, one for each functional subunit, with its catalytic and regulatory sites. Thus, to write the partition function for the whole molecule, begin by writing the partition function for one functional subunit in the T and the R conformational states. Note that ATP binds only to the inhibitory site in the T state, but it binds to the catalytic site in both the T and R states, with the same binding constant K_A.
(c) Next, derive the isotherm for binding of F6P to PFK-1, to describe the data in Table 7.3. Begin by calculating the fraction of F6P sites occupied using the derivative method (shown in Equation 6.21). Now, we want the fraction of the active complex only, not all states with F6P bound. Note, however, that the experiment was performed with 1 mM ATP. Using the partition function for a single ATP catalytic site, calculate the fractional occupancy of catalytic sites by ATP (use K_A from Table 7.5). (This calculation is valid because ATP binds independently (of ATP or F6P) to the catalytic site.) The result of this calculation should tell you that almost all subunits have ATP bound to their catalytic sites if [ATP] = 1 mM. Therefore, the fractional occupancy of the F6P sites that you calculated with Equation 6.21 is also the fraction of active complex, $E_R \cdot F6P \cdot ATP$. Finally, pay attention to concentration units: they have to match the units used for the binding constants! Plot the F6P binding isotherm on the same graph of the experimental data. The curve should fit the data very well.
(d) Now, derive the isotherm for binding of ATP, to describe the data in Table 7.4. This time, we want to calculate the derivative of the partition function with respect to [ATP] using Equation 6.21. However, unlike in the previous case, the result does not directly give us the concentration of the active complex because the binding isotherm obtained includes ATP binding to the T state, which is inactive. Therefore, you have to discard some terms in the numerator of the binding isotherm. On the other hand, with 2 mM F6P present, essentially all forms in the R state have F6P bound. Verify this by calculating the fractional occupancy of F6P sites in the R state using the partition function for a single site in the R state (use K_R from Table 7.5). Plot the binding isotherm on the same graph as the corresponding data for the entire range of [ATP]. Again, the curve should describe the data very well (except for the last point).
(e) Now enlarge the initial part of the plot in the previous question, plotting the data and the curve only up to 0.60 mM ATP. In this range you should obtain a simple Michaelis–Menten plot. Estimate the Michaelis constant K_M for ATP from the graph. To which of the constants in Table 7.5 is K_M related and how?
(f) Explain why the reaction is cooperative with respect to F6P but not with respect to ATP.

7.5 In DHFR, the rate of hydride transfer depends on pH. At very low pH $k_{hyd} \approx 950 \ s^{-1}$. The pH dependence arises because protonation of the active site, with $pK_a = 6.5$, must take place for hydride transfer to occur. Derive the pH dependence of k_{hyd}. Use it to calculate k_{hyd} at pH = 5, 7, 8.5, and 10.

7.6 The drawing below shows four cases of enzyme inhibition. In each case, the figure represents *all possible* states of an enzyme: free, with substrate (S), with inhibitor (I), or with both bound. A *notch* on the enzyme surface means that the binding site is empty but available for binding; the absence of the notch means that the site is unavailable. Identify the type of inhibition shown in each of the four pictures.

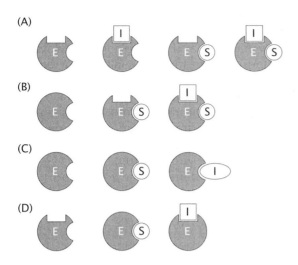

(A)

(B)

(C)

(D)

7.7 You have isolated an unknown enzyme D and determined that one of its substrates is pyrophosphate (PPi). Then you performed a series of experiments to better understand the interaction of PPi with enzyme D.
(a) First, you measured equilibrium binding of PPi to enzyme D by the fluorescence quenching of an intrinsic tryptophan in the enzyme (the Trp fluorescence decreases, or the quenching increases upon PPi binding). The data collected are listed in Table 7.6. Plot and analyze the data to determine the equilibrium dissociation constant (K_d).
(b) Second, you measured the kinetics of binding by stopped-flow fluorescence for several concentrations of substrate PPi. The data obtained at 0.20 mM PPi are listed in Table 7.7. Plot and analyze the data to determine the apparent rate constant for binding at 0.20 mM PPi. Calculate also the characteristic time τ of the kinetics.
(c) You performed the same experiment at several other concentrations of PPi. The apparent (observed) rate constants (k_{app}) obtained are listed in Table 7.8. Complete the table with the k_{app} determined in (b) for 0.20 mM PPi. Then plot k_{app} as a function of [PPi]. Determine the on-rate (k_1) and the off-rate (k_{-1}) constants. Calculate the equilibrium dissociation constant K_d from these data.
(d) Are the equilibrium binding (a) and the kinetic binding (c) data sets consistent with each other? Explain.

Table 7.6 Equilibrium binding of pyrophosphate (PPi) to enzyme D measured by fluorescence quenching (arbitrary units, a.u.)[a]

[PPi] (mM)	Fluorescence quenching (a.u.)
0.05	10.9
0.10	19.2
0.20	31.8
0.40	49.5
0.80	58.5
1.20	60.2
1.60	68.7

[a]The source of the data is given in the Solutions Manual, not to reveal the answer.

Table 7.7 Binding kinetics of pyrophosphate ([PPi] = 0.20 mM) to enzyme D measured by stopped-flow fluorescence (arbitrary units, a.u.)[a]

Time (s)	Fluorescence (a.u.)
0.0022	100.7
0.0029	94.3
0.0039	84.9
0.0068	71.7
0.0093	58.7
0.0154	41.2
0.0190	32.3
0.0273	23.7
0.0409	22.7
0.0474	17.1
0.0600	22.0
0.0643	20.2
0.0790	15.9
0.0912	20.6
0.1084	15.0
0.1163	20.4
0.1264	19.6
0.1472	14.8
0.1691	20.6
0.1896	15.2

[a]The source of the data is given in the Solutions Manual, not to reveal the answer.

7.8 H,K-ATPase is an integral membrane protein that transports H^+ in exchange for K^+ coupled with ATP hydrolysis. The compound imidazo[1,2-α]pyridine (abbreviated here as IAP) is a competitive inhibitor of K^+ transport. In order to understand the mechanism of the competitive inhibition, and to map the binding site of IAP, several mutants of H,K-ATPase were prepared, changing the amino acids residues located close to the presumed binding site of IAP. Table 7.9 shows kinetic data for the transport of K^+ by H,K-ATPase measured by the rate of formation of the product (Pi, inorganic phosphate) of ATP hydrolysis, with no inhibitor, and in the presence of two different concentrations of IAP for each of the three protein mutants (A, B, and C).
(a) Generate a Michaelis–Menten plot for each of the three mutants, including the reaction with no inhibitor and the two IAP concentrations (three graphs, with three plots each). Determine K_M and V_{max} in each case, in the absence and in the presence of the inhibitor, by fitting the Michaelis–Menten equation to each data set.

Table 7.8 Apparent rate constant for binding of pyrophosphate (PPi) to enzyme D as a function of [PPi][a]

[PPi] (mM)	k_{app} (s^{-1})
0.20	—
0.40	116
0.60	132
0.80	161
1.00	172
1.20	187

[a]The source of the data is given in the Solutions Manual, not to reveal the answer.

Table 7.9 Kinetics of K^+ transport by H,K-ATPase[a,b]

	Mutant A				Mutant B				Mutant C		
	v (μmol Pi/mg/hr)				v (μmol Pi/mg/hr)				v (μmol Pi/mg/hr)		
$[K^+]$		[IAP] (μM)		$[K^+]$		[IAP] (μM)		$[K^+]$		[IAP] (μM)	
(mM)	0	0.25	0.50	(mM)	0	2.0	5.0	(mM)	0	0.25	0.50
1.0	21.1	6.4	5.4	0.5	7.8	8.2	4.8	0.7	26.9	9.2	6.5
2.0	30.9	16.2	10.5	0.7	11.4	—	—	1.0	33.4	15.3	10.6
3.5	34.4	21.3	15.5	1.0	15.1	11.2	8.1	2.0	48.1	20.5	12.3
5.0	35.8	24.8	19.6	2.0	21.3	15.1	11.4	3.3	74.7	30.0	17.0
10.0	38.7	30.7	25.6	3.3	26.1	17.8	15.1	6.7	87.3	36.5	25.6
20.0	39.7	34.8	30.8	5.0	30.4	—	—	13.5	94.8	46.0	38.9
33.0	40.5	36.7	32.8	10.0	35.4	24.0	17.2	33.0	103.0	68.5	53.5
				13.5	37.1	—	—				
				20.0	37.1	26.0	21.0				
				33.0	39.1	29.5	19.3				

[a]The rate (v, in μmol Pi/mg/hr) of formation inorganic phosphate (Pi) from ATP hydrolysis is measured with no IAP, or in the presence of two concentrations of IAP for each protein mutant (A, B, C).
[b]The source of the data is given in the Solutions Manual, not to reveal the answer.

(b) Next, transform each plot into a Lineweaver–Burk plot. Again, you should obtain three graphs, one for each mutant, with three plots each. Determine the type of inhibition by IAP for each mutant from the Lineweaver–Burk plots. Determine also K_M and V_{max} from linear fits of the Lineweaver–Burk equation to the data. (Hint: exclude the points at the lowest substrate (S = K^+) concentration (largest 1/[S]) in each plot from the fits because those points have a large error. When you do the linear fits, it will become very apparent that most of the points at high 1/[S] clearly fall off the best-fit line for the lower 1/[S].)

(c) Determine the inhibitor dissociation constants (K_I and K_I') for each mutant.

7.13 FURTHER READING

General

Abeles RH, Frey PA & Jencks WP (1992) Biochemistry. Jones and Bartlett.

Cornish-Bowden A (2012) Fundamentals of Enzyme Kinetics, 4th ed. Wiley.

Frey PA & Hegeman AD (2007) Enzymatic Reaction Mechanisms. Oxford University Press.

Fersht A (1999) Structure and Mechanism in Protein Science. Freeman.

Jencks WP (1987) Catalysis in Chemistry and Enzymology, 2nd ed. Dover.

Kyte J (2006) Structure in Protein Chemistry, 2nd ed. Garland.

Transition State Theory

Arrhenius S (1889) Uber die Reaktionsgeschwindigkeit bei der Inversion von Rohrzucker durch Sauren. *Z Phys Chem* 4:226–248.

Eyring H (1935) The activated complex in chemical reactions. *J Chem Phys* 3:107–115.

Eyring H (1935) The activated complex and the absolute rate of chemical reactions. *Chem Rev* 17:65–77.

Hammond GS (1955) A correlation of reaction rates. *J Am Chem Soc* 77:334–338.

Kramers HA (1940) Brownian motion in a field of force and the diffusion model of chemical reactions. *Physica* 7:284–304.

General Principles of Enzyme Catalysis

Amyes TL & Richard JP (2012) Specificity in transition state binding: The Pauling model revisited. *Biochemistry* 52:2021–2035.

Berg OG & von Hippel PH (1985) Diffusion-controlled macromolecular interactions. *Annu Rev Biophys Biophys Chem* 14:131–160.

Bruice TC (2002) A view at the millennium: The efficiency of enzymatic catalysis. *Acc Chem Res* 35:139–148.

Fischer E (1894) Einfluss der configuration auf die wirkung der enzyme. *Ber Dtsch Chem Ges* 27:2985–2993.

Hammes GG (1964) Mechanism of enzyme catalysis. *Nature* 204:342–343.

Hammes GG (2002) Multiple conformational changes in enzyme catalysis. *Biochemistry* 41:8221–8228.

Jencks WP (1987) Economics of enzyme catalysis. *Cold Spring Harb Symp Quant Biol* 52:65–73.

Koshland DE Jr (1960) The active site and enzyme action. *Adv Enzymol* 22:45–97.

Koshland DE Jr & Neet KE (1968) The catalytic and regulatory properties of enzymes. *Annu Rev Biochem* 37:359–411.

Kraut J (1988) How do enzymes work? *Science* 241:533–540.

Knowles JR (1991) Enzyme catalysis: Not different, just better. *Nature* 350:121–124.

Lichtenthaler FW (1994) 100 Years "Schlüssel-Schloss-Prinzip": What made Emil Fischer use this analogy? *Angew Chem Int Ed Engl* 33:2364–2374.

Pauling L (1948) Nature of forces between large molecules of biological interest. *Nature* 161:707–709.

Entropy of Binding and Intramolecular Reactions

Bruice TC & Pandit UK (1960) The effect of geminal substitution ring size and rotamer distribution on the intramolecular nucleophilic catalysis of the hydrolysis of monophenyl esters of dibasic acids and the solvolysis of the intermediate anhydrides. *J Am Chem Soc* 82: 5858–5865.

Bruice TC & Pandit UK (1960a). Intramolecular models depicting the kinetic importance of "fit" in enzymatic catalysis. *Proc Natl Acad Sci USA* 46:402–404.

Daum S, Lücke C, Wildemann D & Schiene-Fischer C (2007) On the benefit of bivalency in peptide ligand/Pin1 interactions. *J Mol Biol* 374:147–161.

Jencks WP (1975) Binding energy, specificity, and enzymic catalysis: the Circe effect. *Adv Enzymol Relat Areas Mol Biol* 43:219–410. Reprinted as appendix in Jencks WP (1987) Catalysis in Chemistry and Enzymology, 2nd ed. Dover.

Jencks WP (1981) On the attribution and additivity of binding energies. *Proc Natl Acad Sci USA* 78:4046–4050.

Page MI & Jencks WP (1971) Entropic contributions to rate accelerations in enzymic and intramolecular reactions and the chelate effect. *Proc Nat Acad Sci USA* 68:1678–1683.

Thompson RC (1974) Binding of peptides to elastase: Implications for the mechanism of substrate hydrolysis. *Biochemistry* 13:5495–5501.

Zaman MH, Berry RS & Sosnick TR (2002) Entropic benefit of a cross-link in protein association. *Proteins* 48:341–351.

Steady-State Kinetics and Enzyme Inhibition

Briggs GE & Haldane JBS (1925) A note on the kinetics of enzyme action. *Biochem J* 19:338–339.

Cleland WW (1970) Steady state kinetics. In The Enzymes: Kinetics and Mechanism, 3rd ed, vol II (Berg PD, ed), pp 1–65. Academic Press.

Hill TL (1985) Cooperativity Theory in Biochemistry. Springer.

Johnson KA & Goody RS (2011) The original Michaelis constant: Translation of the 1913 Michaelis-Menten paper. *Biochemistry* 50:8264–8269.

King EL & Altman C (1956) A schematic method of deriving the rate laws for enzyme-catalyzed reactions. *J Chem Phys* 60:1375–1378.

Lineweaver H & Burk D (1934) The determination of enzyme dissociation constants. *J Am Chem Soc* 56:658–666.

Michaelis L & Menten ML (1913) Die Kinetik der Invertinwirkung. *Biochem Z* 49:333–369.

Transient Kinetics

Gutfreund H (1995) Kinetics for the Life Sciences. Cambridge University Press.

Serine Proteases

Carter P & Wells JA (1988) Dissecting the catalytic triad of a serine protease. *Nature* 332:564–568.

Fastrez J & Fersht AR (1973) Demonstration of the acyl-enzyme mechanism for the hydrolysis of peptides and anilides by chymotrypsin. *Biochemistry* 12:2025–2034.

Fersht AR & Requena Y (1971) Equilibrium and rate constants for the interconversion of two conformations of α-chymotrypsin. The existence of a catalytically inactive conformation at neutral pH. *J Mol Biol* 60:279–290.

Gutfreund H & Sturtevant JM (1956) The mechanism of chymotrypsin-catalyzed reactions. *Proc Natl Acad Sci USA* 42:719–728.

Gutfreund H & Sturtevant JM (1956a) The mechanism of the reaction of chymotrypsin with *p*-nitrophenyl acetate. *Biochem J* 63:656–661.

Hare M, Su C-T, Frolow F et al. (1991) γ-Chymotrypsin is a complex of α-chymotrypsin with its own autolysis products. *Biochemistry* 30:5217–5225.

Hartley BS & Kilby BA (1954) The reaction of *p*-nitrophenyl esters with chymotrypsin and insulin. *Biochem J* 56:288–297.

Hedstrom L (2002) Serine protease mechanism and specificity. *Chem Rev* 102:4501–4523.

Dihydrofolate Reductase and Conformational Dynamics

Benkovic SJ & Hammes-Schiffer S (2003) A perspective on enzyme catalysis. *Science* 301:1196–1202.

Fierke CA, Johnson KA & Benkovic SJ (1987) Construction and evaluation of the kinetic scheme associated with dihydrofolate reductase from *Escherichia coli*. *Biochemistry* 26: 4085–4092.

Fan Y, Cembran A, Ma S & Gao J (2013) Connecting protein conformational dynamics with catalytic function as illustrated in dihydrofolate reductase. *Biochemistry* 52: 2036–2049.

Hammes GG, Benkovic SJ & Hammes-Schiffer S (2011) Flexibility, diversity, and cooperativity: Pillars of enzyme catalysis. *Biochemistry* 50:10422–10430.

Hammes-Schiffer S & Benkovic SJ (2006) Relating protein motion to enzyme catalysis. *Annu Rev Biochem* 75:519–541.

Olsson MHM, Parson WW & Warshel A (2006) Dynamical contributions to enzyme catalysis: Critical tests of a popular hypothesis. *Chem Rev* 106:1737–1756.

Sawaya MR & Kraut J (1997) Loop and subdomain movements in the mechanism of *Escherichia coli* dihydrofolate reductase: Crystallographic evidence. *Biochemistry* 36:586–603.

Liu CT, Wang L, Goodey NM et al (2013) Temporally overlapped but uncoupled motions in dihydrofolate reductase catalysis. *Biochemistry* 52:5332–5334.

Wang Z, Singh P, Czekster CM et al (2014) Protein mass-modulated effects in the catalytic mechanism of dihydrofolate reductase: Beyond promoting vibrations. *J Am Chem Soc* 136:8333–8341.

Wan Q, Bennett BC, Wilsone MA et al (2014) Toward resolving the catalytic mechanism of dihydrofolate reductase using neutron and ultrahigh-resolution X-ray crystallography. *Proc Natl Acad Sci USA* 111:18225–18230.

Liu CT, Francis K, Layfield JP et al (2014) *Escherichia coli* dihydrofolate reductase catalyzed proton and hydride transfers: Temporal order and the roles of Asp27 and Tyr100. *Proc Natl Acad Sci USA* 111:18231–18236.

Calculation of the Excess Heat Capacity Curve of Protein Thermal Denaturation

Here we give the complete derivation of the excess heat capacity function for protein unfolding. In the text, we wrote the heat capacity function as (shown in Equation 4.18)

$$C_p(T) = C_p^{\text{base}}(T) + \Delta C_p(T). \tag{A.1}$$

The *excess* part is $\Delta C_p(T)$, which gives rise to the peak in the DSC experiment; $C_p^{\text{base}}(T)$ is the baseline. To obtain $\Delta C_p(T)$, we begin by rewriting Equation 4.23,

$$\Delta C_p(T) = \frac{d\Delta H(T)}{dT}, \tag{A.2}$$

and substitute $\Delta H(T)$ by its expression given in Equation A.3,

$$\Delta H(T) = f_{\text{D}}(T)\Delta H^{\text{o}}, \tag{A.3}$$

where ΔH^{o} is the enthalpy change of the transition and $f_{\text{D}}(T)$ is the fraction of denatured protein, to obtain

$$\Delta C_p(T) = \Delta H^{\text{o}}\frac{df_{\text{D}}(T)}{dT}. \tag{A.4}$$

Appendix A Excess Heat Capacity Curve Now, $f_{\text{D}}(T)$ changes with temperature (T),

$$f_{\text{D}}(T) = \frac{K(T)}{1 + K(T)}, \tag{A.5}$$

because the equilibrium constant for unfolding K depends on temperature. Substituting Equation A.5 into Equation A.4, we obtain

$$\Delta C_p(T) = \Delta H^{\text{o}}\frac{d}{dT}\left(\frac{K(T)}{1 + K(T)}\right). \tag{A.6}$$

To calculate the derivative of the temperature-dependent term in Equation A.6, we use the chain rule,

$$\frac{d}{dT}\left(\frac{K}{1 + K}\right) = \frac{dK}{dT}\frac{d}{dK}\left(\frac{K}{1 + K}\right). \tag{A.7}$$

The last derivative is

$$\frac{d}{dK}\left(\frac{K}{1+K}\right) = \frac{1}{(1+K)^2}. \tag{A.8}$$

The derivative of K with respect to T is obtained from the van't Hoff equation (A.9),

$$\frac{d\ln K}{dT} = \frac{\Delta H^o}{RT^2}. \tag{A.9}$$

From calculus we know that the derivative of a logarithm is

$$\frac{d\ln K}{dT} = \frac{1}{K}\frac{dK}{dT}, \tag{A.10}$$

so the van't Hoff equation can be written as

$$\frac{1}{K}\frac{dK}{dT} = \frac{\Delta H^o}{RT^2}. \tag{A.11}$$

Rearranging, we have

$$\frac{dK}{dT} = \Delta H^o \frac{K}{RT^2}. \tag{A.12}$$

Finally, substituting Equations A.8, A.11, and A.12 into Equation A.6, we obtain

$$\Delta C_p(T) = \frac{(\Delta H^o)^2}{RT^2}\frac{K}{(1+K)^2}, \tag{A.13}$$

which is Equation 4.24 in the main text.

The equilibrium constant itself, in Equation A.13, varies with temperature, and so does the Gibbs energy difference (ΔG^o) between the denatured and the native states, according to

$$K = e^{-\Delta G^o/RT} \tag{A.14}$$

and

$$\Delta G^o = \Delta H^o - T\Delta S^o. \tag{A.15}$$

Since $\Delta G^o = 0$ at $T = T_m$, Equation A.15 yields

$$\Delta S^o = \frac{\Delta H^o}{T_m}. \tag{A.16}$$

Substituting this result into Equation A.15, the Gibbs energy change upon unfolding can be written as

$$\Delta G^o = \Delta H^o\left(1 - \frac{T}{T_m}\right). \tag{A.17}$$

Finally, using Equation A.17, the equilibrium constant (shown in Equation A.14) can be written as

$$K = \exp\left(-\frac{\Delta H^o}{R}\left(\frac{1}{T} - \frac{1}{T_m}\right)\right). \tag{A.18}$$

This expression is now substituted for K in Equation A.13, thus yielding the peak function $\Delta C_p(T)$ of the heat capacity curve.

Calculation of the Average Helicity from the Partition Function

In this appendix we derive Equation 5.19, which we use in Chapter 5 to calculate the average helicity (θ) from the partition function (Q). We used a similar equation in Chapter 6 to calculate the average fraction of sites occupied by ligands on a protein (binding isotherm). Let us begin by rewriting Equation 5.19 as

$$\frac{1}{N}\frac{d\ln Q}{d\ln K} = \theta, \tag{B.1}$$

where K is the equilibrium constant. The average helicity is the average number of helical residues ($\bar{n}$) divided by the total number of residues (N),

$$\theta = \frac{\bar{n}}{N}.$$

Any of the partition functions that we have used can be written as a series in powers of K,

$$Q = \sum_{n=0}^{N} \Omega_{N,n} K^n, \tag{B.2}$$

where $\Omega_{N,n}$ is the degeneracy of the term in K^n, or its multiplicity: the number of different ways of having n helical residues in a chain of N residues. In the case of independent residues,

$$\Omega_{N,n} = \frac{N!}{n!(N-n)!}.$$

The probability of observing a chain with n helical residues is the statistical weight of that term divided by the partition function,

$$p(n) = \Omega_{N,n} K^n / Q, \tag{B.3}$$

and the total probability must add to 1,

$$\sum_{n=0}^{N} p(n) = 1.$$

The average number of helical residues ($\bar{n}$) is just the weighted average of the contribution of each state,

$$\bar{n} = \sum_{n=0}^{N} p(n)\, n. \tag{B.4}$$

Substituting Equation B.3 into Equation B.4, we obtain

$$\bar{n} = \sum_{n=1}^{N} \Omega_{N,n}\, nK^n / Q, \tag{B.5}$$

where the term with $n = 0$ drops out because of multiplication by n (which is zero). Now we calculate the derivative of the partition function (shown in Equation B.2) with respect to K (note that $\Omega_{N,n}$ does not depend on K),

$$\frac{dQ}{dK} = \sum_{n=1}^{N} \Omega_{N,n}\, nK^{n-1}, \tag{B.6}$$

then multiply by K,

$$K\frac{dQ}{dK} = \sum_{n=1}^{N} \Omega_{N,n}\, nK^n, \tag{B.7}$$

and divide by Q,

$$\frac{K}{Q}\frac{dQ}{dK} = \sum_{n=1}^{N} \Omega_{N,n}\, nK^n / Q. \tag{B.8}$$

The right-hand side of Equation B.8 is just $\bar{n}$ (shown in Equation B.5). Recall from calculus that for any f,

$$d\ln f = \frac{df}{f}.$$

Thus, the left-hand side of Equation B.8 is

$$\frac{d\ln Q}{d\ln K}.$$

Therefore, dividing both sides of Equation B.8 by N gives us Equation B.1.

Solution of Rate Equations for a Two-State System

Here we derive the solution of rate equations for two-state protein folding, or any two-state system in general:

$$D \underset{k_u}{\overset{k_f}{\rightleftharpoons}} N. \tag{C.1}$$

We will solve this problem by the *matrix method*. Other methods can be used for this simple case. However, for more complicated cases they quickly become much more difficult, whereas the matrix method does not, with just a little more algebra. If you are not familiar with the method, don't let the formalism scare you. If you follow through the derivation, you will see that the mathematics (mostly algebra) are easy. All that is required is to remember matrix multiplication. To simplify the notation, we will use D and N to represent both the species (denatured and native) and their concentrations.

The rate of change of the concentrations of D and N with time (t) for the reversible reaction $D \rightleftharpoons N$ of Equation C.1 is given by the *rate equations*

$$\frac{d\,D}{d\,t} = -k_f D + k_u N \tag{C.2}$$

$$\frac{d\,N}{d\,t} = +k_f D - k_u N. \tag{C.3}$$

We can write these equations as a matrix multiplication: we multiply the matrix of the rate constants by the *column vector* of the concentrations [D, N],

$$\frac{d}{d\,t}\begin{bmatrix} D \\ N \end{bmatrix} = \begin{bmatrix} -k_f & +k_u \\ +k_f & -k_u \end{bmatrix}\begin{bmatrix} D \\ N \end{bmatrix}. \tag{C.4}$$

In symbolic notation, we can write Equation C.4 as

$$\frac{d\,\mathbf{X}}{d\,t} = \mathbf{MX}, \tag{C.5}$$

where $\mathbf{M}$ represents the matrix of rate constants and $\mathbf{X}$ represents the column vector of concentrations. The solution of this equation is a series of exponential functions of time, just like the solution of the standard differential equation $dx/dt = mx$ is of the form $x(t) = e^{mt}$.

Now, there are certain numbers λ, called *eigenvalues*, and certain vectors **X**, called *eigenvectors*, that satisfy the matrix equation

$$\mathbf{MX} = \lambda\mathbf{X}. \tag{C.6}$$

To find the solution of Equation C.4, we need to find the eigenvalues and eigenvectors of the matrix **M**. In general, a 2×2 matrix has two eigenvalues, which we call λ_1 and λ_2, and two eigenvectors, which we call $\mathbf{X_1} = [X_1, Y_1]$ and $\mathbf{X_2} = [X_2, Y_2]$. The eigenvalue λ_1 is associated with the eigenvector $\mathbf{X_1}$ and the eigenvalue λ_2 is associated with the eigenvector $\mathbf{X_2}$.

To solve this problem, let us first rearrange Equation C.6 to

$$(\mathbf{M} - \lambda\mathbf{I})\,\mathbf{X} = 0, \tag{C.7}$$

where **I** is the *identity matrix*:

$$\begin{bmatrix} 1 & 0 \\ 0 & 1 \end{bmatrix}.$$

If we expand Equation C.7, we obtain

$$\begin{bmatrix} -k_f - \lambda & k_u \\ k_f & -k_u - \lambda \end{bmatrix}\begin{bmatrix} X \\ Y \end{bmatrix} = \begin{bmatrix} 0 \\ 0 \end{bmatrix}. \tag{C.8}$$

For Equation C.8 to have a *nontrivial solution*—that is, other than the zero vector,

$$\begin{bmatrix} X \\ Y \end{bmatrix} = \begin{bmatrix} 0 \\ 0 \end{bmatrix}, \tag{C.9}$$

which is always a solution—we must require that the determinant of the matrix $(\mathbf{M} - \lambda\mathbf{I})$ be zero:

$$\begin{vmatrix} -k_f - \lambda & k_u \\ k_f & -k_u - \lambda \end{vmatrix} = 0. \tag{C.10}$$

Equation C.10 is called the *characteristic equation* of the eigenvalue problem. Now this is just an algebraic equation. The determinant is

$$(-k_f - \lambda)(-k_u - \lambda) - k_f k_u = 0, \tag{C.11}$$

from which we obtain a simple quadratic equation,

$$\lambda^2 + (k_f + k_u)\lambda + k_f k_u - k_f k_u = 0, \tag{C.12}$$

whose two solutions are the two eigenvalues

$$\lambda_1 = 0 \tag{C.13}$$

$$\lambda_2 = -(k_f + k_u). \tag{C.14}$$

To obtain the eigenvectors, we substitute each eigenvalue (one at a time) in Equation C.8, perform the matrix multiplication, and solve the system of algebraic equations for [X, Y]. With the eigenvalue λ_1, we will obtain its associated eigenvector $\mathbf{X_1} = [X_1, Y_1]$; with the eigenvalue λ_2, we will obtain its associated eigenvector $\mathbf{X_2} = [X_2, Y_2]$.

The results are, for $\lambda_1 = 0$,

$$\mathbf{X_1} = \begin{bmatrix} X_1 \\ Y_1 \end{bmatrix} = \begin{bmatrix} 1 \\ k_f/k_u \end{bmatrix}, \tag{C.15}$$

and for $\lambda_2 = -(k_f + k_u)$,

$$\mathbf{X_2} = \begin{bmatrix} X_2 \\ Y_2 \end{bmatrix} = \begin{bmatrix} -1 \\ 1 \end{bmatrix}. \tag{C.16}$$

The general solution of our problem is a linear combination of the products of the eigenvectors by exponential functions of time determined by the corresponding eigenvalues:

$$\begin{bmatrix} D(t) \\ N(t) \end{bmatrix} = \alpha_1 \begin{bmatrix} X_1 \\ Y_1 \end{bmatrix} e^{\lambda_1 t} + \alpha_2 \begin{bmatrix} X_2 \\ Y_2 \end{bmatrix} e^{\lambda_2 t}, \tag{C.17}$$

where α_1 and α_2 are constants to be determined from the *initial conditions*—that is, the initial concentrations of D and N. Thus, our final solution is obtained from the corresponding lines of Equation C.17,

$$D(t) = \alpha_1 X_1 e^{\lambda_1 t} + \alpha_2 X_2 e^{\lambda_2 t} \tag{C.18}$$

$$N(t) = \alpha_1 Y_1 e^{\lambda_1 t} + \alpha_2 Y_2 e^{\lambda_2 t}. \tag{C.19}$$

All that is left is to determine the constants α_1 and α_2 from the initial conditions. To do so, we substitute the two eigenvalues and the two eigenvectors in Equation C.17,

$$\begin{bmatrix} D(t) \\ N(t) \end{bmatrix} = \alpha_1 \begin{bmatrix} 1 \\ k_f/k_u \end{bmatrix} e^{0t} + \alpha_2 \begin{bmatrix} -1 \\ 1 \end{bmatrix} e^{-(k_f+k_u)t}, \tag{C.20}$$

and solve this system of equations for α_1 and α_2 setting $t = 0$.

To examine a concrete example, suppose that the initial condition is that *in the beginning* ($t = 0$) all the protein is denatured. Let us express concentrations as the fractions of the two states (instead of molarity). Initially, $D(0) = 1$ and $N(0) = 0$. In vector notation, these initial concentrations are represented by the column vector

$$\begin{bmatrix} 1 \\ 0 \end{bmatrix}.$$

Therefore, from Equation C.20 at $t = 0$,

$$\begin{bmatrix} 1 \\ 0 \end{bmatrix} = \alpha_1 \begin{bmatrix} 1 \\ k_f/k_u \end{bmatrix} + \alpha_2 \begin{bmatrix} -1 \\ 1 \end{bmatrix}, \tag{C.21}$$

where both exponentials have dropped out because $e^0 = 1$. We thus obtain the system of equations

$$1 = \alpha_1 - \alpha_2 \tag{C.22}$$

$$0 = \alpha_1(k_f/k_u) + \alpha_2 \tag{C.23}$$

whose solution is

$$\alpha_1 = \frac{k_u}{k_f + k_u} \tag{C.24}$$

$$\alpha_2 = \frac{-k_f}{k_f + k_u}. \tag{C.25}$$

Finally, we can substitute these results into Equation C.20 and rearrange slightly to write the solutions for $D(t)$ and $N(t)$ as

$$D(t) = \frac{k_u}{k_f + k_u} + \frac{k_f}{k_f + k_u} e^{-(k_f + k_u)t} \tag{C.26}$$

$$N(t) = \frac{k_f}{k_f + k_u} \left(1 - e^{-(k_f + k_u)t}\right). \tag{C.27}$$

These equations are identical to Equations 5.56 and 5.57 in the main text, if you divide both the numerators and denominators of the fractions by k_u, replace the ratios k_f/k_u by the equilibrium constant for folding K_F, and write the sum in the exponents as $k = k_f + k_u$.

INDEX